SYSTEMS ANALYSIS AND DESIGN

An Organizational Approach

Raymond McLeod, Jr.
Texas A & M University

The Dryden Press
Harcourt Brace College Publishers
Fort Worth   Philadelphia   San Diego   New York   Orlando   Austin   San Antonio
Toronto   Montreal   London   Sydney   Tokyo

Acquisitions Editor	Richard J. Bonacci
Developmental Editor	Stacey Fry
Project Editor	Michele Tomiak
Art Directors	Beverly Baker, Sue Hart
Production Manager	Marilyn Williams
Photo & Permissions Editors	Steven Lunetta, Julia Stewart
Publisher	Elizabeth Widdicombe
Director of Editing, Design, and Production	Diane Southworth
Copy Editors	Carolyn Sweeney, Larry Bromley
Compositor	Clarinda Company
Text Type	Palatino
Cover Image	© 1993 Tim Brown, Tony Stone Worldwide

Address for Editorial Correspondence
The Dryden Press, 301 Commerce Street, Suite 3700, Fort Worth, TX 76102

Address for Orders
The Dryden Press, 6277 Sea Harbor Drive, Orlando, FL 32887
1-800-782-4479, or 1-800-433-0001 (in Florida)

ISBN: 0-03-055154-4

Permissions and other credits appear following the Index.

Printed in the United States of America

3 4 5 6 7 8 9 0 1 2 9 8 7 6 5 4 3 2 1

The Dryden Press
Harcourt Brace College Publishers

To Martha

The Dryden Press Series in Information Systems

Arthur Andersen &
Co./Flaatten, McCubbrey,
O'Riordan, and Burgess
Foundations of Business Systems
Second Edition

Arthur Andersen &
Co./Boynton and Shank
*Foundations of Business Systems:
Projects and Cases*

Anderson
*Structured Programming Using
Turbo Pascal: A Brief Introduction*
Second Edition

Brown and McKeown
*Structured Programming with
Microsoft BASIC*

Coburn
Beginning Structured COBOL

Coburn
Advanced Structured COBOL

Dean and Effinger
*Common-Sense BASIC:
Structured Programming with
Microsoft QuickBASIC*

Federico
WordPerfect 5.1 Primer

Goldstein Software, Inc.
*Joe Spreadsheet, Macintosh
Version*

Goldstein Software, Inc.
Joe Spreadsheet, Statistical

Gray, King, McLean, and
Watson
*Management of Information
Systems*
Second Edition

Harrington
*Database Management for
Microcomputers: Design and
Implementation*
Second Edition

Janossy
*COBOL: An Introduction to
Software Engineering*

Laudon and Laudon
*Business Information Systems: A
Problem-Solving Approach*
Second Edition

Laudon, Laudon, and Weill
The Integrated Solution

Lawlor
Computer Information Systems
Third Edition

Liebowitz
*The Dynamics of Decision
Support Systems and Expert
Systems*

McKeown
Living with Computers
Fourth Edition

McKeown
*Living with Computers with
BASIC*
Fourth Edition

McKeown
Working with Computers
Second Edition

McKeown
*Working with Computers with
Software Tutorials*
Second Edition

McKeown and Badarinathi
*Applications Software Tutorials:
A Computer Lab Manual Using
WordPerfect 5.1, Lotus 1-2-3,
dBASE III PLUS and dBASE IV*

McKeown and Leitch
*Management Information
Systems: Managing with
Computers*

McLeod
*Systems Analysis and Design: An
Organizational Approach*

Martin
*QBASIC: A Short Course in
Structured Programming*

Martin and Burstein
Computer Systems Fundamentals

Mason
*Using IBM Microcomputers in
Business: Decision Making with
Lotus 1-2-3 and dBASE III PLUS*

Millspaugh
*Business Programming in C for
DOS-Based Systems*

O'Brien
The Nature of Computers

O'Brien
*The Nature of Computers with
Productivity Software Guides*

Parker
*Computers and Their
Applications*
Third Edition

Parker
*Computers and Their
Applications with Productivity
Software Guide*
Third Edition

Parker
Productivity Software Guide
Fourth Edition

Parker
*Understanding Computers and
Information Processing: Today
and Tomorrow*
Fifth Edition

Parker
*Understanding Computers and
Information Processing: Today
and Tomorrow with BASIC*
Fifth Edition

Robertson and Robertson
*Microcomputer Applications and
Programming: A Complete
Computer Course with DOS,
WordPerfect 5.1, Lotus 1-2-3,
dBASE III PLUS (or dBASE IV)
and BASIC*

Robertson and Robertson
*Using Microcomputer
Applications* (A Series of
Computer Lab Manuals)

Roche
Telecommunications and Business Strategy

Simpson and Tesch
Introductory COBOL: A Transaction-Oriented Approach

Sullivan
The New Computer User

Swafford and Haff
dBASE III PLUS

Lab Manual Series from Electronic Learning Facilitators, Inc.
The DOS Book
The Lotus 1-2-3 Book
Stepping Through Excel 4.0 for Windows
Stepping Through PageMaker 5.0 for Windows
Stepping Through Windows 3.1
Stepping Through Word 2.0 for Windows
Up and Running with Harvard Graphics 1.03 for Windows
Up and Running with PageMaker 5.0 for Windows
Up and Running with WordPerfect 5.2 for Windows
Up and Running with Quattro Pro 1.0 for Windows
Up and Running with Microsoft Works 2.0 for Windows
Up and Running with Lotus 1-2-3 Release 1.1 for Windows
Up and Running with Paradox 1.0 for Windows
Up and Running with DOS 6.0
Up and Running with Paradox 4.0 for DOS
Up and Running with Microsoft Works 3.0 for DOS
Up and Running with Excel 4.0 for the Macintosh
Up and Running with Word 5.1 for the Macintosh
Up and Running with PageMaker 5.0 for the Macintosh
Up and Running with Microsoft Works 3.0 for the Macintosh
Working Smarter with DOS 5.0
Working with WordPerfect 5.0
Working with WordPerfect 5.1

Martin and Parker
Mastering Today's Software Series
Texts available in any combination of the following:
Microcomputer Concepts
Extended Microcomputer Concepts
Disk Operating System 5.0
Disk Operating System 6.0
Windows 3.1
WordPerfect 5.1
WordPerfect 5.2 for Windows
WordPerfect 6.0
Lotus 1-2-3 (2.2/2.3)
Lotus 1-2-3 (2.4)
Quattro Pro 4.0
dBASE III PLUS

dBASE IV (1.5/2.0)
Paradox 4.0
BASIC

Martin, Series Editor
Productivity Software Guide Lab Manual Series
Disk Operating System (DOS)
Windows 3.1
Word Processing with WordPerfect 5.1
Word Processing with WordPerfect for 5.2 Windows
Spreadsheets with Lotus 1-2-3
Spreadsheets with Quattro Pro 4.0
Database Management with dBASE III PLUS
Database Management with dBASE IV
Database Management with Paradox 4.0
A Beginner's Guide to BASIC

Harcourt Brace College Outlines

Kreitzberg
Introduction to BASIC

Kreitzberg
Introduction to Fortran

Pierson
Introduction to Business Information Systems

Veklerov and Pekelny
Computer Language C

One or more courses in systems analysis and design have always been required of both undergraduate and graduate students majoring in information systems. In these courses the students learn the tools and techniques that they will use when they become systems analysts.

Like everything else in the computer field, the processes of analyzing and designing systems are constantly changing. Although the systems theory, which underlies every aspect of systems work, has remained relatively constant, the technologies and methodologies are much different today than just a few years ago.

Although it would be possible to incorporate many of the current trends in a textbook on systems analysis and design, three appear to stand out as absolute requirements. They are the movement to end-user computing, increasing organizational influence, and use of computer-aided software engineering (CASE).

The Need to Recognize End-User Computing

The one change in the computer field that has the potential for affecting systems analysis and design more than any other is end-user computing (EUC), the situation in which users perform some or all of the work in developing their own systems. One of the many impacts of EUC has been a shift in the systems analyst's workload. The analyst is being relieved of many of the simpler applications and is assuming responsibility for a higher concentration of more advanced applications as well as a greater involvement in consulting activity.

College courses and textbooks in systems analysis and design must accommodate end-user computing, not only recognizing that it exists, but also preparing future information specialists to function in such a setting.

The Need for an Organizational Focus

The modern systems analysis and design course should also recognize another reality. This is the fact that systems development is no longer an activity aimed at meeting needs of *individual* users, or even the needs of *organizational subunits*, without considering the needs of the entire organization. Today's systems must be designed to meet the needs of the *organization*.

This organizational influence is reflected in the currently high level of interest in topics such as competitive advantage, strategic planning for information resources, information resources management, ethical computer use, and security of the information resources investment. Systems analysts must be aware of this organizational influence and be able to work within it.

The Need to Incorporate State-of-the-Art CASE Tools

During the forty-odd-year history of business systems, the process of development has evolved through four stages of refinement—traditional, structured, toolkit, and CASE.

The Traditional Approach   The first computer-based systems were developed by following a bottom-up approach and were relatively unstructured. Users played minor roles in actual system development, leaving the work to the information specialists. As the analysts developed the systems, they produced a lengthy, complex narrative called a functional specification. The functional specification was difficult not only for users to understand but also for programmers to follow.

Structured Revolution   The second phase of systems development can be labeled the structured revolution, and it spanned the period from the early 1970s to early 1980s. Once a firm successfully implemented its accounting systems, the demand for more complex and larger systems emerged. The increased complexity of newer systems demanded many more lines of code, which, in turn, increased the size of the development team many-fold. Coordination of the development effort and communication between team members became difficult, and the solution was seen as "structured" systems.

The first evidence of structured techniques came in the form of structured programming, which was followed by graphical analysis and design tools such as data flow diagrams, entity-relationship diagrams, structure charts, and Warnier-Orr diagrams. The main goals of these graphical tools were to allow a top-down approach to system development, to enhance communication, and to simplify the maintenance process.

Using the top-down approach, the system was first defined at a general, overview level, and then successive refinement occurred until the bottom, primitive-level functions were clearly defined. The primitive-level modules were as independent from one another as possible. This structure allowed maintenance programmers to make changes to a single, simple module without having to worry about creating errors in other modules.

The Toolkit Approach   The structured approach stuck, and the next step was to automate the structured tools. This effort began during the early 1980s and continues to this day. The automated tools are expected to:

1. Ensure that the graphical tools are used in a consistent and standard way.

2. Increase analyst and programmer productivity.

3. Provide for an accessible repository of data definitions for the developing system.

4. Allow for rapid maintenance and system enhancement.

The name *toolkit* comes from the fact that the various automated tools are not integrated. They are merely a collection or bag of tools. The next evolutionary step was to integrate these tools into a comprehensive package.

CASE (Computer-Aided Software Engineering)   The CASE approach amassed world-wide support in the late 1980s and continues to draw more attention than any topic in the systems field. CASE applies the formal methods of engineering to software design.

Tools exist to support various stages of software development, but they seldom are integrated and seldom cover the entire development cycle. A special category of CASE tools, called I-CASE (Integrated CASE), supports development from the initial strategic planning stages to the maintenance of the operational system.

I-CASE is the current state-of-the-art in software development. Students in the systems analysis and design course should understand that this approach will likely be the working environment for pursuing a career as a systems professional.

Most analysis and design textbooks up to this point have been constructed around the toolkit approach, using a middle-CASE software package to teach the tools. Middle-CASE means that support is restricted to the design phase in the system development life cycle. The beginning phases of business strategic planning, business area analysis, information systems strategic planning and architecture, and enterprise data modeling are neglected in these texts, with the tools being seen as ends in themselves. *Systems Analysis and Design: An*

Organizational Approach, on the other hand, teaches the full development life cycle, including the methodologies, the tools, and the underlying theoretical constructs.

This Text Satisfies All Three Needs

The three needs of recognizing end-user computing, organizational influence, and I-CASE are all met by *Systems Analysis and Design: An Organizational Approach.* After completing a course that utilizes this text, the graduate should be able to use state-of-the-art CASE tools to develop computer-based systems for users, and also provide assistance to users as they develop their own systems. Such computer-based systems would support the organization's strategic objectives.

The Text Meets Current and Future Needs

Shortly before work on the manuscript began, The Dryden Press conducted a mail survey of instructors of the systems analysis and design course. Responses were received from 652 instructors who provided insights as to current techniques and future needs. This valuable information formed the basis for the entire project. The result is a text that supports a wide variety of teaching environments, including both four-year and two-year schools, and both business and computer science departmental settings.

In addition to the survey, valuable suggestions were also received from reviewers who reviewed preliminary and refined versions of the manuscript. These reviewers include Kirk Arnett of Mississippi State University, Brother William Batt, FSC, of Manhattan College, Jane M. Carey of Arizona State University West, Carl Clavadetscher, Dakota State University, Joey George of the University of Arizona, Robert T. Keim of Arizona State University, Paul Licker of the University of Calgary, Nilakantan Nagarajan of the University of Florida, Julio C. Rivera, Ludwig Slusky of California State University, Los Angeles, and Maureen C. Thommes, Bimidji State University.

We responded to this important feedback and produced a text that we believe more closely meets the needs of the introductory systems analysis and design course than does any other text on the market.

Organization of the Text

The text consists of fifteen chapters, organized into six parts. There are also ten technical modules, and an ongoing case that evolves over ten scenarios.

All chapters are tied to state-of-the-art CASE technology by means of explanations that relate the technology to the chapter contents. In addition, each chapter contains an example of how CASE technology can be applied at that particular developmental phase.

Part One—Development of Computer-Based Information Systems   This part consists of two chapters and a technical module. Chapter 1 provides an introduction to systems work and describes career opportunities. Chapter 2 describes the organizational framework within which modern systems work is performed. Technical Module A describes a high-level data modeling tool, entity-relationship diagrams.

Chapter 1 lays the important foundation for the study of systems work, regardless of the course approach. Chapter 2 is used when the instructor wants to provide the students with the organizational setting within which the work is performed. Technical Module A is positioned after Chapter 1 for a purpose. When the instructor wants to take a data modeling approach, the module can be assigned during the first week of class, and students can begin using the tool in a laboratory setting as early as the second week.

Part Two—Systems in Business   This part views business operations in systems terms and consists of four chapters and one technical module. Chapter 3 presents fundamental systems concepts, and Chapter 4 applies the concepts to the organization, or firm. Chapter 5 explains the influence of the environment on the firm's activities. Chapter 6 explains how computer-based information systems are used to manage the firm as a system. Chapter 6 explains the major applications of computers in business—data processing, management information systems, decision support systems, office automation systems, and expert systems.

Technical Module B describes a popular tool for modeling the firm's processes—data flow diagrams.

Chapters 3, 4, and 5 are used when the course is to include a solid foundation of systems theory. Chapter 6 is used when the instructor wants the students to understand how the type of system influences systems work.

Part Three—Systems Methodologies   A methodology is a recommended way of doing something, and there are quite a few in the systems field. The basis for all of the methodologies is the systems approach to problem solving, which is the topic of Chapter 7. Chapter 8 applies the systems approach to the development and use of a computer-based system in the form of methodologies such as the system life cycle, prototyping, and James Martin's Rapid Application Development (RAD). Technical Module C rounds out this part of the text by describing the tool used to model the firm's data in a detailed fashion, the data dictionary.

Instructors will include Chapter 7 when they want students to understand the theoretical basis for the methodologies. Most instructors will want to include Chapter 8 since it provides the most popular framework for studying systems development, the system life cycle.

Part Four—Systems Analysis   This part of the text describes the process of understanding the users' needs so that they can be satisfied by a new or improved system. Chapter 9 is devoted to the first phase of the system life cycle, explaining how projects are planned and controlled. Chapter 10 is devoted to the second phase, systems analysis.

Technical Module D describes two types of network diagrams: Critical Path Method (CPM) and Program Evaluation and Review Technique (PERT). These diagrams often form the basis for project planning and control. Technical Module E describes some popular approaches to economic justification that are commonly applied early in the system life cycle.

Part Five—Systems Design   This part describes how a new or improved system is designed. Chapter 11 deals with the design process, and Chapters 12 and 13 address two special design issues. Chapter 12 describes how controls are built into systems to ensure that they perform as intended, and Chapter 13 addresses systems security.

The instructor will use Chapter 12 when the need for systems controls is to be emphasized and will use Chapter 13 when students are to understand how system designs can contribute to security.

Technical Module F describes graphical user interface (GUI) design, and Technical Module G describes flowcharting.

Part Six—Systems Implementation, Audit, and Maintenance   This part completes the coverage of the system life cycle. Chapter 14 explains how systems are implemented, and Chapter 15 describes what happens after cutover—system use, audit, and maintenance. Two technical modules describe tools that can model systems processes in detail. Technical Module H explains structured English, and Technical Module I explains action diagrams. Technical Module J provides an example of a user manual that helps the user learn to use the new system.

Modular Design Facilitates Course Tailoring

The modular design of the text enables it to be used by the instructor to assemble exactly the right combination of chapters and technical modules. The *Instructor's Guide* describes six different combinations, and many more are possible.

Incorporation of Experiential Activity

The systems analysis and design course is perhaps the most difficult of the computer courses to teach because it is necessary to bring the real world of business into the classroom. This can be accomplished in several ways, and one of the most popular is the case problem.

The text includes two types of case problems. The **ongoing case** is based on the computer operations of the Harcourt Brace publishing firm in Orlando, Florida, and Bellmawr, New Jersey. As the students read the chapter material, they can see how it is applied in a real business setting. In addition, each chapter concludes with one or two short cases. These **end-of-chapter cases** enable the student to apply chapter material to solve a variety of problems.

In addition to the cases, the text also includes other opportunities for the students to put the material to use. Each chapter concludes with **end-of-chapter questions,** many of which require the student to apply the material in an innovative way. Also included are **end-of-chapter problems** and **end-of-technical module problems** that require the application of the tools and methodologies. In addition, there are **end-of-chapter discussion topics** that can focus the attention of the entire class on issues that are yet to be resolved.

A Grounding in the Literature of Systems

Modern systems work is a craft that has evolved over the years as a result of the contributions of many people. These people have sought to establish a written history and culture upon which to build the technology and to apply the methodologies and tools. The text is grounded to this rich systems literature.

Footnotes provide references to the literature where appropriate, and a **selected bibliography** is included at the end of each chapter, consisting of classics as well as leading-edge references. The footnotes and bibliography point to sources of additional information on selected topics as a means of gaining a full comprehension of modern systems work.

The Text Is a Systems Workshop

Because of its strengths in experiential activities and its solid grounding to the literature the text is a *self-contained workshop,* which enables the students to experience the challenges and rewards of systems work. Should the instructor prefer to stay within the bounds of the text, a successful course can be achieved. However, the instructor who wishes to incorporate additional learning tools has an opportunity to do that by using part or all of the complete package.

A Complete Package

Systems Analysis and Design: An Organizational Approach is supported by the following:

- An **Instructor's Guide** that contains suggestions for teaching the systems analysis and design course, a sample syllabus, and comprehensive lecture notes, keyed to textbook figures. Also included are solutions to all of the end-of-chapter questions and problems, as well as material that relates to the more controversial discussion topics, and suggested solutions to the cases.

- **Overhead transparencies** that consist of fifty two-color acetates of the most important textbook figures.

- **Transparency masters** that enable the instructor to use classroom projections of all textbook figures referred to in the lecture notes, plus many more.

- A **test bank** that contains true/false and multiple-choice exam questions, plus mini-quizzes for each chapter. The test bank is available in both a hardcopy and a computerized form.

- **Laboratory manuals** that provide the instructor with a choice of CASE tools for the students to use in applying the material from the text. One version supports the Visible Analyst from Visible Systems Corporation and is available for both DOS and Windows environments. Another supports the Information Engineering Workbench (IEW) from KnowledgeWare.

  The Dryden Press is also developing laboratory manuals to support the Information Engineering Facility (IEF) from Texas Instruments and Excelerator from Index Technology—for both DOS and Windows.

  The text is designed so that use of a lab manual can begin immediately following chapter 1. This organization enables the students to engage in hands-on, experiential activity *beginning in the second week,* stimulating student interest early on and maintaining it throughout the course.

- **Two videotapes** that support the Harcourt Brace ongoing case by providing a visual tour of both the Orlando computer project and the warehouse operation at Bellmawr. Students not only read about the project, but are able to see the actual facilities and meet the participants.

By taking advantage of this complete package, the instructor can assemble a course that captures the interest of the students and enables them to use the most modern systems technologies, methodologies, and tools.

A Team Effort

I use the term *we* rather than *I.* Some students find this strange since I am the sole author. I follow the *we* strategy mainly so that students will understand that they are a part of the effort. I like to think that the author, the instructor, and the students all follow the same path together.

At the same time, I also recognize that I do not do all of the work. The writing and publishing of a textbook is a team effort, and one of the rewards is the opportunity to work with the many fine people that a publisher such as Dryden has to offer. My main contact at Dryden has been Richard Bonacci, my editor. Richard has gone that extra mile to ensure that we produce a book that meets the needs of the instructors who teach the systems course and the students who take it. Also playing key roles at Dryden were Stacey Fry, Developmental Editor, and Michele Tomiak, Project Editor. I also recognize the support of Kevin Cottingim and Scott Timian, Marketing Managers, who guided us through some rough spots and planned the marketing campaign. From my very first introduction to Dryden, I was impressed with everyone's professionalism. That impression has persisted.

Acknowledgments

Many people outside of Dryden have contributed to this effort as well. Some are from academia, and some from industry.

Much of my academic help comes from my colleagues at Texas A & M University. We have one of the top MIS faculties in the country, and professors Joobin Choobineh, Michael Chung, Jim Courtney, George Fowler, Bill Fuerst, David Paradice, Arun Sen, Marietta Tretter, and Ajay Vinze were always ready to help when needed. Two doctoral students, Hae-ching Chang and Choong Kim, and a masters student, Paul Davis, contributed as well.

I am especially indebted to Joobin Choobineh for preparing Technical Module A on entity-relationship diagrams. Jane Carey of Arizona State University West also contributed

Technical Module I on action diagrams and wrote the CASE segments that appear in each chapter.

Persons from industry have also provided important material. I am indebted to the management of Harcourt Brace for opening the doors of their Orlando and Bellmawr facilities to me for the purpose of developing the ongoing case. I specifically recognize Mark Arak, Neil Aronow, Michael P. Banks, Michael W. Byrnes, Bob Evanson, Alan Fox, Anne Hogan, Ann Johnston, Jenny R. Jolinski, Ira Lerner, Sal Leonetti, Suzanne McNulla, David J Mattson, John Misuira, Cathy Oliva, William Presby, David Renk, and Michael J. Tolen. I also acknowledge the help from Debbie Sandoval of Comshare; Susan Benjamin of Rollerblades; and John Hoffman of Skill Dynamics™, an IBM company.

In writing the chapter on systems security, I was assisted greatly by Dahl Gerbrick and Daniel Faigin of ACM's Special Interest Group on Security, Audit, and Control—SIGSAC. Dahl put me in contact with three of the nation's top security experts—Ken Cutler of American Express, Thomas C. Jones of Phoenix-based Lemcom Systems, Inc., and Daniel E. White of Ernst & Young.

I also want to acknowledge the role played by my systems analysis and design students. Much of the material in this text was developed in class, with valuable suggestions and insights coming from the students. I especially recognize the systems class of the Spring 1992 semester, which used the initial version of the text manuscript. In that class, Melissa Woodlan and Greg Liddle contributed the basic structure of the office automation model described in Chapter 6. Students in my Spring 1993 MIS classes also contributed. Darren Stokes and Tom Kozelsky provided examples of graphical user interfaces, and Cynthia Caldarola provided the user manual that appears in Technical Module J.

I hope that you, the instructor and students, like the book. Please let us know. Our plan is that the book will be around for a long time and will get better with each edition. We can do that only with your help.

I have enjoyed working on the book and have learned a lot in the process. May you experience the same rewards in using it in class and applying it in your job.

Raymond McLeod, Jr.
College Station, Texas
January, 1994

CONTENTS

PART 3   SYSTEMS METHODOLOGIES   253

PART 5   SYSTEMS DESIGN   457

PART 6   SYSTEMS IMPLEMENTATION, AUDIT, AND MAINTENANCE   625

Development of Computer–Based Information Systems

Managers achieve and maintain control over their organizations by establishing policies and defining procedures. When these policies and procedures are incorporated into computer programs, they are called applications *or* systems. *For example, an organization's computer-based payroll system carries out the policies and procedures that management has defined for paying the employees. Any organization uses many such systems, each dealing with a particular facet of the operations.*

The first organizations to install computers more than 40 years ago realized that they should hire computer specialists to do the work. Three types of specialists then existed—systems analysts, programmers, and operators. Systems analysts worked with managers and other employees to identify the necessary computer applications and then describe how the processing would be performed. Programmers used programming languages to transform the systems analysts' descriptions into computer code. Finally, operators ran the programs on the computer to achieve the desired outputs.

This use of computer specialists has continued, but changes have been made. Database administrators, network managers, and other new specialist occupations have been added. Also, some users of the computer output are now doing some or all the work of the computer specialists—a practice called end-user computing.

Chapter 1 focuses on systems analysts. It describes the organizational setting that influences their work, explains what they do, identifies the types of organizations that hire them, projects future changes that will affect the career path, and identifies the keys to success. Systems work is challenging, exciting, and rewarding, and this book prepares you to perform it, either as a systems analyst or as an end user. Chapter 1 provides the important introduction.

Chapter 2 describes the setting within which modern systems work is performed. The first computer-based systems were intended to reduce clerical costs.

*Over time, the emphasis shifted to providing information for decision making. Recently, orga-
nizations have recognized the strategic value of information, using it to gain a* competitive
advantage *in their marketplaces.*

*Management learned that if the organization is to use information to gain a competitive
advantage, it must engage in long-range planning to ensure that the necessary information
resources are available when needed. This planning has been termed* strategic planning
for information resources, *or SPIR. SPIR encompasses all information-related resources,
including hardware, software, data, information specialists,* and *users.*

*The firm's strategic plan for its information resources is developed by top-level executives
as they engage in strategic planning for the entire firm. The computing unit is represented in
this planning effort by its own executive, who often has the title* chief information officer,
or CIO.

*The strategic planning must address end-user computing. Management must provide the
users with the information resources they need but must impose controls to ensure that all
computer use in the organization is coordinated.*

*These four concepts—competitive advantage, strategic planning for information resources,
the chief information officer, and end-user computing—are interrelated. The organization
integrates the concepts in an activity named* information resources management, *or IRM.
IRM is such a new concept that most firms are still working to achieve it. The 1990s may
well be the age of information resources management. Chapter 2 describes the organizational
setting for systems analysis and design activity, performed by systems analysts and users
alike.*

*The ongoing case involving Harcourt Brace & Company publishers opens Part One. The
related problem comes at the end of Chapter 1, under the heading "Consulting with Harcourt
Brace." To best solve the Harcourt Brace problems, first read the scenario and then look ahead
to the problem statement. An understanding of the problem will guide you as you read the
chapter material.*

I—COMPANY OVERVIEW

This case is based on research at the publishing firm of Harcourt Brace & Company from one recent summer through the following spring. If you were to visit Harcourt Brace today, you would find a different situation. For one thing, the company has experienced the executive turnover you would expect in a dynamic business environment. Also, the description of the computers used in the case deviates somewhat from the systems actually used. The two main characters, Sandi Salinas and Russ March, are fictitious, as are some of the other employees mentioned. Despite these deviations from fact, the case is a learning tool that provides a good snapshot of the challenges and opportunities facing new systems analysts in a large organization.

Take the Role of a Consultant

As you read the Harcourt Brace case, take the role of an Orlando computer consultant called in by top management to evaluate the computer operation. Be alert to things that Harcourt Brace is doing well, and be especially critical of things that you think call for improvement.

Assignments at the ends of some chapters require you to apply the text material to the case. *Before* you read a chapter, look at the end for a section titled "Consulting with Harcourt Brace." Read this case assignment, and keep it in mind as you read and study the chapter.

At the end of the last installment, you are asked to critique the entire computing operation. Prepare for this task by keeping two lists as you read each Harcourt Brace installment—one for "Things Harcourt Brace Is Doing Right," the other for "Areas for Harcourt Brace Improvement."

By being aware of your case responsibilities, you will be in an excellent position to respond professionally.

Harcourt Brace and the Book Publishing Business

Harcourt Brace is one of only three publishers in the world that offers all book products to all markets. Its product line includes elementary and high school (elhi) books, college books, professional books, trade books, and medical books. The company is headquartered in Orlando, Florida; has editorial offices, sales offices, and distribution centers throughout the United States and Canada; and has foreign subsidiaries in the United Kingdom and Japan.

Similar to many other publishers, Harcourt Brace increased the volume of its business by acquiring other publishing firms. One acquisition was CBS Educational and Professional Publishing in 1987, which increased sales revenue by 75 percent. Another 1987 event had an even more profound effect on the company. British investor Robert Maxwell attempted to buy a controlling interest in Harcourt Brace, and the corporation's battle to retain control placed the company under severe financial strain. This strain was relieved in 1991 when General Cinema acquired Harcourt Brace.

Harcourt Brace performs all its production and marketing except printing and binding. The printing is performed by printing companies such as R. R. Donnelley and Rand McNally, and the binding is performed by bindery firms. Before a Harcourt Brace book is purchased, it follows a life cycle that begins with the author's manuscript. The manuscript is reviewed by experts in the field and is edited at a Harcourt Brace editorial office. Then, the printing is done, often using the author's edited diskettes to set the type. Printed pages and covers are assembled by the bindery, and the completed book is shipped to a Harcourt Brace distribution center. Book stores place mail and telephone orders to Orlando. The orders are processed by the computer, and the books are shipped from the distribution center. This entire life cycle can span several years.

Headquarters Organization

Approximately 1,500 employees work at the Orlando headquarters. Most are employed in the customer services, accounting, and management information system (MIS) departments. The customer services department processes customer orders, answers customer questions, handles returned books, and provides other such services. Accounting consists of major systems such as billing, inventory, accounts receivable, cost, and general ledger. Most of the customer service and accounting systems are computer based, and the processing is performed by the MIS department.

The MIS Department

The MIS department is organized as shown in Figure HB1.1. Mike Byrnes, the vice-president of MIS, has responsibility for the entire Orlando computing operation. Although he does not have the title, Mike performs the functions of a chief information officer, or CIO. Reporting to Mike are three directors who are responsible for production support, technical support, and data center operations.

Production Support

All the systems analysts, programmers, and managers who develop and maintain applications are located in the production support area. The applications are organized into three groups—order processing systems, financial systems, and COPS development. COPS stands for Customer Order Processing System. One financial system consists of prewritten software that was acquired from a software vendor named Management Systems of America, commonly known as MSA.

COPS is a mammoth systems development project that is intended to result in one distribution system for the entire company. One problem caused by the acquisition of other publishing firms is the retention of their computer systems. Before long, the acquiring firm finds itself using a hodgepodge of systems that process data in different ways. COPS is intended to solve the problem and meet the needs of all the operating groups within Harcourt Brace. Cutover has been scheduled for five months from now.

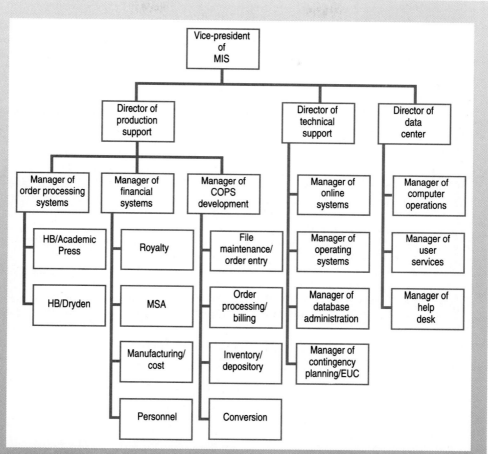

Figure HB1.1
Organizational chart of the MIS department.

Technical Support

The specialists in online data communications systems, operating systems, database administration, contingency planning, and end-user computing form the technical support staff. Contingency planning ensures continuing processing in the event of a disaster.

Data Center

The main computing facility is called the data center. The staff consists of operations personnel, support personnel who provide services to users, and a help desk.

Information Resources

Harcourt Brace's main information resources consist of information specialists, hardware, software, and data. It is Mike Byrnes's responsibility to use these resources efficiently and effectively

Information Specialists

The production support area has two main types of information specialists—systems analysts and programmers. Both specialties have multiple rungs in the career ladder. For example, new systems analysts are given the title of junior systems analyst. They are promoted first to systems analyst and then to senior systems analyst.

A group manager directs each group of specialists. Specialists from the technical support area lend expertise on subjects such as database and data communications. And, an auditor from the internal auditing department is assigned to each group. The COPS order processing and billing group, for example, consists of a group manager, a senior systems analyst, a systems analyst, two programmers, a database administrator, a network specialist, and an internal auditor—eight people in all. Figure HB1.2 is a job description of the systems analyst position.

Also found in the technical support area are junior systems programmers, systems programmers, and senior systems programmers. Other specialists include an end-user analyst, a database administrator, and a senior database administrator.

The data center titles are typical for that area of operations—junior operators, operators, and senior operators. There is also a library supervisor, a scheduler, a production coordinator, a help desk analyst, and several data clerks.

Hardware

The computing equipment at Orlando consists of the configuration illustrated in Figure HB1.3.

The main processor is an Amdahl 5890-300E. This is a large-scale mainframe with 196 megabytes (MB) or million bytes of primary storage, and a speed of 44 MIPS (millions of instructions per second). An IBM 9370 CPU serves as a front-end processor, connecting the host Amdahl to the data communications network. An IBM 3174 CRT control unit links the IBM 9370 to users' terminals at headquarters, and an IBM 3725 provides the same linkage with computing equipment at other Harcourt Brace locations around the world.

The databases are maintained in twenty-seven IBM 3380 direct access storage devices (DASDs). Data is archived onto magnetic tape by twelve STC 4480 tape units that read and write tape cartridges. Also in use are three STC 4674 tape units that use tape reels.

Printed output is produced on an IBM 4248 impact printer at 4,000 lines per minute and an IBM 3800 laser printer at 215 pages per minute.

Software

The software library includes more than twenty-five application systems that are either operational or under development. These application systems handle activities such as

Basic Purpose of the Job

Provide systems analysis support. Apply analytical skills to determine the direction work should take. Assume leadership responsibilities for a small section of a project.

Job Specifications

Minimum education required for job entry:

College degree in Computer Science or Business.

Amount and type of experience required for job entry:

4 years programming experience (Language (COBOL).

Technical Responsibilities (70% of Time)

A. Defines the causes of problems in existing programs and systems.

B. Designs solutions to defined problems.

C. Works efficiently with the database, flat files, and direct access files.

D. Works directly with the user community in the design of enhancements and corrections to their programs.

E. Translates functional specifications into program specs, and communicates this process to others.

F. Creates user and production documentation for all systems and procedures for which responsible.

G. Creates test data.

Administrative & Managerial Responsibilities (10% of Time)

A. May provide direction to programmers.

B. Works directly with users to obtain and clarify functional specifications.

Professional Development Responsibilities (20% of Time)

Receives appropriate technical training as skill level and projects demand.

B. Develops business knowledge required to support projects.

Figure HB1.2
A job description for the systems analyst position.

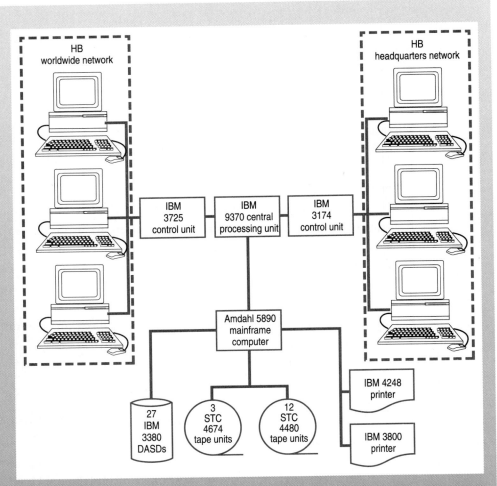

Figure HB1.3
The Harcourt Brace computer configuration at Orlando.

royalty accounting, sales reporting, and forecasting. Also included in the software library are more than fifty system software packages, such as the MVS operating system, the CICS/VS online transaction processor that handles database and data communications applications, the DB2 database management system, and language translators such as COBOL and PL/I. In addition, several software productivity tools are used. Examples are the METHOD-1 methodology and the DESIGN-1 and INSTALL-1 CASE (computer-aided software engineering) tools, from Andersen Consulting.

Data

The DASD units provide more than 148.7 gigabytes (GB), or billion bytes, of database storage. Access to this database is controlled by the DB2 database management system. Data is also stored on more than 5,000 magnetic tape reels and more than 14,000 tape cartridges.

MIS Leadership

Leadership of Harcourt Brace's MIS operations is provided primarily by Mike Byrnes, his three directors, and the managers of the computer specialist groups. These managers provide day-to-day leadership to ensure that the annual objectives of the MIS department are met. Longer-term leadership is provided by Byrnes and other high-level executives.

A special committee has been formed to oversee the COPS implementation program, which is the current focus of most of the executive interest. This COPS Review Committee consists of Michael Banks (controller), John Misura (vice-president of distribution), and Mike Byrnes. The committee meets periodically to check on COPS progress.

Systems Assurance

The COPS review committee is not the only group outside MIS that influences its operations. Guidance also comes from the Harcourt Brace internal auditors. Many firms call in their internal auditors *after* implementing a system, hoping to obtain auditor approval that the system is processing the data in the desired manner. These firms often learn they must redesign the systems to meet the auditors' requirements. Better is Harcourt Brace's approach of calling in the auditors *before* beginning the project and benefiting from their guidance throughout the project. It is a healthy situation, where the Harcourt Brace auditors are members of the application systems and development groups.

Putting Harcourt Brace in Perspective

Harcourt Brace's use of the computer is typical of U.S. firms. The MIS staff keeps busy staying abreast of changing technology and methodology, developing new systems, and keeping existing systems up to date. The computer plays a key role in the firm's activities, and the MIS staff has earned the respect of managers and employees on all levels for the professional way they have provided computer support. Harcourt Brace has used the computer well, but like all companies, has many challenges yet to meet.

CHAPTER 1

Introduction to Systems Analysis and Design

Learning Objectives

After studying this chapter, you should:

- Know how the term *system* is used in the computer field
- Be able to distinguish between physical and conceptual systems and understand what determines the value of each
- Understand what a computer-based information system is and what major subsystems it contains
- Have an awareness of how computer-based information systems are developed in a top-down fashion
- Appreciate that the system belongs to the user and that the systems analyst serves a supportive role
- Recognize that, in a practical sense, everyone is a systems analyst
- Be aware of changes that are affecting the systems analysis profession
- Know the keys to successful systems work and how to become a systems analyst

Introduction

A system consists of several elements that work together to achieve an objective. Systems can be classified in many ways, but the two basic classes are *physical* and *conceptual*. Physical systems are tangible. Conceptual systems represent physical systems. The firm is a physical system, and management uses conceptual systems to manage the firm. A physical system is valuable for what it is. A conceptual system is valuable for what it represents. A computer is a physical system, but its programs and data are conceptual systems that represent the physical processes and resources of the firm.

All the business applications performed on the computer are termed the *computer-based information system, or CBIS.* The CBIS consists of five subsystems: the data processing system, the management information system, decision support systems, office automation, and expert systems.

Conceptual systems are developed within a top-down framework that begins with enterprise planning and modeling, strategic business planning and strategic information planning, and system life cycles. Life cycle participants include users and information specialists who communicate in a chainlike fashion as the five phases of the life cycle evolve. The processes and data of the existing system and the new system are documented using tools such as data flow diagrams, structured English, and the data dictionary. The user controls the development process and system use. The computer specialists provide technical assistance.

All employees in an organization can be regarded as systems analysts, studying their systems and making changes to improve performance. This broad view is especially appropriate today because much systems work is being performed by persons outside the computing organization, a phenomenon called *end-user computing*. End-user computing is having three effects on systems work. First, it is making more complex the systems that the systems analyst must design. Second, it is requiring the analyst to become more of a consultant. Third, it is causing organizations to locate systems analysts in user areas in addition to the central computing facility.

If you want to do systems work, as a systems analyst or as a user, you must first acquire the necessary knowledge and skills. This text will supplement your existing computer knowledge and prepare you for a systems analysis career.

What Is a System?

The term *systems age* could be used to describe our modern world, for this is truly the age of systems thinking. Each day, government officials talk about *economic systems*, military leaders talk about *weapons systems*, city planners talk about *public transportation systems*, and advertisers talk about *shaving systems*. We hear the word *system* so often that we scarcely pay attention to it and most likely fail to appreciate its full meaning.

The term is especially popular in computing. Both government and industrial organizations use *electronic computing systems* to process their data, and the applications that the computers perform are called systems—*payroll systems, decision support systems, expert systems,* and so on.

Any dictionary offers several definitions of the term *system*. However, in the computer field, a **system** is a group of elements that work together to accomplish an objective. The computer is a good example of a system because it comprises thousands of electronic components and mechanical parts used to process its data. Similarly, the payroll system comprises hundreds or thousands of processing steps built into programs that compute the employees' pay. The same can be said for the decision support systems and expert systems. They all consist of many components that contribute to achievement of specific objectives.

The two key elements that define a system are that it (1) is composed of multiple elements and (2) is intended to accomplish an objective.

Physical and Conceptual Systems

Although an infinite variety of system types seems possible, systems can all be grouped into two basic classes—physical systems and conceptual systems.

A **physical system** consists of tangible elements. An excellent example is the firm. This book uses the term *firm* to mean an organization of *any* type—governmental or industrial. The systems principles described here apply to all types of organizations.

All buildings, employees, machines, materials, and money within the firm are its elements. All these elements are applied so that the firm can meet objectives such as profit, return on investment, and market share.

A **conceptual system,** on the other hand, does not exist physically. It *represents* a physical system. The inventory system is an example. Data and information stored in the computer's storage units represent the physical inventory items in the warehouse. The inventory system can print or display reports that describe the

Figure 1.1
The firm is a physical system.

Figure 1.2
Computer contents can be viewed as conceptual systems.

inventory. Persons in the firm, such as managers, can view the reports to monitor the status of the inventory. In this way, a manager does not have to go to the warehouse to determine whether a particular item is in stock. Assuming that the report is accurate, the manager has only to view the report.

The same situation applies to the data and information that relate to other physical systems. For example, the employee data and information represent the firm's employees, and the customer data and information represent the firm's customers. The various files of computer data and programs can be viewed as conceptual systems representing physical systems. The data describes the physical resources of the firm, and the programs describe the processes that the firm performs on the resources.

System Value

The value of a physical system is determined by the physical nature of the system's components and how they are coordinated. For example, a computer with components that enable it to make one million calculations per second has greater value than a computer that can make only a half-million calculations. Similarly, a skilled and experienced employee is more valuable than one just learning the ropes. The point to remember is that a physical system is valuable for what it *is*.

A conceptual system's ability to accurately reflect a physical system is of special value to managers who cannot always be on the scene to observe the physical system. A conceptual system is valuable for what it *represents*.

Our Interest Is Conceptual Systems

You might think that the computer, as a physical system, would be uppermost in the minds of the **information specialists,** those employees who provide computer expertise to the entire firm full time. We use the name **information services** for the organizational unit where the information specialists are located.

Information specialists *are* interested in computers as physical systems. However, once they understand the technology, their interest usually shifts to the many applications the computer can perform. Usually, when information specialists use the term *system*, they are speaking of conceptual systems. In this book, we are primarily interested in conceptual systems.

The Computer-Based Information System

The composite of all the computer-based conceptual systems of the firm is known as the *CBIS*, or *computer-based information system.* During the period of computer use in business, which began in the mid-1950s, the CBIS has evolved to a composite of the five subsystems illustrated in Figure 1.3. A **subsystem** is simply a system within a system. The CBIS subsystems frequently interact and overlap with each other.

Data Processing System   The *data processing system*, sometimes referred to as the *DP system* or the *transaction processing system*, processes the firm's accounting transactions. Included within the DP system are the payroll system, and the inventory system.

Management Information System   The *management information system*, or *MIS*, consists of all the conceptual systems that provide information to managers throughout the firm or within a subunit, such as a management level or a func-

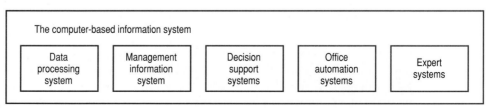

Figure 1.3
The computer-based information system.

tional area. Examples are the executive information system and the marketing information system.

Decision Support Systems A *decision support system (DSS)* provides information to one manager or a small group of managers. Because decision support systems are tailored to individual managers' needs, a firm has many decision support systems.

Office Automation Systems An *office automation system* uses electronic circuits and devices to facilitate communications within the firm and between the firm and its environment. Popular examples are word processing, electronic mail, and electronic calendaring.

Expert Systems An *expert system* is an application of artificial intelligence, or AI, that enables the computer to play the role of consultant. *Artificial intelligence* has been described as the activity of providing computers "with the ability to display behavior that would be regarded as intelligent if it were observed in humans."[1] An expert system is programmed with the logic that an expert employs in solving a particular type of problem. The expert system solves the problem in basically the same manner as the expert.

How Are Systems Created?

A firm's computer-based systems are the result of many people who work together, providing a variety of expertise over a long time. *One* approach to this process is illustrated in Figure 1.4. It consists of enterprise planning, enterprise data modeling, strategic business planning, strategic planning for information resources, and system life cycles. It is a top-down process, beginning on the executive level and ending with work by managers and nonmanagers on lower levels. Not all firms use such a top-down process, but the trend is in this direction.

Enterprise Planning

Executives take an introspective look at the the firm and its position in its environment. An **executive** is a person on the top organizational level who influences long-range strategy, policies, and procedures. The introspective look by the executives is called **enterprise planning** or **enterprise modeling.** The term **enterprise** means the entire organization. Increased competition, often coming from firms with headquarters in other countries, is causing more firms to engage in enterprise modeling and to do so more often.

The formal, written description that the executives produce, using words and graphics, is called an **enterprise model**. It describes the activities that the firm, or enterprise, should perform.

[1]Clyde W. Holsapple and Andrew B. Whinston, *Business Expert Systems* (Homewood, IL: Irwin, 1987): 4.

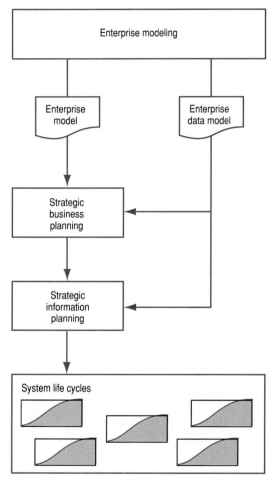

Figure 1.4
The top-down approach to systems development and use.

Enterprise Data Modeling

While developing the enterprise model, executives pay attention particularly to the data that will be required for the firm to perform its processes. This attention recognizes data as an important resource. The description of the firm's data needs is called an **enterprise data model.**

Both the enterprise model and the enterprise data model depict where the firm wants to be. The enterprise model specifies the processes needed. The enterprise data model specifies the data needed. Like snapshots, the models capture the firm's status at some future time.

Data Modeling

Data modeling is the process of constructing a graphic representation of the data contained in an information system. The origin of data modeling can be traced to database design. Understanding the data requirements and how the database must be constructed is crucial to the analysis, design, and implementation of an information system.

An entity-relationship diagram (ERD) is a good data modeling tool. An ERD can illustrate the logical components of a **database schema,** a view of a firm's data resource at the enterprise level, or a **subschema,** a view of the data for a particular application or user. An ERD contains entities, relationships, and attributes. A single enterprise data model may contain hundreds of entities. Some of these diagrams take up the entire wall space in a room.

Figure 1.5 contains a partial enterprise data model for a retailer. This enterprise data model can be read as a series of sentences in the following manner. Entities are in bold print, and relationships are underlined. Many **items** are produced-in a **style.** **Items** are available-in many **sizes. Stores** stock many **items. Workers** work at a **store. Wholesalers** sell **labels. Items** are sold-under a **label.**

Details of how to draw entity-relationship diagrams are included in Technical Module A. However, the notation used in the technical module varies slightly from that used in Figure 1.5. There are many variations in ERD style.

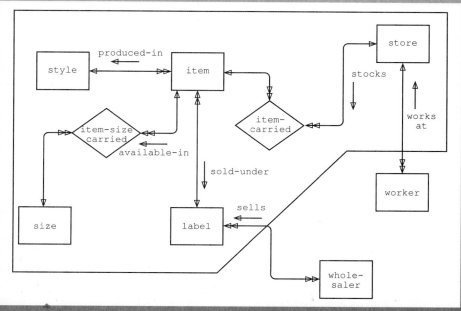

Figure 1.5
A portion of an enterprise data model for a retailer.

Strategic Business Planning

The enterprise model can provide the basis for the firm's strategic business planning. The **strategic business plan** is a formal statement that identifies the actions the firm should take to achieve the status pictured by the enterprise model. For example, if the enterprise model specifies that the firm provide a new customer service in the future, the strategic business plan specifies how that service capability will be achieved.

Strategic Information Planning

The *strategic plan for information resources,* a subset of the strategic business plan, identifies (1) the volume and types of information resources needed to execute the strategic business plan, and (2) the ways those resources will be applied. As an example, the strategic plan for information resources might recognize the need to install a new mainframe computer two years from now and specify the major accomplishments leading to that installation.

All top-level executives contribute to the strategic plan for information resources; however, the executive in charge of information services plays the key role. This executive is often given the title of **chief information officer,** or **CIO.**

The strategic plan for information resources includes resources in user areas as well as in information services.

System Life Cycles

The strategic plan for information resources specifies each computer-based system that the firm will require during the time covered by the plan. The development and use of each system is called the *system life cycle,* or *SLC.* Figure 1.6A identifies the five phases in the SLC: planning, analysis, design, implementation, and use. The curved line represents the cumulative nature of the systems work during the life cycle. Figure 1.6B shows the life cycle as a circular pattern. This pattern recognizes that a system's life cycle is repeated when the system must be redesigned.

Life Cycle Participants   Different types of people can participate in the system life cycle. Always present is the user. A **user** is someone who uses an existing system or has the potential for using a future system. Users can be managers or nonmanagers. Most life cycles involve multiple users.

Information specialists usually participate in the system life cycle, although there may be exceptions, such as when end-user computing is followed. In **end-user computing,** or **EUC,** users perform some or all of the work required to develop computer-based systems for their own use. In some cases, users perform all the development work. In other cases, users will perform only a part, leaving the remainder to the information specialists.

End-user computing is gaining popularity. However, most of the firm's conceptual systems, certainly the large, complex ones, are still being developed by information specialists working closely with users.

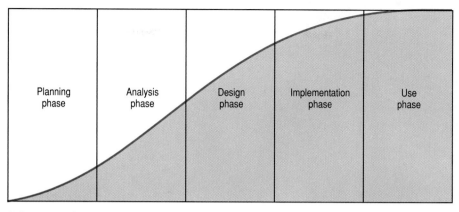

| Planning phase | Analysis phase | Design phase | Implementation phase | Use phase |

A A sequence of steps

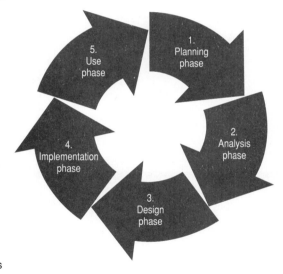

B A repetitive process

Figure 1.6
The system life cycle as a sequence of steps and a repetitive process.

The Communication Chain   The system life cycle can span months or even years, and during that time the participants must communicate. The pattern of that communication is called the *communication chain*. It is illustrated in Figure 1.7. The user is on the left, and the computer is on the right. The task is to communicate the user's problem to the computer so that the computer can send problem-solving information back to the user. When the user pursues end-user computing and communicates directly with the computer, the communication paths are those shown at the top and bottom of the figure.

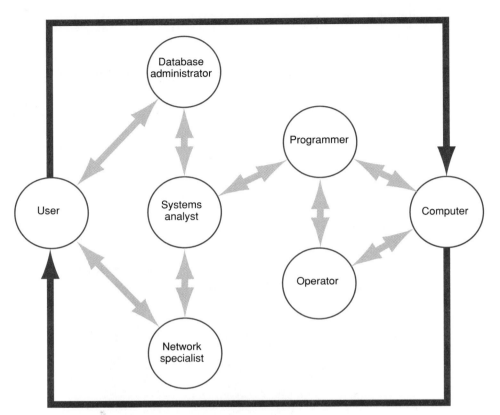

Figure 1.7
The communication chain.

When information specialists are involved, they serve as intermediaries between the user and the computer. The key intermediary is the systems analyst. A
systems analyst is an information specialist who analyzes existing systems to define the information needs and then designs new or improved systems to meet
those needs. When the system requires a change in the firm's collection of stored
computer data, or database, an information specialist called a **database administrator,** or **DBA,** can be involved. Likewise, when the project involves a system that
uses data communications circuits, a **network specialist** can participate. Other
specialists are the well-known **programmer,** who codes and tests the program,
and the **operator,** who runs the job on the computer.

The systems analyst, assisted by the database and data communications specialists, communicates with the user about the problem definition. The analyst
also communicates with the programmer about the new system design. The programmer and the operator communicate about the operating instructions. Each
communication link is accomplished orally and in writing. The programmer

communicates with the computer while coding the program, and the operator communicates with the computer while running the job.

The Planning Phase   Determining that a problem exists, deciding on a system solution, and identifying the steps necessary to put the system into use are all part of **systems planning.**

One product of the planning phase is a planning document such as a network diagram, which identifies the work to be done, who will do it, and when it will be done. Figure 1.9 is a *network diagram* that illustrates the activities required to develop a payroll system. The arrows represent the activities. The circles link the activities into the required sequence.

The Analysis Phase   When planning is completed, the existing situation, or system, must be studied. This study is called the **systems analysis.** The systems analyst, working closely with the user, usually performs it.

The analyst documents the processes and data of the existing system, using various documentation tools. Many existing systems are documented with a diagram called a *system flowchart,* drawn when the system was implemented, and maintained in a documentation file. Figure 1.10 shows a system flowchart of a payroll system. The flowchart symbols represent processes and data. The system

Figure 1.8
Oral communication between the user and the systems analyst forms the foundation for systems analysis and design.

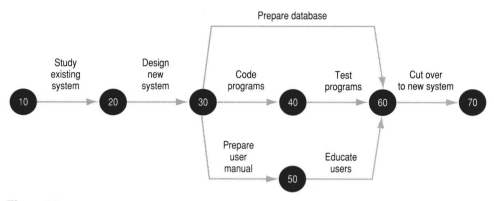

Figure 1.9
A network diagram showing the activities required to develop a payroll system.

flowchart is unique among the documentation tools because it shows the technology used. In the figure, for example, the rectangles represent computer programs and the circles with tails represent magnetic tape files.

Originally, flowcharts and other documentation were prepared by drawing the symbols on paper using a plastic template. Today, software systems called *Case tools* can do the job. **CASE** stands for **computer-aided software engineering**. CASE tools relieve the systems analyst of much work and produce better documentation.

Upper-CASE Technology and the Top-Down Approach to Systems Development and Use

CASE originated in the early 1980s, growing out of the CAD (computer-aided design) software that had been so successful in increasing the productivity of design engineers. All CASE tools generally provide a graphics capability, such as data flow diagrams, entity-relationship diagrams, flowcharts, and structure charts, and contain data repositories that hold details of data elements, data structures, data flows, and process logic. Most CASE tools also provide consistency and completeness checking to ensure the functionality of the resulting systems.

CASE tools can provide varying support during the system life cycle. Some provide support during the early phases in the form of enterprise data modeling, strategic planning, and business area analysis. These CASE tools are referred to as **upper-CASE tools.** Others, called **middle-CASE tools,** provide support during the middle phases as the existing system is analyzed and the new system is designed. Still others, called **lower-CASE tools,** provide support during the latter phases as the new system is implemented.

The most promising of the CASE tools support all phases of the system life cycle. These full-cycle tools are referred to as **integrated CASE tools,** or simply **I-CASE.** The system life cycle support provided by I-CASE consists of the following:

- Enterprise Planning and Modeling
    - A. Perform business strategic planning
    - B. Conduct business function and process identification analysis
    - C. Build the organization chart
    - D. Perform strategic information systems and architecture planning
    - E. Set out quality initiatives
    - F. Build the enterprise data model

- Systems Analysis
    - A. Define the requirements specification
    - B. Define the input/output specification
    - C. Perform process identification and decomposition
    - D. Define data items
    - E. Prepare decomposition diagrams

- Systems Design
    - A. Engage in data modeling
    - B. Perform process definition
    - C. Perform input/output design
    - D. Define systems controls and security issues

- Systems Implementation
    - A. Generate computer code
    - B. Generate database definitions
    - C. Perform testing

- Support and Maintenance
    - A. Prepare documentation
    - B. Engage in systems redesign

- Project Management
    - A. Perform network analysis
    - B. Establish a function/task hierarchy
        1. Work plans
        2. Task dependency relationships
        3. Task tracking
        4. Allocation and chargeback

In each of the following chapters, the related phases of the I-CASE technology will be explained in the form of boxed inserts such as this one and the earlier one that described data modeling. An example of the activities that occur in each system life cycle phase will be included in each chapter to link the chapter material to the I-CASE technology.

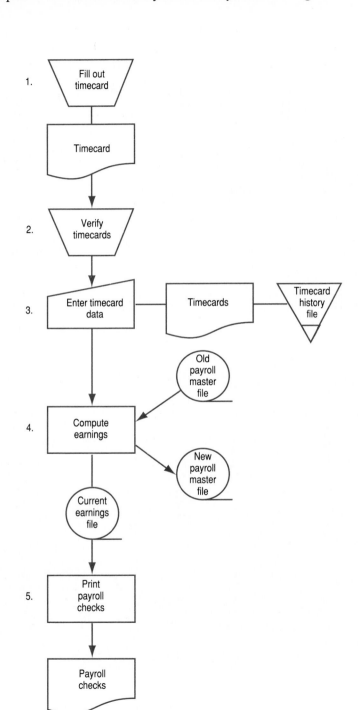

Figure 1.10
A system flowchart of an existing payroll system.

The Design Phase   Once analysis is completed, the new or improved system must be designed. This effort of specifying how the new system will work is called **systems design.** This task, too, is primarily the responsibility of the systems analyst. In designing the system, the systems analyst considers various solutions and documents the best one with tools such as data flow diagrams, structured English, and the data dictionary. Figure 1.11 shows how the processes of the new payroll system can be documented with a *data flow diagram,* or *DFD.* Each upright rectangle represents a process, and the arrows illustrate the flow of data from one process to another. The square represents the origin and destination of data flows, and the open-ended rectangle represents a data file.

Structured English is best suited to the detailed documentation that forms the basis for computer code. Figure 1.12 shows the logic and arithmetic that the new payroll system will use to compute the gross pay for the current pay period.

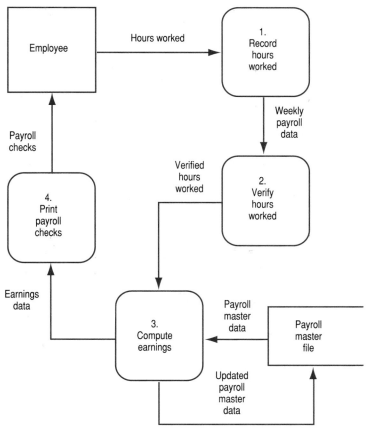

Figure 1.11
A data flow diagram of the main processes of a new payroll system.

```
Compute Current Gross Pay
IF (HOURS.WORKED greater than 40)
  THEN
    Compute OVERTIME.HOURS by subtracting 40 from
      HOURS.WORKED
    Compute OVERTIME.EARNINGS by multiplying
    OVERTIME.HOURS  times 1.5 times HOURLY.RATE
    Compute REGULAR.EARNINGS by multiplying 40
    times  HOURLY.RATE
    Compute CURRENT.GROSS.PAY by adding
    OVERTIME.EARNINGS  to REGULAR.EARNINGS
  ELSE
    Compute CURRENT.GROSS.PAY by multiplying
    HOURLY.RATE  times HOURS.WORKED
End IF
```

Figure 1.12
Structured English of the logic and arithmetic required to compute current gross pay.

Structured English had its roots in *pseudocode,* which means *similar to code.* Structured English is a formal form of pseudocode.

A tool that the systems analyst uses to document data is the *data dictionary.* Figure 1.13 is an example of a data dictionary description of the data element current gross pay.

The Implementation Phase   Once designed, the new system must be transformed from documentation into a working system of computing equipment, programs, people, and data. This transformation is the **systems implementation.** Programmers, computer operators, database specialists, and network specialists do most of the work. When the new system is ready, it is put into use—a process called *cutover.*

The Use Phase   The user uses the new system, perhaps for several years. During this time, changes are made to correct errors, keep the system current, or improve the outputs. These changes are called **systems maintenance.** They

DATA ELEMENT DICTIONARY ENTRY

Use: To describe each data element contained in a data structure

DATA ELEMENT NAME: *CURRENT . GROSS . PAY*

DESCRIPTION: *The amount of an employee's earnings for the current pay period, including regular and overtime earnings*

TYPE OF DATA: *Numeric*

NUMBER OF POSITIONS: *8*

NUMBER OF DECIMAL PLACES: *2*

ALIASES: *None*

RANGE OF VALUES: *000,000.00 to 999,999.99*

TYPICAL VALUE: *1,200.00*

SPECIFIC VALUES: *None*

OTHER EDITING DETAILS: *None*

Figure 1.13
A data dictionary description of the current gross pay data element.

require the contributions of all the information specialists. The maintenance activity continues until an overhaul—a new system—is in order.

Systems evolve within the organizational framework imposed by the enterprise data model and the strategic plan for information resources. This framework ensures that all participants work toward the same end. Keep in mind that the user can work alone throughout the cycle or can work with information specialists, who provide varying degrees of support.

Systems Belong to Their Users

During the system life cycle, the systems analyst and other information specialists take their direction from the user. *The system is the user's system.* Two reasons account for this ownership. First, the system is developed to help solve a problem for the user. Second, in many firms, information services charges the user for system use.

System ownership means that the user makes all the key decisions. The systems analyst provides the specialized knowledge necessary to *advise* the user of decision alternatives. The analyst advises, but the user decides.

Everyone Is a Systems Analyst

Although only certain employees have the title of systems analyst, they are not the only employees who do systems work. In fact, you can take the view that *everyone* in the firm is a systems analyst to a certain degree.

All managers are systems analysts. They are responsible for their systems of people and other resources and are constantly analyzing their systems to make them function properly. After identifying problems, the managers consider alternate solutions, implement the solution that appears to be the best, and follow up to ensure that the solution works.

Nonmanagers such as secretaries, factory workers, and clerical employees are also systems analysts. It is easy to picture factory workers organizing their work areas so that the work flows smoothly. The same scene is easy to envision for a secretary or clerical employee.

Everyone in an organization is responsible for some type of system, even if that system includes only that one person. Each person, from the president to a shipping clerk, is expected to keep his or her system working as it should.

Our Interest Is Computer-Based Systems

Having recognized the broad nature of systems work, we are going to restrict our focus to computer-based systems for the remainder of the text. Our purpose is to prepare you to analyze and design such systems.

We will assume that you are planning on a career as a systems analyst. However, there are two points you should understand. First, the analysis and design principles used by systems analysts can also be used by users. Because of this, the material is as valuable to a career in management as it is to a career as an information specialist. Second, the principles that we establish for computer-based systems apply equally well to noncomputer-based systems.

The Impact of End-User Computing on Information Services

We have recognized that end-user computing (EUC) can exert a strong influence on the system life cycle. However, this influence goes beyond the users' doing a portion of the systems work. EUC is exerting a sweeping influence on the information services organization. The work performed by the information specialists is increasing in complexity, more emphasis is being placed on consulting, and information resources are being distributed throughout the firm.

More Complex Systems

Although some end users can develop complex systems, they generally prefer to concentrate on the simpler ones. Many of the end-user applications are electronic spreadsheets. Others are database management systems such as dBASE and Rbase, which are used to retrieve information from the database. Few end users use conventional programming languages or specialized languages for performing sophisticated operations.

The more difficult systems are left to the information specialists. Although this might appear disadvantageous for the information specialists, it is the main justification for their existence. *Somebody* must assume responsibility for the complex systems, and that assignment goes to information services. On the positive side, the increasing complexity challenges information specialists to perform at a higher level and provides a higher level of satisfaction in meeting that challenge. Systems analysts and other information specialists are professional problem solvers who have the view that no system problem is too tough to handle.

More Emphasis on Consulting

When users do not pursue EUC, they perform mainly a control function during the system life cycle, as shown in Figure 1.14. In this scenario, the user does the planning but turns the analysis, design, and implementation over to the information specialists. The heavy line in the figure shows how the systems workload shifts during the life cycle. This is the *traditional approach to the system life cycle,* which has characterized systems development and use throughout most of the computer era.

Today's user, on the other hand, does development work but does not shut out the specialists altogether. In this scenario, illustrated in Figure 1.15, the systems

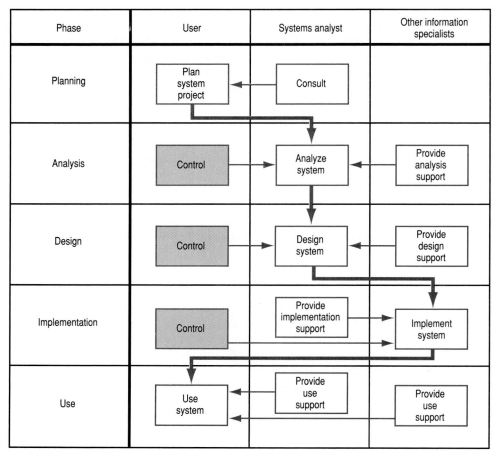

Figure 1.14
The traditional approach to the system life cycle.

analyst provides planning, analysis, design, and use expertise during those phases of the life cycle. Other information specialists can also provide consulting support, as shown. This is the *consultant approach to the system life cycle*. Consulting may well be the dominant role of the analyst of the future.

In a third scenario, the user does all the work alone—a complete dedication to end-user computing. Figure 1.16 illustrates the *end-user computing approach to the system life cycle*.

Firm-wide Distribution of Information Resources

Until EUC emerged, most of the firm's information resources were located in information services. These resources included the computing equipment, the

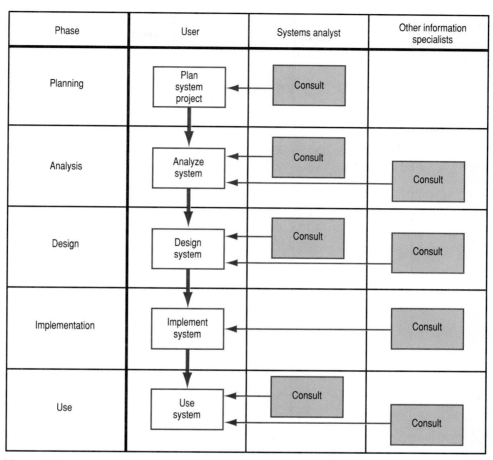

Figure 1.15
The consultant approach to the system life cycle.

software, the database, the information specialists, and the facilities. Users in the various functional areas provided the data, and information services processed the data and provided the users with the information output.

This concentration of information resources within information services no longer prevails—a fact that is evident when you tour firms and see personal computers (PCs) in practically all the offices. These firms have distributed some of their information resources—the PCs, the PC software, and the PC data.

Before long, a user area, such as the marketing division, decides that it needs more computing power than the PCs offer. It requests a minicomputer (mini). The users may also decide it would be more convenient to have their own systems analysts and programmers to develop their systems and their

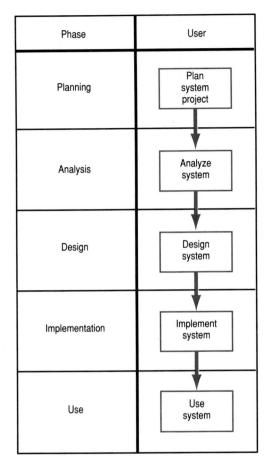

Figure 1.16
The end-user computing approach to the system life cycle.

own operators to operate their mini. Soon, the division's computing unit has the potential to operate independently of information services. At the same time, other satellite computing units crop up in areas such as finance, manufacturing, and human resources.

This distribution trend has been underway in firms for years. If the trend is carried to the extreme, information services *could* eventually have only three primary responsibilities:

- to develop complex, organization-wide systems that the users do not wish to develop themselves
- to provide data communications networks that link the various satellite computing units

- to provide storage for enterprise models, strategic plans, programs, and data in a *repository*, which is available to users throughout the firm

The trends toward more complex systems, more emphasis on consulting, and firm-wide distribution of information resources mean that systems analysts will be located throughout the firm and that the analysts in information services will function differently than before. Systems analysts will still be needed in the organizational framework of tomorrow, even though users will do much of the systems work. Systems analysis will continue to be an excellent career opportunity.

The Keys to Being a Successful Systems Analyst

To become a successful systems analyst, you should acquire as much knowledge and learn as many of the appropriate skills as you can. Systems work, unlike many other professions, demands a wide variety of knowledge and skills.

Systems Analysis Knowledge

Ideally, a systems analyst is computer literate and is knowledgeable in business fundamentals, systems theory, information use in problem solving, systems development, and systems modeling.

Computer Expertise   The analyst must know the fundamentals of computer processing to design computer-based systems. The term **computer literacy** is often used to describe a basic knowledge; however, the knowledge that the systems analyst must have goes far beyond this introductory level. The analyst must have an ability to program, a knowledge of the various computer architectures and how they can be combined to meet particular needs, and an understanding of computer concepts.

Business Fundamentals   Even when working outside the private business sector, such as for a governmental or not-for-profit organization, the analyst should be familiar with the business fundamentals such as economic influences, the nature of free enterprise, and accounting practices and terminology. These fundamentals apply to organizations of all types.

With the trend toward distributed information resources, the future systems analyst would be wise to augment the business fundamentals with specialized knowledge of the activities of the main functional areas—marketing, manufacturing, finance, and human resources. More specialized knowledge of particular industries such as health care, banking, and transportation would also be an asset. You cannot design systems for an organization unless you understand the activity of that organization.

Systems Theory   The systems view that distinguishes between physical and conceptual systems is an example of systems theory. This theory applies to firms, their resources, and their procedures. It provides the pattern for the analyst to follow in performing the analysis and design. The theory tells the analyst which elements must be present and how they must work together for the system to meet its objectives. We elaborate on this systems theory in Part Two.

Information Use in Problem Solving   Modern systems work emphasizes the design of computer-based systems that provide users with problem-solving information. Examples are management information systems, decision support systems, and expert systems. To design such systems, the analyst must have an **information literacy,** an understanding of how information is used in solving problems. Information literacy encompasses the acquisition, application, and sharing of information.

The Systems Development Process   The analyst must thoroughly understand the top-down approach to systems work, described earlier in the chapter. The organizational influences of the enterprise models and strategic plans are explained in Chapter 2. Parts Three through Six are devoted to the system life cycle.

Systems Modeling Capability   The systems analyst should be able to model existing and new system designs, using a variety of documentation tools. The "technical modules" provide an introduction to the most popular tools. The laboratory manual, which accompanies the text, provides an opportunity to learn a CASE tool.

As you can see, you will acquire much of the required knowledge from the text. You will build on this knowledge as you continue your education and pursue your career.

Systems Analysis Skills

Systems work requires many skills, but four stand out as being especially important. They are communications, analytical ability, creativity, and leadership.

Communications   The systems analyst must be able to communicate with other system life cycle participants, as illustrated in the communication chain. During planning and analysis, most communication is oral. As the problem becomes better defined, the analyst transcribes the oral communications. Such written documentation initially takes the form of informal notes. Ultimately, it is refined into proposals, reports, diagrams, and user manuals.

Campus recruiters invariably emphasize the importance of communications skills.

Analytical Ability   The systems analyst must be able to gather data on a problem, sort that data, and identify the cause of the problem. Some of the data may not relate to the problem or may be distorted. Often, user personalities and company politics make it difficult to come to grips with the problem. To function in this environment, the analyst must have an analytical ability. A person who enjoys puzzles and logic problems should enjoy systems work.

Creativity   An experienced analyst is more likely to develop a good solution than is an inexperienced analyst. This is because solutions that proved successful in the past can be repeated. However, the analyst should be creative in developing new solutions when the situation demands. Systems theory provides the basis for developing creative solutions.

Leadership   The systems analyst plays a leadership role when working with users. The systems analyst serves as a *change agent,* stimulating the user to implement systems that incorporate new technologies or methods. In a sense, the systems analyst is a salesperson, convincing the user that the proposed system has the potential for solving the problem.

The systems analyst can also apply leadership skills when working with other information specialists. First, the analyst can apply the skills while a member of project teams. Soon, the systems analyst will be promoted to management, perhaps managing all the systems analysts in the firm. Leadership is the key to career advancement.

Figure 1.17
The systems analyst is a leader.

It is difficult to excel in all these skill and knowledge areas. Even experienced systems analysts continually work to keep their knowledge current and their skills sharp. You never stop learning as a systems analyst. This, combined with the ever-changing nature of the problems that you face, is the main reason the work is so exciting.

How to Become a Systems Analyst

Three types of firms hire systems analysts. These types are *computer-using firms*, such as Federal Express and American Airlines; *consulting and accounting firms*, such as EDS and Arthur Andersen; and *hardware vendors*, such as IBM and DEC. Figure 1.18 is a diagram showing the career paths that exist within these types of firms. Computer-using firms employ systems analysts in information services and user areas. The analysts help the firms solve their problems. Consulting and accounting firms have systems analysts work with their clients to solve client problems. Hardware vendors employ systems analysts as part of marketing teams that sell their customers on hardware solutions and then assist them in implementing the solutions.

Computer-using firms recruit college graduates to begin as programmers or programmer-analysts. A **programmer-analyst** does programming work and analysis work. Some of the firms hire graduates with bachelor's degrees. Some require a master's. Some of the firms expect a very high grade-point average. Some are less demanding. Some of the firms require a major in information systems or computer science. Some do not. Career advancement takes the form of promotion to management positions within information services or in other areas of the firm where the analyst has established a reputation as a problem solver.

Consulting and accounting firms also recruit college graduates to begin work at the programmer-analyst level. These firms have career paths that eventually lead to becoming a partner of the firm.

Hardware vendors recruit college graduates to begin work as systems analysts without an apprenticeship as programmers or programmer-analysts. Some vendors do not require a major in information systems because they have their own formal training programs.

You will note that persons who perform well as analysts for consulting, accounting, and hardware firms are often given management opportunities within computer-using firms.

Steps to Take Now

While in college, you can take several steps toward a career in systems analysis. You can:

- major in information systems to learn the tools of problem solving
- minor in an application area such as marketing, management, or finance, where you can apply the tools

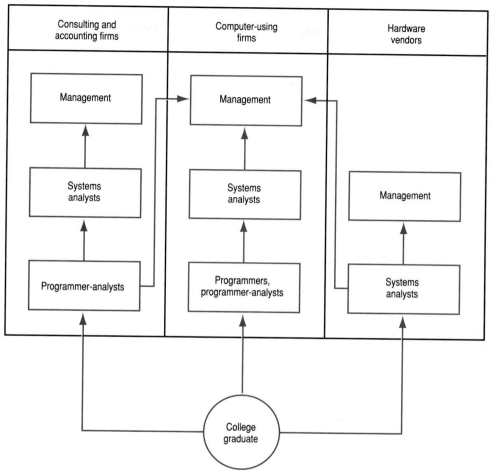

Figure 1.18
Systems analysis career paths.

- take elective courses that contribute to the knowledge and skills identified earlier
- keep your grade-point average high, especially for courses in your major
- develop your leadership skills in student organizations
- continually develop your communications skills by doing your best work on term projects and team presentations
- gain computer-related job experience through summer jobs or co-op programs

These are only a few strategies to consider. Talk with your instructor, your academic advisor, your placement officer, and your parents, and get their advice.

Summary

A system is a group of elements that work together to accomplish an objective. A physical system is tangible, and a conceptual system represents a physical system. A firm can be regarded as a physical system, as can its employee and machine resources. Managers use conceptual systems to manage physical systems. The value of a conceptual system is determined by the accuracy with which it represents its physical system.

Business organizations use the computer in many different ways, but a framework consisting of five major subsystems can represent all the uses. The framework is called the computer-based information system (CBIS). The subsystems are the data processing system, the management information system, the decision support systems, office automation systems, and expert systems.

Computer-based systems are created by following a process that begins at the executive level. The executives engage in enterprise planning to define the activities the firm should perform. This description, called the *enterprise model*, can consist of a definition of the data that the firm needs to perform the processes. The data definition is called the *enterprise data model.*

The enterprise model and enterprise data model facilitate strategic planning. First, strategic business planning spells out the steps to take in achieving the position described by the enterprise model. Then, strategic information planning, identifies the computer-based systems necessary to support the strategic business plan. The strategic plan for information resources builds on the enterprise data model.

With this organizational framework in place, each system project follows a series of steps called the *system life cycle*. These steps are grouped into phases of planning, analysis, design, implementation, and use. The systems analyst plays a key role in the system life cycle, working with the user and other information specialists. Since it is the user's system that is being developed, the user manages the activity.

Everyone is a systems analyst in a certain sense. Everyone manages a system and is expected to keep that system performing at peak efficiency. The principles used in analyzing and designing computer-based systems apply equally well to other types of systems, such as manual systems, and can be applied by users as well as information specialists.

The work of the systems analyst is changing because of many factors. One factor is end-user computing. Because of end-user computing, systems are becoming more complex, greater emphasis is being placed on consulting, and information resources are being distributed throughout the firm.

If you want to pursue a career in systems analysis, you should gain the required knowledge and skills. To meet the knowledge requirements, you should achieve computer literacy, learn the fundamentals of business, become knowledgeable in systems theory, know how information is used in solving problems, understand the systems development process, and know how to model systems in both a graphic and narrative form. You should also be skilled in communications, analytical ability, creativity, and leadership. You can take certain steps in college to enhance your opportunity to become a systems analyst and perform successfully in your career.

Key Terms

Information specialist

Information services

Enterprise planning, enterprise modeling

Chief information officer (CIO)

User

End-user computing (EUC)

Systems analyst

Systems planning

Systems analysis

Systems design

Systems implementation

Systems maintenance

Computer literacy

Information literacy

Key Concepts

A system as a group of elements with a common objective

How a conceptual system represents a physical system

The computer-based information system (CBIS)

The top-down process of systems development

How the enterprise model is a snapshot of the way the firm should look in the future

How the enterprise data model defines the firm's future data needs

How the strategic business plan is a motion picture of the activities necessary to achieve the status described by the enterprise model

The strategic plan for information resources as a subset of the strategic business plan

The system life cycle (SLC)

The pervasive nature of systems work

The communication chain

How tools are used throughout the system life cycle, to facilitate performance of the various tasks

Questions

1. An ad for a razor with a replaceable double blade refers to it as a shaving system. Is it really a system? What are its elements? What is the objective?

2. What are the two basic types of systems? What determines the value of each?

3. What is a CBIS? How many does a firm have? What are its five subsystems?

4. Which CBIS subsystem incorporates artificial intelligence?

5. Which CBIS subsystem emphasizes communications?

6. Which CBIS subsystem represents the firm's accounting system?

7. Which CBIS subsystem is intended to meet the general information needs of all the firm's managers?

8. Which CBIS subsystem is intended to meet the specific information needs of users working individually or in small groups?

9. What does the word *enterprise* mean? Distinguish between an enterprise model and an enterprise data model.

10. Explain the relationship between the strategic business plan and the strategic plan for information resources.

11. What two things are identified in the strategic plan for information resources?

12. Who develops the strategic plan for information resources?

13. What are the five phases of the system life cycle?

14. Why is the SLC represented by a circle?

15. Which information specialists work directly with the user during the system life cycle?

16. Name a tool that you would use to document processes.

17. Name a tool that you would use to document data.

18. Cutover marks the transition from one system life cycle phase to another. Name those two phases.

19. What three effects does end-user computing have on information services?

20. A finance division manager obtains a PC and *Lotus* and learns how to use them by reading the manuals and asking questions of a friend who is a systems analyst. With this knowledge, the manager produces the firm's monthly income statement. Is this an example of end-user computing? Explain your answer.

21. The text says that if the trend to firm-wide distribution of the information resources continues, information services may be left with three main responsibilities. What are these responsibilities?

22. Name six types of knowledge that the systems analyst should possess. Name four skills.

23. What three types of firms hire persons to become systems analysts? Which type starts you off as a systems analyst, rather than a programmer or programmer-analyst?

Topics for Discussion

1. Explain why your college or university is a system.

2. Is a computer a physical system, a conceptual system, or both?

3. Explain why a surgeon is a systems analyst. Do the same for a lawyer. The president of your college or university. A professor. A student.

4. How will the trend toward EUC open up more management opportunities for systems analysts outside information services?

5. Which is more important for a systems analyst—computer literacy or information literacy? What about a manager?

Problems

1. Go to the library and locate the earliest available article containing the term *end-user computing* in its title. Read the article and summarize it in a written report. Your instructor will specify the required length and format.

2. Write a paper titled "How I Can Develop the Four Systems Analysis Skills While in College."

Consulting with

Harcourt Brace

(Use the Harcourt Brace scenario preceding this chapter to solve this problem.)

Assume that you are Mike Byrnes, the vice-president of MIS, and that the president, Richard Morgan, has called you to his office to explain that he has recently become interested in end-user computing. He has read several articles on the subject and wants to know what effect, if any, it might have on Harcourt Brace's hiring policies. If non-MIS employees will be developing their own systems, what types of knowledge and skills should they have? Harcourt Brace will look for those characteristics during employment interviews. Think the matter over, and write a one-page memo. First, list the types of knowledge and then identify the one you feel is the most important. Do the same for the skills. Support your choices with good arguments. Your instructor will specify the memo format to use.

Case Problem
Alpha Foods

Alpha Foods is a producer of frozen foods, located in Watsonville, California. Alpha has been using computers for accounting and production since the early 1960s. It has a staff of 125 information specialists. Each year, these specialists spend more time maintaining existing systems. A study conducted last year by Sterling Bowerman, the CIO, revealed that systems maintenance occupied 65 percent of the specialists' time. Two reasons exist for this more prominent role of maintenance. First, the number of installed systems increases each year. Second, more and more of the new systems are being implemented by users—end-user computing (EUC). Bowerman estimated that half of last year's new systems were implemented by users with little or no assistance from the information specialists. Although the users are becoming more self-sufficient, Alpha has had no organized program to cultivate EUC. EUC just happened.

This increasing popularity of EUC bothers Bowerman, who is a member of Alpha's executive committee, which consists of the president and other vice-presidents. Although EUC is an issue that is being addressed by the entire committee, Bowerman has a special interest. He fears that as the users continue to develop their own systems, the information services unit—and Alpha— will lose control of the information resources.

The growth of EUC is not the only trend worrisome to Bowerman. Of almost equal importance is the increasing turnover of information specialists—especially systems analysts. The specialists become bored with the maintenance work and want to develop more new systems. After working at Alpha for a year or two, most leave for firms with a lighter maintenance workload. This turnover is costly in two respects. First, considerable expense is incurred in recruiting and training replacements. Second, because the experience level is continually eroded by turnover, the specialists often create flawed systems that must be modified earlier than necessary.

Alice Scott, the Alpha president, has notified the executive committee that next week's meeting will kick off with a discussion of the turnover problem. Scott has asked Bowerman to begin the meeting by summarizing the situation.

ALICE SCOTT: Thanks, Sterling. That's a good summary. Do any of you have any suggestions as to how we could solve this problem?

STERLING BOWERMAN: I've given it a lot of thought. I would like to see us put together a staff of information specialists who could develop practically all the systems that Alpha needs. I know that some users will insist on developing their own, but that will cost us in the long run in terms of lost control. A strong information services unit is a real resource for the company. Rather than see it dribble away, I think we should make it as good as possible. What I suggest is that we go after the very best college graduates—meet the Big Six accounting firms and the IBMs head-on in terms of salary. If we provide outstanding support to our users, there will be less need for end-user computing. As I see it, that would solve our turnover problem in terms of new-system development. Those analysts would be challenged by the requirement of providing top-quality support and would be paid accordingly. As for the maintenance people, I think that, by simply increasing their salaries, we'll get them to stay longer.

ALICE SCOTT: It sounds like you are making a plea for more money—higher salaries for both new-system personnel and maintenance personnel?

STERLING BOWERMAN: That's right. If we're going to have a strong information resource, we have to pay for it.

ALICE SCOTT: Mayme, you've been awfully quiet. Human resources is one of the biggest computer users. What do you think?

MAYME VANCE (VICE-PRESIDENT OF HUMAN RESOURCES): I agree with Sterling that we need to upgrade the quality of our systems analysts. I compare the ones from information services to those from our consulting firm and the hardware vendors, and it's like night and day. Our people just don't stack up. But I see the analysts doing a different type of work than Sterling describes. Human resources is trying to become self-sufficient, and we can do most of our work ourselves. What we need is help on the more difficult systems problems. We need analysts who are knowledgeable in human resource systems and who are up to speed on the latest technologies.

ALICE SCOTT: Thanks, Mayme. Bill, how is it from marketing's standpoint?

BILL STUART (VICE-PRESIDENT OF MARKETING): You know what you might do, Sterling, is recruit two kinds of analysts. The ones who will be working with the users in developing new systems could be of the highest quality. Of course, we would have to pay them a premium salary. The others, who will be maintaining existing systems, wouldn't have to have all the interpersonal skills. They just update documentation, don't they?

STERLING BOWERMAN: Oh, it's more than that. An analyst who is given the job of maintaining a system has to interface with the user to understand what changes are needed. The steps in maintaining an existing system are basically the same as for developing a new one.

MAYME VANCE: I would really hate to see us lower our standard of quality anywhere in information services. If we follow your advice, Bill, someone who is hired for maintenance would never have the chance to work on new systems. Isn't that right?

BILL STUART: Oh, there is always the chance of promotion. Maybe we would have a desperate need for a new-system analyst and wouldn't be able to find one on the open market.

MAYME VANCE: Oh, great. Then we're stuck with a substandard analyst developing our new system. I can't get too excited about that possibility.

ALICE SCOTT: Well, we certainly do have a lot of diverse solutions, don't we? Why don't we move on to another topic and think about this some more. When we get together next week, we'll try to reach a conclusion.

Assignments

1. List the advantages and disadvantages of the three solutions. Do not limit your analysis to the discussion by the executive committee. Use material from the chapter or other sources that the executives might not have discussed.

2. Which solution do you like best? Support your answer, briefly referring to the advantages and disadvantages identified in Assignment 1.

3. Can you think of an alternative that was not presented? If so, briefly describe it and identify its advantages and disadvantages. Do you think it is better than the one chosen in Assignment 2?

Selected Bibliography

Brancheau, James C., and Wetherbe, James C. "Key Issues in Information Systems Management." *MIS Quarterly* 11 (March 1987): 23–45.

Carlyle, Ralph Emmett. "Careers in Crisis." *Datamation* 35 (August 15, 1989): 12–16.

Edelman, Franz. "The Management of Information Resources—A Challenge for American Business." *MIS Quarterly* 5 (March 1981): 17–27.

Franz, Charles R. "User Leadership in the Systems Development Life Cycle: A Contingency Model." *Journal of Management Information Systems* 2 (Fall 1985): 5–25.

Frenkel, Karen A. "Women & Computing." *Communications of the ACM* 33 (November 1990): 34–46.

Ginzberg, Michael J., and Baroudi, Jack J. "MIS Careers—A Theoretical Perspective." *Communications of the ACM* 31 (May 1988): 586–594.

Green, Gary I. "Perceived Importance of Systems Analysts' Job Skills, Roles, and Non-Salary Incentives." *MIS Quarterly* 13 (June 1989): 115–133.

Huff, Sid L. "Information Systems Maintenance." *Business Quarterly* 55 (Autumn 1990): 30–32.

Kerr, Susan. "The New IS Force." *Datamation* 35 (August 1, 1989): 18ff.

Leitheiser, Robert L., and Wetherbe, James C. "Service Support Levels: An Organized Approach to End-User Computing." *MIS Quarterly* 10 (December 1986): 337–349.

Leitheiser, Robert L., "MIS Skills for the 1990s: A Survey of MIS Managers' Perceptions." *Journal of Management Information Systems* 9 (Summer 1992): 69–91.

Moad, Jeff. "Why You Should Be Making IS Allies." *Datamation* 36 (May 1, 1990): 26ff.

Munro, Malcolm C.; Huff, Sid L.; and Moore, Gary. "Expansion and Control of End-User Computing." *Journal of Management Information Systems* 4 (Winter 1987–88): 5–27.

Navathe, Shamkant B. "Evolution of Data Modeling for Databases." *Communications of the ACM* 35 (September 1992): 112–123.

Rockart, John F.; Ball, Leslie; and Bullen, Christine V. "Future Role of the Information Systems Executive." *MIS Quarterly* Special Issue (1982): 1–14.

Scheer, August-Wilhelm, and Hars, Alexander. "Extending Data Modeling to Cover the Whole Enterprise." *Communications of the ACM* 35 (September 1992): 166–172.

"The 'Chief Information Officer' Role." *EDP Analyzer* 22 (November 1984): 1–12.

Wasson, Roger E. "Organizing for Future Technologies." *Datamation* 36 (April 1, 1990): 93–95.

TM
A

Entity-Relationship Diagrams

An **entity-relationship diagram,** called an **ER diagram** or simply an **ERD,** identifies the entities that are described by data and the relationships that exist among them. The technique was introduced by Peter Chen in a 1976 journal article.[1] Here, we present a more modern version of this technique.

Entities

An **entity** is a person, organization, place, object, or event that is so important to the firm that its attributes must be recorded and saved in a database. In most cases, an entity is an *environmental element,* such as a customer or supplier, a *resource,* such as the firm's inventory or accounts receivable, or an event recorded as a *transaction document,* such as a sales order or invoice. Entities are represented in an ER diagram with rectangles, as shown in Figure A.1. The rectangles are labeled with the entity names, which are usually singular nouns.

Relationships

A **relationship** is an association that exists among entities. It is most often illustrated with a diamond, as shown in Figure A.1. In the first example, a relationship exists between a customer and a sales order. A customer orders items from the firm by filling out and mailing a sales order. The customer and the sales order are entities, and the relationship is the act of placing the order. This relationship is read as "Customer *places* sales order." It can also be read in reverse sequence: "Sales order *is placed by* customer."

Attributes (Data Elements)

Each entity has distinguishing characteristics, called **attributes.** For example, a customer has some attributes such as name, number, address, and phone number. The attributes are actually data elements maintained for the entity. Although an

[1]Peter Chen, "The Entity-Relationship Model—Toward a Unified View of Data," *ACM Transactions on Database Systems* 1 (March 1976): 9–36.

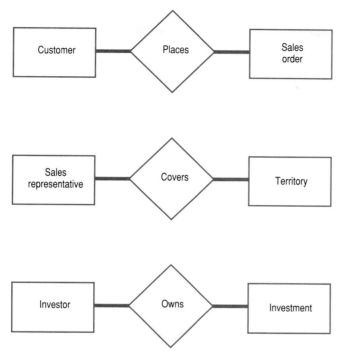

Figure A.1
Examples of entities and relationships.

entity may have several attributes, only some are of interest to a particular information system. For instance, a person has, among others, the attributes of age, height, weight, checking account number, savings account number, debits and credits to these accounts, daily balances, and so on. The attributes age, height, and weight are of fundamental importance to a medical patient information system but irrelevant to a bank customer information system. On the other hand, the checking account number, savings account number, debits and credits to these accounts, and daily balances are crucial to the bank customer information system.

A particular occurrence of an attribute is called an **attribute value.** As an example, a customer name is ABC Company, customer number is 12872, and customer phone is 345-1212. The attribute values are simply the values assigned to the data elements.

The two types of attributes are identifiers and descriptors. An **identifier** uniquely identifies the entity. For example, the identifier of a customer entity is the customer number. On the other hand, a **descriptor** provides information but not identification. Some examples are a customer's name, address, and phone number.

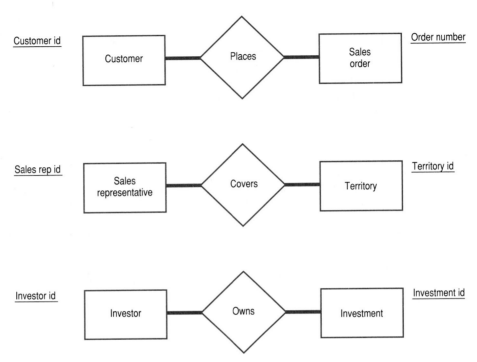

Figure A.2
Identifiers can be shown next to their entities.

Identifiers are often included in ER diagrams as underlined entries next to the entities, as shown in Figure A.2. Descriptors may or may not be shown.

Attributes may also apply to relationships. A relationship attribute is a function of all the entities that participate in the relationship. For example, consider a relationship between students and courses, as shown in Figure A.3. The grade of a student in a particular course is a function of *both* the student and the course—not of the student or the course alone. Therefore, the attributes of the relationship Takes in the figure are Student id, Course id, and Grade.

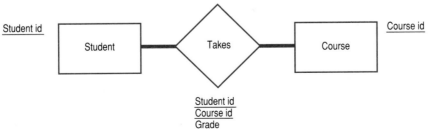

Figure A.3
Attributes of a relationship.

Cardinalities

The term **cardinality** denotes the constraints that can be expressed through a relationship between entities. ER diagrams contain two types of cardinalities: minimum and maximum. For brevity, they are referred to as **min-cardinality** and **max-cardinality.** Figure A.4 illustrates some examples of cardinalities. The min-max cardinalities are shown by symbols separated by a colon (:). The min-cardinality is the symbol to the left of the colon, and the max-cardinality the symbol to its right. The allowable symbols are digits and the letter *m*, where *m* stands for *many*.

The min-cardinality expresses an *existence constraint*—the minimum number of times that the entity can exist within the relationship. For example, in the upper example in Figure A.4, the min-cardinality of 0 for the Customer entity means that a customer can exist in the database even if that customer is not related to an order. The max-cardinality, on the other hand, expresses a *participation constraint,* the number of times that the entity can participate in the relationship. The max-cardinality of *m* for Customer means that a customer can be related to many orders. Similarly, the min-max cardinality for the Order is 1:1,

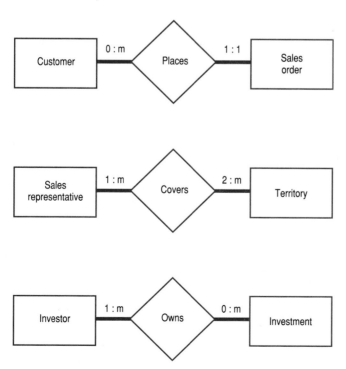

Figure A.4
Examples of specifications of cardinalities.

TM
A

where the min-cardinality of 1 means that an order must be related to at least one customer. An order cannot exist on its own. The max-cardinality of 1 means that an order cannot be associated with more than one customer.

In the middle example in Figure A.4, the 1:m cardinality for Sales representative means that the rep must be related to at least one territory and can be related to many. A sales rep without a territory cannot exist in the database. The 2:m cardinality for Territory means that a territory must be covered by at least two sales reps and can be covered by many.

In a similar manner, in the lower example in Figure A.4, the 1:m cardinality for Investor means that the investor must be related to at least one investment and can be related to many. The 0:m cardinality for Investment means that an investment can exist in the database without being related to an investor and that an investment can be related to many investors.

An entity with a min-cardinality of 0 is called a **strong entity.** That means that it can exist on its own, without being related to the other entity. An entity with a min-cardinality of 1 or more is called a **weak entity** because it cannot exist on its own.

An Example of Cardinality

The ways in which the cardinalities constrain database contents can best be seen in an example. Assume that the firm has four sales reps—John, Kathy, Joe, and Kim—and that there are three territories—Houston, Dallas, and Austin. As shown in Figure A.5, John covers Dallas, Kathy covers Austin, and Joe covers Houston. This database violates the cardinalities because the min-cardinality of 1 for Sales representative means that each rep must cover a territory, and Kim does not. In a similar manner, the min-cardinality of 2 for Territory means that each territory must be covered by at least two sales reps, and that condition is not met.

If the territory assignments are modified as shown in Figure A.6, then the cardinalities are satisfied. Each rep is related to a territory, and each territory is covered by at least two reps.

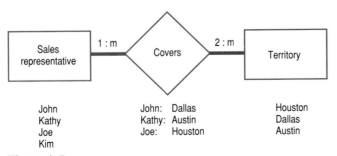

Figure A.5
Example of a database that does not conform to the cardinalities.

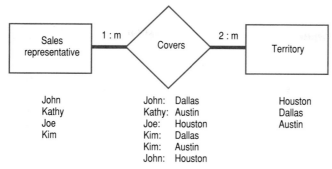

John   John: Dallas   Houston
Kathy   Kathy: Austin   Dallas
Joe   Joe: Houston   Austin
Kim   Kim: Dallas
                Kim: Austin
                John: Houston

Figure A.6
Example of a database that conforms to the cardinalities.

Types of Relationships

All the examples presented so far have been of the binary relationship type. A binary relationship is one between exactly two entities. Other types of relationship not only are possible, but in fact occur quite often. These include unary, ternary, id-dependencies, and is-a relationships.

Unary Relationships

A **unary relationship** is a one between occurrences of the same entity type. For example, consider the hierarchical supervisory relationships among employees of a company. Let John be the president of the company; Mary and Joe be its vice-presidents; Kathy and Richard be supervised by Mary; and Jim and George be supervised by Joe. Only one kind of entity exists, namely, Employee. Even so, the occurrences of these entities are related to one another by a supervisory relationship. The ERD for this situation is demonstrated in Figure A.7. Here, employees are related to one another by the Supervises relationship. This diagram is read as "Employee supervises Employee." The min-cardinality of Employee in its superior role is 1, which means that, to be a superior, an employee must supervise at least one employee. The max-cardinality of m means that an employee playing the

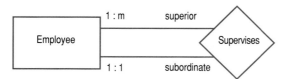

Figure A.7
A unary relationship.

TM
A

superior role may supervise many employees. Similarly, the min-max cardinality of 1:1 for an employee playing a subordinate role means that a subordinate employee must have exactly one supervisor employee. Note that, for unary relationships, the **role** an entity plays in a relationship *must* be specified. The roles in the example are "superior" and "subordinate."

Ternary Relationships

A **ternary relationship** is a relationship among three entities. For instance, in Figure A.8, the relationship Uses exists between the three entities Employee, Software tool, and Project. This relationship can be read as "Employees use software tools in projects." Although rare, it is possible to have relationships among more than three entities. You should, however, avoid creating these relationships. Some experts discourage the creation of even ternary relationships. Techniques can be learned for converting ternary and larger relationships to binary relationships. A discussion of such techniques can be found in a more specialized database or data modeling book.

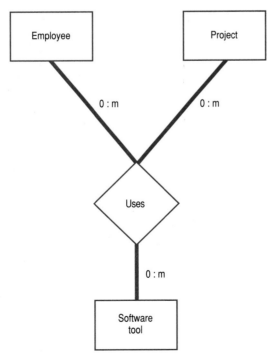

Figure A.8
A ternary relationship.

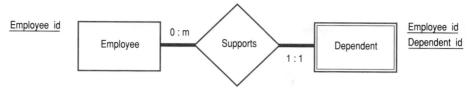

Figure A.9
An id-dependency relationship.

Id-Dependency Relationships

An **id-dependency relationship** (pronounced "eye dee") is created when an entity has no unique identifier of its own. A related entity's identifier must be used to make the dependent entity identifiable. The entity that is not self-identifiable is the id-dependent entity and is represented by a double rectangle, as shown in Figure A.9. This figure models the relationship among employees and their dependents. A dependent does not have its own unique identifier. It has a composite identifier consisting of the employee id and dependent name. The minimum cardinality of an id-dependent entity must be greater than 0.

Is-a Relationships

Generalization and specialization can be represented through **is-a relationships.** Two examples appear in Figure A.10. In the top example, customers are subdivided into current customers and potential customers. The diagram is read by the conjunct "Current customer is a customer and potential customer is a customer." It is not necessary to repeat the word *customer* in both the Current and the Potential rectangles. The Current and Potential entity types are *specializations* of the Customer entity type. Conversely, the Customer entity type is the *generalization* of the Current and Potential entity types. The generalized entity is called the **super-type,** and each specialized entity is called a **sub-type.** In the second example, the generalization of the Management, Line, and Staff entities forms the Employee entity type.

　　The min-max cardinalities are not stated for the generalization hierarchies. It is implicit that the sub-types must always have a min-max cardinality of 1:1. Therefore, the sub-types are weak entities; their existence depends on the existence of their super-type.

An Entity-Relationship Diagram

The following example demonstrates the creation of an ERD. The ERD is intentionally restricted to fit on one page. For larger applications, the ERD may span several pages.

TM
A

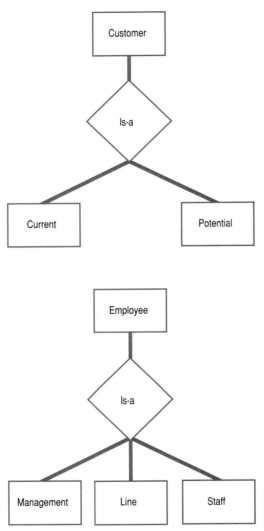

Figure A.10
Two Examples of an is-a relationship.

BuildSolid is a retail furniture store. Customers place sales orders for purchasing furniture and accessories. For each customer, a customer number, name, address, and phone number must be maintained. Each sales order contains an order number, date, information on the customer who placed the order, total before tax, tax, total after tax, customer payment, and balance. Figure A.11 shows a paper sales order.

Each sales order contains one or more order lines. Each order line contains an item number, item name, item description, item quantity (ordered), item unit price, and

```
┌──────────────────────────────────────────────────────────────┐
│                          SALES ORDER                           │
│  ───────────────────────────────────────────────────────────  │
│                                                                │
│  Order number: 1286        Customer number:    673479          │
│                                                                │
│  Order date: 1-19-94       Customer name:      John Adams      │
│                                                                │
│                            Customer address:   2915 Poplar St. │
│                                                Houston, TX 77845│
│                                                                │
│                                                                │
│                            Customer phone:     409-555-1212     │
│                                                                │
│   Item      Item       Item                       Unit          │
│  Number     Name     Description   Quantity       Price    Extension │
│                                                                │
│   9043      chair    dining chair      6         105.99    635.94   │
│   9044      table    dining table      1        1541.99   1541.99   │
│   1893      vase     flower vase       1          45.99     45.99   │
│                                                                │
│                                                                │
│                              Total before tax:    2223.92       │
│                              Tax:                  183.47       │
│                                                                │
│                              Total after tax:     2407.39       │
│                                                                │
│                              Payment:             1000.00       │
│                                                                │
│                              Balance due:         1407.39       │
│                                                                │
```

Figure A.11
An example of a sales order of BuildSolid Furniture Store.

extension. The extension is the quantity multiplied by the unit price. Furniture is delivered to the customer along with a shipment bill. Each shipment bill contains a unique shipment number, shipment date, and the order number that generated the shipment. More than one shipment bill may have to be generated for the same order because items of a single order might be shipped at different times. Similar to the order lines, each shipment bill contains one or more shipment lines. The entries on a shipment line are the item number, item name, item description, and item quantity.

Each piece of inventory is identified by a unique item number, item name, item description, item unit price, quantity on hand, reorder point, purchase order quantity, and quantity on order. Reorder point is the quantity at which a purchase from a manufacturer is requested. Purchase order quantity is the quantity that must be ordered each time. BuildSolid obtains its inventory from various suppliers, or manufacturers. For each manufacturer, a unique name, address, phone number, and contact person must be maintained.

The Development of the Entity-Relationship Diagram

If the ER diagram is to provide a basis for the enterprise data model, it is prepared as part of the enterprise modeling process, and a special task force does the work. On the other hand, if the diagram represents the data of a single system, the project team prepares it. Either way, eight steps are required, and the database administrator, database specialist, and systems analyst play key roles. These steps are iterative and not necessarily performed in the order shown. Analysts go through several iterations before obtaining a satisfactory diagram.

Identify the Entities   First, you identify those persons, organizations, places, objects, or events that you want to describe with data. A good starting point is to examine the firm's existing forms, reports, files, and other documents. Augment this list with user input. Using the order processing example, the following entities can be identified in the first iteration: Customer, Sales order, Shipment bill, Inventory, and Manufacturer.

Identify the Relationships   The second step is to relate the entities in the manner of a subject and an object. One entity is the subject and another is the object. A connecting verb describes the relationship.

Prepare a Rough ER Diagram   Third, sketch a rough ER diagram, and try to arrange the symbols so that the relationships read from left to right or top to bottom. Figure A.12 shows the appearance of the diagram at this point.

Identify Data Elements   The fourth step is to make a list of the data elements that will be maintained for each entity or relationship, underlining the identifiers. The list for Figure A.12 appears in Table A.1.

Perform a Data Analysis   A formal procedure exists for analyzing the data elements for each entity and identifying the arrangement that represents the best logical database design. This process is called **data analysis,** and it uses a technique called normalization to eliminate redundant elements and make the structure as flexible and efficient as possible. **Normalization** consists of converting the data to a series of normal forms—first normal form (1NF), second normal form (2NF), third normal form (3NF), and so on. You begin by putting the data into 1NF and then proceed to the higher levels in sequence. In most cases, 3NF is as far as you go.

First Normal Form When data is in **first normal form,** it has no repeating elements. When we mapped the data elements to the entities in Table A.1, we identified repeating elements for three entities. These repeating elements are the ones that *occur n times.* They recognize that a customer can make several payments, that a sales order can contain multiple inventory items, and that a shipment bill can contain several line items. We solve the problem of repeating groups by creating a new entity and a new relationship for each entity that contains a data element occurring *n* times. Figure A.13 is the modified diagram with three new entities—Customer payment, Order line, and Shipment line. For example, multiple order line items exist, but each has only a single item

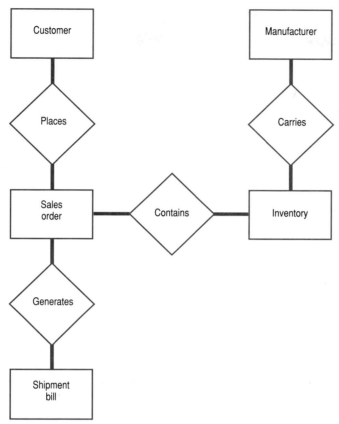

Figure A.12
A rough ERD for BuildSolid Furniture Store.

number, quantity, unit price, and extended price element. The same situation exists for the Customer and the Shipment bill entities.

Table A.2 lists the data elements for the entities in the figure. This table contains no "occurs *n* times" phrases. We have normalized the data to the first normal form. However, the table shows that the newly added entities have two data elements as identifiers. For example, the Customer payment entity is identified by both the Customer number and the Payment date. The same situation exists for the Order line and Shipment line entities. Identifiers that consist of multiple elements are called **composite keys.**

Second Normal Form Data is in **second normal form** when the descriptor elements rely on the entire composite key for identification. If a descriptor element

Table A.1   Initial Data Elements for Each Entity

Customer Entity

Customer number

Customer name

Customer address

Customer phone

Payment date   (occurs n times)

Payment amount   (occurs n times)

Balance   (occurs n times)

Sales Order Entity

Order number

Order date

Customer number

Item number   (occurs n times)

Item name   (occurs n times)

Item description   (occurs n times)

Item quantity   (occurs n times)

Item unit price   (occurs n times)

Item extension   (occurs n times)

Total before tax

Tax

Total after tax

Payment amount

Order balance

Shipment Bill Entity

Shipment number

Shipment date

Shipment amount

Order number

Item number   (occurs n times)

Item name   (occurs n times)

Item description   (occurs n times)

Item quantity   (occurs n times)

Inventory Entity

Item number

Item name

Item description

Item unit price

Quantity on hand

Reorder point

Purchase order quantity

Quantity on order

Manufacturer

Manufacturer name

Manufacturer address

Manufacturer phone number

Manufacturer contact person

depends on *part* of the key, then the entity is not in second normal form. As an example, consider the Order line entity. This entity has a composite key of Order number and Item number. This entity *is not* in second normal form because three of its descriptor elements (Item name, Item description, and Item unit price) can be obtained from the Inventory entity using the item number alone. For similar reasons, the Shipment line entity is not in second normal form. To transform an entity to second normal form, all its descriptor elements that can be obtained from another entity must be omitted. In Table A.3, both the

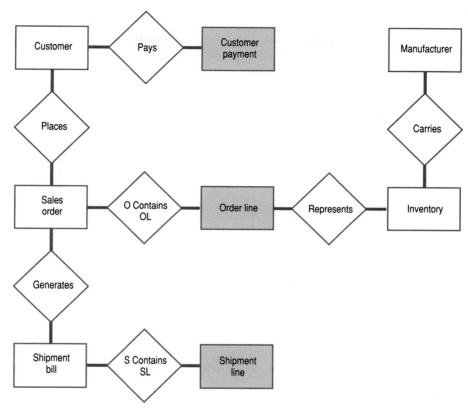

Figure A.13
The modified ERD for BuildSolid Furniture Store.

Order line and Shipment line entities are in second normal form. The Order-item quantity depends on *both* Order number and Item number for identification. Likewise, the Shipment-item quantity depends on both the Shipment number and Item number for identification.

Third Normal Form Data is in **third normal form** when descriptor elements are not dependent on other descriptor elements *in the same entity* for the assignment of values. We have such a problem in the Order line entity (see Table A.2). Item extension can be computed from Item quantity and Item unit price. The solution is to eliminate the Item extension element. Similarly, for the Sales order entity, the descriptor elements Total after tax and Order balance can be omitted, as they can be computed from other descriptor elements. Table A.4 shows these two entities in third normal form.

Table A.2   New Entities and the Modified Data Elements for Each Entity

	Changes from Table A.1
Customer Entity Customer number Customer name Customer address Customer phone	Data elements with "occurs n times" are eliminated.
Customer Payment Entity Customer number Payment date Payment amount Balance	New entity is created from the "occurs n times" data elements.
Sales Order Entity Order number Order date Customer number Total before tax Tax Total after tax Payment amount Order balance	Data elements with "occurs n times" are eliminated.
Order Line Entity Order number Item number Item name Item description Item quantity Item unit price Item extension	New entity is created from the "occurs n times" data elements.
Shipment Bill Entity Shipment number Shipment date	Data elements with "occurs n times" are eliminated.

Continued

Table A.2   New Entities and the Modified Data Elements for Each Entity   Continued

Shipment amount
Order number
Shipment Line Entity   New entity is created from the "occurs *n* times"
Shipment number   data elements.
Item number
Item name
Item description
Item quantity

Inventory Entity   No changes.
Item number
Item name
Item description
Item unit price
Quantity on hand
Reorder point
Purchase order quantity
Quantity on order

Manufacturer   No changes.
Manufacturer name
Manufacturer address
Manufacturer phone number
Manufacturer contact person

Eliminate Duplicate Data Elements if Necessary   Most data elements in an ER diagram must be unique, with one exception. The identifiers may be duplicated in different entities. This enables the users to easily identify the related entities. An example is Order number, which is duplicated in three entities. Two types of duplications, however, do require resolution.

The first type is when two or more different data elements are called by the same name. In this case, the most common practice is to prefix each duplicate data element name with the first (or the first few) letters of its entity. Consider Item quantity in Table A.2, which is a descriptor data element of both the Order line and Shipment line entities. In Order line it refers to the quantity ordered. In Shipment line it refers to the quantity shipped. The two quantities are not

Table A.3   The Order Line and Shipment Line Entities in Second Normal Form

Order Line Entity
<u>Order number</u>
<u>Item number</u>
Item quantity
Item extension

Shipment Line Entity
<u>Shipment number</u>
<u>Item number</u>
Item quantity

necessarily the same. To solve this problem, in Table A.5, the Item quantity of the Order line entity is prefixed by O, and the one in Shipment line is prefixed by S.

The second type of duplication is when the same data element is called by two or more different names. In this case, one must be omitted. In our example, the data elements Balance of the Customer payment entity and Order balance of the Sales order entity are the same. See Table A.2. Both reflect the customer's balance at a point in time. Since all customer payments, including payments upon placing an order, are recorded in the Customer payment entity, Order balance must be eliminated from the Sales order entity. These changes are reflected in Table A.6.

Table A.4   The Order Line and Sales Order Entities in Third Normal Form

Order Line Entity
<u>Order number</u>
<u>Item number</u>
Item quantity

Sales Order Entity
<u>Order Number</u>
Order date
Customer number
Total before tax
Tax

Table A.5   Use Prefixes to Solve the Problem of Duplicate Names with Different Meanings

Order Line Entity
Order number
Item number
O-item quantity

Shipment Line Entity
Shipment number
Item number
S-item quantity

Table A.6   The Final Entity and Data Elements List

	Changes from Table A.2
Customer Entity Customer number Customer name Customer address Customer phone	No changes.
Customer Payment Entity Customer number Payment date Payment amount Balance	No changes.
Sales Order Entity Order number Order date Customer number Total before tax Tax	Total after tax and Payment amountand balance are omitted.

Continued

Table A.6   The Final Entity and Data Elements List   Continued

Order Line Entity Order number Item number O-item quantity	Item name, Item description, Item unit price, and Item extension are omitted. Item quantity is prefixed by O.
Shipment Bill Entity Shipment number Shipment date Shipment amount Order number	No changes.
Shipment Line Entity Shipment number Item number S-item quantity	Item name and Item description are omitted. Item quantity is prefixed by S.
Inventory Entity Item number Item name Item description Item unit price Quantity on hand Reorder point Purchase order quantity Quantity on order	No changes.
Manufacturer Manufacturer name Manufacturer address Manufacturer phone number Manufacturer contact person	No changes.

Identify Min-Max Cardinalities   Identifying the min-max cardinalities requires careful attention. These cardinalities reflect the most important business rules of the enterprise, and their identification requires detailed discussion with the users. Figure A.14 is the final ER diagram after the identification of min-max cardinalities. Note that id-dependent entities are identified by double boxes.

Review the ER Diagram with Others and Refine   As a final step, the members of the task force or project team review the ER diagram with other users (and perhaps members of top management) to ensure that it represents a true picture of the firm's data.

TM
A

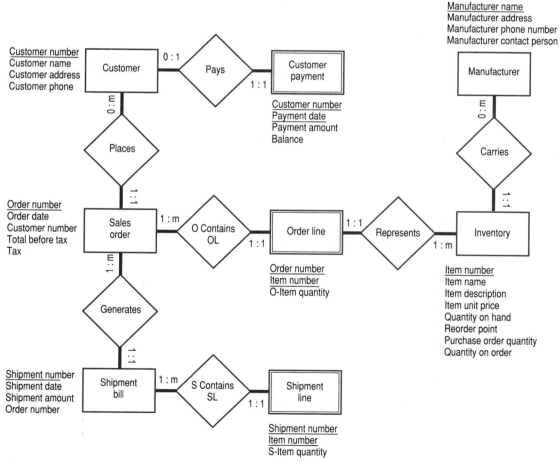

Figure A.14
The final ERD for BuildSolid Furniture Store.

Putting the Entity-Relationship Diagram in Perspective

The popularity of the ER diagram as a systems analysis and design tool recognizes the important role played by the firm's data resource. With the data in computer storage, systems can be developed that provide virtually any information users might need. The ER diagram can provide the basis for:

- the design of a relational database
- documenting the firm's data in a more detailed fashion with the data dictionary
- documenting the processes with tools such as data flow diagrams and system flowcharts

These capabilities explain the popularity of the entity-relationship diagram in the design of modern computer-based systems.

Problems

Draw an ER diagram, including a list of attributes for the following problems.

1. Customers prepare sales orders that specify the firm's products they want to purchase. Each customer is identified by a customer number, and each product by a product number. Each sales order is identified by a sales order number and can specify multiple products. Each customer is described by a name, address, and phone number. Each product is described by a product name, unit price, balance on hand, reorder point, order quantity, and quantity on order. Each sales order describes both the customer and the products. Hint: Do not regard the firm as an entity.

2. Modify the ERD in Problem 1 to include data involved with the ordering of replenishment products from the firm's suppliers. Buyers in the firm's purchasing department obtain products from suppliers by means of purchase orders. Each buyer is identified by a buyer number and described by a name, a phone number, and a product line speciality code. Each supplier is identified by a supplier number and described by name, address, and phone number. Each purchase order can specify multiple products and is identified by a purchase order number and describes both the supplier and the products. Hint: Do not regard the purchasing department as an entity.

3. The Woodstone Corporation owns a large apartment complex by the same name. The owners want to develop a database to keep track of each individual unit, the leases, the lessees (tenants who sign the leases), and workers who perform maintenance or housekeeping jobs on the units.

   For each worker, the worker id, name, address, phone number, and worker's specialty must be maintained. For each apartment unit, a unit number, number of bedrooms, number of baths, street rent, and street deposit must be kept. (Street rent and street deposit are the rents and deposits that will be charged for a new lease. These amounts may or may not be the same as the previous rent and deposit of the lease for that unit.) A worker may work on many units, and the same unit may be worked on by many workers. The work on a unit is documented by recording the worker's id, the unit number, a work description, the date required, and the date performed.

Each lease contract has a unique lease number, rent, deposit, move-in date, and move-out date. A unit may not have more than one lease on it, and a lease is associated with exactly one unit. Lessees sign the leases. Each lessee is identified by his or her name. In addition, the occupation and the previous address of the lessee must be maintained. A lessee must be on at least one lease but can be on many leases. Similarly, a lease can be associated with many lessees but must be associated with at least one lessee.

Lessees generate transactions. Each transaction is identified by a composite key consisting of the lessee's name and a transaction id. In addition, date, amount, and a description of the transaction is maintained. A lessee generates at least one transaction (at the time of signing the lease and paying the deposit) and at most, many. Each transaction is associated with exactly one lessee.

4. Get-Well-Soon is a veterinary clinic owned by a few veterinarians in a small rural town. The vets would like to set up a database to keep track of their operations. These operations relate to clinics' employees, various animals the veterinarians attend to, the animals' illnesses, and the animals' owners.

Employees are categorized into two groups: doctors and staff. For each employee a social security number, name, address, phone number, and salary are maintained. In addition, a specialty must be kept for doctors, and a title must be kept for staff.

Each animal is identified by a unique id. In addition, the kind (e.g., horse), the gender, and the age of the animal is also maintained. If there is a birth, it is required that the mother be related to the offspring. An offspring has the same attributes as any other animal. Therefore, some animals are related to others through birth.

Doctors attend to animals. A doctor may attend to more than one animal, and an animal may be attended to by more than one doctor. For each attendance, the doctor's id (i.e., employee id), the animal id, the date of the visit, the diagnosis, the prescription, and the charge is specified.

For each owner of an animal, an account number, an address, a phone number, and the balance in the account is maintained. An animal can be owned by exactly one owner, but an owner may own zero or more animals.

CHAPTER 2

The Organizational Influence On Computer Use

Learning Objectives

After studying this chapter, you should:

- be familiar with the current strategy for gaining competitive advantage through information and be able to predict additional strategies for the future
- understand the reciprocal relationship that exists between the firm's strategic plan and its resources
- be aware that strategic plans exist for functional areas such as information services
- understand what the strategic plan for information resources should include
- appreciate how information management has evolved since the early days of the computer
- understand the concept behind the CIO title
- better understand end-user computing—who engages in in it, what applications are performed, its potential benefits and pitfalls, and how firms are dealing with it
- know what is meant by the information resources management (IRM) concept and how it integrates the concepts of competitive advantage, strategic planning for information resources, the CIO, and end-user computing

Introduction

As firms participate in a global marketplace, they seek a favorable position relative to their competitors—a status called competitive advantage. Competitive advantage can be achieved in many ways, but much attention currently is being focused on the use of information. The firms that have achieved competitive advantage have concentrated on improving information flows to and from their customers. This strategy will persist in the 1990s but other information-oriented strategies will be added.

As a firm's executives set their sights on achieving competitive advantage, they develop a strategic business plan that provides the long-range guidelines. They select those strategies for which the firm will be able to focus the necessary resources. Each functional area also has its own strategic plan. The term *strategic planning for information resources,* or *SPIR,* refers to the development of a strategic plan that guides the use of all the firm's information resources.

Firms have always recognized the need to manage their information resources, but the process of information resource management has evolved from viewing the computer as a financial tool to viewing it as a strategic organizational resource. The person whose main responsibility is information resource management is often called the *CIO,* for *chief information officer.* More important than the title, however, is the status the title implies.

The information resources are no longer concentrated in information services but are scattered throughout the firm because of end-user computing. End users have varying degrees of computer skills and attempt to implement only certain types of systems. Although end-user computing is generally beneficial, it has potential pitfalls. Firms follow several strategies in controlling their end-user computing activities.

The term *information resources management (IRM)* describes computer use in firms where information is regarded as a strategic resource, the manager of information services is a top-level executive, and information services is a major functional area. IRM is beginning to emerge as a concept that provides the overall setting for computer use in organizations.

Competitive Advantage

In Chapter 1, we recognized that executives engage in enterprise modeling and strategic planning to define where the firm is to be at some future time and what must be done to get there. As the executives engage in this planning, they seek to gain a competitive advantage. The term **competitive advantage** has received widespread attention in the world of business during recent years. It can be defined as the favorable position of a firm in its marketplace in relation to its competitors.

A competitive advantage can be gained by following many strategies. Some are obvious, such as high-quality products, low prices, superior service, and fast delivery. One strategy that has not been quite as obvious is the use of information.

In the 1980s, firms began to realize that information could be just as effective as other resources in helping them get an edge on competitors. The reasoning is logical: if the firm's managers have better information than their competitors have, the managers can make better decisions than can their competitors. The better decisions lead to better problem solutions.

Prevailing Competitive Advantage Strategy

The firms that have emerged as leaders in using information to gain competitive advantage have all followed the same general strategy. They have strengthened the information flows with their customers, making it easier for the customers to buy the firm's products and use its services.

Three firms that pioneered this customer-oriented strategy are American Airlines, American Hospital Supply, and McKesson Drug.

American Airlines   During the early years of the computer, American Airlines saw the potential value of a computer-based reservation system and implemented their Sabre system. Sabre gave American's reservation agents up-to-the-minute flight status information that gave American a position of market leadership. However, American went one step farther, making the flight schedules available to travel agents. This strategy increased the likelihood that the travel agents would recommend American's flights rather than their competitors'.

American Hospital Supply   American Hospital Supply sells products to hospitals and other health care organizations. American Hospital makes ordering easy by letting customers access the American Hospital computer system. Customers do not have to make telephone calls and prepare purchase orders. They simply key the order data in at their terminals, and the data is transmitted directly to the American Hospital computer. The computer advises whether the items are available. If they are, the order is filled. This strategy of achieving competitive advantage by improving information flow from and to its customers enabled American Hospital to triple its sales volume with practically no increase in its support staff.[1]

McKesson Drug   The story at McKesson Drug is just as impressive. McKesson supplies drug stores with pharmaceutical products. McKesson also permits customers access to its computer system. This direct entry of sales orders has enabled McKesson to eliminate 250 clerks who previously took orders and typed purchase forms.[2] McKesson also improved the information flow *to* its customers. When the customers advise McKesson of the prices they will charge for the purchased items, McKesson prepares price labels for them. The labels are enclosed with the items so that the customers can easily attach the labels when the items are received.

In each of the preceding examples, the firm makes its computer system available to its customers—travel agents, hospital administrators, or pharmacists. The customers can easily assess the firm's ability to provide the needed products and services, and make their purchases accordingly. When a firm establishes such a close relationship with its customers, it is difficult for other firms to compete.

Future Competitive Advantage Strategy

Firms will continue to pursue a customer-oriented competitive advantage strategy. However, information linkages will be made with other elements in the firm's environment to achieve even better results. In the future, we can expect

[1]H. Russell Johnston and Michael R. Vitale, "Creating Competitive Advantage with Interorganizational Information Systems," *MIS Quarterly* 12 (June 1988): 161.
[2]Ibid.

Figure 2.1
American Airlines achieved a competitive advantage with its computer-based reservation system.

firms to seek competitive advantage by strengthening information flows with suppliers, the financial community, the government, and competitors.

Information Links with Suppliers   If a firm cannot obtain the raw materials needed to produce its products, no sales will be made. The firm should arrange for direct access to its suppliers' computer systems so that it can place orders quickly and easily.

Information Links with the Financial Community   Similarly, a firm must maintain strong relations with members of the financial community, such as banks, insurance companies, and credit unions. The financial community can keep the firm informed of the state of the economy and can have funds available when the firm needs them.

Information Links with the Government   A firm can also use information from the government to remain current on legal constraints and the economic influences in the marketplace. Is the federal government likely to take an action, such as changing the interest rate, that will affect the economy? Will the government enact legislation, such as that relating to tariffs on imported goods, that will affect the firm's sales? Incoming flows of government information keep the firm alert to these influences.

Incoming Links of Competitor Information   More firms are recognizing the necessity of establishing formal systems to gather competitive information. Computer-based intelligence systems enable firms to not only react quickly to competitive activity, but—in some cases—to anticipate that activity.

In the coming years, more firms will implement electronic information flows with multiple environmental elements, as shown in Figure 2.2. Note that all are two-way flows except for the one-way flow from competitors. The intent of these flows will be to make maximum use of information as a strategic resource.

The Relevance of Competitive Advantage to Systems Work

Information services once exercised almost total control over which systems would be implemented on the computer. Now, that determination is being influenced by the firm's executives. The systems projects that contribute to competitive advantage have the best chance of receiving approval by top management.

Strategic Planning

Many firms have an **executive committee** of top-level managers who make the decisions that have long-term, strategic implications. The executive committee consists of the president and the vice-presidents of the functional areas such as finance, human resources, information services, marketing, and manufacturing. The committee meets periodically, perhaps weekly, to address strategic issues. The executives engage in formal strategic planning on perhaps a quarterly or an annual basis. The planning process considers both environmental and internal influences, as pictured in Figure 2.3.

The Environment Influences the Plan   An organization exists to serve its environment. Business organizations are established to provide products and

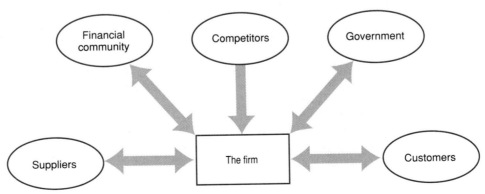

Figure 2.2
Future competitive advantage strategy will feature information flows with multiple environmental elements.

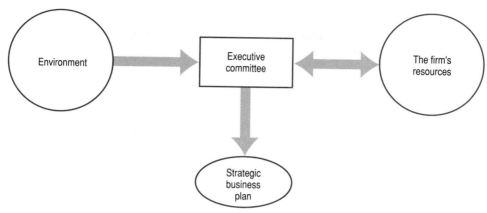

Figure 2.3
The executive committee considers both environmental and internal influences when developing the strategic business plan.

services to their customers and to provide a return on investment to their owners and stockholders. The environment also imposes constraints. For example, competitors seek to take customers away from the firm and the government enacts legislation to ensure ethical business practices. The strategic business plan reflects these environmental influences.

The Plan Influences the Firm's Resources   Once the strategic business plan is in place, the firm must acquire the resources needed to meet the strategic objectives. Take, for example, a strategic plan that includes the opening of a new plant. All the facilities, machines, materials, and personnel required for the plant must be put in place. The firm, therefore, acquires resources in response to the strategic plan.

The Resources Influence the Plan   The executives do not blindly select business strategies. First they appraise the firm's ability to successfully achieve the strategies. The executives know that achievement will be determined by the resources that can be directed at the strategy. For example, they know that a strategy of adding a new product line requires appropriate production facilities and a knowledgeable sales staff, among other resources.

A reciprocal relationship therefore exists between the strategic business plan and the firm's resources. This relationship is illustrated by the two-way arrow in the figure. To develop good strategic plans, the executives must be in tune with both their current and future resources.

Strategic Plans for the Functional Areas

A firm is not restricted to a single strategic plan. Each functional area can develop its own strategic plan to support the firm's plan. A strategic marketing plan, a strategic manufacturing plan, and a strategic plan for information resources are examples of functional plans.

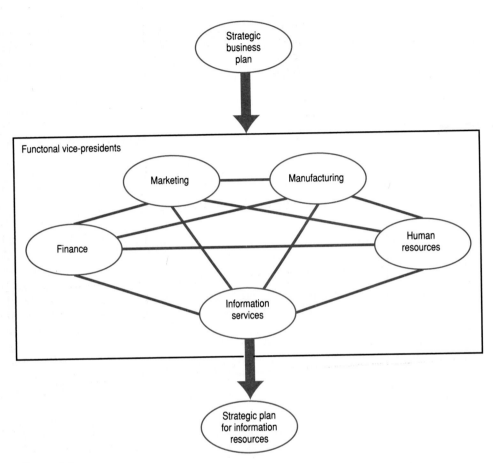

Figure 2.4
The vice-presidents of the functional areas work together when developing their strategic plans.

Each functional plan identifies the resources needed to support its plan and describes how those resources will be acquired. The functional plans coordinate the use of all functional resources.

The functional plans are developed by the executives *working together*, as illustrated in Figure 2.4. For example, the information resources plan is developed by the information resources executive working with the other executives and the information services managers. The term *SPIR* for *strategic planning for information resources*, describes this activity of specifying the guidelines for managing the firm's information resources so that they support the firm's strategic objectives.

Content of the Strategic Plan for Information Resources

Each firm tailors its strategic plan for information resources to its own needs, paying attention to those areas most critical to meeting the firm's strategic objectives. The plan content can vary considerably from one firm to the next. As a minimum, however, the format should consider each CBIS subsystem and the information resources needed to accomplish each subsystem's objectives. This structure is shown in Figure 2.5.

The executives estimate the level of information resources required for each CBIS subsystem. These resources include personnel, hardware, software, data and information, and facilities.

Personnel   The strategic plan for information resources focuses mainly on identifying the future information needs of the users. The executives address the important subject of end-user computing and specify the number and types of information specialists who will be required to work with the users.

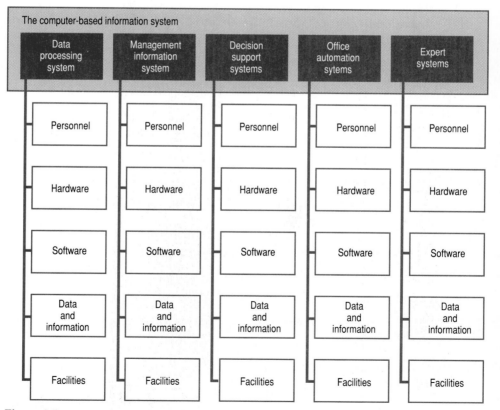

Figure 2.5
The strategic plan for information resources specifies the types of information resources needed by each CBIS subsystem.

Hardware   The executives also select the computing equipment, or hardware, that will provide the platform for the firm's computer-based systems. For example, they might commit to a particular IBM mainframe architecture. They also pay attention to microcomputers, specifying who will approve their acquisition, whether the equipment will be purchased or leased, and whether only particular brands will be allowed.

Software   The strategic plan for information resources specifies the types of computer programs, or software, needed. The firm includes its *make-or-buy policy*, which states whether the software will be acquired by custom programming, purchase of prewritten packages, or a combination of the two.

Data and Information   After identifying the information needed by the users, the strategic plan for information resources describes the data needed to produce that information. The basis for this description is the enterprise data model, developed in the enterprise planning phase.

Facilities   All the preceding information resources are housed in facilities that consist of computer rooms, tape libraries, offices, classrooms, and the like. Facilities housing hardware can be very expensive, with separate temperature and humidity controls, fire detection and extinguishing systems, backup power, and security features.

Putting Strategic Planning in Perspective

The strategic plans for the firm and its information resources provide a positive, stabilizing influence on systems work. The plans minimize the likelihood that the information specialists and computer users will waste their efforts developing misdirected systems. All the systems support the firm's strategic objectives.

Using CASE to Conduct Strategic Business Planning

Upper-CASE tools are CASE tools that support the beginning phases of the system development life cycle—the planning and analysis phases. The activities supported during these phases include strategic business planning, strategic planning for information resources, and business area analysis. The typical outputs from these phases are a business plan, an information strategy plan, and the documentation of a business area under study.

Before planning can begin, enterprise data modeling should be performed to define the data relevant to the organization and how that data interacts. Once the enterprise data model is constructed, all the data and the dependencies can be viewed as a whole. This is an important first step in the construction of a business plan.

A formal business plan has three benefits:

1. It allows data sharing, which ensures consistency among applications.

2. It ensures that the system fits into the strategic plans of top management.

3. It helps reduce costs.

In some situations where no prior formal business strategy exists in the organization, efforts to create a business plan for the proposed information system result in an organizational understanding that the planning process is important and must occur. One of the planning methodologies currently used to undertake business systems planning was devised by IBM and is called *Business Systems Planning (BSP).* The four main steps in BSP are:

1. Define the business objectives of the systems.
2. Define the business processes.
3. Define the business data.
4. Define the information architecture.

Any systems project being cosidered for development must fit into the business mission of the organization. If an executive committee exists and it has chosen this project because of its strategic or operational merit, then the committee need only bring the merit and rationale for the project to the attention of the project team. Understanding why an information system is being developed and how the proposed system fits into the overall information strategy of the firm is important to project team morale, motivation, and effectiveness.

Acceptable objectives for an information system project include, but are not limited to, the following:

1. Improving the accuracy of data input and thereby increasing the quality of information output
2. Reducing CBIS cost by eliminating redundant and unnecessary processes and human intervention
3. Integrating CBIS subsystems to achieve data consistency and increased responsiveness
4. Upgrading customer services to increase customer satisfaction and possibly revenue
5. Increasing throughput (the number of transactions that can be processed within a specified time) or input/output
6. Increasing CBIS reliability
7. Protecting the CBIS from security violations
8. Increasing the usability of the CBIS

Potential objectives of an individual information system must be aligned with the overall objectives of the firm. It is important that the mission statement and business objectives of the organizational unit be well established and understood by all those involved in determining the goals of the specific information system. CASE tools that support the initial stages of the development life cycle force organizations to adhere to a methodology, or recommended approach. This methodological adherence results in better planning, better analysis, and a system that meets organizational and user needs.

The Evolution of Information Management

When firms first installed computers in the 1950s, managers had a vague appreciation of the potential value but felt powerless to apply the new technology. The solution was to assemble staffs of computer specialists and turn the computers over to them. This was done, and firms formed computer departments.

The Computer as a Financial Tool

Because the early computers were used for accounting, the computer departments were located in the finance area with the accounting department. It was common for the manager of the computer department to report to the vice-president of finance, as pictured in Figure 2.6.

Over time, news of the computer successes spread throughout the firm and managers in other areas began experimenting with computer use. Computer specialists sought to apply their electronic tool in new ways, and began working with managers throughout the firm.

Many nonaccounting applications were difficult to computerize, but some successes were realized. Production control, inventory control, and other manufacturing applications proved especially adaptable. It did not take the executives long to realize that the computer had a potential use outside accounting.

The Computer as an Organizational Resource

As the number of computer applications grew during the early 1960s, staffs of computer specialists grew too large to be housed within the finance function.

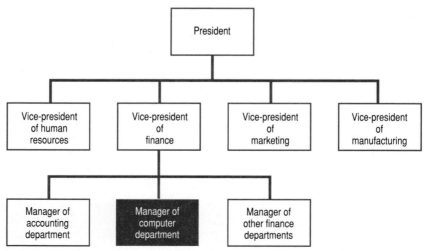

Figure 2.6
The computer department as a part of the financial function.

Many firms decided to give the computer department its own organizational status. The underlying assuption was that the computer department would then become a service unit for the entire firm, not just the finance function. Organizational structures such as the one pictured in Figure 2.7 were common in the 1960s. This structure recognized the computer unit as a major functional area of the firm. Note, however, that the manager of the computer unit was not a vice-president, as were the managers of the other functional areas.

The Computer Manager as a Vice-President

In the 1970s, the status of computing in firms gradually improved and the managers of the larger computer units in the more progressive firms were promoted to vice-president. Also, the increased emphasis on information, rather than data, prompted many firms to adopt the name *MIS department* for their computer units. In most cases, however, the decisions made by these vice-presidents of MIS were limited to technical issues relating to computer use. Figure 2.8 reflects this exclusion of the vice-president of MIS from the strategic decision making of the other executives.

The Computer Manager as a Key Executive

In the 1980s, it became obvious that the computer manager should participate in the firm's strategic planning. This fact became clear when the other executives recognized that information could be used for competitive advantage.

Today, many firms not only have given the computer manager the title of vice-president but have included that person in solving strategic problems of all kinds. Figure 2.9 reflects this arrangement.

The CIO Concept

Some firms have bestowed the title of **CIO,** for **chief information officer,** on the person who manages information services. This title is in keeping with CEO (chief executive officer), CFO (chief financial officer), and COO (chief operating officer).

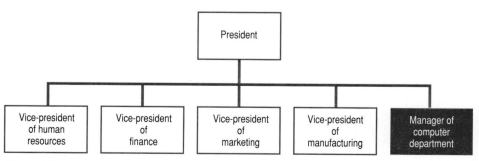

Figure 2.7
The computer department as a separate organizational unit of secondary importance.

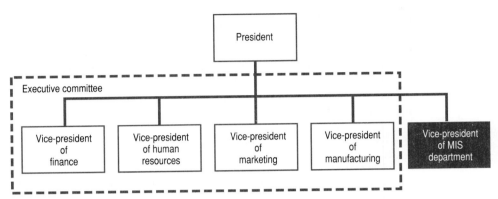

Figure 2.8
Information services led by a vice-president who is excluded from strategic decision making.

Adoption of the CIO title usually indicates that the person who holds the title has achieved executive status and that information services is recognized as one of the major organizational units. However, there are exceptions. Some information services managers function as executives without having the CIO title. Also, some managers with the title do not perform as executives.

The importance of CIO is not in the title, but in the concept. The **CIO concept** implies that the manager of information services is recognized as an executive whose status equals that of managers of the other functional areas.

Time is not the Determining Influence

Before we conclude this description of the evolution in information management, one important point should be made. Although computer use *in general* has

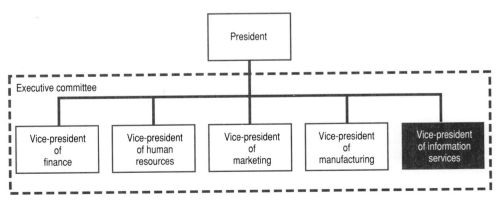

Figure 2.9
Information services led by a vice-president who is considered a key executive.

evolved as described, in many firms the computer manager is *not* a vice-president and information services is *not* a separate functional area. The evolution of computer use is not linked solely to the calendar. More important than time are the attitudes of the executives who determine the status of information services in the firm.

End-User Computing

In Chapter 1, we saw that end-user computing (EUC) dramatically affects the location of the firm's information resources. Systems work is no longer the exclusive property of information services. To fully understand the proper role of EUC in an organization, we must learn more about it. What types of persons engage in EUC, and what are their capabilities? Which applications lend themselves to EUC? What are the potential benefits of EUC? What are its potential pitfalls? How are firms coping with the EUC approach to computer use? Because the objective of this chapter is to address organizational issues, we answer these questions in the following sections.

Types of End Users

To learn more about EUC, two MIT researchers, John Rockart and Lauren Flannery, conducted a study in the early 1980s. They identified six types of end users.[3] An **end user** is a user of computer output who does some or all the work of developing the systems that produce that output.[4] The types of end users are illustrated in Figure 2.10, with percentages indicating the proportions found by the researchers.

- **Nonprogramming End Users** These end users have computer capabilities limited to making selections from menus displayed on their screens, and following prompt messages to get the information they need.

- **Command-Level Users** These end users can use the commands of various prewritten software packages to perform operations beyond those provided by the menus.

- **End-User Programmers** These end users can use programming languages to produce software for their own needs. The end-user programmers also often develop programs for the nonprogramming end users and command-level users in their areas.

[3]John F. Rockart and Lauren S. Flannery, "The Management of End User Computing," *Communications of the ACM* 26 (October 1983): 776-784.

[4]We use the term *end user* in this chapter, but will not make that distinction in the remainder of the text. After this chapter, people will be referred to as *users* regardless of whether they do any developmental work.

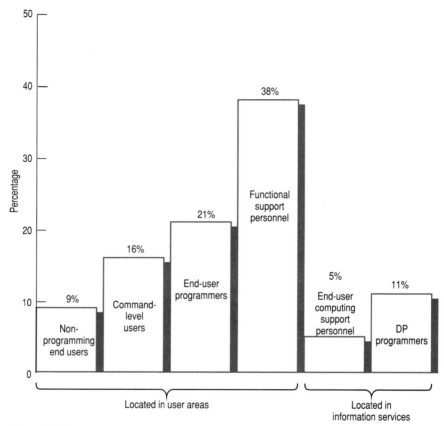

Figure 2.10
Types of end users as identified by rockart and flannery.

- **Functional Support Personnel** These end users have the same capabilities as the information specialists but are not in the information services unit. Rather, they are assigned to user areas.
- **End-User Computing Support Personnel** These end users are information specialists assigned to information services but who specialize in helping users develop their own systems. This group performs in much the same manner as consultants.
- **DP Programmers** These end users are programmers who are a part of information services but who do work for users for a contract fee.

One would expect the situation today to be quite different from that found by Rockart and Flannery. For example, one would expect a higher proportion of nonprogramming end users and less of a need for a programming capability. However, the research contributes three important findings.

- The typical end user does not exist. Various types exist, depending on their degree of computer skills. Some end users are self-sufficient. Others depend almost entirely on other end users and information specialists.

- Some end users have the same computer capabilities as do information specialists in the information services unit.

- Some end users are located in information services. These are information specialists dedicated to user support.

The Rockart and Flannery classification does not consider the knowledge and skills necessary to do systems analysis and design work. Rather, the classification emphasizes programming and computer use. Before the programming can begin, someone—either the systems analyst or the end user—must define the problem, consider various ways to use the computer, select the best way, and define how the computer will be applied.

Which Applications Lend Themselves to End-User Computing?

Because they lack the required skills, because they have better ways to spend their time, or because of some other reason, users limit the kinds of applications they develop. Typically, they develop applications that solve their own problems and the problems of their units. As shown in Figure 2.11, end users have addressed the simpler MIS and DSS applications and the office automation applications, such as word processing and electronic calendaring, that increase personal productivity. The simpler MIS applications include functional subsets of the MIS, such as marketing information systems and human resource information systems. The simpler DSS applications use database management systems to retrieve database contents. They also use electronic spreadsheets for various types of data analysis and modeling.

As end users become more computer literate, we can expect progress beyond the simpler systems. More complex MIS and DSS applications are certain to migrate to user areas, and some day the accounting department might assume responsibility for data processing. It is unlikely, however, that users will attempt to implement complex decision support systems, office automation applications aimed at increasing organizational productivity, the overall MIS for the firm, and expert systems. Nobody is certain how far end users are willing to go in developing their own systems, but there is no question that the trend is in that direction.

Potential Benefits of End-User Computing

Well-planned and well-organized end-user computing can benefit the firm and the users in the following ways:

- **EUC can make computer processing readily available throughout the firm.** The users, rather than information services, determine whether their

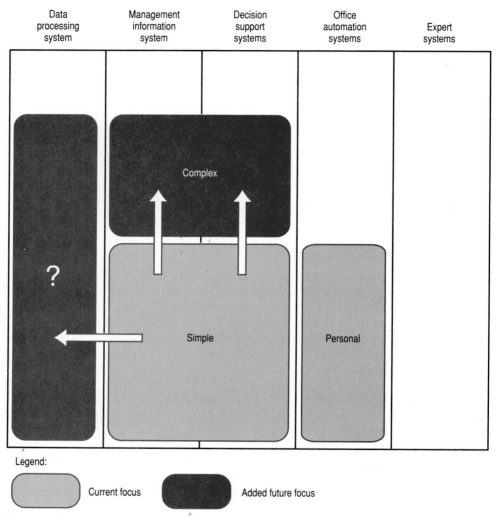

Figure 2.11
CBIS subsystems that lend themselves to end-user computing.

applications go on the computer. Areas of the firm that have generally been ignored by information services in the past, such as human resources, become frequent computer users.

- **EUC increases the likelihood that user needs will be met.** Users understand their information needs better than anyone else. When information specialists work with users, a communications gap can exist. This problem is solved or minimized when users do their own computing.

- **EUC enables users to enjoy computer benefits sooner.** In some firms, information services has so much work that getting a job on the computer can take several years. The queue of jobs awaiting computer development is called the **application backlog.** By not having to rely on information specialists, users can implement their systems and begin using the outputs immediately.

- **EUC gives information specialists more time for other applications.** When information specialists are freed from the application backlog, they can devote time to the complex applications that the users do not develop.

The combined effect of these advantages is a higher level of computer support for the users.

Potential Pitfalls of End-User Computing

As firms attempt to realize the maximum benefits from end-user computing, management must be aware that potential pitfalls exist:

- **Users might obtain the wrong information resources.** Users not properly educated in hardware and software selection might acquire resources ill-suited to their needs. For example, a user who wants to produce computer graphics might purchase a computer not supported by graphics software. Dissatisfaction caused by the mismatch might give the users negative views of computer use.

- **Users might use information resources ineffectively.** When users determine which applications will be computer based, the assembled information resources might not support the strategic plan. For example, the marketing function might assemble a staff of information specialists to work on marketing applications when a greater need exists to develop a human resources information system.

- **Users might use information resources inefficiently.** Users left to determine their own needs can make unnecessary hardware and software purchases. For example, a user might purchase an electronic spreadsheet package when someone in the office next door has the same software and could share it without violating the licensing agreement.

- **Users might contaminate the firm's database.** Careless users can enter error-ridden data into the database. This data is used by others who assume it is accurate and then produce false or misleading information. Using such information to make critical decisions can be disastrous.

- **Users might cause security breaches.** Some users might be careless in keeping their hardware and software secure. For example, users often leave magnetic disks on their desks overnight. Such irresponsibility opens up opportunities for accidental and intentional destruction or misuse.

These potential pitfalls represent serious threats to successful computer use. All, however, can be prevented or minimized if management recognizes that EUC is here to stay and installs the appropriate controls and provides the necessary support.

How Firms Cope with End-User Computing

Firms plan and control end-user computing in the following ways:

- **Emphasize user participation in information management.** More and more firms are engaging in information management in a top-down manner, with a strong user participation at each level. Figure 2.12 shows how the executive committee establishes the strategic guidelines, which are used by the functional vice-presidents as they develop the strategic plan for information resources. Implementation of this plan is accomplished by the **MIS steering committee,** a group of upper-level managers who make key decisions and settle disputes concerning computer use. The leaders of the various systems project teams report to the MIS steering committee.

  Usually, the CIO is a member of the executive committee and the MIS steering committee. However, the CIO seldom chairs either group; that role usually is played by an executive or a manager from a user area. The project teams also have a high concentration of users. In fact, some firms have policies that users, not information specialists, serve as project team leaders.

- **Set policies and standards concerning computer use.** The MIS steering committee sets policies that govern procurement and use of information resources. For example, the committee authorizes certain brands of microcomputers for purchase and establishes procedures that prescribe how computer systems are to be designed. The standards ensure that all systems are compatible.

- **Provide user education and training programs.** Information services offers courses and individual assistance to give users the skills and knowledge necessary to develop and use their own systems.

- **Establish information centers.** An **information center** is an area in the firm where information resources are made available to users. Rather than obtain their own hardware and software, users share the resources of the information center. A help desk staff assists users who encounter difficulty.

Figure 2.14 shows the key role played by the MIS steering committee in controlling the information resources. The resources are located in three areas: the firm's central computing facility, user areas, and one or more information centers. These resources are used by users who rely on information services for all of their systems development work, and by end users who have varying developmental capabilities.

Putting End-User Computing in Perspective

Of all the changes in computing during the past several years, end-user computing has had the most dramatic effect. For the first time, people other than

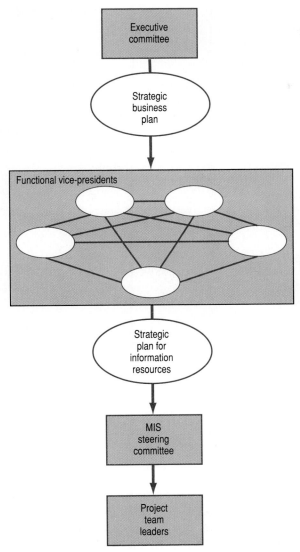

Figure 2.12
Users play key roles in controlling the firm's computing activities at all organizational levels.

information specialists are developing their own applications. This activity means that end users should have systems analysis and design capabilities. It is not sufficient for end users to simply be able to use an electronic spreadsheet or a database management system. The end users go through the same steps and use many of the same tools as the systems analysts in developing the applications.

Figure 2.13
An information center is reserved for the exclusive use of end users.

This requirement becomes more crucial as systems grow in size and complexity and affect multiple organizational units.

The Information Resources Management Concept

Thus far, this chapter has discussed four major topics: competitive advantage, strategic planning for information resources, the CIO concept, and end-user computing. The four topics become tightly intertwined in a concept called information resources management.

Information resources management, or **IRM,** describes the status of computing in a firm in which the executives devote the same attention to managing information as to managing the physical resources. Five conditions exist in such a firm:[5]

[5]For other views of IRM, see Tor Guimaraes, "Information Resources Management: Improving the Focus," *Information Resources Management Journal* 1 (Fall 1988): 10–21.

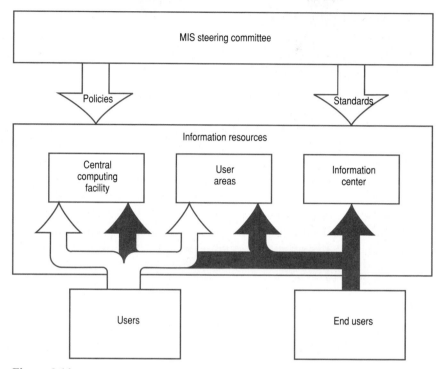

Figure 2.14
Policies and standards by the MIS steering committee control the use of all the firm's information resources.

- The executives recognize that information can be used to achieve a competitive advantage.
- The executives consider the information resources when developing the strategic plan for the firm.
- The manager of information services and the other executives participate equally in developing the strategic plan for the firm.
- The manager of information services participates with the other executives and the MIS staff in developing a strategic plan for information resources.
- The strategic plan for information resources provides for the controlled expansion of end-user computing in the firm.

When IRM is practiced, information services is regarded as a major functional area of the firm. Also, the manager of information services is considered equal to the managers of the other functional areas and has an equal voice in decision making. Having the CIO title is not required.

IRM as a Unifying Concept

The process of pursuing IRM is a series of actions that embraces the major concepts we have addressed and is illustrated in Figure 2.15. The numbered descriptions below correspond to the numbers in the figure.

1. The executive committee develops a strategic business plan that reflects environmental and internal influences.

2. Each functional area develops its own strategic plan to support the strategic plan of the firm. The strategic plan for information services specifies how *all* the firm's information resources will be used.

3. The MIS steering committee, consisting of the CIO and other upper-level managers, uses the strategic plan for information resources to guide their policies and standards for controlling the firm's computing activities.

4. *Users* use information resources in the central computing facility and user areas to satisfy their information needs. They rely on information specialists to develop the systems. The specialists can be located in information services or in user areas as functional support personnel. *End users* use information resources in these same two locations, plus the information center, to engage in end-user computing.

This is the organizational framework within which all information systems work is performed. Information specialists and users alike should understand the setting, because it both imposes constraints and provides opportunities. The firm looks to the information systems for support in achieving the strategic objectives. All information systems, whether developed by information specialists or users, should be designed to meet that expectation.

Using CASE to Prepare Organization Charts

All CASE products support some form of graphic diagramming. Symbols such as rectangles and lines are stock. Figure 2.16 is an organizational chart for Rag City, a small retail chain for women's fashions, that has been prepared by a CASE tool.

The organizational chart is an important step toward understanding the nature of the business and how the functional areas are compartmentalized into areas and levels.

Summary

Today's firms seek a competitive advantage, which can be achieved by meeting customers' needs better than their competitors. A recent innovation in competitive advantage strategy has been the use of information. Firms have established two-way linkages with their customers to simplify order placement. Attention

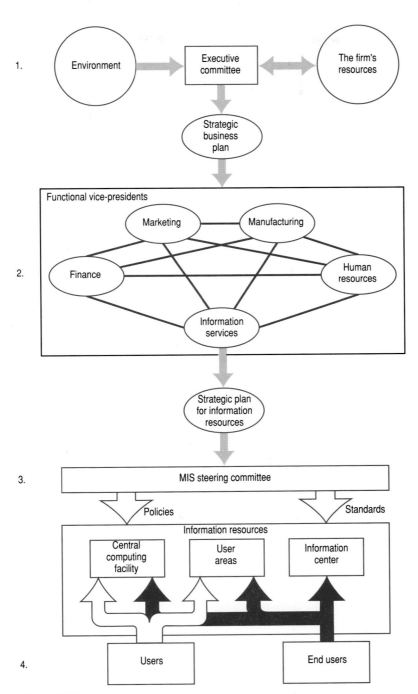

Figure 2.15
An IRM model.

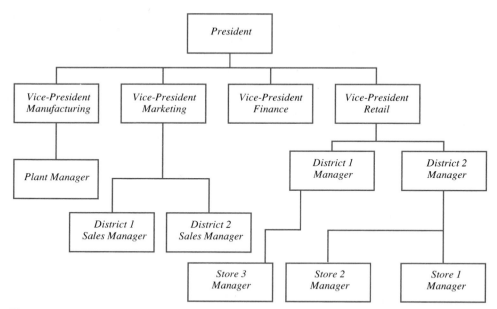

Figure 2.16
A CASE-generated organizational chart.

will continue to be directed at customers, but other environmental elements also will be included. The elements offering the best opportunity for two-way flows are suppliers, the financial community, and the government. Incoming flows can be established from competitors.

The long-range plan that the executive committee develops is called the *strategic business plan*. While developing this plan, executives are aware of the resources they have available. Once set, the plan influences the acquisition of additional resources needed to meet the plan.

Once the strategic business plan is set, similar plans are made for each functional area. The functional plan for information services is called the *strategic plan for information resources*. This plan is developed by the vice-president of information services, working closely with executives in the other functional areas. One way to divide the strategic plan for information resources is to specify the information resources required by each of the major CBIS subsystems. The information resources should include personnel, hardware, software, data and information, and facilities.

It has always been clear to the firm's executives that the information resources must be managed. This management has evolved in stages. The first computer units were part of the finance function, and then they were given separate status. Many managers of the computer units have been promoted to vice-president, but not all have been given a role in making the firm's strategic decisions. Some information services managers have the title of CIO. The title, by itself, means nothing. What is important is the concept that the title represents. The CIO concept recognizes the information services manager as an executive on the same level as the managers of the firm's other functional areas.

The most dramatic changes in computer use are a result of end-user computing. It is important to understand that there are different types of end users, distinguished by their varied computer literacy and organizational location.

Thus far, these end users have only developed systems aimed at solving their own problems and those of their units. They have not addressed organizational systems such as data processing, MIS, complex decision support systems, and expert systems. The responsibility for developing the more complex systems and those systems aimed at supporting the entire firm should remain with information services.

EUC will continue to flourish. It spreads the computing resource throughout the firm, increases the likelihood that the computer output will meet the users' needs, and enables the users to enjoy computer benefits sooner. EUC also benefits information specialists by giving them more time for the complex applications.

If not controlled, EUC can result in the wrong mix of information resources, ineffective and inefficient use of the resources, an inaccurate database, and loss of security. Firms are dealing with the potential pitfalls by establishing user-oriented control groups, setting policies and standards governing computer use, providing users with education and training, and establishing information centers.

The information resources management concept reflects a positive attitude of the firm's executives toward information as a strategic resource. IRM integrates the concepts of competitive advantage, SPIR, the CIO, and EUC. These concepts provide the framework within which all systems work is performed. Information specialists and users must understand this organizational influence so that their systems support the strategic activities of the firm. We focus on these systems in the next part of the text.

Key Terms

Executive committee

Chief information officer (CIO)

Nonprogramming end user

Command-level user

End-user programmer

Functional support personnel

End-user computing support personnel

DP programmer

Application backlog

MIS steering committee

Information center

Key Concepts

Competitive advantage

The reciprocal relationship between the firm's strategic plan and its resources

Strategic planning for information resources (SPIR)

Users as information resources

The CIO concept

End-user computing

Information resources management (IRM) as a concept that integrates competitive advantage, strategic planning for information resources, the CIO, and end-user computing

How IRM influences all systems work

Questions

1. Name three ways that a firm can achieve competitive advantage that do not involve the flow of information.

2. Which element in the firm's environment has received the most attention in prevalent competitive advantage strategy?

3. Who are the American Airlines customers who have access to the Sabre system?

4. Explain how McKesson Drug has established a two-way information link with its customers.

5. Name four environmental elements that will receive attention in the future as a firm works to strengthen its environmental information flows.

6. Which formal group in the firm has responsibility for strategic planning? Who belongs to the group?

7. What is the relationship between the environment and the strategic business plan?

8. What is the relationship between the firm's resources and the strategic business plan?

9. Who develops the strategic plan for information resources?

10. What two groups of people are considered to be information resources?

11. What are the other information resources?

12. How does a make-or-buy policy relate to software?

13. In what area of the firm were the first computer units installed? Why?

14. Name the four phases through which information management has evolved.

15. What must a firm do to show that it supports the CIO concept?

16. Of the six types of end users identified in the text, which are assigned to the information services unit? Which function as information specialists?

17. Which of the following applications do you think would be developed by the firm's credit manager? By the manager of the payroll department? By the sales manager? By information specialists?

    - the firm's electronic mail network that links managers in all functional areas and on all hierarchical levels

    - a report that is produced by scanning the accounts receivable file and identifying poor credit risks

    - a sophisticated mathematical model comprising thousands of equations that is used for forecasting the economy

    - a program that prints payroll checks

    - an expert system that uses artificial intelligence to approve or disapprove a customer's credit application

    - an electronic spreadsheet used to simulate the effect of a change in the sales commission

    - a report that lists the employees who worked more than ten hours of overtime last week

18. What is an MIS steering committee? To whom does the committee report in the organizational hierarchy? Who reports to it?

19. Is an information center another term for the firm's central computing facility? Explain your answer.

20. Which of the potential pitfalls of end-user computing are eliminated or minimized when the firm establishes an information center? Explain your answer.

21. Which five conditions must exist for a firm to reflect the IRM concept?

22. What is the relationship between competitive advantage and strategic planning?

23. What is the relationship between the strategic business plan and the strategic plan for information resources?

24. With whom does the CIO work in strategic business planning? With whom does the CIO work in developing the strategic plan for information resources?

25. What is the relationship between the strategic plan for information resources and end-user computing?

26. Name three places in the firm where hardware can be located. Place an asterisk next to the locations that would provide support to users who are not computer literate.

Topics for Discussion

1. The example in the chapter of an information link with the goverment is an incoming flow of economic and legislative information. What are some examples of an outgoing flow to the government?

2. The chapter mentions the application backlog, but does not explain what caused it. What are your ideas?

3. In addition to the application backlog, what do you think some of the other stimuli to end-user computing might have been?

4. What could the firm do to prevent the database from becoming contaminated by end users who enter erroneous data?

5. What could the firm do to ensure that end users do not breach computer security?

6. Why would the CIO encourage end-user computing? Why would the CIO discourage it?

Problems

1. Assume that you are a systems analyst and your boss asks you to design a questionnaire to be sent through the company mail to all employees using personal computers. Design the questionnaire so that you can determine whether the user is a nonprogramming end user, a command-level user, or an end-user programmer. The questionnaire should include information about the individual, such as department, position, and degree of computer and information literacy. Also include questions about computer use, such as frequency and purpose. The questionnaire responses will enable information services to tailor future support programs to individual users.

2. Go to the library, and find a recent article on the information center. Summarize the article in a written report. Your instructor will specify the length and format.

Case Problem
Anderson's

Anderson's is a leading department store chain in northern California. Its headquarters are in San Francisco. Three distribution centers provide merchandise to its 18 retail stores. A mainframe computer is located at the headquarters and is linked to minicomputers in each store and distribution center. The Anderson's top-level managers include Bill Glass, president; Harold Hall, director of information systems; Chung Kim, vice-president of marketing; Charlie Sims, vice-president of finance; Alice Wingate, vice-president of administration; and Beth Yardley, director of human resources.

Eight years ago, a management consultant studied Anderson's strategic planning system and recommended that an executive committee be formed. This was done, with the membership limited to the president and vice-presidents. The committee wasted no time in developing the initial strategic plan, spanning the period five to ten years in the future. Each year, during September through November, the plan is revised. At the same time, the executives engage in management control—deciding how to implement the strategic plan.

Most strategic planning and management control discussions take place during the committee's weekly meeting. It is early September, and the committee is meeting in Bill Glass's office.

BILL GLASS (PRESIDENT): One item that we must address is how to implement our decision last year to go mail order. As you know, we decided to publish catalogs quarterly, distribute them nationally, and fill the orders from a new distribution facility in Oakland. We decided on this new venture with the idea of gaining at least a 5 percent share of the mail-order market. Are we going to be ready? Kim?

CHUNG KIM (VICE-PRESIDENT OF MARKETING): Well, I don't think we will have any trouble assembling the marketing resources that we need. The mailing list will be no problem, as well. We have access to one of the most complete lists on the market. It is the list of people who have made catalog purchases in the past. If anybody doesn't believe it's a good list, may they be hit on the head with all the catalogs I receive each week. (Everybody laughs.)

CHARLIE SIMS (VICE-PRESIDENT OF FINANCE): What about the mail-order distribution center staff, Kim? Will it be ready?

CHUNG KIM: No sweat. We've got such a good reputation as an employer that staffing won't be a problem. I've been working very closely with Beth (Yardley, director of human resources). We meet every week to look at the situation concerning marketing personnel. We're confident that we'll be able to assemble a top-flight staff. I've told her to follow the same procedure we did when we opened the San Jose distribution center.

BILL GLASS: That sounds good. Does she have a plan underway?

CHUNG KIM: She sure does. Everything is written down—headcounts, schedules, interview dates, the works.

BILL GLASS: Excellent. Charlie, do you see any problems getting the financing to build the new building and buy the equipment we need?

CHARLIE SIMS: None at all, Bill. Our credit rating is tops with every lending institution in the area. I can see only clear sailing.

BILL GLASS: Great. Kim, what about our inventory of stock? Will we be able to offer the same quality of merchandise that we sell in our stores? And, I don't want a lot of backorders. We've got to be able to fill the orders quickly. If we don't, the whole thing goes belly up.

CHUNG KIM: I don't anticipate any problems at all, Bill. The buyers who will order merchandise for the mail-order distribution center will, for the most part, have several years experience with the company. It's the promotion-from-within policy that Beth and I agreed on. And practically all of our suppliers are large-scale operations. They will be able to keep us supplied with what we need.

Bill Glass: They will as long as the computer tells the buyers that it is time to order. Which brings up a question to you, Alice. What about computer support? Harold (Hall, director of information systems) reports to you.

ALICE WINGATE (VICE-PRESIDENT OF ADMINISTRATION): I don't see any problem there. When I told Harold last year about our decision to go mail-order, he was pretty upset. I expected that. That's just his style. He's always complaining that he doesn't have enough people.

CHUNG KIM: Harold has his hands full. He's been working on my marketing information system for more than two years now. I don't think he's ever going to get finished. He claims he doesn't have enough programmers.

CHARLIE SIMS: I've had the same experience with the cash flow model that he's doing for me. Harold has good people, but they're overworked. You have to be patient.

BILL GLASS: Well, we can't be patient when it comes to the mail-order distribution center. When it goes on the air four years from now, the computer operation will have to be in place. Alice, have you given Harold any idea of what level of computer support we will need?

ALICE WINGATE: I've told him what we decided—that the mail-order distribution center will be operated just like the ones for our stores. That's about all he should know, isn't it?

CHUNG KIM: (Interrupting) Well, the systems aren't exactly the same. In the stores, the orders are entered into the cash-register terminals. Everything is automatic. In the mail-order center, the orders will be keyed in by my order takers. The mail-order operation is going to be pretty new from a computer standpoint.

BILL GLASS: That's a good point. Alice, I want you to get with Harold and make certain that he understands just what we expect of him.

ALICE WINGATE: Right, Bill. That's what strategic planning is all about, isn't it?

Assignments

1. In what way is Anderson's doing a good job of strategic business planning?
2. In what way is Anderson's not doing a good job?

3. Should Beth Yardley be a member of the executive committee? What about Harold Hall? Support your answers.

4. Suppose that Bill Glass decides to add only one person to the executive committee. Which person would you recommend? Beth or Harold? Support your answer.

5. Assume that Bill Glass is against enlarging the executive committee. What recommendations would you make to him to solve any problems that exist with the current membership?

Case Problem
Ratliff Stokes Carlyle

Ratliff Stokes Carlyle (RSC) is a large investment firm headquartered in New York City. It has 25,000 employees worldwide. One hundred fifty-five employees are in the human resources (HR) function at the headquarters. Within HR is a section named human resources management systems (HRMS).

Weldon Holley is an MIS professor at a large New York university. He has decided to research human resource applications of the computer. A friend has given him the name of the HRMS manager at RSC. Her name is Geraldine Sullivan, and Weldon has scheduled an interview.

GERALDINE: Well, Mr. Holley, I'm flattered that you picked us for your interview. What is it exactly that you want to know?

WELDON: You can call me Weldon. RSC is one of the more progressive examples of computer use in human resources. I would like to have an idea of what is possible. Then, when I interview other companies, I can see how they stack up. Could you please start by telling me how large your operation is and how it got started?

GERALDINE: Well, five years ago the HRMS unit had only four employees. Although RSC has an information services division with over 1,000 employees, we were generally dissatisfied with the computer support that IS was providing. So, Barbara Wilson, who is the vice-president in charge of HR, decided to increase the size of our HRMS staff. Today, we have thirty-eight employees.

WELDON: That's a big increase. What do these people do?

GERALDINE: The largest section is operations and records. We have nineteen people who do data entry and perform certain control and coordination functions. Then, we have three systems analysts and six programmer-analysts. We also have two project coordinators. The rest are management personnel.

WELDON: Did any of these people come from your IS division?

GERALDINE: Yes, quite a few. In fact, the ones who made the switch regarded it as a promotion, although they stayed at the same level at the same pay. Because of our smaller size, they have a greater sense of importance.

WELDON: That comes as no surprise. If anybody knows how to motivate employees, it should be HR. Tell me, are both IS and HR represented on the RSC executive committee? You do have an executive committee, don't you?

GERALDINE: We do, and HR is represented, but IS isn't. Barbara's boss, Evelyn Taylor, is a member. I think the IS director is four or five levels below the committee.

WELDON: Barbara is not on the committee herself?

GERALDINE: That's right, but Evelyn does an excellent job of looking out for us.

WELDON: I can understand why that arrangement pleases you. But isn't it rather strange that IS isn't represented at a higher level? Doesn't RSC management regard the computer as an important part of the organization?

GERALDINE: Oh, they do, but you've got to remember that this is an investment firm and all the top executives have a financial background. Money's the big thing. The computer is regarded as a support operation. Corporate management looks upon the IS people primarily as technicians.

WELDON: I guess that's to be expected in a financial firm. Does RSC have a strategic plan?

GERALDINE: We sure do. It's all in here. (Geraldine reaches into a desk drawer and pulls out a notebook labeled "Ratliff Stokes Carlyle Strategic Plan." She lays the notebook on her desk.)

WELDON: What about HR? Does it have its own strategic plan?

GERALDINE: Yes. It's in here, too.

WELDON: What about HRMS? Do you have your own strategic plan?

GERALDINE: We sure do. It's in here as well, and it spells out what we would like to accomplish in the next three years.

WELDON: Could you tell me what that is?

GERALDINE: Certainly. We want to accomplish four things. (Geraldine opens the notebook, and places it in front of Weldon. At the top of the page, Weldon reads "HRMS 3-Year Plan." Listed are four items:

• Convert all existing HR applications to DB2

• Replace all PCs with mainframe applications

• Decentralize data entry to the RSC branch offices

• Emphasize end-user computing by providing a database retrieval ability to users)

WELDON: I'm impressed. It looks like a good plan. I guess DB2 is the mainframe database management system that RSC has adopted as the standard, right?

GERALDINE: That's right.

WELDON: Does HR have a mainframe? And, how many micros do you have?

GERALDINE: No mainframe. Only standalone micros—125 of them. There are about forty separate applications installed, which mainly produce reports. A few of the applications enable managers to simulate the effect of decisions. The HRMS programmer-analysts coded about half the programs. The rest are prewritten packages. We've spent about $500,000 on the packages.

WELDON: That's a lot. But your plan says that you intend to get rid of the PCs. What are you going to replace them with?

GERALDINE: Terminals. We want to centralize all the processing on the mainframe. That's a corporate decision made by the executive committee. The feeling is that it will be more economical in the long run and give us better control of the applications. As it is now, everybody can do anything with their computers that they want to. Listen, Weldon (looking at her watch), I've got a meeting in two minutes. I'm going to have to run now. If we need to talk some more, just give me a call.

Assignments

1. Explain what ways RSC has embraced the IRM concept.
2. Explain what ways they have not.
3. Assume that you are a consultant hired by the RSC executive committee to recommend changes in their use of computers. Briefly describe your recommendations.

Selected Bibliography

Alavi, Maryam; Nelson, R. Ryan; and Weiss, Ira R. "Strategies for End-User Computing: An Integrative Framework." *Journal of Management Information Systems* 4 (Winter 1987–88): 28–49.

Amoroso, Donald L., and Cheney, Paul H. "Quality End User-Developed Applications: Some Essential Ingredients." *DATA BASE* 23 (Winter 1992): 1–11.

Berry, Michael F. "Managing the Multiplatform Environment: A Strategic Approach." *Datacenter Manager* 3 (September 1991): 23ff.

Cheney, Paul H.; Mann, Robert I.; and Amoroso, Donald L. "Organizational Factors Affecting the Success of End-User Computing." *Journal of Management Information Systems* 3 (Summer 1986): 65–80.

Clark, Thomas D., Jr. "Corporate Systems Management: An Overview and Research Perspective." *Communications of the ACM* 35 (February 1992): 61–75.

Clemons, Eric K. "Evaluation of Strategic Investments in Information Technology." *Communications of the ACM* 34 (January 1991): 22–36.

Fleck, Robert A. "Information as a Competitive Weapon." *Information Executive* 3 (Spring 1990): 42–46.

Fuller, Mary K., and Swanson, E. Burton. "Information Centers as Organizational Innovation: Exploring the Correlates of Implementation Success." *Journal of Management Information Systems* 9 (Summer 1992): 47–67.

Grant, Robert M. "The Resource-Based Theory of Competitive Advantage: Implications for Strategy Formulation." *California Management Review* 33 (Spring 1991): 114–135.

Harrison, Allison W., and Rainer, R. Kelly, Jr. "The Influence of Individual Differences on Skill in End-User Computing." *Journal of Management Information Systems* 9 (Summer 1992): 93–111.

Hershey, Gerald L., and Eatman, John L. "Why IS Execs Feel Left Out Of Big Decisions." *Datamation* 36 (May 15, 1990): 97–99.

King, William R. "Strategic Planning for Information Resources: The Evolution of Concepts and Practice." *Information Resources Management Journal* 1 (Fall 1988): 1–8.

Lederer, Albert L., and Sethi, Vijay. "Critical Dimensions of Strategic Information Systems Planning." *Decision Sciences* 22 (Winter 1991): 104–119.

Lederer, Albert L., and Sethi, Vijay. "Root Causes of Strategic Information Systems Planning Implementation Problems." *Journal of Management Information Systems* 9 (Summer 1992): 25–45.

Lindsey, Darryl; Cheney, Paul H.; Kasper, George M.; and Ives, Blake. "TELCOT: An Application of Information Technology for Competitive Advantage in the Cotton Industry." *MIS Quarterly* 14 (December 1990): 347–357.

Mockler, Robert J. "Computer Information Systems and Strategic Corporate Planning." *Business Horizons* 30 (May–June 1987): 32–37.

Moynihan, Tony. "What Chief Executives and Senior Managers Want From Their IT Departments." *MIS Quarterly* 14 (March 1990): 15–25.

Newman, William A., and Brock, Floyd J. "A Framework for Designing Competitive Information Systems." *Information Executive* 3 (Spring 1990): 33–36.

Porter, Michael E. "How Competitive Forces Shape Strategy." *Harvard Business Review* 57 (March–April 1979): 137–144.

Porter, Michael E., and Millar, Victor E. "How Information Gives You Competitive Advantage." *Harvard Business Review* 63 (July–August 1985): 149–160.

Reid, Richard A., and Bullers, William I., Jr. "Strategic Information Systems Help Create Competitive Advantage." *Information Executive* 3 (Spring 1990): 51–54.

Sabherwal, Rajiv, and King, William R. "Decision Processes for Developing Strategic Applications of Information Systems: A Contingency Approach." *Decision Sciences* 23 (July–August 1992): 917–943.

Systems in Business

Chapter 1 explained that the firm is a physical system and that management uses a conceptual system to manage the physical system. Thinking in systems terms is known as taking a systems view. Both the manager and the systems analyst benefit from taking a systems view of the manager's area of responsibility. The systems view makes it easier to identify the parts of the manager's system that are not functioning properly and to identify ways of correcting the weaknesses. Part Two prepares you to take a systems view so that you can solve systems problems.

Part Two consists of four chapters. Chapter 3 presents fundamental systems and management concepts that provide the basis for viewing the firm as a managed system. Chapter 4 describes the firm as a system, Chapter 5 describes the environmental system of the firm, and Chapter 6 explains the computer-based information system.

Chapter 3 consists of two major topics—systems theory and management theory. You recall from Chapter 1 that systems theory is an important area of knowledge for the systems analyst. Management theory helps you understand how information is used in problem solving, another of the analyst's knowledge areas.

In Chapter 4, we build the firm as a system, one element at a time. The physical system of input, transformation, and output is explained first. Then, the conceptual system of management, the information processor, and standards is presented as the means of controlling the physical system. The chapter concludes with examples of firms as systems.

In Chapter 5, we continue our description of the firm's environment. We recognize additional environmental elements, and explain how resources flow between the firm and each element. We conclude by recognizing the trend to interorganizational systems that consist of multiple firms linked by electronic information flows.

Chapter 6 explains how the computer can provide the basis for the conceptual systems that managers use to manage physical systems. It examines the fundamental approaches to computer processing and the ways the manager can obtain information from the computer. The chapter concludes with a description of the five subsystems of the CBIS—the data processing system, the management information system (MIS), decision support systems (DSSs), office automation (OA) systems, and expert systems.

II—THE COMPANY IN ITS ENVIRONMENT

(Use this scenario in solving the Harcourt Brace problem at the end of Chapter 5.)

Harcout Brace is a physical system that exists within a larger environmental system. The Harcourt Brace management serves as the control mechanism of the firm—setting objectives, engaging in long-range strategic planning, and monitoring performance. In terms of Harcourt Brace's activities, upper-level management is most interested in the COPS project. Although COPS is essentially an activity of the MIS department, the firm has created an environment that permits end-user computing to develop within certain constraints.

The Harcourt Brace Environment

The most important elements in the Harcourt Brace environment are the customers, the stockholders, and the financial community. All Harcourt Brace activities are aimed at meeting customer needs and providing stockholders a good return on investment. The company must maintain good relations with the financial community to maintain a solid financial position.

Other external organizations also play a key role in the operations. These other elements include the suppliers and competitors. Most of the activities that are performed by the accounting, customer service, and MIS departments in Orlando are concerned with the flow of products from the suppliers and to the customers, in a competitive environment. These three elements—suppliers, customers, and competitors—directly affect Harcourt Brace's computer-based systems.

Suppliers

The supplier situation at Harcourt Brace is different from that of a typical manufacturer. Because Harcourt Brace does not do its own printing, you would expect that it does not purchase its own raw materials. That is not always true. To keep costs low, Harcourt Brace, like most publishers, purchases the paper used to print its books. In a company the size of Harcourt Brace, annual paper costs can be as high as $50 million. The paper companies ship the paper to the printers, who store it until it is used to print the book pages. The pages are shipped to the binderies, who bind the books and ship them to Harcourt Brace.

Other materials provided by suppliers include book covers, cardboard shipping cartons, and other shipping supplies.

Customers

A variety of organizations and individuals buy Harcourt Brace'st books. Elementary and high-school books are bought by school boards and school districts. The decisions to buy college books are made by professors, and the books are ordered by college bookstores. The professional, trade, and medical books are bought by bookstores, individuals, and libraries.

Most customer contacts are made by the Harcourt Brace sales representatives, who are assigned specific territories. The sales representatives call on school board members, school administrators, professors, bookstore managers, librarians, and so on.

The customer services department at Orlando also interacts with the customers—most often with the bookstores.

Competitors

Harcourt Brace competes with several publishing firms in each market. In the college market, the competitors include firms such as Macmillan/McGraw-Hill and Irwin. In many cases, the differences among the competing publishers' books are slight and the performance of the sales representatives and customer service personnel decides which publisher gets the order.

The competition is carried on at the highest ethical level. One reason is that many of the people in the various companies know one another. Moving from one publisher to another occurs frequently. This is especially true at the management level. Large

Figure HB2.1
A Harcourt Brace sales representative making a sales call on a librarian.

publishers such as Prentice-Hall often furnish managers for smaller publishers. Also, the editors and sales representatives often see one another at trade shows, book fairs, and customers' offices.

As an example of the openness with which the publishers compete, the editors typically exchange copies of new texts so that each will know what is on the market.

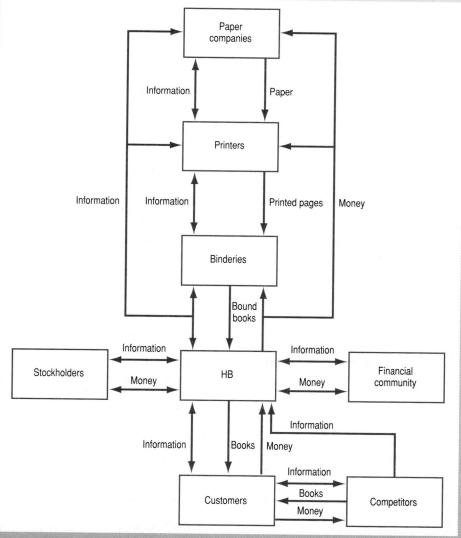

Figure HB2.2
Harcourt Brace and its environment.

Figure HB2.2 illustrates Harcourt Brace's position in its environment. The arrows show the flow of resources among Harcourt Brace and the various elements. Management seeks to maintain these flows at the proper rate to keep the system in equilibrium.

Strategic Planning

The president and vice-presidents make a five-year projection of activities every eighteen months. At the time of this case,the strategic plan was aimed at achieving two primary objectives. One was to return the company to solid financial footing. The other was to implement the COPS system.

In addition to the strategic plan and objectives, the corporate operating budget imposes a guideline for future operations. Each year, the budget for the coming three years is set. This budget planning takes place in July and August .

End-User Computing

Harcourt Brace has neither encouraged nor discouraged end-user computing. If a user has a legitimate need to access computer resources, those resources are made available within certain constraints. One constraint is a PC committee in the MIS department that must approve any user requests for personal computers.

A tour of the headquarters building and the other sites gives the impression that the PC committee is not trying to keep users from obtaining PCs. Although not everyone in the company has a PC, PCs do exist in large numbers. The typical arrangement is to locate a PC, or a network of PCs, in an area convenient to several employees.

The most dramatic example of end-user computing is the SLS system developed by Dave Mattson, manager of packaging at the Bellmawr, New Jersey, distribution center. SLS stands for stock locator system. It is a set of PC programs that determines the location of stock in the warehouse. When books arrive from the bindery, someone must decide where in the warehouse to store them. SLS provides the answer. The system works so well that it has recently been implemented in the Elk Grove, Illinois, distribution center. The plan is to implement it in the other four distribution centers as well. SLS is a system development effort that required about seven years to complete. For all purposes, it was accomplished independently of the MIS department.

While end-user computing abounds in operations and lower management, the opposite is true at the executive level. None of the executives has a computer or a terminal in the office. Instead, executives rely on subordinates for information. One manager was asked what he does when something on the monthly operating report indicates a problem or potential problem. The manager answered, "I get on the phone and try to find out what is going on, why, and what the remedy might be."

Corporate Intelligence

A good example of the environmental focus of Harcourt Brace strategic management is the work of Jenny Jolinski, the assistant director of information management. Jenny's job is to scan the environment for information of potential interest to Harcourt Brace and

funnel that information to the right person. She is especially alert to the interests of the executives. Jenny has established an information profile for each and has built a search structure that is stored in her PC, which is networked to several commercial databases. Each day, the search structure is executed automatically and any items of interest are retrieved and printed. This activity is called *selective dissemination of information*, or *SDI*.

The commercial databases contain information from newspapers, magazines, and TV. For example, the Nexus database contains articles from the *New York Times*, and the Burrelle's Broadcast Database contains the texts of TV news programs. Harcourt Brace pays $10,000 to $12,000 per year for these online database services. The cost may seem high but the databases provide vital environmental information and relieve the managers of the time-consuming task of searching for it.

Personal Profile—Jenny R. Jolinski, Assistant Director, Information Management

Jenny Jolinski received a Bachelor of Arts in Behavioral Sciences from Southern Colorado State College and a Master of Science in Library Sciences from the University of North Carolina at Chapel Hill. She has been with Harcourt Brace since 1985. Her hobbies include tennis and softball. She recently obtained the status of certified records manager from the Institute of Certified Records Managers. She attributes the success of her career to being able to look beyond the trees to the forest. She is willing to view change as a challenge and a broadening experience. Jenny believes that the key to Harcourt Brace's success will be the ability to manage information efficiently and effectively. Jenny recognizes the need to develop an enterprise-wide information policy.

Systems and Management Concepts

Learning Objectives

After studying this chapter, you should:

- understand the hierarchical nature of systems
- know the types of resources that flow through the firm
- know the guidelines for designing conceptual systems
- be able to distinguish between open systems and closed systems, open-loop systems and closed-loop systems
- know why managers are needed in a firm
- be familiar with the more widely known explanations of what managers do
- understand the influence of the computer on management work
- know where managers are located in a firm's organizational structure and understand how the location influences information needs

Introduction

Because systems exist on multiple levels, all employees in a firm can perceive their organization in systems terms. By taking a systems view, they recognize that resources flow to the system from its environment, flow through the system, and then flow back to the environment. Some resources are physical. Others, conceptual resources, represent the physical resources. When designing systems that consist of conceptual resources, the objective is to accurately reflect the current status of the physical system.

All the systems in business interact with their environments. However, some systems can control their own actions and others cannot. Management plays the key role in systems control. It receives information about the status of the physical system, compares that status with the standards of system performance, and changes the physical system as required.

Several management theories have sought to provide an understanding of what managers do. All the theories recognize that information is needed and that the computer has satisfied this need in part. Attention has also been given to the effect of the manager's organizational location on information needs. An understanding of this management theory is of value to managers and systems analysts alike in developing information systems.

System Levels

Systems can exist on several levels. The highest-level system is the **supersystem,** or **suprasystem.** A **subsystem** is a system within a system. For example, if the firm is the system, the national economy is the supersystem and the departments in the firm are subsystems. However, if you direct your attention to one department, the department becomes the system, the firm becomes the supersystem, and any sections in the department become subsystems.

The concept of a system fits a unit on any organizational level. The president can view the firm as a system, and the supervisor of the mail room can view that unit as a system. This universal applicability of the systems concept is why managers, systems analysts, and others with organizational responsibilities find it helpful to think in systems terms. Thinking in systems terms is also called taking a **systems view,** which means regarding the organization as an integration of elements working toward an objective.

System Resources

In Chapter 1, we recognized that the firm is a physical system. The firm consists of physical elements, which can be seen and touched. *The elements composing the physical system are physical resources.*

Management uses conceptual systems to manage the physical systems. *The elements composing the conceptual systems are conceptual resources.*

Physical Resources

The many different physical resources that a firm can assemble can be classified into four types: material, machines (including buildings and land), money, and personnel. These four types exist in all firms.

Material Resources   When thinking of material resources, we tend to think of raw materials used in production. That is perhaps the best example, but we should also recognize that nonmanufacturing firms use materials. For example, banks obtain printed checks from printing companies, restaurants obtain food products from food wholesalers, and newspapers obtain newsprint from paper mills. All the diskettes, magnetic tape cassettes, and paper forms used in a computer installation are material resources.

Machine Resources   Some machines perform specialized functions and are found only in certain types of organizations. Examples are factory machine tools used to transform material resources into products, and presses used to print and collate newspaper pages. Other machines perform functions that can be applied in many situations. Examples are pocket calculators, telephones, and computers.

Figure 3.1
Some firms have automated their warehouses as a way to improve their materials management.

As with the material resources, we use the term *machine* in a broad sense, applying it to machines of all types. We include the firm's buildings and land in this category.

Money Resources   Money is a unique physical resource in two respects. First, it is used to buy all the other physical resources. Second, money itself usually does not flow through the firm. Instead, something representing money, such as checks or electronic signals, is used. Only at the retail level does cash actually change hands, and even there it is giving way to other forms of payment.

Personnel Resources   If you ask managers which physical resource is the most valuable, they will almost invariably reply, "My personnel." This is because the personnel apply the other physical resources to perform the activities of the firm. They use the money to obtain materials and then use machines to transform those materials into finished products.

It is management's responsibility to manage the available resources to meet the firm's objectives.

Conceptual Resources

Conceptual systems are composed of two types of conceptual resources—data and information.

Figure 3.2
A firm's personnel are its most valuable resource.

Data   A firm's **data** consists of facts and figures that are relatively meaning-less to the user. Typically, data exists in such volume that analysis is difficult. To be meaningful, the data must be transformed into information by an information processor.

Information   A firm's **information** is processed data that is meaningful to the user. The **information processor,** which transforms the data into information, can be any kind of mechanism that does the job. Humans, as well as computer and other data processing devices, are information processors.

Guidelines for Designing Conceptual Systems   When designing a concep-tual system to provide information for managing a physical system, keep in mind three points:

Figure 3.3
Supermarket checkout scanners enable the conceptual system to be updated as transactions occur.

- You cannot produce the information without the required data.
- The system should always provide information—not data. When you provide data, you require the user to transform it into information.
- The system should not provide too much information, creating a situation called **information overload.** Too much information can be just as harmful as too little.

The primary objective is to design a conceptual system that is a *mirror image* of its corresponding physical system. The conceptual system should reflect changes in the physical system as quickly as possible after they occur. A conceptual system that functions in this manner is a valuable resource to the user.

The Relationship of the System to Its Environment

Physical systems and their accompanying conceptual systems exist in an environment. Systems theorists have coined two terms that describe the relationship of a system to its environment. These terms are *open system* and *closed system*.

Open Systems

An **open system** interfaces with its environment in some way. Because we have recognized that systems are composed of resources, we will use resource flows as the connections between a system and its environment. All physical and conceptual resources flow from the environment into the firm and from the firm back to its environment as shown in Figure 3.4. A business firm is a good example of an open system.

Closed Systems

The term *closed system* exists in theory only. A **closed system** would be one not connected to its environment by resource flows. Since a system relies on its environment for its life-giving properties, it cannot exist long when these properties are taken away. For that reason, no true closed systems exist. Scientists attempt to create closed systems for experimentation in laboratories, but those are artificial cases. We are interested only in open systems.

The Ability of a System to Control Itself

Systems theorists have also coined terms for systems that can and cannot control their own operations. The terms are *closed-loop system* and *open-loop system*.

Closed-Loop Systems

To control itself, a system must contain a control mechanism. This mechanism compares the actual system performance with the expected system performance.

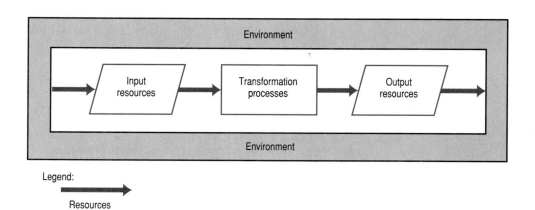

Legend:

Resources

Figure 3.4
Resources flow into a firm from its environment, and then flow back to that same environment.

When the system does not perform as expected, the control mechanism issues signals intended to bring the performance in line with the desired level.

Figure 3.5 shows a system with a control mechanism, a flow of information from the system's output element to the control mechanism, and a flow of signals from the control mechanism to the system's input element. A component that contains the system performance standards is included.

The flow of information and signals between the system's output, control mechanism, and input forms a **feedback loop.** The system is called a **closed-loop system.** It includes a control mechanism and a feedback loop, and can adjust its operation to the desired level.

A Thermostatically Controlled Heating Unit is a Closed-Loop System   You control a heating unit by adjusting the thermostat. When the temperature drops below the set temperature, the thermostat issues a signal that turns on the heater until the temperature rises to the desired level. When that level is reached, the thermostat issues another signal that turn off the heater. Figure 3.6 illustrates how the desired temperature setting, the thermostat, and the feedback loop control the heating unit.

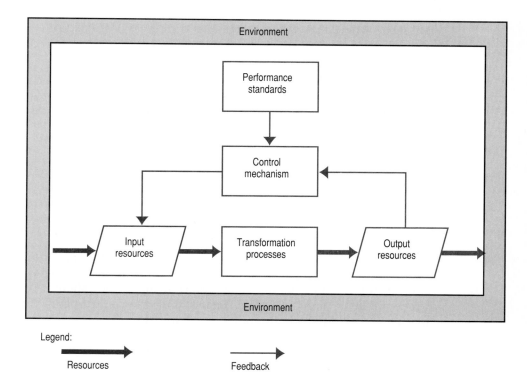

Figure 3.5
A system with elements that enable it to control itself.

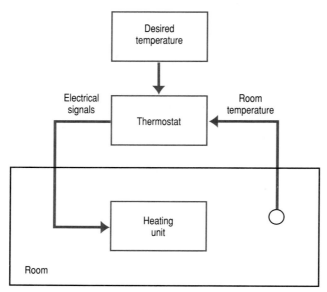

Figure 3.6
A heating unit is controlled by the thermostat.

A Firm is a Closed-Loop System   Managers achieve control by setting objectives and making decisions that cause the objectives to be met. This controlling action is diagrammed in Figure 3.7. Information that describes the firm's operations is compared with the objectives. When the operations do not measure up to the objectives, the managers make decisions that change the operations.

Understand that the firm pictured in the figure is also an open system. Its input resources come from the environment and its output resources go back to the same environment. *A firm is both an open system and a closed-loop system.*

Open-Loop Systems

A system that has no control mechanism and feedback loop, is called an **open-loop system.** Perhaps the feedback loop is missing, or perhaps a gap somewhere prevents information and decisions from flowing as they should. Such a system has no ability to monitor its performance and make adjustments. If adjustments are made, they are made by something external to the system—usually another system.

An automobile is a good example of an open-loop system. Although it might include much sophisticated circuitry, including one or more computers, it cannot drive itself. *You* must provide the control mechanism and feedback loop.

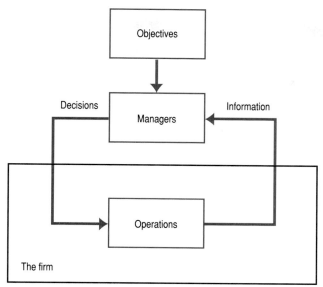

Figure 3.7
A firm is controlled by the managers.

You monitor the performance and make the necessary adjustments with the steering wheel, gear shift lever, and brake and accelerator pedals. By itself, your car is an open-loop system, but you and your car together represent a closed-loop system.

The Need for Managers

Managers serve as the control element of the physical system of the firm. Managers decide where the firm should go and then ensure that it gets there.

A firm's success depends on the ability of its managers to set good objectives, to know whether the objectives are being met, and to quickly and decisively solve problems before a situation gets out of hand.

What Managers Do

Managers have existed for centuries, but not until the early 1900s were formal studies conducted to catalog what managers do. The first descriptions were developed primarily by Frenchman Henri Fayol and American Frederick W. Taylor. Modern contributions have been added during the computer era by Robert N. Anthony and Henry Mintzberg.

Fayol's Management Functions

In a 1916 book, Henri Fayol identified five *management functions,* or activities, that *all* managers perform. The functions are:

- *Planning* what the unit will accomplish. This function includes setting the objectives.
- *Organizing* the firm's resources so that the firm's objectives can be met.
- *Staffing* the units with the personnel needed to accomplish the objectives.
- *Directing* the personnel as the work is done.
- *Controlling* the personnel so that the work helps accomplish the objectives.

Although Fayol primarily was interested in managing the personnel resource, we have recognized that other resources are also important. The modern manager recognizes the need to manage all resources, including data and information.

Taylor's Exception Principle

Frederick W. Taylor, an industrial engineer who lived at the same time as Fayol, recognized that managers have so much responsibility that they cannot be involved in everything. Taylor advocated that managers establish ranges of acceptable performance and become involved only when performance falls outside the ranges. Taylor used the name *exception principle* for this belief that managers should become involved only when performance is exceptionally good or exceptionally bad. The managers try to correct the bad performance and take advantage of the good performance. Today, Taylor's principle is known as *management by exception.*

Anthony's Planning and Control Systems

In 1965, Robert N. Anthony, a Harvard professor, saw a need for a framework that could organize all the material being compiled on management activities.[1] His framework included three main categories—strategic planning, management control, and operational control. Each of these categories can be regarded as *management systems.*

- **Strategic Planning Systems** A **strategic planning system** is the process of deciding on organizational objectives; on changes in the objectives; on the resources used to attain the objectives; and on the policies that govern the acquisition, use, and disposition of the resources. This type of system was

---

[1]Robert N. Anthony, *Planning and Control Systems: A Framework for Analysis* (Boston, MA: Harvard University Graduate School of Business Administration, 1965).

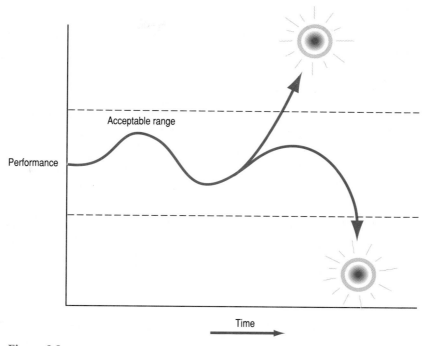

Figure 3.8
Managers become involved when performance goes outside the acceptable range.

the subject of Chapter 2. Executives use it to develop the strategic business plan and its strategic plan for information resources.

- **Management Control Systems** A **management control system** is the process by which managers ensure that resources are obtained and used effectively and efficiently to accomplish the organization's objectives.

- **Operational Control Systems** An **operational control system** is the process of ensuring that specific tasks are carried out effectively and efficiently.

Although Anthony did not intend his three categories to define the three levels of management, he felt that a relationship did exist. Top-level managers engage in strategic planning, middle-level managers perform management control, and lower-level managers provide operational control.

Mintzberg's Managerial Roles

While a doctoral student at the Massachusetts Institute of Technology (MIT), Henry Mintzberg gathered data on five executives that provided additional insight into what managers do. Mintzberg observed each executive for two weeks

Figure 3.9
Mintzberg's three categories of managerial roles: informational role (top); interpersonal role (bottom left); decisional role (bottom right).

and identified ten *managerial roles* that *all* managers play. The roles can be grouped into three categories—interpersonal, informational, and decisional.[2]

In fulfilling the *interpersonal roles*, the manager attends to basic interpersonal relationships. The interpersonal roles are:

- *Figurehead*—performs ceremonial duties, such as attending an employee's wedding
- *Leader*—hires and fires employees, provides motivation, and aligns individual needs with the goals of the organization
- *Liaison*—makes contact with persons outside the manager's own unit

By combining the three *informational roles*, the manager serves as the nerve center of the organization. The informational roles are:

- *Monitor*—gathers information
- *Disseminator*—shares information with subordinates
- *Spokesperson*—shares information with superiors and other persons outside the manager's unit

The manager's *decisional roles* require decisions that commit the unit to new courses of action. The decisional roles are:

- *Entrepreneur*—makes long-term improvements to the unit
- *Disturbance handler*—handles unexpected issues
- *Resource allocator*—decides who in the unit will get which resources
- *Negotiator*—settles disputes within the unit and between the unit and its environment

Figure 3.10 shows how the roles fit into the flow of information to and from the manager. The circles represent the people and organizations who provide and receive information, the arrows depict the flows, and the rectangles represent the roles. The figurehead role is not included because it typically does not directly involve communication or decision making. Instead, it sets the stage for those key activities.

Mintzberg's roles provide a good basis for designing computer-based systems. The roles recognize that information is an important part of the manager's activities and that the information is used in making decisions.

How the Computer Affects Management Work

Data, which is used to produce management information, existed before the computer era. However, the data typically was gathered days or even weeks after the

[2]Henry Mintzberg, "The Manager's Job: Folklore and Fact," *Harvard Business Review* 53 (July–August 1975): 49–61.

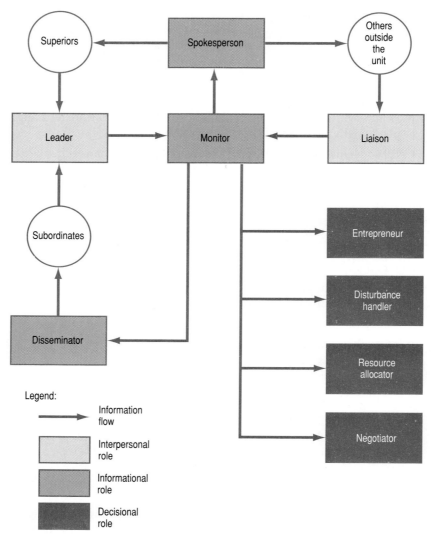

Figure 3.10
Mintzberg's managerial roles are linked by information flows.

transactions, and the machines that processed the data operated at tortoise-like speeds compared to today's machines. For these reasons, the conceptual systems that used the data did not provide management with a good picture.

The computer makes possible accurate and timely data gathering and quick transformation of the data into problem-solving information. The computer affects management work as described by Fayol, Taylor, Anthony, and Mintzberg.

Computer Support for the Management Functions

The computer has not provided equal support to each of Fayol's five functions. Figure 3.11 shows the results of a study that indicates that planning and control receive the most support.[3] In this example, directing receives some support, but organizing and staffing receive little.

In fairness to the designers of today's systems, the low level of support for organizing, staffing, and directing has not been intentional. These are difficult applications to perform on the computer. As such, they represent almost unlimited opportunities to systems developers of the future.

Some successes are finally being achieved in these more difficult application areas. For example, the human resources information systems (HRISs) being

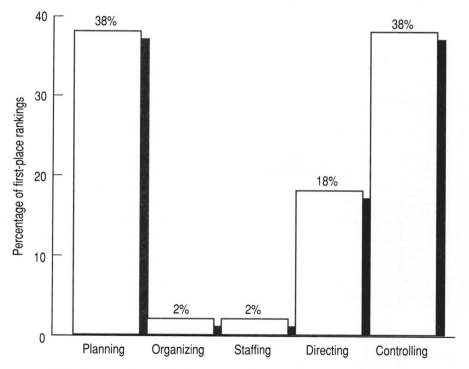

Note: Because of rounding, percentage figures do not total to 100 percent.

Figure 3.11
Computer support for management functions.

[3]In the study, marketing managers ranked the management functions in terms of computer support. The graph shows the percentage of first-place rankings for each function. For more information, see Raymond McLeod, Jr. and John C. Rogers, "Marketing Information Systems: Their Current Status in Fortune 1000 Companies." *Journal of Management Information Systems* 1 (Spring 1985): 57–75.

developed in many firms are aimed at increasing the support for the organizing and staffing functions.

Computer Influence on Management by Exception

The computer can compare transaction data, as it is gathered, to a range of acceptable performance that the manager has defined. When the data values fall outside the range, the computer can notify the manager. For example, the computer can print an **exception report** that calls the manager's attention to the activity and perhaps advises the manager of the appropriate response.

Figure 3.12 illustrates an exception report that identifies customers who are late in paying their bills. Customers whose payments are 30–60 days past due will receive reminder notices in the mail, customers in the 60–90 days category will receive telephone calls, and customers whose payments are more than 90 days past due will be called upon in person by the firm's sales representatives. In this manner, management uses information from the conceptual system to influence the activities of the physical system.

Computer Influence on Planning and Control Systems

The computer supports the managers on each organizational level. As executives use the strategic planning system, they receive reports and outputs from mathematical models. The executives use this information to forecast the environmental situation and to understand the current and projected status of the firm's resources. The management control systems used by middle-level managers also employ mathematical models. These systems develop the operating budgets that allocate the firm's resources to the various units for the next several years. The operational control systems used by lower-level managers consist of reporting systems that keep the managers current on what is happening in their areas.

Until recently, top-level managers did not receive as much support as did managers on lower levels. This situation is changing because of the currently high interest in executive information systems.

Computer Influence on Managerial Roles

The computer enables the manager to better perform the managerial roles by facilitating the flow of information along the paths in Figure 3.10. The manager obtains the information needed to make decisions and help others do their jobs.

You can see a strong connection between the various theories of what managers do and the design of computer-based systems. This connection does not come by accident. System designers have made it a point to understand the theories so that their systems can provide the required support. The theory helps the designers, the systems analysts and the managers, understand *why* the systems must function as they do.

```
                                                              Page 8
                  AGED ACCOUNTS RECEIVABLE REPORT
                     AS OF SEPTEMBER 30
```

----------CUSTOMER----------		CURRENT	30-60	60-90	OVER 90	TOTAL
NUMBER	NAME	AMOUNT	DAYS	DAYS	DAYS	AMOUNT
12383	COATS AND KELLEY	1,003.10	20.26			1,023.36
13972	COBB AUTOMOTIVE	181.90				181.90
13999	COCHRAN HOME CARE	445.19				445.19
14109	COLEMAN AND SONS		153.26	114.14	11.12	278.52
14238	COLLAZO MFG. CO.	367.94	101.74			469.68
15330	COLLEGE APTS.				419.73	419.73
16267	COLOGERO TILE CO.	24.12	122.81			146.93
16329	COMPUTER ACCESS	26.30				26.30
16419	COMPUTER LEARNING		49.42			49.42
16527	CONOCAST CONSTR.	31.29	192.52			223.81
16667	CRESTVIEW CREAMERY	217.82				217.82
17002	CRIPPLE CREEK CONDO	106.95				106.95
17003	DAN'S BODY SHOP		723.80			723.80
19520	DAVID, ANTHONY	1,140.23				1,140.23
19665	DAVIS AND DAVIS	21.93	1.94			23.87
19719	DEAN'S DONUTS	1.10	476.93	174.96		652.99
20011	DEFENSIVE DRIVING	35.87	35.95			71.82

Figure 3.12
An exception report.

Where Managers are Found

In addition to understanding what managers do, it is important to understand where they are located in the organization. The location influences information needs. Managers are found on different organizational levels and in different functional areas.

Management Levels

We have recognized the hierarchical nature of the management structure of an organization. Although a large organization can have many management levels, only three levels typically are used for classification. These levels are the top, middle, and lower.

- Top-level managers include the board of directors (if the firm is a corporation), the president, and the vice-presidents. These are the executives, the managers who engage in the firm's strategic planning as described in Chapter 2. The CIO is a top-level manager.

- Middle-level managers, such as plant superintendents, the sales manager, and the director of purchasing, are persons responsible for large units within the firm. In the information services unit, the managers who report directly to the CIO—such as the managers of systems analysis, programming, and operations—are considered middle-level managers.

- Lower-level managers manage smaller units in the firm. Examples are shop supervisors, sales office managers, the credit manager, and the payroll section supervisor. Within information services, the project team leaders are lower-level managers.

How do you draw the line separating the levels? One approach is to use the person's orientation toward the future. This orientation is called the manager's **planning horizon.** Figure 3.13 shows the planning horizons for the three management levels.

Top-level managers have a planning horizon of roughly five or more years in the future. They must plan the firm's activities for that period. Middle-level managers plan for the period that ranges from the current year to the point at which top-level planning takes over. Therefore, middle-level managers have a planning

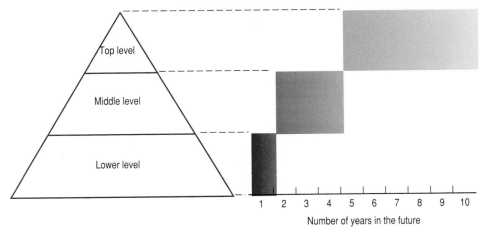

Figure 3.13
Managers can be classified based on their planning horizons.

horizon of approximately one to five years. Lower-level managers plan their units' activities so that they meet the objectives for the current year.

Although some managers plan far into the future, they do not ignore what is happening now. *All* managers, even the CEO, are *very much involved* in the day-to-day activities of the firm.

Influence of Management Level on Information Needs

The management levels are important to the design of computer-based systems. They influence the types of information the managers need and the ways that information is presented. Figure 3.14 shows that top-level managers require more information about the firm's environment than do lower-level managers. This external orientation is based on the environment's strong influence on the firm's long-range operations. Similarly, managers on the lower levels require more internal information. The reason for this internal focus is that the firm's daily activities are carried out by employees who report directly to the lower-level managers.

Note that upper-level managers require some internal information and that lower-level managers require some environmental information.

Figure 3.15 shows how management level also affects the degree of detail required. The higher you are in an organization, the more you rely on summary information. The wider the span of responsibility, the more difficult it becomes to deal with all the details.

You should understand that these preferences for information source and level of detail are believed to apply to *most* managers. However, no two managers are alike and their differences affect their use of information. Many top-level

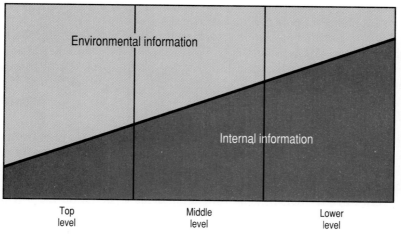

Figure 3.14
Influence of management level on information topic.

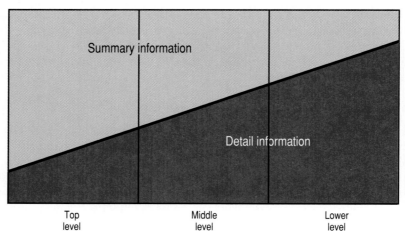

Figure 3.15
Influence of management level on the degree of information detail.

managers prefer to work with details, and many lower-level managers have an environmental focus.[4] The variability in information needs and preferences is one reason that systems work is so challenging and exciting.

Functional Organizational Areas

The most popular approach to organizing a firm's *personnel resources* is based on the functions, or the work, that they perform. This type of organization is known as a **functional organization structure.** Practically all firms have marketing, human resources, finance, and information services functions. Production organizations also have manufacturing functions.

The *functional organization structure* also involves grouping the *nonhuman resources* into the functional areas. For example, the personnel in each functional area have the materials, machines, money, and information that they need to do their jobs.

The Influence of Functional Area on Information Needs

The managers in the various functional areas have different information needs, depending on the primary resources that they manage and the environmental elements with which they interact. These principal information needs are shown in Table 3.1.

Notice that managers in all functional areas receive information about money. All managers pay attention to reports that measure their abilities to manage their

[4]John F. Rockart and Michael E. Treacy, "The CEO Goes On-Line," *Harvard Business Review* 60 (January–February 1982): 82–88.

Table 3.1   Information Needs of Functional Area Managers

Functional Areas

		Marketing	Manufacturing	Human Resources	Finance	Information Services
Physical Resources	Personnel	■		■		■
	Materials	■	■			
	Machines		■			■
	Money	■	■	■	■	■
Environmental Elements	Customers	■				
	Suppliers		■	■		
	Stockholders or Owners				■	
	Labor Unions		■	■		
	Government	■	■	■	■	■
	Financial Community				■	
	Local Community			■		
	Competitors	■				

monetary resources. Also, all managers obtain governmental information that applies to their areas.

   Marketing Information   Marketing managers need information that answers questions about their personnel, such as which sales representatives are selling and which are not. The managers also need information about finished products. The marketing managers want to know which products are selling and which are not. Marketing managers need similar information concerning the firm's customers—who is buying, who is not, which products certain customers are buying, and so on. Marketing managers also need information about their competitors.

   Manufacturing Information   Manufacturing managers need to know about the productivity and efficiency of the manufacturing employees and machines.

They also need information about the raw materials and the use of the various production machines. In addition, manufacturing managers must remain current on supplier status and labor union relationships.

Human Resources Information   Managers in the human resources area are interested primarily in the flow of personnel into, through, and out of the firm. Two environmental elements exert a strong influence on this personnel flow— labor unions and the local community.

Finance Information   Because the finance function is interested mainly in the money flow, special attention is paid to the stockholders or owners and the financial community—the environmental elements that provide funds and offer investment opportunities.

Information Services Information   Managers in information services particularly like to stay current on the status of their personnel, the information specialists. The managers also monitor the status of their machines, the hardware and software. The hardware and software suppliers also receive attention.

These are the information topics of primary interest to the managers. However, the managers are not blind to other information that might have potential value. Managers are usually quick to acquire *any* information that they believe will help them or others in their units.

Using CASE to Perform Business Function Analysis and Conduct an Information Needs Analysis

One of the major tasks in the early stages of system development that is supported by upper-CASE tools is the analysis of the firm's basic activities. The resources of a business organization are commonly grouped into functional areas. A **functional area** is a formal grouping of personnel and other resources according to a major category of work that is performed. A function that is typically found in retail stores such as Rag City is merchandising. Merchandising is concerned with all the firm's sales activities.

An analysis of any functional area will identify multiple functions being performed. A **function** is a group of activities that, when taken as a whole, represent one aspect of the mission of the enterprise. In Rag City, merchandising consists of such functions as buying, inventory, display, sales, returns, and advertising.

The activities performed within each function can be divided into processes. A **process** is a business activity that has related inputs and outputs. The name of a process usually consists of a verb and an object. In Rag City, you would find such processes as *Prepare purchase order*, *Receive stock*, and *Conduct credit check*.

CASE tools can play a key role in gaining an understanding of the business functions early in the system development life cycle. A popular technique in documenting information needs is to divide the functional areas by system types, producing a type of documentation known as an **information needs matrix**. Figure 3.16 is an example,

	Marketing	Manufacturing	Purchasing	Finance Accounting	Administration
Strategic Planning	New Product R&D Service	Capacity Process alternatives	Sourcing Commodity forecasting	Accounting policy Tax model	Resource Systems R&D
Operational Planning	Product mix Pricing	Production plans Labor MRP	Contracts Vendor	Cash flow Capital expenditure Tax management	Systems development R&D budget Staffing
Control Reporting	Budget vs actual Sales force Performance	Scheduling Quality control Process control	Vendor performance Materials availability	Financial statements Funds Tax reserve	Expenses Utilization MBO
Transaction Systems	Order entry Billing	Goods in process Machine utilization Rework Inventory	Inventory Back orders Lead time Scheduling	A/P A/R Payroll General ledger	Skills inventory Legal Environmental

Figure 3.16
An information needs matrix.

produced by a CASE tool, that identifies the system types as strategic planning, operational planning, control reporting, and transaction systems. The functional areas include marketing, manufacturing, purchasing, finance and accounting, and administration.

Using CASE to Build a Decomposition Diagram

One of the simplest diagrams supported by CASE is the decomposition diagram. In a **decomposition diagram,** a high-level task is decomposed, or divided into its lower-level constituent activities. At each successively lower level, the activities represent greater and greater detail. An organizational chart, such as the one developed in Chapter 2, is an example of a simple decomposition diagram.

Decomposition diagrams can be arranged in a horizontal or vertical format, or they can take the form of an *action diagram.* We describe action diagrams in Technical Module I.

Figure 3.17 is a decomposition diagram of the Rag City inventory system. The rectangles represent activities within the treelike structure. This example shows four main branches—sales, receipts, returns, and physical inventory. The high-level task at the

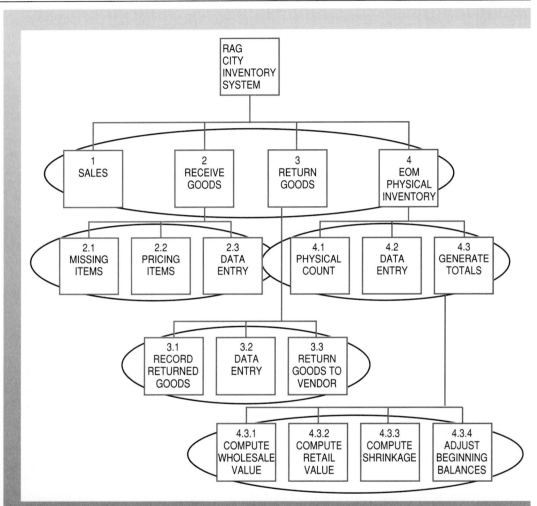

Figure 3.17
A decomposition diagram.

top is the **root,** and the activities at the lowest level are called **leaves.** The activities in between are called **nodes.** When two activities are arranged in a hierarchy, the one on top is the **parent,** and the one on the bottom is the **child.** The root is always a parent, and the leaves are always children.

The children of one parent compose a **sibling family.** In this example, the circles indicate the sibling families that will be identified as processes in data flow diagrams as the system documentation proceeds, later in the system development life cycle.

Virtually all CASE tools support decomposition diagrams in one form or another.

Putting Systems and Management Concepts In Context

Taking a systems view lets you compare an existing organization to a recommended or ideal pattern. The organization must permit physical resources to flow at the desired rate. It should also include a conceptual system consisting of management, objectives, and feedback information. If any of these conditions is missing, the system will not function as it should.

Understanding management theory will help you develop information systems for managers. The theory explains why certain information is needed. The only word of caution is that the theory is believed to apply to *most* managers, but not *all*. The theory provides a good starting point, but each manager has unique needs.

Summary

Although systems can exist on many levels, names have been coined for only three levels. The system under study is called the *system*. The system immediately above is the *supersystem*, or *suprasystem*. The system immediately below is the *subsystem*. Anyone on any organizational level can view his or her unit as a system. This is known as taking a systems view.

Physical systems consist of physical resources—material, machines, money, and personnel. Managers manage these physical resources by using conceptual resources—data and information. Data consists of facts and figures that are relatively meaningless until transformed into information by an information processor, such as a human or a computer. When designing an information system, be sure you have the needed data. Then, provide the user with information rather than data. Do not, however, provide too much information.

A system that interfaces with its environment is called an *open system*. *Closed systems* exist only in artificial environments, such as laboratories. A system that can control itself is called a *closed-loop system*. A business firm is an example of an open, closed-loop system.

Managers are the control element in the firm as a system. According to Henri Fayol, all managers plan, organize, staff, direct, and control. They apply these management functions to all types of resources—physical and conceptual.

Frederick W. Taylor recognized that managers must budget their time effectively, concerning themselves only with issues that warrant their attention. He called this selective focus the *exception principle*. It has been renamed *management by exception*.

Robert N. Anthony is often credited with naming the three management levels, but his terminology was intended to describe levels of management systems. Strategic planning systems address long-term objectives, management control systems allocate resources to achieving these objectives, and operational control systems ensure that the resources are applied correctly.

Henry Mintzberg is the most recent theorist. The management roles he has defined explicitly recognize the use of information. Two role categories, interpersonal and informational, are concerned with receiving and routing information. The third category, decisional, has to do with applying information in making decisions.

Management's use of the computer as an information system can be related to all the theories. For example, of Fayol's five management functions, the computer has best supported planning and controlling. The computer excels as a device for providing Taylor's exception information and for facilitating Anthony's three levels of management systems. It also facilitates the flow of information that connects Mintzberg's managerial roles.

Besides addressing what managers do, management theory addresses the location of managers in an organization. Managers can be found on three levels—top, middle, and lower—and in functional areas. All organizations have finance, human resources, and marketing functions. Manufacturing organizations also have a manufacturing function.

Management level and functional area both influence information needs. Top-level managers generally prefer environmental information in a summary form. Lower-level managers prefer internal information in a detailed form.

All functional managers need information about money and the government, but other resources and environmental elements are especially important to certain functions. Marketing managers need information on personnel, material, customers, and competitors. Manufacturing managers need information on personnel, machines, materials, suppliers, and labor unions. Human resources managers need information on personnel, suppliers, labor unions, and the local community. Finance managers need information on stockholders or owners and the financial community. Information services managers need information on personnel, machines, and the suppliers of those machines.

To design information systems for use by managers, you should understand management theory as well as systems.

Key Terms

Supersystem, suprasystem

Subsystem

Systems view

Data

Information

Information processor

Information overload

Open system

Closed system

Feedback loop

Closed-loop system

Open-loop system

Strategic planning system

Management control system

Operational control system

Exception report

Planning horizon

Key Concepts

Levels of systems

How physical and conceptual resources flow from the environment to the firm, through the firm, and back to the environment

Achievement of system control by means of a control mechanism, performance standards, and feedback

Management functions

Management by exception

Management systems tailored to organizational levels

Managerial roles

Hierarchical levels of managers

Functional organization structure

How organizational level and functional area influence information needs

Questions

1. What is the name of the highest system level? What is the name of a system within a system?

2. What is a systems view? Who in an organization can benefit from taking such a view?

3. Which class of physical resources includes buildings and land?

4. Which physical resource usually does not actually flow through a firm?

5. Which physical resource would a manager likely believe to be the the most important? Why?

6. Are data and information the same? Explain.

7. What is an information processor?

8. Which three guidelines should you keep in mind when you design conceptual systems?

9. What is a system called that interacts with its environment? How is the interface achieved?

10. What name is used to describe a system that can control itself? That cannot control itself?

11. Which elements are necessary for the firm to control itself?

12. Which resources are being managed when a manager performs Fayol's management functions?

13. Why would a manager apply Taylor's exception principle?

14. What did Robert N. Anthony contribute to management theory?

15. Which of Mintzberg's three categories of management roles is concerned with receiving and passing along information? With using information to solve problems?

16. Which of Fayol's management functions have been best supported by the computer?

17. Which of Fayol's management functions appear to be targets of the human resource information system?

18. A particular type of report is an example of Taylor's theory. What is it called?

19. What is a mathematical model used for in a strategic planning system? In a management control system?

20. Explain what can be used to distinguish between managers on different management levels.

21. What kind of information is a top-level manager supposed to prefer? A lower-level manager?

22. In what form is a top-level manager supposed to prefer information? A lower-level manager?

23. Which physical resource is of interest to managers of all functional areas?

24. Managers of which functional area are most interested in information about customers? About competitors? About suppliers? About the financial community?

Topics for Discussion

1. Although it is not mentioned in the chapter, management attempts to speed the flow of certain physical resources through the firm and attempts to slow down the flow of others. Which would be speeded up? Which slowed down?

2. Would a firm ever be considered an open-loop system?

3. Why would a manager want to be alerted when performance rises above the range of acceptable performance?

4. The sample exception report in the chapter has separate columns for the various exceptions. In what other ways can reports be designed to facilitate management by exception?

5. Give examples of how the physical resources are organized functionally.

Problems

1. Read the Boulding article identified in the chapter bibliography, and write a paper titled "How Boulding's Systems Theory Relates to Business." Your instructor will advise you concerning format and length.

2. Read the Duncan article identified in the chapter bibliography, and write a paper titled "How Duncan's Systems Theory Relates to Business."

Case Problem: Blue Bonnet Motor Homes

Blue Bonnet Motor Homes is located in Hempstead, Texas, on the outskirts of Houston. Hempstead is a small town, but Blue Bonnet derives most of its business from buyers who drive from Houston for the low prices. In its TV ads, Blue Bonnet claims to be the lowest-priced motor home dealer in the Houston area.

About three months ago, Blue Bonnet installed a personal computer to handle inventory. A manual system had been used, but it was always causing problems. The salespersons would convince a customer to buy a certain style and color motor home, check the manual inventory records to ensure that the motor home was in stock, and then learn that the vehicle had already been sold. The manual inventory records did not accurately reflect the Blue Bonnet inventory.

Blue Bonnet sales manager Bubba White convinced owner Henry Bailey to replace the manual system with a computer. A PC was bought and installed, along with a software package designed especially for automobile dealerships. Carol Olsen, who had been keeping the manual inventory records, was trained to perform the data entry. A systems analyst, Jennifer Gordy, was hired to develop additonal applications. Responsibility for the computer operation was given to Sunny Popp, manager of the accounting department.

One bright summer day, Bubba walks into Henry's office.

Bubba White (sales manager): Henry, it's still happening. The inventory records still aren't right. Twice yesterday the computer said we had vehicles that weren't on the lot. We went out and found they had been sold. And, the same thing happened this morning. I thought you ought to know.

Henry Bailey (owner): I'm glad you told me. This is pretty discouraging. We spend all that money on the computer and the people to operate it, and it's no better than the manual system. I'm going to talk to Sunny about this.

(Henry phones Sunny, and asks her to come to his office. Sunny is there in a few minutes, and Henry explains what has happened.)

HENRY: Sunny, do you have any idea what's going on? I thought that computers were supposed to be so accurate.

SUNNY POPP (ACCOUNTING MANAGER):   I think I do. The salespersons aren't letting us know when they make a sale. When we receive a shipment of motor homes from the manufacturer, we immediately update our inventory records to show the new stock. But when the sales are made, the salespersons either are late in letting us know or forget to notify us altogether. The result is that the records indicate that a vehicle is in stock when it really isn't. As you know, the salespersons are supposed to fill out a Vehicle Sold slip and personally carry it to the computer room so that Carol can update the computer records. We decided on that procedure so that the computer records would be accurate up to the minute. If the salespersons would follow the procedure, I don't think we would have any problems.

(Henry thanks Sunny, and she leaves. Henry calls Bubba on the phone, and explains what Sunny has said.)

BUBBA: It sounds like Sunny is just trying to cover up for her own operation and put the blame on sales. Listen, Henry, our sales reps have more important things to do than fill out a bunch of paperwork. They're salespersons, not clerks. If we can't get this problem solved without turning our reps into paper shufflers, I'm going to tell them to just forget the sales slips altogether.

Assignments

1. Explain the Blue Bonnet situation in systems terms, using systems theory terminology.
2. What has caused the problem?
3. What should Henry do to solve the problem?

Selected Bibliography

Boulding, Kenneth E. "General Systems Theory—the Skeleton of Science." *Management Science* 2 (April 1956): 197–208.

Brewer, Stanley H., and Rosenzweig, James. "Rhochrematics and Organizational Adjustments." *California Management Review* 3 (Spring 1961): 72–81.

Cashman, James F., and Seers, Anson. "Teamwork: An Open Systems Process Analysis." *Journal of Management Systems* 3 (Number 3, 1991): 41–50.

Debons, Anthony; Horne, Esther; and Cronenweth, Scott. *Information Science: An Integrated View.* (Boston: G. K. Hall & Co., 1988.)

Duncan, Otis Dudley. "Social Organization and the Ecosystem." *Handbook of Modern Sociology,* edited by Robert E. L. Faris. (Chicago: Rand McNally, 1964): 36–45.

Forrester, Jay W. "Industrial Dynamics: A Major Breakthrough for Decision Makers." *Harvard Business Review* 36 (July–August 1958): 37–66.

Gurbaxani, Vijay, and Whang, Seungjin. "The Impact of Information Systems on Organizations and Markets." *Communications of the ACM* 34 (January 1991): 59–73.

Hopeman, Richard J. *Systems Analysis and Operations Management.* (Columbus, OH: Charles E. Merrill, 1969.)

Johnson, Richard A.; Kast, Fremont E.; and Rosenzweig, James E. *The Theory and Management of Systems.* 2nd ed. (New York: McGraw-Hill, 1967.)

Pavett, Cynthia M., and Lau, Alan W. "Managerial Work: The Influence of Hierarchical Level and Functional Specialty." *Academy of Management Journal* 26 (Number 1, 1983): 170–177.

Schoderbek, Peter P.; Schoderbek, Charles G.; and Kefalas, Asterios G. *Management Systems: Conceptual Considerations.* 4th ed. (Homewood, IL: BPI/Irwin, 1990.)

von Bertalanffy, Ludwig. "General System Theory: A Critical Review." *General Systems* 7 (1962): 1–20.

von Bertalanffy, Ludwig. "General System Theory: A New Approach to the Unity of Science." *Human Biology* 23 (December 1951): 302–361.

Weinberg, Gerald M. *An Introduction to General Systems Thinking.* (New York: John Wiley & Sons, 1975.)

CHAPTER 4

The Firm As a System

Learning Objectives

After studying this chapter, you should:

- know two approaches to taking a systems view of a firm
- better understand systems theory
- know what elements must exist for the firm to function effectively as a system
- be able to describe any type of firm in systems terms

Introduction

Viewing a firm as a system enables you to detect and solve problems that threaten the firm's ability to operate properly. Two basic approaches have been taken to achieving a systems view. One approach views the firm in terms of its energy. The other approach views it in terms of its resources. We favor the resource view and recognize the need for a conceptual system to control the physical resources.

The conceptual system consists of management, an information processor, and standards of performance. An intricate network of data and information flows connects management and the information processor with the physical system and the environment.

The *general systems model of the firm* shows how all the system elements fit together and how the resources flow. Managers and systems analysts alike can benefit from using this model as a blueprint of ideal system performance. When the manager and the systems analysts evaluate the manager's system, they look for eight criteria that affect the firm's ability to operate efficiently and effectively.

All types of firms can be represented by the general systems model.

Two Main System Views

Systems theorists have taken several approaches in explaining the firm as a system, but two approaches are especially well suited to systems analysis and design. One view recognizes the firm's ability to store energy. The other view recognizes the firm's ability to transform resources.

The Firm as an Energy System

Social psychologists Daniel Katz and Robert L. Kahn refer to the organization as an *energetic input-output system*.[1] As Katz and Kahn explain, an organization such as a business firm receives energy from its environment in the form of labor and materials. The energy is used to create products that are sold to the environment, and the money received from the sales is used to purchase more energy. Because the firm makes a profit, it can purchase more energy than it needs. The excess energy is stored. Figure 4.1 shows the flows of energy, products, and money, and the storage of the excess energy.

The Firm as a Transformation System

Another view recognizes that the firm's resources are transformed as they flow through the firm. This is the view taken in the late 1960s by Richard J. Hopeman, a

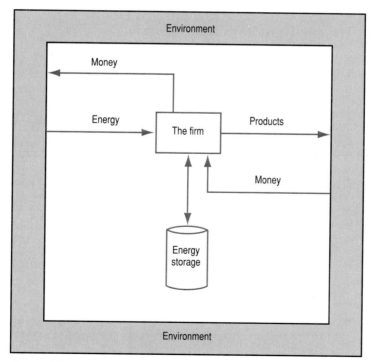

Figure 4.1
An energetic input-output system.

[1]Daniel Katz and Robert L. Kahn, *The Social Psychology of Organizations* (New York: John Wiley & Sons, 1966): 14–29.

professor of production operations and management at Syracuse University.[2] Hopeman's view provides the basis for the approach taken in this text.

Figure 4.2 shows this resource flow. The firm's physical system has the task of transforming the input physical resources into forms needed by the environment.

Transformation Systems Seek a Steady State   As the resources flow through the firm, management seeks to maintain the proper rate of flow. If the resources flow too slowly, the firm will not have enough resources to perform its necessary activities. For example, if a supplier is late in shipping raw materials, the production process must be delayed until the materials arrive. Or, if the firm cannot fill customer orders immediately, the customers will go elsewhere.

Resources that flow too quickly can have equally disastrous consequences. For example, high personnel turnover can leave the firm short of experienced workers.

The managers manage the resource flows for the purpose of achieving a **steady state,** an equilibrium produced by a balanced flow of input and output resources.

Key Components in the Transformation System

The view we have taken in this text emphasizes the importance of resource flows. We assert that management's task is to manage the flow of physical resources, and

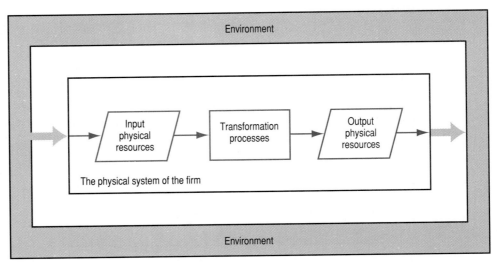

Figure 4.2
A resource transformation system.

[2]Richard J. Hopeman, *Systems Analysis and Operations Management* (Columbus, OH: Charles E. Merrill, 1969): 125–150.

we recognize that management needs a conceptual system of data and information to manage these resources. In the sections that follow, we examine more closely each element of the firm as a system.

Management Uses an Information Processor for Control

Figure 4.2 illustrates an open-loop system that has no means of controlling its own operations. Such control can be achieved by means of a conceptual system. Two elements in the conceptual system are management and an information processor. These elements are added in Figure 4.3.

An **information processor** is any device, computer or noncomputer, that transforms data into information and makes information available to its users. As shown in the figure, the information processor gathers data and information from the input, transformation, and output elements of the physical system. This data and information describes the firm's current status. The information processor either provides the information to management immediately or stores it for possible future use.

The two-headed arrow connecting management and the information processor represents a two-way communication. Management provides information to the information processor, and receives information from it. For example, a manager enters decisions into the information processor for use in a mathematical model, and the information processor provides the manager with the model output.

The addition of the information processor to the physical system enables management to get up-to-the-minute information on the physical system's status. The data is gathered as transactions occur and is processed by the information processor immediately upon receipt.

Performance Standards are Needed for Control

Knowing what is happening in the firm is not sufficient. Management must also know whether the activity is acceptable. To determine this, management establishes standards of acceptable performance. Figure 4.4 shows how management uses the standards.

Management evaluates system performance by comparing the information processor output with the standards. The information processor output describes **actual activity**—what the system is accomplishing. The standards describe **desired activity**—what the system should accomplish.

For example, the management of a shoe factory might desire the production of 900 to 1,100 pairs per day. The 900 to 1,100 pairs represent a **performance range** that identifies both the lower and upper ends of an area of acceptable performance. Management is satisfied when performance remains in the range.

Assume the information processor reports that 875 pairs were produced yesterday. What action, if any, does management take? Management must determine whether the 875 pairs are acceptable. If such a production rate lasts only 1 day,

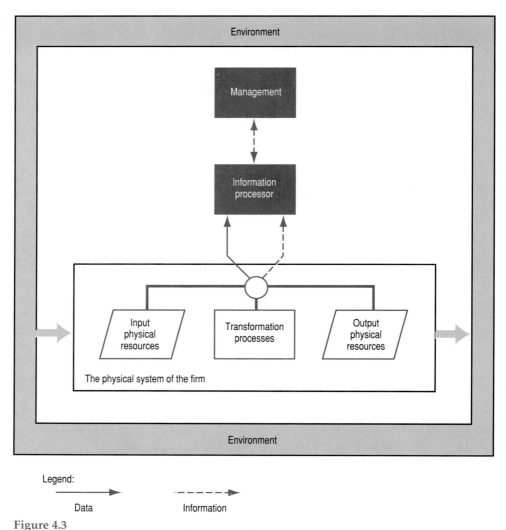

Legend:

——————→
Data

- - - - -→
Information

Figure 4.3
The information processor provides management with information describing the performance of the physical system.

management might regard it as a chance variation and do nothing. However, if the output continues for several days, management might decide that action is needed to raise the actual performance level. Management takes action only when appropriate. This is an example of *management by exception,* which we described in Chapter 3.

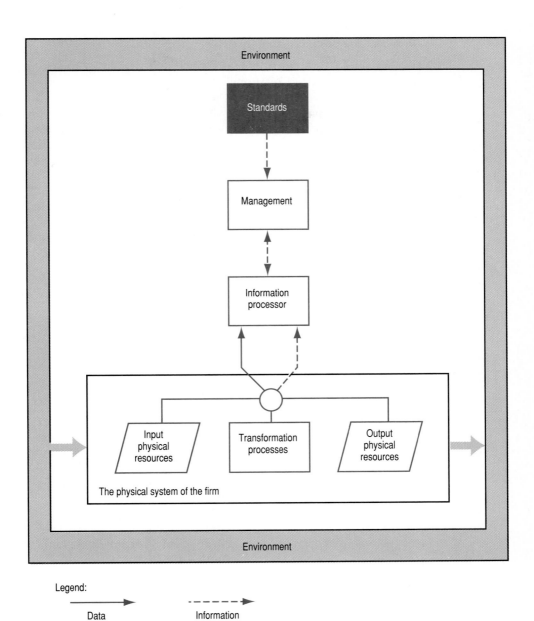

Figure 4.4
Performance standards are intended to direct the firm toward its objectives.

Where the Standards Come From

Management sets standards that, when met, will enable the firm to meet its objectives. Therefore, a direct relationship exists between the firm's objectives and its standards.

The objectives tend to be broad statements of desired performance. The standards are more specific. For example, a firm might have an objective of providing quality products. A standard to support that objective would be a limit on the percentage of products returned by dissatisfied customers. Table 4.1 shows some objectives and their corresponding standards.

It is common for more than one standard to support an objective. This way, several activities can help achieve the objective. The standards should be expressed quantitatively so that it is easy to see whether they are met. Notice that the first standard in the table specifies a specific number of days. This is much better than saying "Respond quickly."

The Information Processor Can Perform a Control Function

If the performance standards are made available to the information processor, as shown in Figure 4.5, the information processor can relieve the manager of much

Table 4.1   Objectives and Their Corresponding Standards

Objectives	Standards of Performance
Operate in an ethical manner	Respond within 3 days to all customer requests for data on their individual credit ratings Have no more than 1 percent of invoices vary from quoted charges by more than 10 percent
Achieve a global operation	Open five sales offices in Southeast Asia by July 1, 1995 Update foreign currency price lists within 5 days of changes to U.S. currency price lists
Meet customer needs	Fill at least 98 percent of customer orders without backorders Operate a customer hotline 24 hours a day
Provide a return on investment to the owners	Pay dividends each quarter Maintain common stock price of $90–$125
Provide a good place to work	Keep employee turnover below 5 percent per year Keep time lost due to accidents below 0.5 percent of the total hours worked
Operate efficiently	Keep scrap costs below $25,000 per month
Innovate technology	Invest at least $2,500,000 in R&D each fiscal year

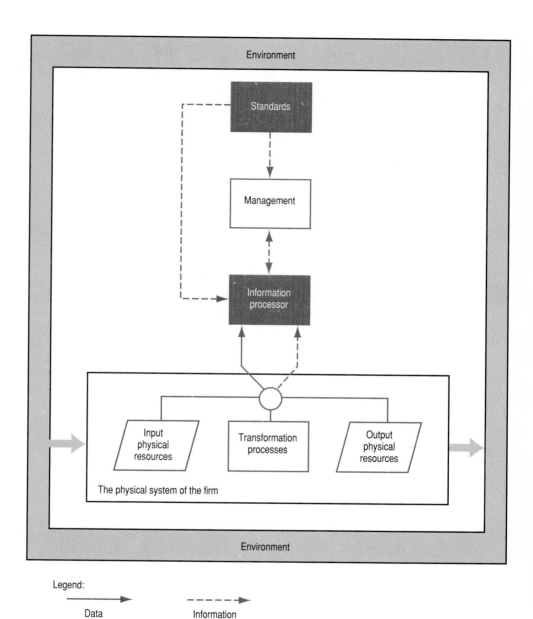

Figure 4.5
Making the performance standards available to the information processor facilitates control of the physical system.

time-consuming monitoring. The information processor can compare the actual and desired performance levels and notify the manager when action might be needed. For example, the information processor can display a message on the manager's computer screen, stating that yesterday's production fell below the acceptable range.

You can see that adding the flow of the standards information to the information processor facilitates management by exception.

Decisions Must Be Transmitted to the Physical System

When management determines that corrective action is needed, a communications network transmits decisions to the physical system so that changes can be made. In our shoe factory example, the problem might be in the input area, and management could decide to add buyers in the purchasing department. Or, the problem might be in the transformation area, and management could decide to replace some aging production machines. If the problem is in the output area, management might use the computer to determine the best delivery route.

In transmitting the decisions, management has historically used memos, letters, telephone calls, face-to-face conversation, and other conventional communication media. These conventional communications are represented by the arrow leading from management to the physical system in Figure 4.6. The communications are directed to those lower-level managers and employees responsible for making the needed changes in the physical system.

Recently, management has begun to use office automation (OA) systems such as word processing, electronic mail, and teleconferencing to communicate decisions. We discuss OA in detail in Chapter 6. OA's significant contribution to system management is the use of the computer in all three segments of the feedback loop. Previously, the computer was useful only for the first two segments—gathering data from the physical system and transforming the data into information.

The arrow leading from the information processor to the physical system in Figure 4.6 reflects the use of OA for decision communication. Management makes a decision and enters it into the information processor. The information processor then electronically relays the decision to the appropriate element of the physical system.

Information Also Flows Directly to Management

Not all information must flow to management through the information processor. Information can flow directly to management from the physical system. As an example, a production employee can telephone the plant manager to report a loss of electrical power, or a clerk in the receiving department can personally alert an inventory manager to the arrival of a particular shipment.

By the same token, some information can flow directly to management from the environment. A banker, for example, can send a newsletter to the vice-president of finance, warning of a projected economic downturn.

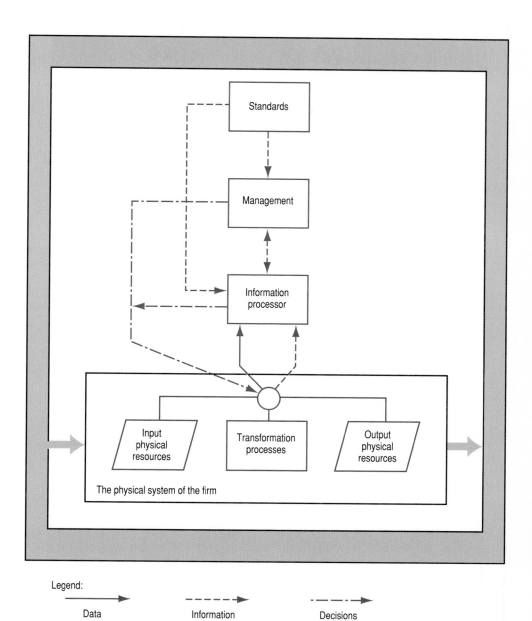

Figure 4.6
Management achieves changes in the physical system by means of decisions.

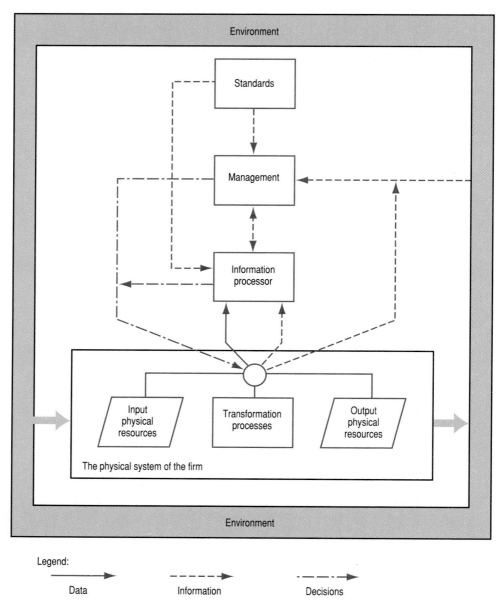

Figure 4.7
Information flows to management from the physical system and the environment.

These direct information flows are illustrated by the arrows in Figure 4.7 that lead to management from the physical system and the environment.

Conceptual Resources Also Flow Through the Firm

In addition to transforming physical resources to meet environmental needs, the firm transforms conceptual resources for the same purpose. The environment provides the firm with data, the firm transforms the data into information, and the firm makes information available to the environment. For example, a manufacturing firm conducts marketing research to measure consumer interest in a new product and then makes the findings available to its retailers.

The information processor performs the transformation of the conceptual resources. In Figure 4.8, the information processor gathers data and information from the environment and transforms the environmental data into information. Management uses the information to remain current on environmental conditions.

The information processor also provides data and information to the environment. For example, data on employee social security deductions is made available to the federal government. An example of an outgoing information flow is an income statement made available to a bank for obtaining a loan.

Management also has a responsibility to provide information to the environment. The president, for example, might call the local mayor to pass along the news that the firm will expand operations and offer more jobs to local residents.

The incoming and outgoing flows of data and information, which involve management and the information processor, are shown in Figure 4.8.

Using CASE to Determine Strategic Uses of Information Systems

CASE tools support a particular methodology, or way of doing something. Although methodologies differ from tool to tool, they drive the order, nature, and output of the CASE products.

Analysis and design of information systems can be viewed as a pyramid structure. At the top of the pyramid is *strategic planning*. So, the first step in systems analysis and design is to develop a strategic plan. This is extremely important even though it does not require as much time and effort as do the activities on the lower levels. The second level from the top is *business area analysis*, the study of the functions and processes performed within each functional area. The third level from the top is *system design*, and the bottom level is the implementation of the system that is often called *system construction*.

The key to this pyramid view of analysis and design is its top-down nature, which enables the user or information specialist to view the entire system in a broad and understandable way. The viewer who wants a more detailed understanding can navigate down through the system from the general to the more specific levels. This top-down format provides the user with a *mental model* of the system that is easy to understand and learn.

The objectives of the strategic information planning process include not only identification of information systems projects for the next three to five years, but also an estimation of the resources required to develop these projects. Any project, regardless of complexity, requires a data repository, a stable methodology, and tools, such as code generators, graphics support, screen generators, and project management software. All I-CASE products support these components. Planning must also be accomplished for system capacity and communications capability and should be done on both a project and an organizational basis.

The General Systems Model of the Firm

Figure 4.8 is the **general systems model of the firm.** It is a graphic model that shows the intricate feedback loop and control mechanism that enable the physical system to maintain a steady state. The model illustrates how the standards, management, information processor, and communications channels work together to keep the physical system on course.

It might appear from the model that the information processor separates management from the physical system. The situation is the opposite. The information processor provides a window through which management can monitor the performance of the physical system. As the physical system grows beyond a single geographic site, it becomes impossible for management to monitor by direct contact and observation. The information processor provides the monitoring help that management needs.

The Model Serves as a Guideline

Viewing the firm as a system makes it easier to identify the characteristics that are key to successful performance. The systems framework serves as a guideline for both the manager and the systems analyst.

The Manager Uses the Model

When you manage an organizational unit, whether it be an entire firm or a subsidiary area, you must evaluate that system. The evaluation helps you determine whether the system needs improvement, and, if so, how you might achieve that improvement. A systematic evaluation is the most effective.

One systematic approach is to view the organizational unit as a system that should fit the general systems model. Compare the unit to the guideline, looking for the necessary elements. When the elements are missing or not performing

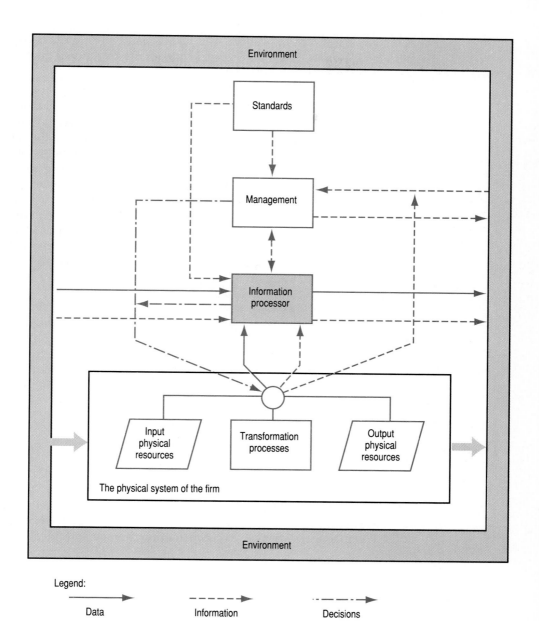

Figure 4.8
The general systems model of the firm.

correctly, you know that improvement is needed and which parts need attention. The systems view is a logical way to analyze your organizational unit.

The Systems Analyst Uses the Model

If you become a systems analyst, you will use the general systems model in basically the same way as the manager. The difference is in the area of system responsibility. The manager is responsible for the day-to-day working of the system, and the analyst is responsible for improving the system. When assigned to study a system with the intent of possibly making improvements, you can use the model as a guideline of how the system ought to function. When you spot defects, you know that those areas must be improved or replaced.

Although systems analysis frequently requires direct contact with the physical system, you typically do not attempt to make changes to that physical system. The changes are made by persons skilled in analyzing and designing physical systems. **Industrial engineers,** called **IEs,** are systems analysts who specialize in physical systems. Most of the IE's activity is directed toward the manufacturing function.

The systems analyst from information services typically works with only the conceptual framework represented by the standards, management, information processor, and communications network. As we can see from the model of the firm as a system, this conceptual framework can represent an intricate flow of data, information, and decisions. The complex nature of the firm's conceptual systems is a major reason why systems work is so challenging and rewarding. When you improve or design a system, you can take satisfaction in having performed a difficult task that requires specialized knowledge and skills.

Criteria for Good Systems Performance

Whether a manager or a systems analyst, look for the following criteria when evaluating a business system:

- **A Good Management Team** Management performs the critical control function, so be sure you have enough managers with the necessary knowledge and skills. The managers should have information literacy and, ideally, computer literacy.

- **Good Standards** The firm must know where it wants to go. The objectives satisfy this need. The performance standards specify the level of activity required for the firm to meet its objectives.

- **A Balanced Flow of Resources Through the Physical System** The personnel, material, machine, and money must flow through the physical system at the desired rate.

- **Data Gathering Throughout the Physical System** Data must be gathered from the input, transformation, and output areas so that management can monitor the resource flows.

- **Data and Information Gathering from the Environment** Management monitors the environment to expedite the flows of resources into and out of the firm. Management seeks to remain aware of what is currently happening in the environment, as well as what might happen in the future.

- **Transformation of Data into Information** Data from the firm and its environment must be transformed into information by an information processor. The information is used by management and the environment. Some information is made available immediately. Some is stored for later use.

- **Standards Made Available to the Information Processor** When the information processor has access to the standards, it can perform much of the monitoring for the manager.

- **A Network for Communicating Management Decisions** A communications network enables management to communicate decisions to the physical system quickly and effectively. Management selects the communication medium, either conventional or electronic, based on the needs and the available resources.

These criteria seem very logical and it is easy to assume that they exist in all firms. However, it is difficult to achieve them all and to keep them current. Management's job is to ensure that each desired element exists. It is a full-time and ongoing job. The systems analyst's job is to assist the manager in this important activity.

Examples of Firms Viewed as Systems

When we think of firms as systems, we typically think of the physical systems, which are the most obvious. We see the smoke stacks of a manufacturer, the showroom of an automobile dealer, the home office building of an insurance company, and the delivery trucks of a service organization. If we go behind the scenes, we can see all the elements of the model of the firm as a system, including the elements that compose the conceptual system.

A Manufacturer as a System

It is easiest to see a manufacturer as a system. You can stand in the production area and see the personnel and machines transforming raw materials into finished products. If you follow the motions of the factory workers, you will see that they frequently stop to fill out paperwork or enter data into a computer terminal. Figure 4.9 shows a factory worker entering data into a data collection terminal. The data describes the status of the current job—start time, completion time, number of units produced, and so on. The photograph illustrates how data on the transformation process is gathered.

The data that the workers provide to the information processor is used by management to monitor production. Not all the managers are located in the production area. The ones on the shop floor are lower-level supervisors who

Figure 4.9
Data collection terminals in the production area keep management informed concerning the transformation process.

monitor their areas mainly by observation. The upper-level managers are located elsewhere, in offices in other parts of the building, or perhaps hundreds or thousands of miles away, relying on the information processor for monitoring help.

If you talk with the manufacturing managers, you will be impressed with their knowledge of their performance standards. The lower-level supervisors know how many units their areas are to produce in a day or even an hour. Upper-level managers can show you formal statements of longer-term production goals.

The managers in the production area are also concerned about their environment. They are aware of the impact labor unions and suppliers can have on their operations. Managers in other areas of the firm are also concerned about the environment. Finance managers monitor the health of the economy, and marketing managers want to keep the finished products flowing to their customers.

In a well-managed manufacturing firm, you can find examples of each element of the general systems model of the firm.

A Retailer as a System

When you shop in a retail store such as a supermarket, you are an active participant in the output element of the system. The bar code scanners at the checkout counters provide the store's computer with data that describes the output.

If you are allowed to tour the storeroom at the rear of the store, you will see huge stacks of merchandise from suppliers. You may even see a truck unloading more merchandise. Many stores use hand-held devices to scan bar codes on the shipping cartons as they are unloaded, as illustrated in Figure 4.10. The photograph illustrates how data describing the input of physical resources is gathered. The scanning of the input, combined with the scanning of the output, enables store management to track the flow of merchandise, the material resource, through the store.

The system input and output occurs during the day, but transformation takes place at night. If you visit a supermarket during the early morning hours, you will see stock clerks bringing merchandise from the storeroom and placing it on the shelves. This is the transformation process. The items are removed from their cartons and displayed so that you can easily select them as you shop.

Your tour of the supermarket may also include a look at the store computer. In addition to maintaining the status of the physical system, the computer makes

Figure 4.10
Supermarkets scan barcodes on cartons of merchandise received from suppliers.

information available to store management. Like the manufacturing firm, the retail store has performance standards, and store management uses the information to see how well the standards are being met. Management is also aware of the environment. Management wants to keep the products flowing smoothly, and to do this, they must monitor their suppliers, competitors, and customers. If you doubt this environmental awareness, take note of the sales, special promotions, and even remodeling that goes on when a new supermarket opens in the area.

The retail store, like the manufacturer, contains all the elements of the general systems model of the firm. Store management is very aware of these elements.

A Wholesaler as a System

A wholesaler is similar to a retailer. For both firms, the transformation consists of making the products available to the customers. For the wholesaler, the customers are retail stores.

A wholesaler's physical resources consist primarily of one or more warehouses and a fleet of delivery trucks. Merchandise is delivered to the warehouses by the manufacturer's trucks and is stored until it is shipped by the wholesaler's trucks. Like retail managers, wholesale managers want to keep the materials flowing smoothly. Computers determine the best time to order replenishment merchandise so that stock is available to fill customers orders but the warehouse does not become overstocked. Computers can also schedule deliveries so that the routes are covered in the least number of miles. This fleet scheduling, illustrated in Figure 4.11, is an example of the output portion of the physical system.

Because the wholesaler's activity is concentrated in the warehouse area, it is easy to see that these operations form a system. If you stand in the storage area and look out on the shipping and receiving docks, you can see it all—the input, the transformation, the output, and the role of conceptual systems.

An Insurance Company as a System

It is not as easy to see the flow of resources through an insurance company, but the flow is there. The personnel flow is there, with some loyal employees remaining with the firm for many years. Likewise, the machine flow is there. Insurance companies make excellent use of machines—especially computers.

Computers handle the flow of materials and money. The materials flow in the form of paper documents. Prospective policyholders fill out application forms that flow into the company, and policy forms flow out. The money flow is just as visible. Policyholders mail premium payments to the company, and the company makes investments and pays claims.

What transformation does the insurance company perform? Essentially, its products are intangibles. They do not exist in a physical form. They provide financial security. The insurance company transforms premium income into policyholder security.

Many operations in an insurance company are computerized. For automobile policies, for example, the computer calculates the premium using characteristics

Figure 4.11
The computer can determine truck routes that minimize delivery cost.

of the auto, the driver, and the geographic area. In addition to computing the premiums, the computer prints the policies. Figure 4.12 is an example of an automobile policy's cover page. This document represents the flow of information from the firm to its environment—in this case, to the firm's customers.

Not all the money an insurance company receives comes from the policyholders. Much of the premium income is invested in stocks and bonds with the intent of producing additional revenue in the form of capital gains and dividends. The company relies on the returns from the investments to pay a large portion of its claims and provide interest and dividends to its stockholders and owners. Most insurance companies have investments departments that monitor the economy and decide which investments to make. The computer can assist in this monitoring and investing activity. Figure 4.13 shows a screen containing financial information provided by the Dow Jones News/Retrieval Service. A firm can subscribe

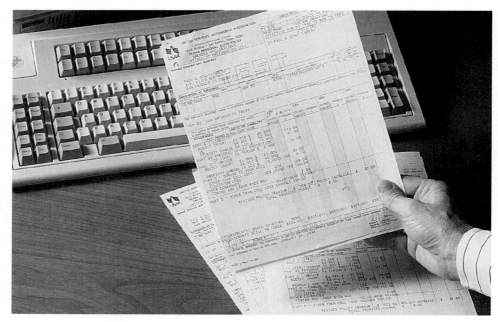

Figure 4.12
The computer can provide information to the environment.

to the service so that its managers can have up-to-date financial information at their fingertips. The managers enter requests on their terminal keyboards, and the news is displayed on their screens. This is an example of the input of environmental information in the general systems model of the firm.

Dow Jones Business Newswires	//WIRES

hon

Business Newswires-HON	HEADLINE PAGE 1

1	PI	10/24	*(HON) Up 6%; Buying Interest Seen in Wake of (ATK) Sp
2	PR	10/24	MEYER RESIGNS FROM HONEYWELL
3	PI	10/24	Honeywell Inc (HON) P-Alert: +3 1/4 on 114,300
4	DJ	10/23	Alliant Computer Sues Honeywell; Charges Infringement
5	DW	10/18	Moody's Upgrades Honeywell Inc; Senior Debt to Bail
6	FF	10/16	Latest 8K Filings
7	DW	10/16	Honeywell Earnings -2-
8	DW	10/16	Honeywell 3rd Qtr Net Cont Op $2.30 A Shr Vs $1.

4

Figure 4.13
The computer can provide information from the environment.

Figure 4.14
The computer facilitates communication of management decisions.

A Hospital as a System

A hospital, like an insurance company, is a service organization. The hospital combines the skills of personnel (the doctors and nurses), machines (such as X-ray machines), and materials (medications) to cure the ailments of their customers, the patients.

Hospitals make extensive use of computers. Most of the accounting activities are computerized—as is evident when you check out. The computer can immediately print a bill that details each service the hospital provided. Hospital managers, called administrators, also use the computer for problem solving. Computer programs can schedule the use of facilities, such as operating rooms, and ensure that patients receive the right medications and foods. Hospital adminstrators also use the computer to communicate the problem-solving decisions. For example, an administrator can prepare a work schedule that is displayed on terminals in the nurses' areas. Figure 4.14 illustrates this decision flow from management to the physical system.

All firms can be viewed as systems. All the elements of the model of the firm as a system are there, connected by flows of data, information, and decisions. This way of viewing a firm is perhaps the most valuable orientation that a systems analyst can have. It provides the basis for solving systems problems.

Using CASE to Draw a Staffing Matrix for an Individual Project

Many I-CASE tools generate matrix-type diagrams that can be used to estimate overall systems personnel requirements determined in the systems plan. Such a diagram, called a **staffing matrix,** begins with the level of staffing required for each individual project and each type of personnel.

The goal of this type of matrix tool is to help determine the level of recruitment that will be needed to staff all phases of systems development. Figure 4.15 is an example. The letter *P* stands for Programmer, and the letter *A* stands for Analyst. Other categories of personnel can be included, such as database administrators, data communications specialists, and users who are assigned to projects.

Task	1994 Q1		1994 Q2		1994 Q3		1994 Q4		1995 Q1		1995 Q2		1995 Q3		1995 Q4	
	P	A	P	A	P	A	P	A	P	A	P	A	P	A	P	A
Document existing system	1	3														
Automate Delivery Tickets	1		1	2	1	2										
Develop Procedures and training	1	1	1	2												
Code Modules					1	1	1	2	5	1	5	1	5	1		
Enhance Code							2	1	2	1	2	1				
Install new system									1	4		4		4		4
Totals																
Gross Requirement	3	4	2	4	2	3	3	3	8	6	7	6	5	5	0	4
Available	4	4	4	4	4	4	4	4	4	4	4	4	4	4	4	4
Excess	0	0	0	0	0	0	0	0	4	2	3	2	1	1	0	0
Recruitment	0	0	0	0	1	1	1	1	0	0	0	0	0	0	0	0

Figure 4.15
A project staffing matrix.

Summary

One systems view recognizes the ability of the firm to store energy. Energy, in the form of labor and materials, is acquired from the environment and applied to production. Finished products are sold to customers, and the money received is used to buy more energy. Excess energy is stored so that the firm will not experience an energy rundown.

Another basic systems view emphasizes the transformation of resources as they flow through the firm. When the firm's management takes this view, they seek to achieve and maintain a steady state, a balanced flow of resources into and out of the firm.

Management is the key element in controlling the resource flow, but the information processor also plays an important role. The information processor keeps management advised of the status of the physical system by gathering data from the input, transformation, and output elements, and transforming that data into information.

The information from the information processor tells management what the physical system is accomplishing, but management must also know what the system should accomplish. The objectives of the system must be stated in the form of performance standards that are compared with the actual performance. Management or the information processor can make the comparison. Management determines whether changes should be made to the physical system. When appropriate, management makes decisions that achieve the desired changes.

The information processor does not serve as a filter through which all of management's information must travel. Information can go directly to management from the physical system and from the environment.

Data and information also flow to the information processor from the environment and from the information processor back to the environment. Management has a responsibility to provide information to certain environmental elements.

The graphic model that illustrates how the elements of a firm are integrated to form an open, closed-loop system is called the general systems model of the firm. It applies to all firms in a general way, and it illustrates the firm's operations as systems processes.

The value of the general systems model is its use as a guideline. The manager evaluates the firm in terms of the model. The model helps the manager identify weaknesses and the locations of those weaknesses. The systems analyst uses the model in the same way, but applying it to the manager's system. The systems analyst is a professional problem solver who uses the model as a means of coping with the intricate nature of flows of data and information through the information processor.

When you compare your system or a manager's system to the model, you look for a good management team, good standards, a balanced flow of physical resources, data gathering throughout the firm and its environment, a transformation of data into information, standards incorporated into the information processor, and a means for communicating decisions to the physical system. A problem exists when any of these criteria is not met.

All types of organizations can be viewed as systems. Some, such as manufacturers, retailers, and wholesalers, specialize in the flow of tangible products to their customers. Others, such as insurance companies and hospitals, provide intangible products. All use conceptual systems to enable management to manage the resource flows.

Key Terms

Steady state

Information processor

Actual activity

Desired activity

Industrial engineer (IE)

Key Concepts

How a firm can store energy to keep from running down

Management as the activity of ensuring that the resources flow through the firm at the proper rate

The three elements that compose the conceptual system—management, an information processor, and standards

How the feedback loop in the firm as a controlled system consists of three segments—data, information, and decisions

Office automation as a means of facilitating the flow of information between the manager and the physical system of the firm

The general systems model of the firm

Questions

1. What are the two main system views of the firm?
2. How does a firm keep from running out of energy? Give an example.
3. How do managers achieve steady states in systems?
4. What is an information processor? What function does it perform for management?
5. What two types of activity does one compare to engage in management by exception? Where does the information on the two types of activity come from?
6. Are standards the same as objectives? Explain.
7. Why make the standards available to the information processor?

8. To which elements in the physical system does the manager transmit decisions?

9. Which CBIS subsystem has made it possible for the computer to be used in all three segments of the feedback loop?

10. Management receives information from three main sources. What are they?

11. To whom does management provide information? Data?

12. Who uses the general systems model of the firm?

13. What special type of systems analyst solves problems that occur in the physical system?

14. What eight criteria must be met for a firm to operate as an efficient system?

15. What type of terminal does a factory worker use to clock on and off the job? In what system element does that activity take place?

16. What role does optical character recognition play in managing the flow of material resources through a retailing firm such as a supermarket?

17. What flow in the model of the firm as a system represents the Dow Jones News/Retrieval Service?

Topics for Discussion

1. Which view of the firm as a system do you like better—an energy system or a resource flow system?

2. Does the resource flow view of the firm ignore energy?

3. What are some examples of *data* flowing to the information processor from the environment? Some examples of *information* flowing to the information processor from the environment?

4. What are some examples of *data* flowing from the information processor to the environment? Some examples of *information* flowing from the information processor to the environment?

5. Explain a college or university in terms of the general systems model of the firm. Do the same for a law firm.

Case Problem
Hoelscher, Nickerson, and Jones

Wow! A job as a systems analyst for a law firm. Who could ask for anything more. Nobody *ever* hears anything about how lawyers use computers. Systems work is challenging enough as it is, but being able to create applications in a field of such untapped potential is more than anyone can ask.

You have just received your degree in information systems and have landed a job with Hoelscher, Nickerson, and Jones, one of the biggest commercial law firms in town. Hoelscher, as the firm is called, represents business firms in a wide variety of litigations—trademark infringement, product safety, pollution, and so on. You head for the office and your first day of work with your worn copy of the general systems model in your coat pocket.

MICHAEL HOELSCHER (PARTNER): Welcome to the firm. This is a big day for us. We've had our 486 for about a year, and have really been pleased with the results. We subscribe to the Westlaw legal information retrieval system. Their database is in St. Paul. We retrieve abstracts and legal summaries using the 486. Everything has gone so well that we decided to expand our applications, and that is why we hired you. We want you to conduct a thorough systems study of our operation and advise us which applications we should add next. We want to use the 486 as an information system to enable us to do a better job. Then, we can add the necessary resources—programmers, operators, and the like. Why don't you ask me some questions, and we'll take it from there.

You: Great. You mention your 486. Are you familiar with computers?

MIKE: Pretty much so. I studied them at school and have learned a lot from my kids. We have a Mac at home and I've fiddled around with spreadsheets and the like. But I feel like an outsider here. Nobody else in the firm knows the least thing about computers. That's one reason I hired you. I wanted someone to talk to. (Mike smiles and you both laugh.)

You: Well, you can count on me for conversation. I love computers. Let me find out some more about the firm. (You remember what your professor said about the general systems model being a good guideline for evaluating a system, and feel your coat pocket to make certain it is still there.) How large is your management team?

MIKE: We really don't have managers as such. We have three partners, twelve other lawyers, and an office staff of six legal secretaries and four clerks. We also have an administrative assistant who helps the partners. We sometimes hire college students part time to do various types of research.

You: OK. (You are somewhat shocked that there are no managers. You wonder whether the general model is as great as your professor said it was.) What about performance standards? How do your people know what they are supposed to achieve?

MIKE: We don't have formal standards. We just assume that all the lawyers know that they are supposed to win their cases, and the secretaries and clerical personnel know that they are supposed to do high-quality work, keeping errors to a minimum.

You: Nothing in writing?

MIKE: Not a word.

You: All right. (You begin to wonder how this firm has stayed in business all these years. No managers and no standards. That's not the picture your professor painted.) Do you have objectives?

Mike: Well, we certainly want to be profitable. We want to return an investment to the owners, who are the partners. We want to provide a good place for our employees to work. We feel like we perform a service for the business community, providing legal service to those who need it. We would like to continue our growth, but we don't want to grow so fast that we lose our reputation as a firm that really cares for its clients and its people.

You: Those sound like good objectives. You say you want to use the 486 as an information tool. Exactly what do you mean by that?

Mike: As you can appreciate, our business is communications. We communicate with everybody—our clients, the D.A., the jury, our sources of information, Washington, ourselves. All this communication is now being done in conventional ways—in person, on the phone, through the mails, by reading the paper and listening to TV and the radio, and so on. We will have to continue these conventional ways, but we would like to take advantage of electronics as well. We would like to use the 486 to better communicate here in the office, *and* with our outside contacts. Most of our commercial clients have huge computer setups. We would like to be able to tie in with them.

(At that moment, Mike's administrative assistant walks in the office, holding *The Wall Street Journal*. He excuses himself for interrupting and asks, "Mike, have you seen the article on air bags? I think it's relevant to the class action suit we're handling. Maybe we ought to talk about it."

Mike excuses himself. You head to your office, holding on to your coat pocket all the way:)

Assignments

Use the general systems model of the firm to evaluate your new firm.

1. Examine each element of the conceptual system, and identify any associated problems.
2. Examine each flow of data and information in the model, and explain how the flows can be improved with the 486. It is not necessary that the 486 be applied to every flow.

Selected Bibliography _____

Brewer, Stanley H., and Rosenzweig, James. "Rhochrematics and Organizational Adjustments." *California Management Review* 3 (Spring 1961): 72–81.

Burch, John G. "Adaptation of Information Systems Building Blocks to Design Forces." *Journal of Management Information Systems* 3 (Summer 1986): 96–104.

Debons, Anthony; Horne, Esther; and Cronenweth, Scott. *Information Science: An Integrated View.* (Boston: G. K. Hall & Co., 1988.)

Forrester, Jay W. "Industrial Dynamics: A Major Breakthrough for Decision Makers." *Harvard Business Review* 36 (July–August 1958): 37–66.

Gurbaxani, Vijay, and Whang, Seungjin. "The Impact of Information Systems on Organizations and Markets." *Communications of the ACM* 34 (January 1991): 59–73.

Johnson, Richard A.; Kast, Fremont E.; and Rosenzweig, James E. *The Theory and Management of Systems.* 2d ed. (New York: McGraw-Hill, 1967.)

Miller, Jeffrey G., and Gilmour, Peter. "Materials Managers: Who Needs Them?" *Harvard Business Review* 57 (July–August 1979): 143–153.

Peters, Tom, and Austin, Nancy. "MBWA (Managing by Walking Around)." *California Management Review* 28 (Fall 1985): 9–34.

Schoderbek, Peter P.; Schoderbek, Charles G.; and Kefalas, Asterios G. *Management Systems: Conceptual Considerations.* 4th ed. (Homewood, IL: BPI/Irwin, 1990.)

Stewart, Thomas A. "The Search for the Organization of Tomorrow." *Fortune* 125 (May 18, 1992): 92–98.

CHAPTER 5

The Environmental System

Learning Objectives

After studying this chapter, you should:

- know the eight elements that exist in the firm's environment
- appreciate that all physical resources flow to the firm from the environment and from the firm back to the environment
- be able to distinguish between primary and secondary flows of physical resources
- know the main types of information that flow between the firm and each environmental element
- be aware of some future developments in information systems that involve the environment

Introduction

We have discussed the firm's environment in each of the previous chapters. It would be impossible to describe the firm in systems terms and not include the environment because all firms are open systems. However, until now, we have not addressed the environment specifically. That is the purpose of this chapter.

The firm exists within an environment of eight other elements—individuals or organizations such as customers, suppliers, and competitors. The firm is connected to its environmental elements by flows of physical and conceptual resources. Management manages the physical resource flows by managing the information flows.

Future information systems will feature electronic data and information linkages between the firm and its environmental elements. In the past, systems developers have concentrated their efforts within their firms. In the future, representatives of multiple firms will have to work together, most likely using new techniques and tools.

Environmental Elements

We have previously recognized several environmental elements. In Chapter 2, we described how competitive advantage can be achieved through information links between the firm and five environmental elements—customers, suppliers, the financial community, the government, and competitors. This text uses an environmental framework that consists of these five elements and three others. Figure 5.1 identifies the eight elements that make up the environment of any firm.[1] This framework is more than 20 years old, but it remains an effective structure. That is the good thing about systems theory; it remains relatively stable. In the dynamic world of computing, some stability is a blessing.

Each of the eight environmental elements are organizations or individuals, and each affects the firm in some way.

Customers   A firm is organized to meet the marketplace's need for products or services. A firm's customers can be individuals, called **consumers** by marketers. Supermarket shoppers are an example. A firm's customers can also be organizations, as when Ace Hardware purchases electric drills from Black & Decker.

Suppliers   The firm obtains its material and machine resources from suppliers. A **supplier,** or **vendor,** provides products and services that firms use in

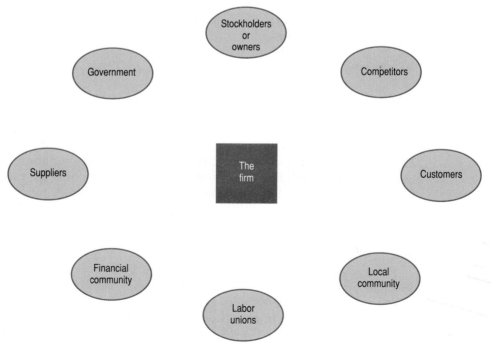

Figure 5.1
The environment of the firm.

[1]Richard J. Hopeman, *Systems Analysis and Operations Management.* (Columbus, OH: Charles E. Merrill, 1969): 79–81.

producing their own products and services. Suppliers are organizations just as is the firm. In fact, the firm is one of the supplier's customers.

Stockholders or Owners   A *stockholder* is anyone who invests in a corporation and can be an organization, such as a mutual fund, or an individual. When the firm is not a corporation, the owners are called proprietors or partners. A *proprietor* is a sole owner. A *partner* shares ownership. In most cases, proprietors and partners are top-level managers, but the money they invest is their own—coming to the firm from the outside.

The ultimate responsibility of the firm is to meet the investment expectations of its owners.

Labor Unions   Many industries have work forces that belong to labor unions. Management can regard the unions as a help or a hindrance. Unions help by providing skilled employees. They can hinder by preventing activities that union officials believe are not in the best interests of the members. Management of a heavily unionized firm must work closely with the union officials to enact policies that are mutually acceptable. Otherwise, the firm faces the threat of a crippling strike or other kind of work stoppage.

Government   Federal, state, and local government can also be viewed positively and negatively. When we think of governmental influence on business, we usually think of constraints—the requirements to report information and pay taxes. However, government is also a source of support for the firm. Government is especially helpful to small businesses, providing specialized expertise and guaranteeing loans. It also makes available a wealth of data and information, often free of charge, so that businesses of all sizes can better understand the economy and their marketplaces. These governmental relationships are common in North America. In other countries, the governmental role can be quite different.

Financial Community   The *financial community* consists of those institutions that make money available to the firm or provide investment opportunities. These institutions include banks, investment firms, insurance companies, credit unions, and any other type of organization that has large accumulations of money. These institutions can lend money to the firm when it is short of cash and provide investment opportunities when the firm has a money surplus.

Local Community   The firm has a responsibility to be a good citizen of its city or town. The community provides fire and police protection, cultural opportunities and education for the firm's employees and their families, and health care facilities. The firm usually pays for these services through local taxes and contributions to local organizations and activities.

A large firm with widely distributed operations has a local community that consists of many cities, perhaps in different states and even different countries. The local community of today's multinational corporation includes the entire world. This worldwide definition of the local community is the view taken by the text.

The local community also serves as a source of personnel. We assume that all nonunion employees come from the local community, although they might be recruited from other geographic areas. Similarly, all nonunion personnel go from the firm back to the local community.

Competitors   The firm seeks to maintain a good relationship with the seven environmental elements just described. However, it does not have such an accommodating attitude toward the competition. Competitors are the enemy. Their objective is to put the firm out of business. For that reason, the manager attempts to make the competitors' life as difficult as possible, while still conducting business ethically. The firm competes by offering better products and services.

Resource Flows

We have seen how resources flow through the firm. All physical resources originate in the environment and return to the environment. Management's task is to create, direct, and control these resource flows. The manager makes certain that the resources flow at the desired volume, quality, and rate. The manager tries to speed up some flows and slow down others.

Personnel Flow

Although we do not always think of personnel as flowing through a firm, they do. All employees come from the environment, are transferred from one area in the firm to another, and eventually return to the environment.

The manager tries to expedite the flow of personnel *into* the firm but delay the flow of personnel *out* of the firm as long as possible. In this way, the firm benefits from the employees' services for the longest period possible.

Material Flow

A firm's materials represent a large investment. After obtaining the materials, the manager wants to put them to use, rather than let them sit idle. Therefore, the manager attempts to speed up the flow of material into and out of the firm.

An example of this speedy flow is an approach to manufacturing popularized by the Japanese, called *JIT*, for *just in time*. The idea is that raw materials should arrive at the plant "just in time" for production. To accomplish this goal, suppliers often locate a distribution center directly across the street from a large customer's plant. This strategy ensures shipment without delay. By speeding the material flow, firms that practice JIT have become tough competitors in the marketplace.

The manager also attempts to speed up the material flow after production. The sooner the customers receive their products, the sooner the firm receives its money.

Machine Flow

The manager treats the machine flow exactly the same as the personnel flow, expediting the incoming flow and slowing down the outgoing flow. The machines are used as long as possible. Maintenance prolongs the life of the machines.

Figure 5.2
College recruiters seek to speed up the flow of personnel into their firms.

Money Flow

It should come as no surprise that the manager wants money to flow into the firm as quickly as possible. Firms develop special systems to expedite this flow. The computer-based systems of the energy companies such as Texaco and Standard Oil are economically justified in reducing the delay, called **float,** in receiving customers' credit card payments. These companies have millions of dollars outstanding at any one time. By reducing the float two or three days, the companies can earn more than enough interest to pay for the computer systems.

You might believe that the firm should hold onto its money as long as possible. However, firms want to put their money to work earning interest. This is accomplished by investing their money without delay. Applying this logic, the manager wants the money to flow out of the firm to an investment opportunity as soon as possible after it is received.

Data and Information Flow

The manager wants data to flow to the information processor, such as the computer, as quickly as possible. The data flow can originate inside or outside the firm and can be expedited through data communications networks. The manager also wants to expedite the flow of information from the information processor to the users. The users can be inside or outside the firm. Again, data communications can be used.

The firm's environment is involved in all the physical and conceptual resource flows. The manager must establish the needed relationships with the environmental elements to make the resources flow at the desired rate.

Environmental Resource Flows

The main distinction between the internal and environmental resource flows is control. The manager has greater control over resources while they are inside the firm. It would be a mistake to say, however, that the manager is at the mercy of the environment. The manager can influence the environmental flows by offering the various environmental elements something in return.

In many cases, the manager can offer information. For example, a department store might want a skiwear manufacturer to replenish stock quickly during the skiing season. To achieve this, the store can give the manufacturer statistics describing how the manufacturer's products are selling. By building such alliances with the noncompetitive elements in the environment, the manager exercises some control on environmental resource flows. This is the fundamental strategy for surviving in a competitive environment.

Primary Physical Resource Flows

Not all resources flow in both directions. Each element is connected to the firm by a combination of physical and conceptual resource flows. Some resources regularly flow in large volumes. They are called **primary physical resource flows.** Figure 5.3 shows the primary flows of the physical resources that connect the firm to its environment.

Primary Personnel Flows   Two-way personnel flows connect the firm to labor unions and the local community. The personnel flow to the firm when hired and to the environment when employment is terminated.

Primary Material Flows   Raw materials, parts, and subassemblies flow from suppliers to the firm. Finished products flow from the firm to the customers.

Primary Machine Flows   The firm obtains machines from suppliers. Eventually, the machines are sold to other firms, returned to suppliers as trade-ins, or returned to the local community as scrap. Customers and even competitors are among the firms to whom the used machines might be sold.

Primary Money Flows   The firm pays for the materials and machines it receives from suppliers. The customers pay for the materials they receive from the firm. Money flows both ways between the firm and the stockholders or owners, as the owners make investments and receive interest, dividends, or other returns on their investments. Money also flows both ways between the firm and the financial

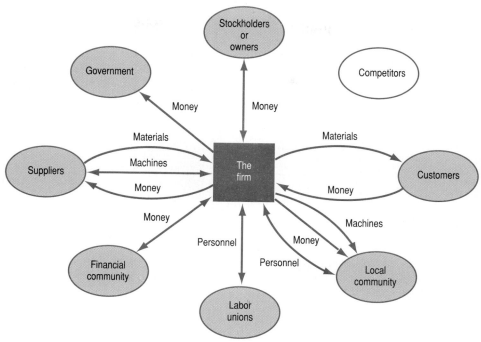

Figure 5.3
Primary environmental resource flows.

community. The community makes loans, and the firm invests surplus funds. The primary money flow between the firm and the government is the one-way flow of tax money to the government. Money also flows one way to the local community, as the firm donates to local institutions and charities such as the United Way.

Note that there is no primary flow of physical resources to or from competitors.

Secondary Physical Resource Flows

Other resource flows between the firm and its environment occur less often and in smaller volume. These **secondary physical resource flows** are illustrated in Figure 5.4.

Secondary Personnel Flows   Secondary personnel flows exist between the firm and its competitors when employees switch firms in the same industry.

Secondary Material Flows   Materials flow from the firm to the suppliers when the firm returns items for credit. Materials flow from the customers to the firm when the customers return items for credit.

Secondary Machine Flows   Secondary machine flows occur rarely, such as when a machine is returned to the supplier because of a defect.

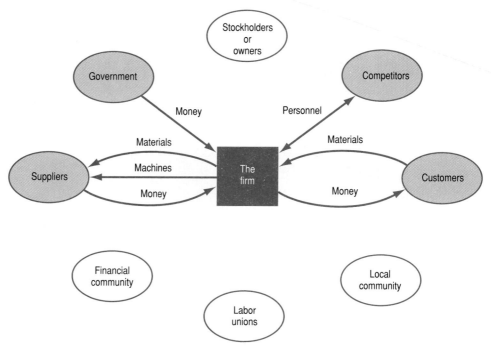

Figure 5.4
Secondary environmental resource flows.

Secondary Money Flows   Returns for credit cause infrequent money flows from the suppliers to the firm and from the firm to the customers. There can also be a secondary money flow from the government to the firm in the form of loans and grant money.

The manager encourages the primary physical resource flows, which establish the desired relationships between the firm and those environmental elements. The manager devotes less time to the secondary flows. With the exceptions of money from the government and personnel from competitors, the manager seeks to minimize the secondary flows. Sound purchasing decisions reduce the need to return materials and machines to the suppliers. Good products and services give customers little reason to return their purchases. And, good employment practices give employees reason to stay with the firm.

Flows of Conceptual Resources

The firm is also connected to its environment by flows of conceptual resources. Data and information are included, but we will refer to everything as information.

Information flows in both directions, as shown in Figure 5.5. The manager seeks to maximize the incoming flows from all the elements. In return, the manager provides outgoing flows to all elements except competition. The firm and the non-competitive elements recognize the benefits of sharing information.

The manager attempts to make all conceptual resource flows primary ones.

Customer Information Flows   Customers give the firm information about the products and services they need. The firm makes the products and services available, and the customers place orders identifying the products that they want to purchase. The firm ships the products, and provides information on product use. The firm sends invoices and statements to the customers, asking for payment.

Supplier Information Flows   The information flows that exist between the firm and its customers also exist between the firm and its suppliers.

Stockholder or Owner Information Flows   A corporation provides its stockholders with the opportunity for formal and informal communications. Formal communications include the annual stockholders meeting and the annual report. Informal communications include letters, telephone calls, and plant tours.

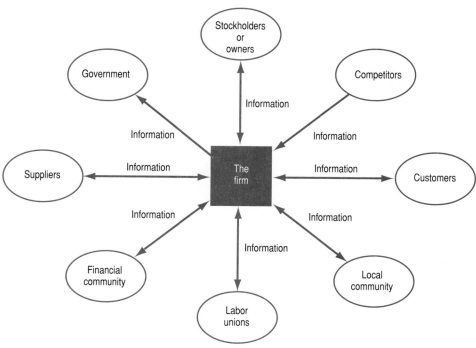

Figure 5.5
Conceptual resource flows.

Figure 5.6
Firms use marketing research as a means of learning about consumer needs.

Before the annual meeting, the firm distributes its annual report to the stockholders. These reports represent a high volume of information. At the annual meeting, the managers make speeches and the owners ask questions and express opinions.

Labor Union Information Flows   The most important information flow with the labor union is the contract between the firm and the union. The contract binds both parties for several years.

The firm normally has a labor relations department that serves as the main information conduit with the union. Most of the information flow is informal, in the form of personal communications.

Government Information Flows   The firm is required to provide the government with data and information that describe the firm's operations. For example, the firm periodically reports the income and social security taxes that are withheld from each employee's pay. The computer plays a key role in this reporting, summarizing large volumes of data.

The government advises the firm of its reporting responsibilities by providing copies of the various rules and regulations that describe the flows. The government provides statistics that are valuable for forecasting activity. As described earlier, the government also provides valuable information to certain types of firms, such as those with limited resources. Much of this information is in the form of brochures and handbooks.

Financial Community Information Flows   The financial community provides periodic reports and newsletters that describe economic conditions. As with government statistics, the firm uses this financial information in planning. The financial institutions also keep the firm updated on how its investments are doing.

Information also flows to the financial community. Financial institutions that loan money to the firm ask for financial statements that support the firm's ability to repay the loans.

Figure 5.7
Stockholders communicate to the firm at the annual stockholders meeting.

Local Community Information Flows   The flows of information between the firm and its local community usually consist of personal communications. The flows also occur in relatively low volume, coming when the community asks for financial support or when the firm wishes to publicize an aspect of its operations.

Competitor Information Flows   Incoming information that describes the activities of the firm's competitors is called **competitive intelligence,** or **CI.** In the past, firms gathered this intelligence informally. Sales representatives would report unusual competititive activity in their territories, and managers would read business periodicals. Today, however, a firm can acquire competitor information by subscribing to services that make the information available on compact disks or access to databases. Figure 5.8 illustrates the type of intelligence available from CI subscription services.

The information flows facilitate the physical resource flows. The information flows also help the firm achieve a competitive advantage.

Future Developments of Firms as Systems

Chapter 2 discussed competitive advantage and described how firms are using their information resources to compete better.

```
News Report 04/20/93 NEW PRODUCTS: Claris Corp., a subsidiary of
Co., announced the introduction of ClarisWorks 2.0, the complete
productivity solution for business, education and home.
ClarisWorks 2.0 includes new capabilities to address the day-to-
day demands of most generalist computer users who produce a range
of documents and need to manage information. A new outlining
feature gives users a choice of seven outline formats including
Harvard, legal and numeric, to easily organize their thoughts.
The price of ClarisWorks 2.0 remains unchanged from the original
version. In the U.S., the suggested retail price is $299.
Upgrades from ClarisWorks 1.0 are $99. A $129 competitive trade-
up price is also being offered through Sept. 30, 1993.

News Report 04/13/93 NEW SERVICE: Apple USA, a division of Co.,
announced Apple Assurance, an enhanced service and support
platform, designed to offer Apple customers greater flexibility
and choice in their service and support options. Apple Assurance
will provide support to help customers setup and begin to use
their Apple products. This support is available for as long as
the customer owns their Apple, hardware, software and networking
products.  Apple USA also announced 24-hour built-in lifetime up-
and-running support and a one-year on-site service warranty for
Apple WorkGroup Server 95.  The warranty includes set-up,
installation, configuration, compatibility and troubleshooting.
```

Figure 5.8
An example of competitive intelligence.

As a firm's executives decide on the appropriate competitive strategy, they can approach their task in a systems fashion. They can view their firm as a subsystem of a larger environmental system—their marketplace. The executives can then follow a two-step strategy to achieving competitive advantage. First, they can transform their firms into *value chains* that enable them to distinguish their products from those of their competitors'. Second, they can transform their firms into *value systems* that include links to the value chains of other firms. Both steps involve the environment.

Value Chains

In a *Harvard Business Review* article, Harvard professor Michael E. Porter and Arthur Andersen partner Victor E. Millar used the term **value chain** to describe the various activities that can provide the firm with product differentiation and a cost advantage.[2]

Figure 5.9 illustrates a value chain. *Inbound logistics* are all materials procurement activities. *Operations* are all production activities. *Outbound logistics* are all distribution activities, such as warehousing and transportation. Porter and Millar also added *marketing and sales* activities and *service* activities. Both activities involve the firm's products and are directed at the customers.

You can see the close resemblance between the Porter and Millar model and the lower portion of the general systems model. The inbound logistics correspond to the input resources element, and the operations correspond to the transform element. The outbound logistics, marketing and sales, and service elements correspond to the output element of the general system model.

Each link in the value chain has a physical component and an informational component. The physical component is the flow of material from one link to the next, from left to right, on its way to the customers. The material flow represents the firm's product and service offerings. The informational component is the two-

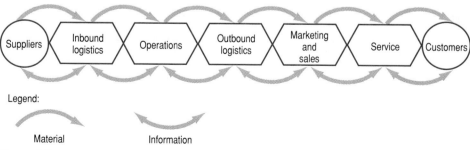

Legend:

Material   Information

Figure 5.9
A value chain.

[2]Michael E. Porter and Victor E. Millar, "How Information Gives You Competitive Advantage," *Harvard Business Review* 63 (July–August 1985): 149–160.

way information flow that connects all links. This information flow serves as the nervous system of the value chain, enabling all the links to work together.

The value chain is a systems concept. It recognizes that product value is a result of all the activities beginning with suppliers and leading to customers. Multiple elements work together to produce the competitive product. The value chain concept also recognizes the necessity of a conceptual information flow to parallel the physical flow. The value chain is an example of applying systems theory to the task of producing high-quality products for the purpose of achieving competitive advantage.

Value Systems

Throughout most of the computer era, systems analysts and managers have concentrated on implementing computer-based systems that improve the informational components of the value chain links. Efficient, computer-based systems for purchasing, production control, distribution, customer billing, and service management have become commonplace.

The next logical step for the analysts and managers is to connect the value chains of the different firms involved. Typically, the firms include suppliers, manufacturers, wholesalers, and retailers. Porter and Millar use the term **value system** to describe the linkage of the value chains of multiple firms. Figure 5.10 illustrates a value system.

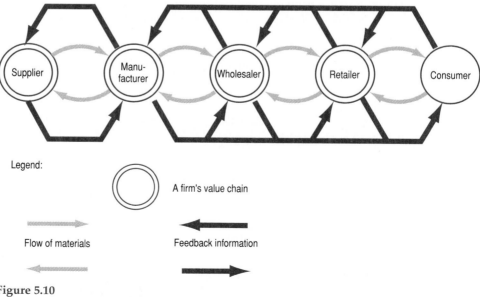

Legend:

A firm's value chain

Flow of materials

Feedback information

Figure 5.10
A value system.

Using CASE to Identify External Systems Boundaries

One of the key activities in the analysis and design of an information system is identifying external entities that interact with the system. The entities serve to establish the system's boundary. One approach to identifying the entities is the entity-relationship diagram, which we have already described. Another approach is the data flow diagram, or DFD. Both ERDs and DFDs are supported by virtually all CASE tools.

DFDs are explained in Technical Module B, and the goal here is not to repeat that explanation but to explain how the DFD can be used to identify external entities of a system. The entities serve as sources and sinks of a system's data. A **source** is an origin of a data flow, and a **sink** is a destination of a data flow.

A DFD provides a dynamic view of the data as it flows through a system. This view can most clearly be seen in the top-level DFD, called the context diagram. A context diagram for the Rag City inventory system is pictured in Figure 5.11. The external system entities are illustrated with squares and consist of suppliers and customers. The diagonal marks on the squares indicate that the entities appear in the diagram multiple times. In this particular diagram, one pair of entity symbols illustrates the sources, and another pair illustrates the sinks.

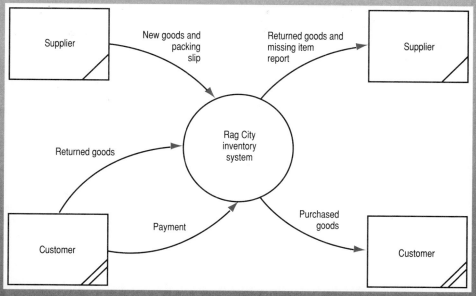

Figure 5.11
Context diagram entities establish system boundaries.

CASE not only supports the physical drawing of the DFD but originates data dictionary entries as the various components are defined. For example, when the data flow labeled *Payment* is drawn, an entry is also placed in the data dictionary as a shell or a complete description. In the case of a shell, the analyst revisits the data dictionary at a later time and completes the entry. If flows, stores, or processes already exist in the data dictionary from another diagram source, CASE will alert the analyst that a data dictionary entry already exists and will list all occurrences of the item.

The figure shows the flow of materials to the final customer, or consumer, and the reverse flow of money. The figure also shows two types of information flow—feedback information and feedforward information. **Feedback information** notifies an upstream firm of activity that *has occurred,* or *is occurring,* downstream. For example, customers notify retailers and perhaps manufacturers of their experiences with the manufacturer's products. **Feedforward information,** on the other hand, notifies a downstream firm, or the consumer, of activity that *is to occur.* For example, the manufacturer notifies wholesalers, retailers, and consumers of new products or changes in the prices of existing products. Feedforward information also gives consumers information they need to properly use their products.

Interorganizational Systems

The term **interorganizational system,** or **IOS,** has been coined to describe the informational portion of a value system. IOSs offer new opportunities for the participating firms to realize a competitive advantage. They will also offer new challenges. For example, questions of system control and security take on new dimensions when system users include persons outside the firm.

Putting the IOS in Perspective

The broadening of systems boundaries to encompass elements in the firm's environment will affect the way that systems are developed. Such systems will require that project teams include representatives from different firms and that the system configurations feature the latest developments in data communications and database management.

This inevitability of change could be worrisome to you as you embark on your career in systems work. However, the problem is not as serious as it might appear. The *fundamentals* of systems work should remain constant because they are based on systems theory. That theory will hold true in the future—even for the IOSs.

Summary

Eight elements make up the firm's environment—customers, suppliers, stockholders or owners, labor unions, government, the financial community, the local community, and competitors. The manager manages the rate at which the resources flow through the firm. Management seeks to slow down personnel and machine flow and speed up material and money flow. Management also strives to speed up the flow of data and information between the firm and all of the environmental elements except the competitors. The manager strives for only an incoming flow of competitive information. The manager can control resource flows within the firm easier than those flows to and from the environment. However, the manager achieves some control over the environmental flows by cooperating with all the elements except the competitors.

Resources that flow frequently and in large volume are primary resource flows. Resources that flow infrequently and in small volume are secondary flows. The manager does not neglect the secondary flows but devotes most attention to the primary flows.

The conceptual resources flow between the firm and the environmental elements, and management stimulates all the flows except the outgoing flow to competitors. Many firms are implementing computer-based competitive intelligence systems that provide the firms with an almost unlimited supply of up-to-the-minute information.

Future information systems will feature linkages between multiple firms. Value chains consist of all the activities between the suppliers and customers and are aimed at achieving product quality. Each link in the chain includes a physical component and a conceptual component. Value systems connect the value chains of multiple firms.

The information component of the value systems is called the interorganizational system, or IOS. The IOS consists of feedforward information and feedback information that keeps the member firms aware of activities throughout the network. Although new methodologies and tools will be used to implement the IOSs, systems theory will provide a continuity in the form of a familiar foundation.

Key Terms

Customer, consumer

Supplier, vendor

Float

Competitive intelligence (CI)

Feedback information

Feedforward information

Interorganizational system (IOS)

Key Concepts

>   The firm's environment as a system of eight elements, or subsystems
>
>   Management as the speeding up and slowing down of resource flows
>
>   How the firm is connected to its environmental elements by resource flows
>
>   The subdivision of resource flows into primary and secondary groupings
>
>   Value chains
>
>   Value systems

Questions

1. What are the eight environmental elements?

2. Which environmental elements could be individuals, as opposed to organizations?

3. What are two names given to organizations that provide the firm with materials and machines?

4. How does the firm communicate information in large volume to its stockholders?

5. What is the most important information flow for labor unions?

6. In what situation does the financial community require an information flow from the firm?

7. Which three environmental elements provide money to the firm?

8. Name three kinds of firms that are included in the financial community.

9. Which two environmental elements provide employees to the firm?

10. Which environmental element is connected to the firm by only a one-way primary physical resource flow?

11. Which secondary resource flows does the manager attempt to stimulate?

12. Which resource flows does the manager strive to speed up? To slow down?

13. Which resource flow have the Japanese become famous for speeding up? What is their approach called?

14. What is float?

15. What is the two-step strategy that firms can follow to achieve a competitive advantage?

16. How does a value chain differ from a value system?

17. Which two resources flow through a value chain?

18. Give an example of feedforward information that an automobile manufacturer would provide to buyers or prospective buyers. Do the same for the manufacturer of drug products such as Excedrin or Tylenol.

19. Which types of firms would you normally find in a value system?

20. What effect will interorganizational systems have on the composition of project teams?

Topics for Discussion

1. Why would an owner, who is president of the firm, be considered an element in the environment?

2. Is there any environmental element that would never constrain the firm in some way?

3. Can you think of a physical resource that does not originate or terminate in the firm's environment?

4. Do all data and information flows originate and terminate in the environment?

Consulting with

Harcourt Brace

(Use the Harcourt Brace scenario prior to Chapter 3 in solving this problem.)

Pretend that you are Mike Byrnes and that Robert Simons, the vice-president of finance, is interested in improving computer use by the managers. Robert asks for your recommendations. He is knowledgeable in systems theory and asks that you express your recommendations in systems terms. Write a memo to Robert, using as many of the systems terms and concepts from Chapters 3, 4, and 5 as you can. One approach would be to key your comments to the general systems model of the firm and the eight environmental elements. Robert is familiar with those structures.

Case Problem
Condor Industries

Roy Cotter stared at the graph on his computer screen for what seemed like an hour and then brought up a series of electronic calendars. Then, he keyed in the following memo:

TO: Markham, McClain, Robinson
FROM: Roy
SUBJECT: Competitive Sales Strategy
DATE: Nov. 12

We to have to do something to counter the plastic bicycle that Saito brought out last week. Let's have a conference Thursday at 2 P.M. I've checked your calendars, and we're all free for at least two hours.

When Thursday afternoon rolled around, all the participants gathered in Roy's office.

ROY COTTER (PRESIDENT): Sammie, I'd like to know why marketing didn't pick up on Saito's new product—the Divine Wind, I think it's called. Your sales reps must not have the foggiest idea what's going on in their territories.

SAMMIE MARKHAM (VICE-PRESIDENT OF MARKETING): I have to admit it doesn't make us look good. We continually stress the importance of feedback, but the reps don't always do it. Maybe they just forget, or maybe they're too busy. There's really no incentive for them to take the time to fill out a call report and mail it to headquarters.

ROY COTTER: Well, we should have known what's going on. I'm sure the Saito reps have been calling on the same wholesalers and retailers we have. If we had just asked, I'm sure our customers would have told us all about it. We have good relations with our customers, don't we?

SAMMIE MARKHAM: The best. But, you know, there's no incentive in it for them. Why should they give us competitive information? We certainly wouldn't want them to give information on us to Saito or any of our other competitors.

ROY COTTER: Well, that's water under the bridge as far as the Divine Wind goes, but I don't want it to happen again. Sammie, I want you to put in place some mechanism that lets us know what is happening at both the wholesale and retail levels. In the future, when a competitor plans to introduce a new product, I want us to know about it before we read it in the papers. Now, Al, we've got to gear up and produce our own plastic bicycle. Is there anything that will hold us back? I would like to have our product in the retailers' stores by June.

AL MCCLAIN (VICE-PRESIDENT OF MANUFACTURING): One problem I can see is the plastic. I don't know what the situation is with our suppliers when it comes to large volumes. We've been buying in pretty small quantitites. I'll have to get my buyers on the WATS line and check with the various suppliers. That will take a while. It's hard to make telephone connections with the people you want to talk with. You call them, and they are out or busy. They call you, and you are out or busy. Telephone tag.

ROY COTTER: It seems to me that you should have better relations with your suppliers. Assuming you get the plastic, can you foresee any other problems?

AL MCCLAIN: Sure. We will have to buy some additional plastic extrusion machines. And, we will need the operators.

Roy Cotter: Can we get them?

Al McClain: I'll have to see. We buy production machines so seldom that we don't try to stay up with what's going on in the market. We don't even know who the suppliers are. We'll just have to get on the telephone. And, we'll have to check with the union to see if there is a pool of extrusion machine operators out there somewhere. I haven't talked with the union guys since the strike two years ago. There's no telling what the situation is.

Roy Cotter: It doesn't sound promising. Will, can you see any problems in getting a new product out?

Will Robinson (vice-president of finance): If Al starts buying new machines and hiring new operators, we do have a problem. Our budget can't handle costs like that. We're talking big bucks. We'll have to go to the bank, and that might take a while. I'm not sure where we can get the best deal. Most of our money sources have dried up. You know we used to do business with savings and loans in Texas, but most of them are history. I don't even know who our contacts are anymore.

Roy Cotter: Gadzooks. How did we get ourselves in this mess?

Assignments

1. What can the marketing function do to ensure that new competitive products are detected as early as possible in the future?

2. What about manufacturing? What can they do to be more responsive when new threats arise?

3. Evaluate finance in the same way. How can Condor be guaranteed a quick source of funds when they are needed?

Case Problem
The Toy Mart

Al Bailey, Gene Starr, and Frank Lent grew up in the same Albuquerque neighborhood. They attended elementary, middle, and high school together. They graduated from the same college, and, more remarkably, they all ended up in the toy industry.

Upon graduation from college, Al opened his own store, the Toy Mart. Eventually, the store grew into the largest retail toy outlet in the area.

Gene did a little job hopping but finally found a home at New Mexico Toys, a wholesaler that supplies all brands of toys to retailers in New Mexico and West Texas.

Frank went to work for an uncle who owns Zoom Toys, a manufacturer that specializes in small plastic toys such as miniature cars and doll house furniture.

The three stayed in touch over the years, and now they play golf practically every Saturday morning. Today, as is their habit, they stop by a Wendy's restaurant on their way home to have a snack and engage in small talk. Only, this time, the talk is not so small.

FRANK LENT (MANUFACTURER): What do you mean, you aren't going to be able to play any more golf for a while?

AL BAILEY (RETAILER): Just what I said. Things are too hectic at the store. You two just don't know how competitive the toy business is getting. It used to be that maybe one or two new hot sellers would come along every few years—you know, hula hoops, Barbie dolls, Frisbees. Then, it picked up to maybe one a year. But, lately it seems like there is something new every week. Maybe it all started with the Ninja Turtles.

FRANK: What's so bad about that? You're complaining about too much business?

AL: Oh, no. The problem is that these items sell out the first day. Then, it's a steady stream of kids and their parents all wanting to buy more. You say you're sorry, and the tears start to flow. And, the kids take it even harder. (Everybody laughs.)

FRANK: Boy, I wish we had your problem of too much demand. Our company has its good-selling items, but not often enough. The toys that have been the really big sellers in the last few years have been pretty evenly distributed among a dozen or so manufacturers. No single manufacturer gets all the business. But, Al, I still don't see your problem. I could take the tears.

AL: When I run out of stock, I can't get more. Every retailer is facing the same situation at the same time. The wholesalers can't fill our orders.

GENE STARR (WHOLESALER): He's right. There's just no way we can keep up with the demand. Like when the Turtles first came out, we could have sold anything, and I mean *anything,* that had a Turtle picture on it. The manufacturers who were licensed to make the Turtle products just couldn't make them fast enough.

FRANK: I can see your point. When we came out with that doll house kit, our production people worked around the clock. We could have sold twice as many if we could have produced them while the demand was still high. But you know Al, we didn't know they were selling so well until you mentioned it that day after golf. I went back and told our marketing people and they were really shocked. By the time we got our production geared up to crank out more stock, much of the demand had died out. But, that's the toy business. Chicken one day, feathers the next.

GENE: It's not quite that bad, Frank. You manufacturers do a lot of advertising that extends the demand for your products. Just look at the Saturday morning cartoons. A good-selling toy will do well for several months. But we wholesalers have the same problem as you manufacturers. We don't know what is selling well either. We just wait until the orders arrive in the mail from the retailers. Did you know that sometimes it takes up to a week for an order to get to us from some little town up in the mountains?

AL: Why don't you stay in closer touch with us retailers? I only see your rep about once a month.

GENE: That's as often as we can call. We only have eight reps, and they cover almost 150,000 square miles. We're the largest toy wholesaler in the area, and we have a couple thousand customers. You're lucky. If you were that retailer in the mountains, you wouldn't see us but about once every six months.

AL: Well, there ought to be a better way.

GENE: Another problem is that we're only one of about six toy wholesalers in our area, and we all carry basically the same product line. Nobody has a monopoly. It's a very competitive business. My company would do anything to get the jump on competition.

AL: Gene, wouldn't New Mexico Toys like to have a bigger share of my business? If I could get good service from you, I wouldn't buy from the other guys. And, Frank, wouldn't Zoom like to capitalize on your hot sellers when they come along?

FRANK: Sure, but we've got the same problem as Gene—too few reps. And, don't forget, our territory is the entire world. I'll bet we sell to a thousand wholesalers.

AL: Frank, let me ask you a question. If you knew how sales of your products were going on a daily basis, wouldn't that help you keep up with demand?

FRANK: Oh, sure. But how am I going to do that? I don't have a crystal ball.

GENE: Me either.

AL: But you do have computers. That's more than I've got. The Toy Mart is a big operation, but we're just a grain of sand on the beach when it comes to the whole toy market.

GENE: Well, we don't have to solve the problem of the whole market. If the three of us could just work together, that would be a step in the right direction.

FRANK: That's for sure.

AL: (Looking at his watch.) Hey, I have to go. Gene, one of your competitors is coming in this afternoon. He has a cellular phone in his car. He's going to show me how he uses it in his business.

Assignments

1. Is there a need for feedback information from Al's retail operation to Gene's wholesaler? From Gene's wholesaler to Frank's manufacturer? Directly from Al to Frank's company? If so, explain.

2. Is there a need for feedforward information? If so, what should it include? Who should provide it? Who should receive it?

3. Can this feedback and feedforward information be provided quickly, say daily, without using a computer? If so, how? How could a computer be used? Would the companies' sales reps play a role in this communication?

4. Whose company—Frank's, Gene's, or Al's—should take the initiative in creating a system that will improve the information flow? Explain your reasoning. Which strategy can this company follow in achieving the cooperation of the other types of companies? How will each type of company benefit?

Selected Bibliography

Brewer, Stanley H., and Rosenzweig, James. "Rhochrematics and Organizational Adjustments." *California Management Review* 3 (Spring 1961):72-81

Chan, K. Caleb, and Thachenkary, Cherian S. "A Close-up Study of Interorganizational Information Systems (IOS) and Competitive Advantage." *Proceedings 1990 Annual Meeting Decision Sciences Institute,* coordinated by Betty Whitten and James Gilbert. (San Diego: 1990.) 1057–1059.

Debons, Anthony; Horne, Esther; and Cronenweth, Scott. *Information Science: An Integrated View.* (Boston: G. K. Hall & Co., 1988.)

Duncan, Otis Dudley. "Social Organization and the Ecosystem." *Handbook of Modern Sociology,* edited by Robert E. L. Faris. (Chicago: Rand McNally, 1964): 36–45.

Elofson, Gregg, and Konsynski, Benn. "Delegation Technologies: Environmental Scanning with Intelligent Agents." *Journal of Management Information Systems* 8 (Summer 1991): 37–62.

Forrester, Jay W. "Industrial Dynamics: A Major Breakthrough for Decision Makers." *Harvard Business Review* 36 (July–August 1958): 37–66.

Frolick, Mark N., and Carr, Houston H. "The Role of Management Information Systems in Environmental Scanning: A Strategic Issue." *Journal of Information Technology Management* 2 (Number 3, 1991): 33–37.

Gilad, Tamar, and Gilad, Benjamin. "Business Intelligence—The Quiet Revolution." *Sloan Management Review* 27 (Summer 1986): 53–61.

Johnson, Richard A.; Kast, Fremont E.; and Rosenzweig, James E. *The Theory and Management of Systems.* 2d ed. (New York: McGraw-Hill, 1967.)

Keon, Thomas L.; Vazzana, Gary S.; and Slocombe, Thomas E. "Sophisticated Information Processing Technology: Its Relationship With an Organization's Environment, Structure, and Culture." *Information Resources Management Journal* 5 (Fall 1992): 23–31.

Schoderbek, Peter P.; Schoderbek, Charles G.; and Kefalas, Asterios G. *Management Systems: Conceptual Considerations.* 4th ed. (Homewood, IL: BPI/Irwin, 1990.)

Wey, Jason Y. J., and Gibson, David V. "The Development and Implementation of Inter-organizational Systems: Considered at Three Levels of Analysis." *Proceedings of the Twenty-Third Annual Hawaii International Conference on Systems Sciences,* edited by Ralph H. Sprague. (Kona, January 1990): 158–164.

Ongoing Case

Harcourt Brace & Company

III—THE COMPANY AS A MANAGED SYSTEM

(Use this scenario in solving the Harcourt Brace problem at the end of Chapter 6.)

Harcourt Brace uses the computer primarily for processing accounting data, but some information is made available to management for use in decision making. This information most often is in the form of periodic reports, but some output comes from mathematical models and decision support systems. Less has been accomplished in the areas of office automation and expert systems.

The Physical System

Harcourt Brace's most obvious physical resources are personnel and material—the people who make possible the flow of books to the customers, and the books themselves. Most administrative employees work at Orlando, and distribution employees work in the distribution centers identified in Table HB3.1. Sales representatives, including regional managers, typically work out of their homes. The machine resources consist primarily of computing equipment, some automated warehouse equipment in the distribution centers, and the sales reps' cars. Of course, money is required to acquire and maintain these physical resources.

Management's attention is focused on the flow of the books to the distribution centers and, from there, to the customers. Harcourt Brace strives to have the right product available for shipment when the customer needs it—an objective that requires much planning. As stated earlier, it takes one or more years to produce a book. Harcourt Brace must anticipate customer needs long before orders come in.

Table HB3.1 **The Harcourt Brace Distribution Centers**

Distribution Center Location	Space (Square Feet)	Number of Employees
Bellmawr, New Jersey	420,000	105
Elk Grove, Illinois	130,000	45
Irving, Texas	200,000	62
Niles, Illinois	125,000	65
San Antonio, Texas	123,000	78
Troy, Missouri	130,000	45
Totals	1,128,000	400

Note: The Niles and Troy personnel figures include administrative employees who work in the customer services departments at those two distribution centers. Orlando provides customer services support for the other centers.

Objectives and Performance Standards

Harcourt Brace's corporate objectives and performance standards, which serve as guidelines for the physical system, begin with the sales forecast. This forecast predicts the number of copies of each title that will be sold during the coming year and serves as a basis for the production schedule. The production schedule provides the basis for the operating budget.

The sales forecast, production schedule, and budget represent the main performance standards for the various organizational units. These standards identify the levels of performance expected of the units and, indirectly, the individual employees.

Organizational Performance Standards

The monthly operating statement and other reports represent standards the organizational units are expected to achieve. Managers look at particular indicators of exceptionally good or poor performance. For example, one manager pays special attention to the cost of sales margin.

Individual Performance Standards

An example of individual performance standards can be found in the customer services TOS (telephone order service) section. Many customers place orders by telephone. The

Figure HB3.1
A TOS operator taking a customer order over the telephone.

AGENT	Signed-in duration	In-coming	Out-going	Interval	On Hold	Wrap-up	Avail-able	Unavail-able
				ALLOCATION OF TIME hr.:min.				
BETTY	8:28	5:09	0:00	0:00	0:21	0:45	1:11	1:02
		60.8%	0.0%	0.1%	4.0%	8.8%	14.0%	12.2%
AMOS	8:24	5:27	0:10	0:01	0:08	0:35	1:27	0:45
		64.9%	0.0%	0.2%	1.7%	7.0%	17.3%	9.0%
HENRY	0:36	0:08	0:00	0:03	0:00	0:00	0:00	0:24
		22.8%	0.0%	9.5%	0.0%	0.0%	0.0%	67.7%
BILL	9:24	5:39	0:00	0:00	0:05	1:05	1:42	0:53
		60.0%	0.0%	0.0%	0.9%	11.5%	18.1%	9.4%
BLANCA	8:24	4:20	0:00	0:01	0:09	1:05	1:13	1:37
		51.5%	0.0%	0.2%	1.8%	12.9%	14.4%	19.3%
Group Total	35:16	20:42	0:00	0:05	0:43	3:30	5:33	4:42
		58.7%	0.0%	0.3%	1.0%	9.9%	15.8%	13.3%

Note: The format of this report is slightly different from the format of the report prepared by the ECD-6000 system.

Figure HB3.2
Performance statistics for telephone order service representative.

volume can be as high as 4,000 calls per day. The calls are received by the 16 TOS representatives in Orlando.

Recognizing a need to know the level of service provided by the TOS reps, customer services management installed a computer-based ECD-6000 system by Alphacon. ECD-6000 is a standalone microcomputer that is tied into the telephone system. It logs statistics concerning each rep's calls. Periodic and special reports can be printed that reflect the performance of each rep and the entire TOS unit. Figure HB3.2 is a daily report that shows how the five reps' time on the job was allocated—the signed-in duration. Management looks at the exceptions, such as Henry being signed in for only 0:36 hours, and seeks to learn why the exceptions occurred.

The Computer-Based Information System

The CBIS consists of five subsystems—the data processing system, the management information system, the decision support systems, the office automation systems, and the expert systems.

Data Processing System

Most Harcourt Brace computer time is spent in processing accounting data. The data processing system consists of the order processing, billing, inventory, and accounts receivable subsystems.

The order processing system processes customer order data received by mail and telephone. The output from the order processing system is the pack lists, which are transmitted to the appropriate distribution centers. Once the orders are filled, data is transmitted back to Orlando where the billing system prepares customer invoices and the inventory system updates the inventory file. The accounts receivable system processes the customer payments upon receipt.

Management Information System

Harcourt Brace's MIS provides the information in the form of reports and output from mathematical models. The managers receive periodic reports and can request special reports. The managers tell the MIS specialists the special information they need, and the specialists use the DB2 query language to produce the reports from data in the database. On occassion, mathematical models simulate a particular aspect of the operation. For example, one model simulates the effects of changes in the sales representatives' territories.

Decision Support Systems

An example of a DSS is the computer program used when Harcourt Brace closed an Orlando warehouse and a warehouse in Petaluma, California. Using the inventory balances of the two warehouses, the program computed the number of pallets and trucks needed to move the items to other warehouses.

Office Automation Systems

Secretaries and clerks at Harcourt Brace use word processing software to prepare letters and memos, and TOS reps use fax machines to transmit rush orders to the distribution centers. Faxes and word processing are the main office automation, or OA, systems. Other OA systems such as electronic mail and electronic calendaring, are not used. The company uses OA for secretarial or clerical tasks, rather than to communicate problem-solving information.

Expert Systems

Harcourt Brace has not actively developed expert systems. It does use a system, however, that has many characteristics of an expert system. The system is the SLS, or stock locator system. When books are received from the bindery, SLS determines where in the warehouse to store them. The determination is influenced by several factors. One is the expected activity of the books. Frequently ordered books are placed in the most accessible locations.

SLS is also used when the items are picked from the shelves. Take, for example, the situation that arises when a customer asks for only a portion of the books stored on a pallet and the area where the books are stored is for pallets only. Where should the remainder of the books be located?

SLS represents the accumulated knowledge of Dave Mattson, a manager at the Bellmawr distribution center, and other Harcourt Brace warehouse specialists. The SLS programs and databases enable the computer to make expert stock location decisions.

Most of top management's energies currently are directed at the COPS (Customer Order Processing System) implementation. However, the managers recognize that the cutover to the new system will not end systems development work. The strength of Harcourt Brace's computer use is in data processing, MIS, and DSS efforts such as SLS. Management is obtaining valuable information from the computer, but a wider application of the computer is possible. Improvements in the COPS system will be needed, as will be other, new systems.

Personal Profile—Michael W. Byrnes, Vice-President of MIS

Mike Burns, a native of Flushing, New York, received a Bachelor of Science degree in marketing and management from Fordham University. His hobbies are golf, walking, and reading mysteries and history books. Mike joined Harcourt Brace in 1968 and believes that the keys to the company's success will be a single image of mission-critical systems; a computer platform that reflects current technology, good forecasting, and marketing research systems; and connectivity of PC and mainframe data. Mike attributes his personal success to patience, good listening skills, and being able to communicate in nontechnical terms.

CHAPTER 6

Computer-Based Information Systems

Learning Objectives

After studying this chapter, you should:
- better understand information resources
- know two ways to organize information services personnel
- know the two basic ways to process data and understand how they affect system performance
- understand the factors that contribute to information value
- know the three basic ways users obtain information from the computer
- better understand the CBIS subsystems

Introduction

Chapter 2 identified the information resources as hardware, software, data and information, personnel, and facilities. Personnel include information specialists, who are often organized into project teams, and users.

The computer has been used to solve a wide variety of problems, yet all the methods of processing fall into one of two categories—batch or online. In batch processing, transactions are accumulated and processed together. In online processing, transactions are processed separately—usually at the time they occur.

Both processing methods transform data into information. Data has value because it is the raw material of information. Information has value because it has the ability to describe a relevant subject accurately and completely, in a timely manner.

Users get information from the computer through reports, outputs from simulations, and communications. Communications, the most recent form of computer output, have been popularized by office automation.

These basic output methods, coupled with the basic processing methods, form the basis for five CBIS subsystems. Each subsystem makes a unique contribution to problem solving. The systems analyst should be familiar with all five subsystems.

The Information Resources

All information resources—hardware, software, data and information, personnel, and facilities—are present in each computer-based information system.

Hardware Resources

The late 1960s and 1970s saw firms centralize their data processing at headquarters sites to take advantage of increases in computer power, capacity, and speed. Employees in outlying locations used terminals to send data to headquarters and to receive information output. The data was transmitted by communications networks, usually operating over telephone lines. The sharing of the central computer was called **timesharing.** It is illustrated in Figure 6.1.

Also during the 1960s, a new breed of smaller computers, called *minicomputers,* or *minis,* was introduced. The larger computers became known as *mainframes.* The

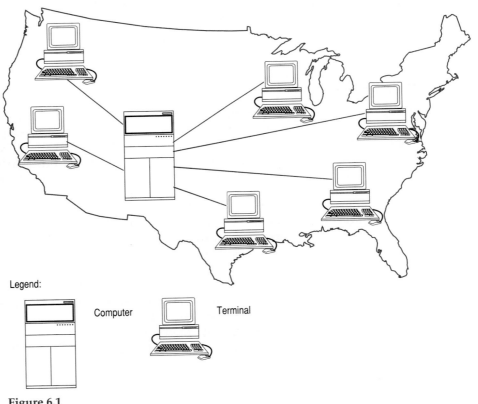

Legend:

Computer   Terminal

Figure 6.1
Timesharing.

lower-priced minis made computer processing available to medium-sized firms. The larger firms also took advantage of the minis, installing them in their outlying locations. Users could then do their computing on the nearby minis or on the mainframe at headquarters. The sharing of computer resources throughout the firm was called **distributed processing.** It is illustrated in Figure 6.2.

The 1980s saw the boom in even smaller computers—the *microcomputers,* or *micros.* These computers, also called *PCs,* for *personal computers,* opened up opportunities for both small and large firms. Small firms could now enjoy the same computer processing as their larger competitors. Large firms could use the micros in distributed processing and make them available to users as standalone systems.

While most of the attention during the 1980s was directed at the small computers, very large and powerful *supercomputers* were being developed primarily for scientific, rather than business, applications.

The four classes of computers—supercomputers, mainframes, minicomputers, and microcomputers—represent the hardware resources. The current trend is to

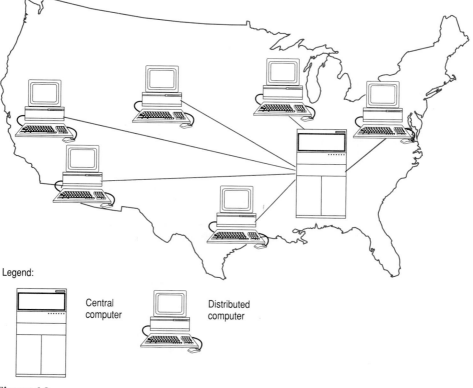

Legend:

Central computer

Distributed computer

Figure 6.2
Distributed processing.

link these resources in a network to facilitate client/server computing. In **client/server computing,** the centrally located, large *server* computer processes the high-volume applications. *Clients* use the distributed, smaller computers as stand-alones or as terminals to access the server.

Software Resources

The several types of software resources are illustrated with the hierarchy diagram in Figure 6.3. For most of the computer era, two main categories of software—system and application—existed.

System software causes the hardware to perform certain fundamental tasks required by all users. Examples of system software are operating systems, translators that translate source language into object language, and utility programs that perform necessary functions such as disk formatting and sorting. **Application software** causes the hardware to perform tasks required by only certain users. Payroll and inventory programs are examples.

The system software and application categories did a good job of classifying all software until a new approach to software development came along in the 1980s.

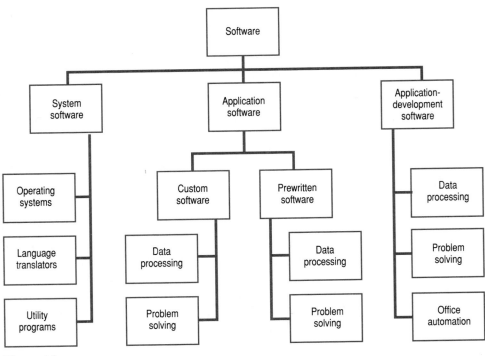

Figure 6.3
Types of software.

We call this category **application-development software,** recognizing that it is easily tailored to user applications. Application-development software includes personal and organizational productivity software, such as electronic spreadsheets, word processors, and graphics packages. The term **fourth-generation language (4GL)** is used for application-development software that facilitates querying the database. Database management systems and database query languages are examples. Most end-user computing has been in application-development software.

System and application-development software are purchased from suppliers such as Microsoft and Borland. Application software might be custom software prepared by the firm's information specialists and users or prewritten software purchased from suppliers. Application software is used to process data or solve problems. Application-development software has these same two uses, plus office automation.

Data and Information Resources

Data and information resources describe what goes on inside the firm and in the firm's environment.

Most data is internal data gathered by the data processing system. *Internal data* provides the details concerning the firm's business activities—its sales orders, payroll transactions, inventory levels, and so on. Some *internal information,* meaningful descriptions of activities inside the firm, is also entered into the CBIS. For example, a supervisor might enter a report citing the reasons for overtime work.

In recent years, *environmental data* has increased in importance. However, top-level managers favor *environmental information,* meaningful descriptions of activities in the firm's environment.

Figure 6.4 shows the types of data and information entering the CBIS. The data can be converted into information, which is immediately provided to users, or it can be stored in the database for later processing.

Personnel Resources

Personnel resources consist of users and information specialists.

Users   CBIS users can be located inside and outside the firm, as shown in Figure 6.5. Internal users consist of managers and nonmanagers. External users consist of all the environmental elements except competitors. All the users receive information. Internal users receive information that they use to solve their problems. Environmental users receive information that facilitates their relationships with the firm.

Information Specialists   Chapter 1 identified the main types of information specialists, and, Figure 1.7 illustrated how they are arranged in a communication chain. The information specialists are located in the information services unit, managed by a vice-president or another manager, who might be called the chief

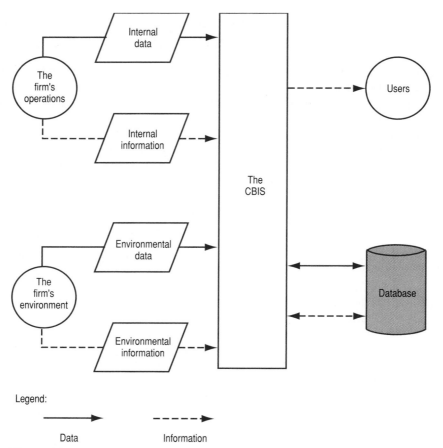

Figure 6.4
The database contains both data and information.

information officer, or CIO. Information specialists traditionally have been organized in a **functional organization structure,** based on the functions they perform,
as shown in Figure 6.6.

- *Systems analysts* study existing systems and design new or improved systems.
  The analysts communicate with users to determine their needs and decide how
  the systems will be applied. The analysts document the new systems and communicate the details to the programmers.

- *Database administrators,* called *DBAs,* have the responsibility of implementing
  and maintaining the firm's computer-based data, or database. In performing
  their work, the database administrators communicate with the users, the systems analysts, and the network specialists.

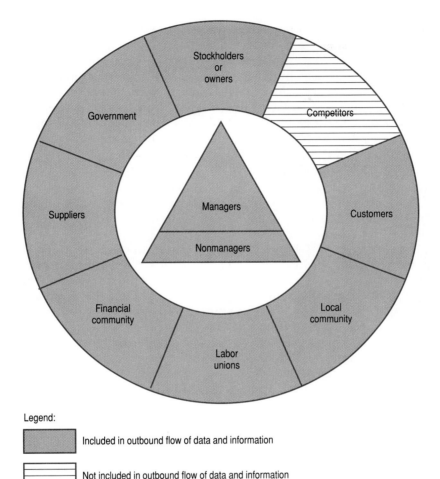

Figure 6.5
Users of the CBIS output.

- *Network specialists,* also called *data communications specialists,* implement and maintain the firm's data communications networks. The network specialists assist the systems analysts in designing the networks.

- *Programmers* use the analysts' system documentation to code and test the programs for the new systems. When the documentation fails to tell the entire story, the programmers communicate orally with the systems analysts and, when the situation demands, with database administrators and network specialists. The programmers also prepare documentation for the operations personnel.

- *Operations personnel* implement the software systems on the hardware. Several categories of personnel are included in computer operations. *Data entry personnel* key data from source documents into keyboard terminals. *Operators* operate the supercomputers, mainframes, and minis. *Schedulers* determine when jobs will be run, and *production control personnel* enforce the schedule. *Input and output control personnel* ensure that all the data received from users is processed and that the outputs are returned to the users. *Library personnel* maintain magnetic tape and disk files in a storeroom when the files are not being used by the computer.

A Life Cycle Organization Structure   Although the functional organization in Figure 6.6 is the traditional way to organize the information specialists, it is only one approach. A modification that exists in many firms is shown in Figure 6.7. We call this the **life cycle organization structure,** because it subdivides the systems analysts and programmers in terms of their responsibilities during the system life cycle.

The *manager of systems development* manages all the systems analysts and pro-grammers who develop systems.The *manager of systems maintenance* manages all the systems analysts and programmers who maintain systems. The operations personnel, database administrators, and network specialists support both groups.

Project Teams   In either structure, information specialists typically work in project teams. A **project team** consists of one or more systems analysts, DBAs, net-work specialists and users, programmers, an internal auditor, and other persons such as consultants who develop or maintain a computer-based system. Each team is led by a project team leader, who reports to the MIS steering committee, as was illustrated in Figure 2.12.

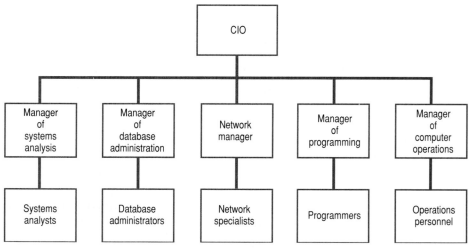

Figure 6.6
A functional organizational structure for information services.

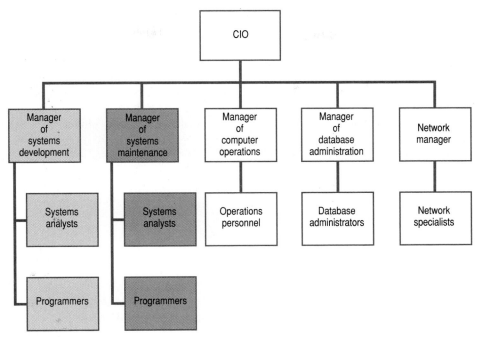

Figure 6.7
A life cycle organizational structure for information services.

Facilities Resources

Facilities include such resources as computer rooms, media libraries, offices, and classrooms. Facilities tend to be quite expensive. A computer room, for example, has its own air conditioning and humidity controls, fire extinguishing equipment, backup power generators, and security features.

Until end-user computing came along, all computing facilities were located in information services. Today, they can be located anywhere. The facilities represent a significant investment and a valuable resource.

Using CASE to Determine System Architecture

One of the upper-CASE tasks is to determine a system architecture. A **system architecture** defines the hardware, software, and data that compose an information system. One class of tools assists the systems analyst with drawing the architecture and also evaluating the performance and reliability of the system, given the requirements. These tools combine the features of CASE with computer performance determination and provide design trade-off results through simulation and other quantitative techniques such as queueing models.

First, the analyst describes the proposed system using qraphical tools to depict the software, hardware, and data components of the design. Then, the response times, capacity, and availability can be assessed using analytic and simulation models. This upper-CASE tool can be brought into play during the early phases of the SDLC by supporting modeling of the existing system. As new designs are proposed, the tool can be used to assess different performances from design to design. These tools start out by capturing the system-level workload and architecture and evaluating them against system-level requirements. Once the system is operational, the tool provides an ongoing mechanism for capacity management and planning.

Fundamental Approaches to Computer Processing

The task of the firm's hardware, software, personnel, and facilities resources is to transform data into information. Data can be processed in two main ways—batch or online. The first computers mimicked the earlier punched card and keydriven machines by processing their data in batches. Magnetic disk storage made it possible to process transactions as they occur, online.

Batch Processing

In **batch processing,** transaction data is accumulated in a batch and all the transactions in the batch are processed through each step of the procedure.

Figure 6.8 is a system flowchart that illustrates batch processing. The first symbol in the figure represents a paper document. The next symbol represents a keying operation using a computer keyboard. Each rectangle represents a program, and the cylinders represent disk storage. The brackets to the right of the sort programs identify the sort keys. The two small circles are called onpage connectors. They are used instead of a long arrow.

In this flowchart, sales transactions are keyed onto a magnetic disk throughout the day in steps 1 and 2. The disk file is a **transaction file,** which contains the data describing one of the firm's activities. In this case, the transaction file describes the firm's sales.

The transactions are held on the disk until the end of the day, when they are processed. The processing requires that three master files be updated with the transaction data. A **master file** is a file that represents an important resource or an element in the firm's environment. The master files in the flowchart include the inventory master file, the customer master file, and the salesperson master file. The inventory master file is maintained in item number sequence, the customer master

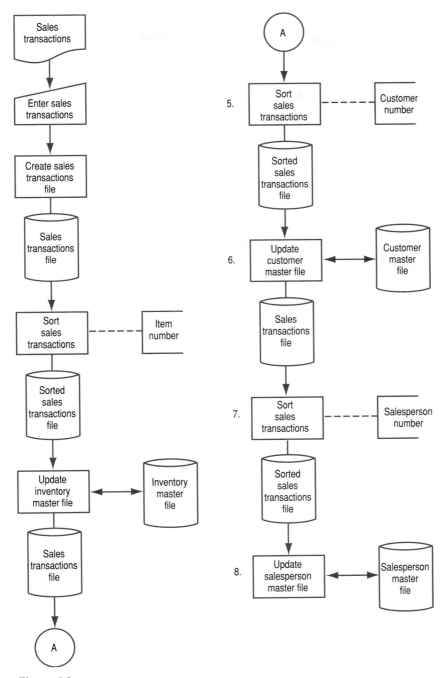

Figure 6.8
Batch processing.

file in customer number sequence, and the salesperson master file in salesperson number sequence. Before each master file can be updated, the transaction file must be sorted into the master file sequence. The sorting occurs in steps 3, 5, and 7, and the files are updated in steps 4, 6, and 8.

The main limitation of batch processing is that it does not provide current data. The master files are current only immediately following file maintenance. Because a current database is so important to many modern computer applications, systems designers have shifted almost exclusively from batch to online processing.

Online Processing

The technique of updating the master files as each transaction is entered is called **online processing.**[1] Figure 6.9 shows how the same three master files can be updated online. Data for a single transaction is keyed into the computer and, while the transaction data is in the primary storage, all three master files are updated. Then, another transaction is entered.

Ideally, the transaction data should be entered as the transaction occurs, but that is not a necessity. A transaction can be held for several hours before being entered. Online processing takes place as long as all the master files are updated from one transaction before another transaction is entered. The file is as current as the most recent transaction that has been processed.

An online system that controls the physical system is called a realtime system. **Realtime** means *right now.* A *realtime system* controls one or more processes as they occur. The key word is *control.* Take, for example, a system used by a department store to enter data into POS (point of sale) terminals as sales are made. If the

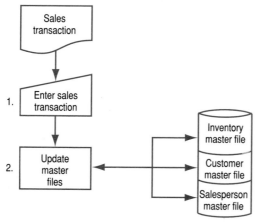

Figure 6.9
Online processing.

[1]The term *transaction processing* is often used to describe the separate handling of transactions. We do not use this term because it is also used to mean the firm's data processing system.

data simply updates the inventory and other files, it is an online system. If, on the other hand, the data is used to check credit to determine whether the sale will be made, the system is a realtime system because the system controls the firm's sales activity.

During the designing of a new computer-based system, several decisions concerning system alternatives must be made. However, none affects system performance more than the choice between batch and online. This choice is *the fundamental design decision* because of its ultimate influence on information value.

Information Value

The value of information to its user depends on its four dimensions—accuracy, timeliness, completeness, and relevance.

Accuracy

The computer has earned a reputation for processing data accurately and for producing accurate information. *Accuracy* is the degree of precision with which the information describes its subject.

It is not always necessary to strive for 100 percent accuracy. When money is involved, as with the firm's payroll or accounts payable, complete accuracy is desired. However, many other applications can tolerate a degree of error. For example, a projection of the coming year's sales revenues provides a marketing manager with only a broad guideline for decision making. The report is not expected to be accurate to the penny.

Timeliness

The computer produces timely information. *Timeliness* refers to whether the information gives the user an understanding of the subject in time to act on the information. Timeliness is the information dimension that has been improved the most by the computer. Earlier technologies could not process the data fast enough to provide timely information.

Completeness

Completeness of information is the dimension that was handled best by precomputer and early computer systems. *Completeness* refers to the thoroughness with which the information describes the subject.

It is important not to go overboard by giving more information than the user can handle, creating a condition called **information overload.** Modern systems seek to overcome the overload problem by enabling the user to specify the level of detail. Executive information systems, for example, feature **drill down,** which allows the user to obtain summary information and then gradually produce the desired degree of detail.

Figure 6.10 illustrates the drill-down technique. First, the manager requests a display of sales for the entire company and sees that the western region is far behind quota. To learn more about the situation in the western region, the manager requests a display of the region's sales by product line. This display shows that the region is having the least success with the VCR line. Finally, to learn whether the problem is common to the entire region or to individual salespersons, the manager requests a display of western region VCR sales by sales representative. This analysis enables the manager to pinpoint the problem so that various solutions can be considered. In this case, representatives Arnold and Tumlinson might be given special training to help boost their VCR sales.

Relevance

Information has *relevance* if it bears on the issue at hand. For example, if a manufacturing manager is interested in increasing production quality, a report showing the number of defects per thousand units produced would be relevant. The relevance dimension imposes severe demands on the systems analyst because users' interests are always changing. Systems designs must be flexible enough to respond to changing user needs.

Accuracy, timeliness, completeness, and relevance all have their cost. Increases in accuracy and timeliness are achieved by increased investments in hardware and software. These costs escalate so rapidly as perfection is approached that users usually are willing to settle for something less than perfection. The cost of completeness and relevance is not so much the cost of the hardware and software but, rather, the time the system designers spend in precisely determining the user's needs.

A key element in system design is finding the balance between what the user needs and what the user is willing to pay.

How Users Obtain Information from the Computer

Problem solvers obtain information in many ways. Some use *oral media,* such as scheduled and unscheduled meetings, tours, telephone calls, and business meals. Others employ *written media,* such as letters, memos, periodicals, and computer output. Some information is provided by *formal systems,* which operate according to a schedule or are documented in writing. Other information is provided by *informal systems,* which function on demand to meet unexpected needs.

In this text, our main concern is with the design of formal and informal computer-based systems. The computer provides information to users in three basic ways—reports, outputs from simulations, and communications.

REGION	YEAR-TO-DATE SALES	YEAR-TO-DATE QUOTA	YEAR-TO-DATE VARIANCE
COMPANY SALES (IN THOUSANDS OF DOLLARS) FOR PERIOD ENDING AUGUST 31			
EASTERN	425,000	410,000	15,000+
MIDWEST	378,500	360,000	18,500+
SOUTHWEST	163,250	160,000	3,250+
MOUNTAIN	150,000	155,000	5,000-
WESTERN	410,000	445,500	35,500-
TOTALS	1,526,750	1,530,500	3,750-

A Company sales

PRODUCT LINE	YEAR-TO-DATE SALES	YEAR-TO-DATE QUOTA	YEAR-TO-DATE VARIANCE
WESTERN REGION SALES BY PRODUCT LINE (IN THOUSANDS OF DOLLARS) FOR PERIOD ENDING AUGUST 31			
CAMCORDER	60,000	63,000	3,000-
CD PLAYER	92,300	95,000	2,700-
TAPE PLAYER	15,000	10,000	5,000+
TUNER, AM/FM	78,200	77,500	700+
TV	103,600	100,000	3,600+
VCR	60,900	100,000	39,100-
TOTALS	410,000	445,500	35,500-

B Western region sales by product line

Figure 6.10
Managers can drill down to lower levels of detail in searching for problems and their causes.

Figure 6.10 *continued*

WESTERN REGION SALES BY PRODUCT LINE AND SALES REPRESENTATIVE (IN THOUSANDS OF DOLLARS) FOR PERIOD ENDING AUGUST 31			
PRODUCT LINE:	VCR		
SALES REPRESENTATIVE	YEAR-TO-DATE SALES	YEAR-TO-DATE QUOTA	YEAR-TO-DATE VARIANCE
ARNOLD, G. N.	5,000	30,000	25,000-
PERKINS, B. J.	50,450	40,000	10,450+
TUMLINSON, M. E.	5,450	30,000	24,550-
TOTALS	60,900	100,000	39,100-

C Western region VCR sales by sales representative

Reports

Over the years, reports have been the most popular way to represent computer output. The two basic types of reports are periodic and special.

A **periodic report** is prepared according to a schedule, perhaps weekly or monthly. It is triggered by a passage of time. The names **repetitive report** and **scheduled report** are also used for a periodic report.

A **special report** is triggered by a request or an event. For exmple, a human resources manager who queries the database for a list of employees with more than ten years of service makes a *request*. An accident that decreases production time is an *event* that can cause a lost time report to be prepared. Another event is an exception to planned performance, as when the manager practices management by exception. An *exception report* is a special report used in management by exception.

Figure 6.11 is a hierarchy diagram that shows the classification of report types. Each type can portray its contents in a printed or displayed form and in a tabular, graphic, or narrative manner.

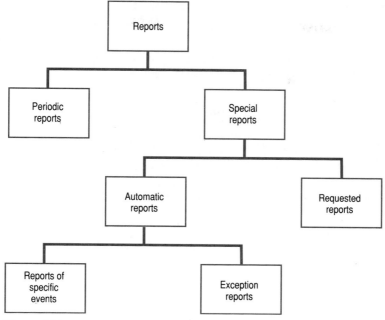

Figure 6.11
Types of reports.

Printed and Displayed Reports   Before computer screens became common-place, reports were always printed on paper. Such output is called a *hard copy*. The popularity of the screens has shifted much of the output to a displayed form. Users prefer the fast and convenient displays. Users can use their screens to scan large volumes of information and then request hard copies of specific portions.

Tabular, Graphic, and Narrative Reports   Traditionally, the computer presented report information in rows and columns, forming a table. Such an arrangement is called a **tabular report.** A report that presents information pictorially is called a **graphic report.** When information is communicated in sentences, the report is called a **narrative report.** A memo prepared with a word processor is an example of a narrative report.

Recently, the tabular, graphic, and narrative presentations have been combined. The narrative briefly explains the tabular and graphic information. This technique has been especially effective in executive information systems (EISs). Figure 6.12 is a screen display produced by Comshare's Commander EIS, which combines the three information formats.

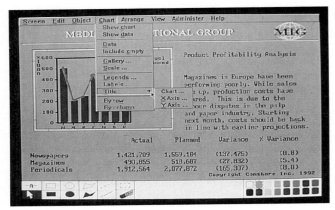

Figure 6.12
Executive information system displays frequently combine tabular, graphic, and narrative information.

Outputs from Simulations

Early designers of computer-based systems quickly recognized the computer's ability to represent business phenomena in the form of mathematical formulas. Any representation of a phenomenon, or entity, by means of one or more mathematical formulas is a **mathematical model.** The act of using such a model is called **simulation.** The model simulates its entity.

A mathematical model can represent an entity such as cash flow or inventory fluctuations. Then, a manager can use the model to project the effect of certain changes. Modeling is the only output method that makes projections for the user.

Figure 6.13 is the output from an inventory model. As with reports, model output can be either printed or displayed. In the example, the days are listed down the left margin. For each day, the model computes the beginning balance, the quantity of receipts (if any) from suppliers, the quantity of sales, the ending balance, and the backorder quantity. A backorder is an order that cannot be filled due to lack of stock. An inventory manager could use the output in the figure to identify the order quantity that produces the lowest average inventory balance without incurring backorders.

The difficulty of model building prevented simulation from being a common output method for much of the computer era. However, the advent of the electronic spreadsheet has enabled persons who are not management scientists or modeling specialists to create models. The model output in Figure 6.13 could have been produced by an electronic spreadsheet as well as a programming or modeling language. A large portion of end-user computing, perhaps the largest portion, involves spreadsheet modeling.

```
                              INVENTORY MODEL

       NUMBER OF DAYS:        10
       BEGINNING BALANCE:    100
       ORDER QUANTITY:        80
```

DAY	BEGINNING BALANCE	RECEIPTS	SALES	ENDING BALANCE	BACK-ORDER QUANTITY
1	100		23	77	
2	77		16	61	
3	61	80	21	121	
4	121		36	85	
5	85		28	57	
6	57		44	13	
7	13		23	0	13
8	0	80	31	36	
9	36		18	18	
10	18		23	0	5

```
       AVERAGE INVENTORY BALANCE:                    47

       TOTAL BACK-ORDER QUANTITY:                            18
```

Figure 6.13
Output from a mathematical model that simulates inventory activity.

Communications

When we explained the general systems model in Chapter 4, we recognized the need for the manager to communicate decisions to the physical system. This communication can be facilitated by office automation, or OA. OA can communicate problem-solving information *to* and *from* the user. We describe this computer application later in the chapter.

The task of the user and the systems analyst is to match the form of output to the problem-solving situation. Reports are effective for distilling large volumes of data into information, mathematical models can project future activity, and communications facilitate the exchange of information with others. These three forms of computer output are combined with noncomputer information to provide the basis for business problem solving.

The Computer-Based Information System

We use the term **computer-based information system,** or **CBIS,** to describe all the business applications performed by the computer. The CBIS contains five subsystems—the data processing system, the management information system, decision support systems, office automation systems, and expert systems.

The Data Processing System

The **data processing system,** or **DP system,** consists of the firm's accounting applications. Practically all firms perform their accounting applications on the computer. A firm has one data processing system, with subsystems dedicated to the various applications such as payroll and inventory.

Characteristics of the Data Processing System   A firm's data processing system has the following characteristics:

- **A required application**—A firm does not choose whether it has a data processing system. Data processing is a required application. The firm's managers need it to maintain control over operations. Environmental elements, such as the government and the stockholders, demand that a firm perform data processing to establish an accountability.

- **Standard procedures**—Data processing procedures are fairly standard from one firm to the next. The accounting fundamentals you learn in college provide the common thread.

- **An internal data focus**—The *primary* purpose of the data processing system is *not* to provide information for problem solving but to process the data that represents the firm's activities.

- **An environmental information focus**—Elements in the firm's environment look to the firm for information describing the firm's activities. This information is provided by the data processing system in the form of purchase orders, invoices, and data in annual reports to stockholders.

- **Some information for internal use**—The data processing system produces *some* problem-solving information in the form of traditional accounting reports, such as the balance sheet and income statement.

- **A historical focus**—Information generated by the data processing system is historical. Although the information can, in some instances, tell what *is* happening, it sheds the most light on what *has* happened.

The preceding characteristics of the data processing system are reflected in the model in Figure 6.14. Data is gathered from the physical system of the firm, is stored in a database, and is transformed, using data processing software, into information for management and the environment.

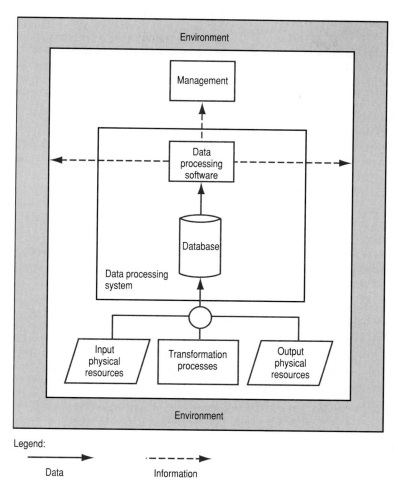

Figure 6.14
A data processing system model.

Keys to the Analysis and Design of the Data Processing System   The three keys to the analysis and design of the data processing system are:

- an understanding of accounting fundamentals
- an understanding of the information needs of the firm's environmental elements
- an ability to design databases that describe the firm's operations

Putting the Data Processing System in Perspective   The DP system provides the basis for the other CBIS subsystems. If a firm does not have a good foundation of accounting data, the other subsystems face an uphill battle.

The Management Information System

After large firms successfully implemented their data processing systems, they began developing computer-based systems explicitly for providing information to managers. The name *management information system,* or *MIS,* was given to these systems. Many of the early MIS efforts met with failure because information specialists knew little about management and managers were unaccustomed to describing their information needs. Some information specialists made unrealistic promises, and some managers had unrealistic expectations. After a while, the more persistent firms overcame these shortcomings and implemented successful MISs.

From 1964 to 1971, the term *MIS* encompassed *all* computer applications aimed at management support. Since, the addition of other CBIS subsystems has given the term a different meaning. We offer a current definition of **management information system,** or **MIS,** as the effort by the entire firm to provide all of its managers and nonmanagers with information for solving all types of problems.

Characteristics of the Management Information System   Current MIS designs have the following characteristics:

- **Organizational Focus** An MIS is intended to meet the needs of all problem solvers in the firm. As such, the decision to implement an MIS is an organizational one, most likely made by the president.

- **All Employees as Potential Users** Although the term *MIS* implies otherwise, not all potential users of MIS output are managers. Nonmanagers also obtain the information they need to do their jobs.

- **A Focus on All Types of Problems** The MIS is not aimed at particular types of problems. It is a broad-based, problem-solving system.

- **An Emphasis on Information** The underlying logic of the MIS is that a person can solve problems when provided with good information. The task of the MIS is to provide the information. The task of the problem solver is to use it.

- **An Emphasis on Reports and Simulations** An MIS provides users with information in the form of reports and outputs from simulations.

These characteristics are reflected in the model of an MIS in Figure 6.15. The database contains data and information provided by the data processing system and the environment. Report-writing software and mathematical models provide organizational problem solvers with information.

Keys to the Analysis and Design of the MIS   The keys to the analysis and design of the MIS are:

- an understanding of organizational structure—how firms are organized, the activities of the various organizational units, and the information needs of persons in each unit

- an understanding of management responsibilities, such as management functions and roles

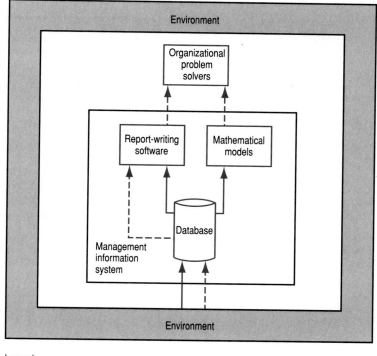

Legend:

→ Data   - - - → Information

Figure 6.15
An MIS model.

- an ability to design periodic reports that reflect the four dimensions of information
- an ability to design databases to facilitate the preparation of special reports
- an ability to design complex mathematical models using programming languages or special modeling languages

Putting the MIS in Perspective   The firm has one MIS. The MIS can have subsystems designed to meet the needs of groups of managers within the firm. For example, the *marketing information system (MKIS)* provides information about the firm's marketing operations, the *manufacturing resource planning (MRP-II) system* provides information about the firm's manufacturing operations, the *financial information system* provides financial information, and the *human resource information system (HRIS)* provides personnel information. In addition, the *executive information system (EIS)* meets the information needs of the firm's upper-level managers.

Figure 6.16 illustrates the subdivision of the MIS into organizational subsystems, an area of systems development currently receiving much attention because of end-user computing.

Decision Support Systems

As firms struggled with their early MIS designs, two professors at the Massachusetts Institute of Technology (MIT) concluded that greater success could be achieved if the information support were more focused. The professors, G. Anthony Gorry and Michael S. Scott Morton, published an article in which they used the term *decision support system (DSS).*[2]

Gorry and Scott Morton identified three types of problems all managers face. In a **structured problem,** it is possible to define algorithms, or decision rules, that allow managers to (1) identify and understand the problem, (2) identify and evaluate alternate solutions, and (3) select a solution. In an **unstructured problem,** none of the three ingredients exists. In a **semistructured problem,** one or two of the ingredients exist.

Structured problems can be solved by the computer. Unstructured problems must be solved by the problem solver working alone. Most problems, however, are semistructured and can be solved by the problem solver using the computer.

The revelation by Gorry and Scott Morton was that a **decision support system,** or **DSS,** can support an individual problem solver during the solving of a

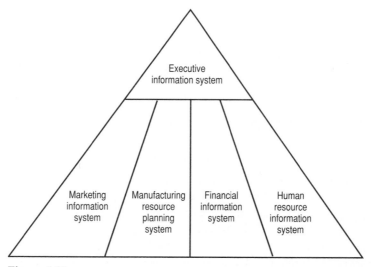

Figure 6.16
Organizational subsystems of the MIS.

[2]G. Anthony Gorry and Michael S. Scott Morton, "A Framework for Management Information Systems," *Sloan Management Review* 13 (Fall 1971): 55-70.

semistructured problem. The scope of the DSS is therefore narrower than that of the MIS. The DSS is aimed at particular problems solved by particular problem solvers.

Group Decision Support Systems   The last few years have seen the DSS concept broadening to include problem solving by groups. The groups vary in size from two or three to a hundred or so. The group members can interact face to face or be linked by a data communications network. The computer-based system that facilitates group problem solving is called a **group decision support system,** or **GDSS.** The software used by a GDSS is called *groupware*.

One GDSS setting is a *decision room,* such as the one pictured in Figure 6.17. Problem solvers use keyboards to enter comments, criticisms, and suggestions, which are displayed on a giant screen. The simultaneous data entry is called **parallel communication.** Because participants can voice their views without being identified, a feature called **anonymity,** information flows more freely than in a normal group setting.

Characteristics of DSSs   The characteristics that distinguish DSSs are:

- an individual or small group focus
- use of all three types of information output—reports, simulations, and communications
- an emphasis on decision making by assisting the manager in going through the problem-solving steps
- a focus on semistructured problems

Figure 6.17
A decision room.

These characteristics are reflected in the model of a DSS in Figure 6.18. The groupware provides a communications ability when problem solvers work in groups.

Keys to the Analysis and Design of DSSs   The five keys to the analysis and design of DSSs are listed below. The first four are also required for the MIS.

- an understanding of management responsibilities
- an ability to design good periodic reports
- an ability to design databases to facilitate the preparation of special reports
- an ability to design complex mathematical models
- an understanding of group dynamics—how people interact in a group setting while solving problems—and how to influence this interaction to achieve the best results

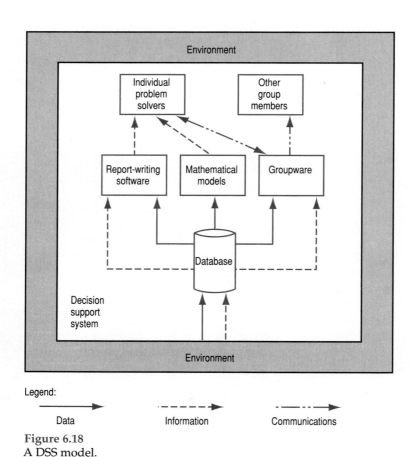

Legend:

→ Data   - - -→ Information   -·—··→ Communications

Figure 6.18
A DSS model.

Putting the DSSs in Perspective   The high level of computer literacy combined with the low cost of hardware and the availability of application-development software stimulates users to develop their own DSSs. For this reason, the continued popularity of DSSs is ensured. As a manager, you can expect to develop your own DSSs. As a systems analyst, you can expect to work with users in developing more complex DSSs and GDSSs.

Office Automation

Office automation got its start in 1964 when IBM invented the Magnetic Tape/Selectric Typewriter (MT/ST), a typewriter with a magnetic tape unit attached. The MT/ST enabled the typist to type a form letter, store it on tape, and then print it many times, each time keying in only the receiver's name and address. Each letter appeared to the receiver to be an original. This was the first example of word processing, which was the first example of office automation. **Office automation,** or **OA,** is the use of electronic devices to communicate between persons in the firm, and between persons in the firm and its environment. Some OA systems involve the computer. Some do not.

Computer-based OA Systems   Eight OA systems involve the computer. They are:

- word processing
- electronic mail
- electronic calendaring
- computer conferencing
- voice mail
- imaging
- videotex
- desktop publishing

Word processing is the use of the computer to prepare written documents.

Electronic mail, or *E-mail,* is the use of networked computers for sending and receiving messages.

Electronic calendaring is the maintenance of appointments calendars in the computer's secondary storage. Users can review other calendars for the purpose of scheduling meetings. Users can also keep portions of their calendars confidential.

Computer conferencing is the use of electronic mail software by a specified group of participants to discuss a particular topic. This use differs from E-mail, which is available to anyone with a terminal, for use in any kind of communications.

Voice mail is the storage of oral communications in secondary storage. Voice mail requires a networked computer with a message encoding and decoding unit. The system is used in the same way as electronic mail, except that the messages are sent and received by telephone.

Imaging is the use of a computer equipped with a laser disk, or compact disk, to store images of documents. The documents are scanned, using an optical character recognition (OCR) device, and the images are converted to a digital form for storage on the disk. The images can then be retrieved and displayed rapidly.

Videotex is the use of a computer to store information in text or graphic form for future retrieval. A firm can store and retrieve material in its own computer, can retrieve material from other firms' computers, or can retrieve material provided by a videotex subscription service such as Dow Jones News/Retrieval Service.

Desktop publishing (DTP) is the use of a computer to prepare typeset copy of the same high quality as that prepared by a printing firm. DTP requires a high-resolution screen and a letter-quality printer.

Noncomputer-based OA Applications   Three OA applications do not require a computer. They are:

- facsimile transmission
- audio conferencing
- video conferencing

Facsimile transmission, or *fax,* is the most popular. Fax technology consists of special copying machines that transmit document images from one location to another, using ordinary telephone circuits. You can also give your microcomputer fax capability by installing a fax board.

Audio conferencing is the use of audio circuits to permit more than two people to engage in two-way communication over long distances. An example of audio conferencing is the conference call that enables a small group of people to communicate, all from their own telephones.

Video conferencing is the use of television equipment to transmit video and audio signals in either a one-way or two-way manner. People at geographically disbursed locations meet in rooms designed specifically for the purpose. Rooms at the transmitting and receiving locations are equipped with video screens, and the rooms at the transmitting locations are equipped with video cameras. Figure 6.19 shows a video conferencing room provided by AT&T.

Characteristics of OA Systems   OA systems have the following unique characteristics:

- an emphasis on communication
- satisfaction of organizational needs and individual needs
- emphasis on informal systems
- an ability to send and receive information
- an increasing capability for obtaining information from the environment
- ease of implementation

Figure 6.20 is an OA model. The rectangle labeled *Other problem solvers* straddles the line separating the firm from its environment, indicating that the other

Figure 6.19
A video conference room.

OA users can reside inside or outside the firm. OA emphasizes communications and information, rather than data.

Keys to the Analysis and Design of OA Systems   The two keys to good OA systems work are:

- an understanding of the role of communication in problem solving

- a knowledge of the available OA systems

Putting OA in Perspective   Many people view OA only as a way to *send* information, failing to recognize its value as a problem-solving tool. This narrow view ignores OA's ability to facilitate two-way communication within the firm and between the firm and its environment.

OA is the only CBIS subsystem that includes a large proportion of informal systems. In fact, *all* the OA systems are most often operated informally. Only when they operate according to a schedule, do they become formal systems. The informal nature of OA appeals to problem solvers.

Expert Systems

Whereas OA has been widely accepted by users but not given full credit as a problem-solving tool, the opposite is true for expert systems. Expert systems have captured the interest of management scientists and information specialists, but have been difficult to achieve.

Expert systems are a subset of artificial intelligence. *Artificial intelligence*, or *AI*, is the activity of giving machines the ability to display behavior that would be

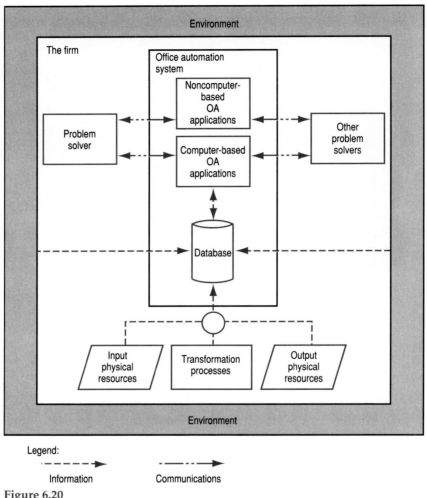

Figure 6.20
An OA model.

regarded as intelligent if it were observed in humans.[3] The term *artificial intelligence* was coined in 1956, but few advances were made until the 1980s.

Artificial intelligence applies computer technology in five areas:

- **Perception** Giving computers the ability to react to visual and auditory signals.

- **Learning** Giving computers the ability to acquire knowledge beyond what has been built into their programs. Much research activity in this area involves *neural nets,* mathematical representations of the cellular activity of the human brain.

[3]Clyde W. Holsapple and Andrew B. Whinston, *Business Expert Systems* (Homewood, IL: Irwin, 1987), 4.

- **Robotics** Providing computer-controlled devices with the ability to perform some human motor activities.

- **Natural Language Processing** Being able to generate computer code from natural-language descriptions of problems to be solved.

- **Expert Systems** Giving computers the ability to function as consultants to problem solvers.

Of the AI areas, expert systems, also called **knowledge-based systems,** have stimulated the most development activity in business. By serving as consultants, the expert systems enable problem solvers to achieve solutions that exceed their own capabilities. For example, a system that captures the knowledge of an expert bank loan officer can be used by less-experienced officers in making better loan decisions.

Characteristics of Expert Systems   Expert systems are characterized by the following:

- an ability to diagnose complex situations

- an ability to handle uncertainty

- an ability to solve problems previously solved only by humans

- an ability to explain solutions

An expert system consists of the four main parts illustrated inside the rectangle in Figure 6.21.

Knowledge about the problem that the expert system is expected to solve is the *problem domain.* For example, the problem domain of the bank loan system would be installment loans. All the expert's knowledge is stored in the *knowledge base.* The knowledge can be stored in various ways, but the most popular is in the form of rules.

A rule is similar to an IF statement in a programming language. For example, an expert system rule might appear as:

```
IF SEASONAL.TREND = "UP" AND ECONOMIC.OUTLOOK = "GOOD"
   THEN SALES.FORECAST = "EXCELLENT"
```

The part of the expert system that analyzes the rules is the *inference engine.* The inference engine examines the rules in a certain pattern, with the objective of reaching a conclusion, called the *goal variable.* In the bank loan system, the goal variable would be the determination of whether to approve the loan.

The user communicates with the expert system by means of the *user interface.* The user enters instructions and information, and the expert system responds with solutions and explanations.

The expert system is developed by one or more experts working with a systems analyst knowledgeable in the development process. The name **knowledge engineer** is commonly used to describe such an analyst. The analyst and expert use the *development engine* to create the expert system. The most fundamental form of the development engine is a programming language. Another form is an expert system with a prewritten inference engine. Such a system is called an *expert system*

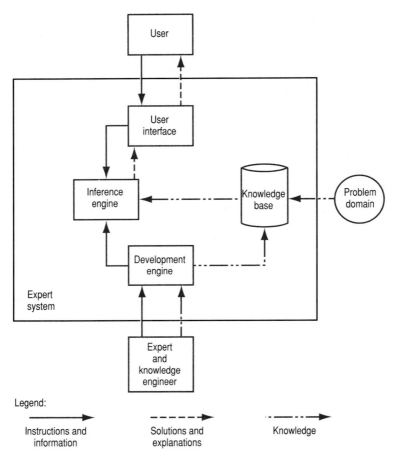

Figure 6.21
An expert system model.

shell. The shell contains all the expert system components except the knowledge base, which the developers must add.

Keys to the Analysis and Design of Expert Systems   Two keys are crucial to the analysis and design of expert systems:

- an understanding of the problem domain
- an ability to extract the expert's knowledge—an activity called **knowledge acquisition**

Putting Expert Systems in Perspective   Expert systems face more difficult challenges than the other CBIS subsystems, and their development requires more time and money. Even though the results have been relatively meager thus far, the opportunities for applying expert systems to business problems seem boundless.

For that reason, a specialization in expert systems would appear to be a good career strategy.

Putting the CBIS in Perspective

We have seen that each CBIS subsystem has unique characteristics and requires a different mix of skills for analysis and design. This aspect is often overlooked. However, as soon as you begin analyzing and designing the various types of systems professionally, the differences will become clear. Throughout the text, as we explain the processes of developing computer-based systems, we will call your attention to the different demands placed on you by each CBIS subsystem.

Using CASE to Draw and Define System Architecture

Figure 6.22 is a top-level hardware drawing generated with a CASE tool that supports system architecture graphics. The figure shows that there are three mainframe processors connected to a communications bus. Two of the processors have secondary storage disks. The diagram shows not only the hardware but also the utilization levels of the hardware. The utilization is determined by entering processing specifications, software specifications, and communication volumes. The tool acts as a simulator to determine the usage levels.

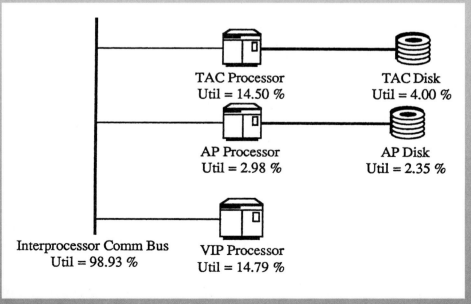

Figure 6.22
A top-level hardware drawing produced by CASE.

Summary

A firm's information resources are a combination of hardware, software, data and information, personnel, and facilities. Hardware comes in the form of supercomputers, mainframes, minis, and micros. Software is of three types—system, application, and application-development. Personnel once consisted only of information specialists but now includes users.

Data is processed two ways on the computer—batch and on line. In batch processing, the batched transactions update master files one at a time. In online processing, all the master files from one transaction are updated before the next transaction is handled. One type of online system, a realtime system, can influence the operation of the physical system by means of rapid feedback.

Information value depends on the information's accuracy, timeliness, completeness, and relevance. Precomputer systems provided the most complete information, but computer systems provide the most accurate and timely information. Relevance is difficult to achieve and maintain because user needs change constantly. Each dimension of information value has a cost, and the systems analyst and user must find the best balance between needs and cost.

Users obtain information from the computer in three basic ways—reports, outputs from simulations, and communications. Reports can be periodic or special, and can be printed or displayed. Reports also provide information in three formats—tabular, graphic, and narrative. A special report that is prepared when performance exceeds preestablished boundaries is called an exception report. Business phenomena can be represented by mathematical models. The process of using such models is called simulation. Models have the unique advantage of being able to project activity. The difficulty of model building has been lessened by electronic spreadsheets. The third output form, communications, is the specialty of office automation.

All the business applications that computers perform are given the name computer-based information system, or CBIS. The CBIS consists of five subsystems. All firms have a data processing system. Most data processing systems are computer-based. The primary purpose of a data processing system is to maintain an accurate database. However, the data processing system provides much information to the environment and some information to the firm's managers.

The firm's management information system is an organizational system that is intended to provide all the firm's problem solvers with information. The recent emphasis on information as an organizational resource has pumped new life into the MIS and its subsystems, which include the human resource information system and executive information systems.

DSSs help individual managers or small groups of managers solve semistructured problems. Reports and simulation outputs provide the information. When a GDSS is used, communications between members of a problem-solving team are improved by groupware.

Office automation systems employ computer and noncomputer technologies to improve the communication flow both inside the firm and between the firm and its environment. Much of the OA appeal is due to its informal nature.

An expert system can function as a consultant and enable the manager to achieve problem solutions not ordinarily expected. An expert system consists of a knowledge base that reflects the problem domain. An inference engine uses programmed logic to analyze the knowledge base. An expert system with a prewritten inference engine is called an expert system shell. A systems analyst, known as a knowledge engineer, and one or more experts use a development engine such as a programming language or a shell to develop the expert system. The systems analyst must understand the problem domain and be skilled in knowledge acquisition. A unique characteristic of the expert system is its ability to explain how it arrived at its conclusion, called the goal variable.

If you are to develop successful computer applications as a manager or an information specialist, you must understand the CBIS subsystems and how each subsystem influences analysis and design.

Key Terms

Timesharing

Distributed processing

Client/server computing

System software

Application software

Application-development software

Fourth-generation language (4GL)

Functional organization structure

Life cycle organization structure

Batch processing

Transaction file

Master file

Online processing

Realtime system

Information overload

Drill down

Periodic report, repetitive report, scheduled report

Special report

Exception report

Tabular report

Graphic report

Narrative report

Mathematical model

Simulation

Data processing system (DP system)

Management information system (MIS)

Decision support system (DSS)

Group decision support system (GDSS)

Parallel communication

Anonymity

Office automation (OA)

Expert system, knowledge-based system

Knowledge engineer

Knowledge acquisition

Key Concepts

Users as information resources

The dimensions of information that give it value

Formal and informal systems

How business phenomena can be represented by mathematical models

The computer-based information system (CBIS)

The management information system (MIS) as an effort to meet information needs of the entire organization

Problem structure

The decision support system (DSS) as an effort to meet decision needs of individual problem solvers

Office automation as a problem-solving tool

The ways that characteristics of human behavior can be captured by artificial intelligence, specifically expert systems

Questions

1. What are the information resources? Place an asterisk next to the resources always located in information services.

2. How does timesharing differ from distributed processing? What technological innovation accounted for the popularity of distributed processing?

3. In client/server computing, who or what is the client? The server?

4. What are three examples of popular application-development software packages?

5. What two things can happen when data and information are entered into the CBIS?.

6. Which environmental element is not a user?

7. Whom would you expect to find on a project team?

8. What key factor distinguishes batch processing from online processing?

9. What is the difference between a transaction file and a master file?

10. What is the difference between online processing and realtime processing?

11. What are the four dimensions of information value? Which one has been improved the most by the computer?

12. When should information be 100 percent accurate?

13. How responsive must a system be to meet the requirement of timeliness? How responsive must it be to qualify as a realtime system?

14. Which information value dimension runs the risk of producing information overload?

15. What is the difference between a formal and an informal system?

16. What are the two types of triggers for special reports?

17. What three types of report formats have been integrated in executive information systems?

18. What unique capability does a mathematical model offer the user?

19. Which CIBS subsystem must the firm have? Who demands it?

20. Why did many early MISs fail?

21. What is happening to rekindle interest in the MIS?

22. What are the three types of problems in terms of their structure? Which type requires the user to work with the computer?

23. Which GDSS feature encourages the participants to enter into the problem discussion?

24. What is the difference between an MIS and a DSS?

25. Who in the firm would most likely make the decision to implement an MIS? A DSS?

26. Which OA applications require a computer? Place an asterisk next to the ones that require multiple computers to be linked in a network.

27. What is the difference between computer conferencing and electronic mail?

28. Which OA application can be either computer-based or noncomputer-based?

29. Which expert system component is missing in an expert system shell?

30. Which two CBIS subsystem types facilitate communications?

Topics for Discussion

1. Which of the eight elements in the firm's environment would insist that the firm have a data processing system? The eight elements are illustrated in Figure 6.5.

2. What computing hardware would be required to shift a data processing system focus from historical to current data?

3. Draw a grid on a sheet of paper. List your school's business core courses down the left margin, and list the CBIS subsystems across the top. Enter checkmarks in the squares to show which courses include material that will be helpful in analyzing and designing the particular CBIS subsystems. Which CBIS subsystem is supported the best (has the most checkmarks)? Can you explain why? Which subsystem is supported the least? Why?

Problems

1. Draw a diagram of a client/server computing network. Use rectangles for the computers, and use cylinders, as in Figure 6.8, for the databases. Label each rectangle with the word *Client* or *Server*. Label each cylinder to identify it as either an organizational or personal database. Use lines to connect the symbols.

2. Read the Gorry and Scott Morton article, and describe the contributions of Anthony and Simon to the DSS concept.

3. Read the Alter article included in the chapter bibliography, and briefly explain each of the six DSS types. For each DSS type, identify which form of computer output (reports, simulation results, or communications) would be used.

Consulting with
Harcourt Brace

(Use the Harcourt Brace scenario prior to this chapter in solving this problem.)

Pretend that you are Mike Byrnes and that Robert Simons, the vice-president of finance, has asked you for a memo that identifies the areas where the computer should be applied in the future. Write a one-page memo, listing at least three areas, in order of priority, and briefly explaining each area.

Case Problem
Baskin Business Forms

Robert Baskin is president of Baskin Business Forms. The company uses its mainframe computer for data processing and management information applications. All data processing systems are performed in a batch manner and provide the data for periodic reports. A database management system enables managers to prepare special reports.

Mr. Baskin recently attended a demonstration of Comshare's executive information system (EIS), which stimulated his interest in obtaining better information from the computer. Upon his return, he asked Michael Raye, the CIO, to assign someone to help him investigate the feasibility of an EIS. Michael assigned Molly Evers, a senior analyst who has enjoyed much success in designing systems for executives. Molly reports to Mr. Baskin's office at the appointed time.

MR. BASKIN: I'd like to tell you about the EIS demo I saw. It was well done. The Comshare people are really pros. The part I liked best was the Dow Jones News/Retrieval Service. It was impressive. All you have to do is key in a code for a company, and you get all its stock information. You can get financial news, too. It's like reading *The Wall Street Journal,* only on the screen.

MOLLY: I'm glad it went well. You know I have been encouraging you to consider some new applications all along. Maybe this can stimulate some ideas. Are you interested in an EIS?

MR. BASKIN: Not the whole thing. I've never been much on using the computer myself. The company couldn't get along without it, but I'd rather my staff do the keyboarding. There's so much of the EIS that I would never use—all the mathematical modeling. I did like the electronic mail feature, though. My brother, who owns a trucking company in Dubuque, says he couldn't live without E-mail.

MOLLY: You know that the Dow Jones service and E-mail are both examples of office automation. Maybe that should be our focus, rather than EIS. There are several more OA applications that could help not only you but the other managers. For example, if you use E-mail to send a memo to someone else, it helps you both. Why don't we take a look at OA, and see what the possibilities are for you? A lot of companies implement their OA applications from the top down, starting with the executive suite. I think a good strategy would be to get you on OA first and then include lower-level managers.

MR. BASKIN: That's fine with me. Can we talk about it now? I'm free until eleven o'clock.

MOLLY: Sure. Let me ask you a few questions so that I can get a better handle on your communications network. Do you have many personal contacts outside the company?

MR. BASKIN: I sure do—many bankers, lawyers, hospital administrators, and so on. I call them on the phone all the time, and they call me. I probably talk on the phone to a dozen different people outside the company every day.

MOLLY: Of course, because you are at the top level of management, you don't have to communicate with other organizational units on the same level. But what about your superiors? Tell me about that communication.

MR. BASKIN: That's a good question. Many people don't realize that a president has superiors—the board of directors. We meet each month. Sometimes it's hard to get everybody together at one time. They are all executives or professionals, and most live in other cities. We've had to cancel some meetings just because everyone wasn't available. Another problem we've had is not everyone being prepared to discuss the issues. We've solved that, though, by faxing an agenda to each member a few days before the meeting. That has worked well.

MOLLY: I'm glad to hear that. Everyone is using fax these days. Do you and the board communicate in any other ways?

MR. BASKIN: We have had telephone conference calls on occasion. That works pretty well. It's especially good when we have to call a special meeting. You know, I think that all the board members use video conferencing in their companies. They were all talking about it at our last meeting. Is that office automation?

MOLLY: It certainly is. That's a thought. Do you and the board ever communicate one on one—say, by telephone or mail?

MR. BASKIN: All the time.

MOLLY: Do you ever prepare any special documents for the board—you know, the high-quality reports that Macintosh is always advertising on TV?

MR. BASKIN: We have all that material prepared in our print shop. Being in printing, we have all the capability we need. And, we should use our own products. If we don't, we have a hard time convincing our customers to use them.

MOLLY: I understand. Now let's talk about how you communicate with your subordinates.

MR. BASKIN: Most of my communications go to the members of the executive committee, which includes the vice-presidents. I rarely communicate below that level.

MOLLY: I'm familiar with the weekly executive committee meetings. I've overheard the executives say that it's helpful when your secretaries send the memo reminding them of the time and the topics. But I'm sure that your secretaries get tired of everyone calling constantly to find out whether you are free. That must take a lot of everyone's time. Now, I would like to ask a few questions about how you use the information for problem solving. First, let's talk about strategic planning.

MR. BASKIN: Molly, I'm sorry. I'm really enjoying this, but its almost eleven and I have to make a conference call to the board. Let's finish this some other time.

Assignment

List the OA applications that Mr. Baskin could use. For each, write one paragraph describing how the application would benefit Mr. Baskin and his company. Identify the type of computer or noncomputer technology that would be required.

Selected Bibliography

Agarwal, Ritu; Tanniru, Mohan R.; and Dacruz, Marcos. "Knowledge-based Support for Combining Qualitative and Quantitative Judgments in Resource Allocation Decisions." *Journal of Management Information Systems* 9 (Summer 1992): 165–184.

Alter, Steven L. "How Effective Managers Use Information Systems." *Harvard Business Review* 54 (November–December 1976): 97–104.

Amaravadi, Chandra S.; Sheng, Olivia R. Liu; George, Joey F.; and Nunamaker, Jay F., Jr. "AEI: A Knowledge-based Approach to Integrated Office Automation." *Journal of Management Information Systems* 9 (Summer 1992): 133–163.

Braden, Barbara; Kanter, Jerome; and Kopsco, David. "Developing an Expert Systems Strategy." *MIS Quarterly* 13 (December 1989): 459–467.

DeSanctis, Gerardine. "Human Resource Information Systems: A Current Assessment." *MIS Quarterly* 10 (March 1986): 15–26.

DeSanctis, Gerardine, and Gallupe, R. Brent. "A Foundation for the Study of Group Decision Support Systems." *Management Science* 33 (May 1987): 593–595.

Durand, Douglas; Floyd, Steven; and Kublanow, Samuel. "How Do 'Real' Managers Use Office Systems?" *Journal of Information Technology Management* 1 (Number 2, 1990): 25–32.

Ellis, C. A.; Gibbs, S. J.; and Rein, G. L. "Groupware: Some Issues and Experiences." *Communications of the ACM* 34 (January 1991): 38–58.

Frappaolo, Carl. "The Promise of Electronic Document Management." *Modern Office Technology* 37 (October 1992): 58ff.

Huber, George P. "Issues in the Design of Group Decision Support Systems." *MIS Quarterly* 8 (September 1984): 195–204.

Jones, Jack William, and McLeod, Raymond, Jr. "The Structure of Executive Information Systems: An Exploratory Analysis." *Decision Sciences* 17 (Spring 1986): 220–249.

Keen, Peter G. W., and Scott Morton, Michael S. *Decision Support Systems: An Organizational Perspective.* (Reading, MA: Addison-Wesley, 1978.)

Klepper, Robert, and Hartog, Curt. "Trends in the Use and Management of Application Package Software." *Information Resources Management Journal* 5 (Fall 1992): 33–37.

McLeod, Raymond, Jr., and Jones, Jack W. "A Framework for Office Automation." *MIS Quarterly* 11 (March 1987): 86–104.

Meyer, Marc H., and Curley, Kathleen Foley. "Putting Expert Systems Technology to Work." *Sloan Management Review* 32 (Winter 1991): 21–31.

Olson, Margrethe H., and Lucas, Henry C., Jr. "The Impact of Office Automation on the Organization: Some Implications for Research and Practice." *Communications of the ACM* 25 (November 1982): 838–847.

Sinha, Alok. "Client-server Computing." *Communications of the ACM* 35 (July 1992): 77–97.

Sprague, Ralph H., Jr. "A Framework for the Development of Decision Support Systems." *MIS Quarterly* 4 (December 1980): 1–26.

Stylianou, Anthony C.; Madey, Gregory R.; and Smith, Robert D. "Selection Criteria for Expert System Shells: A Socio-technical Framework." *Communications of the ACM* 35 (October 1992): 30–48.

Sviokla, John J. "An Examination of the Impact of Expert Systems on the Firm: The Case of XCON." *MIS Quarterly* 14 (June 1990): 127–140.

Viskovich, Fred. "Is Your Accounting System Right for You?" *EDGE* 2 (July–August 1989): 14ff.

Vogel, Douglas R.; Nunamaker, Jay F., Jr.; Martz, William Benjamin, Jr.; Grohowski, Ronald; and McGoff, Christopher. "Electronic Meeting System Experience at IBM." *Journal of Management Information Systems* 6 (Winter 1989-90): 25–43.

Weitzel, John R., and Kerschberg, Larry. "Developing Knowledge-based Systems: Reorganizing the System Development Life Cycle." *Communications of the ACM* 32 (April 1989): 482–488.

Wilkes, Maurice V. "The Long-term Future of Operating Systems." *Communications of the ACM* 35 (November 1992): 23ff.

TM

B

Data Flow Diagrams

The most popular tool for documenting systems processes is the data flow diagram. A **data flow diagram,** or **DFD,** is a drawing that shows how a system's environmental elements, processes, and data are interconnected. DFDs are well suited to the development of structured systems because they can document multiple levels of a top-down design. DFDs typically exist in a hierarchy, as shown in Figure B.1.

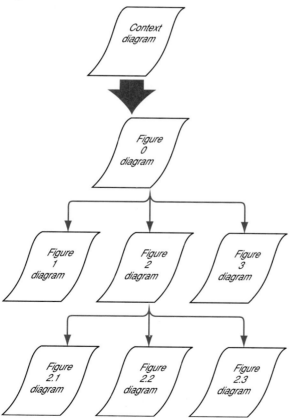

Figure B.1
The hierarchical arrangement of DFDs.

TM

B

The Context Diagram

The DFD on the top level is called the **context diagram** because it describes the system in the context of its environment. The context diagram contains only a single unnumbered process step.

Figure B.2 shows the context diagram for an order entry system. The single circle represents the system. The squares represent the environmental elements. Arrows representing data flows connect the system to its environmental elements. A **data flow** is one or more data elements that travel together.

The Figure 0 Diagram

The data flow diagram on the second-highest level is the **Figure 0 diagram.** It is called a Figure 0 diagram because it shows the major processes of the unnumbered process step of the context diagram. The processes of the Figure 0 diagram are numbered, beginning with 1. To prevent a DFD from becoming too cluttered, limit the number of processes to about seven. Also apply this rule to lower-level DFDs.

Figure B.3 shows the Figure 0 diagram of the order entry system. The system contains three main processes. The first process screens sales orders, determining

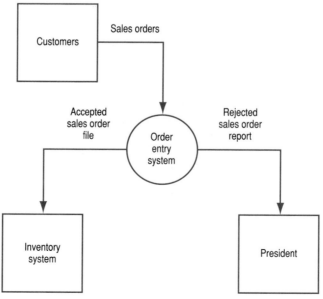

Figure B.2
A context diagram.

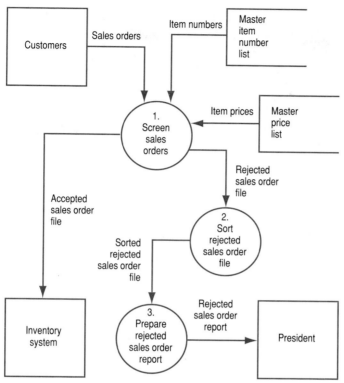

TM

B

Figure B.3
A Figure 0 diagram.

whether to accept or reject them. The second process sorts rejected sales order records. The third process uses the sorted records to prepare a report.

Note that the Figure 0 diagram includes all the environmental elements from the context diagram. Also note that the data flows connecting the environmental elements with the processes are labeled in the same way as they are in the context diagram.

The Figure 0 diagram contains a symbol not included in the context diagram. The symbol is an open-ended rectangle. It represents a **data store,** a place where data is kept. You can think of a data store as a file that is maintained in an up-to-date manner. This system's data stores are master lists of valid item numbers and current prices.

Figure *n* Diagrams

On the next-lower level are DFDs with names such as Figure 1 diagram, Figure 2 diagram, Figure 3 diagram, and so on. The **Figure 1 diagram** documents the major processes of Process 1 of the Figure 0 diagram, the **Figure 2 diagram** documents Process 2, the **Figure 3 diagram** documents Process 3, and so on. We refer to the DFDs on this level as **Figure n diagrams.**

Figure B.4 is an example of a Figure 1 diagram. It documents the first major process of the order entry system, the screening of the sales orders. The screening consists of two processes, numbered 1.1 and 1.2. Process 1.1 verifies the item numbers on the sales orders by comparing them with the numbers on the master item number list. Process 1.2 refers to the master price list to verify that the customers have used the current prices.

Lower-Level DFDs

The top-down decomposition of the system structure can be continued to lower levels. For example, a Figure 1 diagram can be divided into diagrams numbered 1.1, 1.2, 1.3, and so on. In turn, a Figure 1.1 diagram can be divided into diagrams

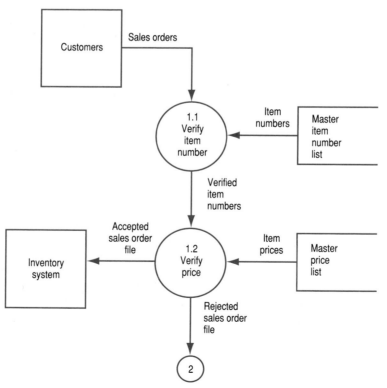

Figure B.4
A Figure 1 diagram.

numbered 1.1.1, 1.1.2, 1.1.3, and so on. However, not all processes require the same level of detail. Some can be adequately documented at a higher level in the hierarchy. The sort process in Figure B.3 is an example. Its documentation in the Figure 0 diagram is sufficient.

Continue documenting with DFDs until you reach a level of detail where the focus is on individual data elements. In the order entry system, that level will likely come when you attempt to document the details of Process 1.1. For that task, use a better-suited documentation tool, such as structured English or program flowcharting.

Leveled and Balanced DFDs

Multiple DFDs used in a hierarchy are referred to as **leveled DFDs.**

Leveled DFDs are developed in a systematic, top-down manner, maintaining consistency from one level to another. This consistency can be seen in the process numbering—Process 1, Process 1.1, Process 1.1.1, and so on. Consistency can also be seen in the data flow names that remain constant from one level to another. For example, the *Sales orders* data flow represents the same group of data elements regardless of the level. Leveled DFDs that have this top-down consistency in processes and data structures are called **balanced DFD.** Your leveled DFDs should always be balanced.

Basic DFD Methodologies

Two basic DFD methodologies exist. They are distinguished by the shape of their process symbols. The **Gane-Sarson methodology** uses an upright rectangle with rounded corners. We use this methodology in Chapter 10 to document an order entry system. Refer to Figure 10.9 and 10.10.

The **Yourdon-Constantine methodology** uses a circle to document a process. We use Yourdon-Constantine for Figures B.2, B.3, and B.4.[1]

The shape of the process symbols is the only significant difference between the two methodologies. Gane-Sarson probably is preferred because the rectangle provides more space for labeling. Your choice of methodology probably will be influenced by which methodology is supported by your CASE tool. Some CASE tools allow you to select between the two methodologies.

---

[1]For more information on the Gane-Sarson approach, see Chris Gane and Trish Sarson, *"Structured Systems Analysis: Tools and Techniques* (Englewood Cliffs, NJ: Prentice-Hall, 1979). For more information on the Yourdon-Constantine approach, see Edward Yourdon, *Modern Structured Analysis* (Englewood Cliffs, NJ: Yourdon Press, 1989).

TM

B

DFD Symbols

DFDs consist of four types of symbols that represent processes, environmental elements, data stores, and data flows. Connector symbols can also be used.

Process Symbols   A **process** is a transformation of data. Data flows into a process, is transformed, and then flows out. The transformation makes the output data different from the input data in some way.

Process symbols usually are labeled with a verb and an object, such as *Process accounting data* and *Edit sales orders*. However, upper-level processes normally regarded as systems can be labeled with system names. Examples are *Data processing system* and *Inventory system*.

Environmental Element Symbols   Systems interface with persons, organizations, locations, and other systems. Each element in the environment *of the system* is called a **terminator.** The names **entity** and **environmental element** are also used, but we use *terminator* because it is often used by CASE tools. Each terminator can be represented by a square or rectangle that is labeled with a noun describing the entity.

Data Flow Symbols   Think of a data flow as "data on the move." The data moves from an environmental element to a process, from one process to another, from a data store to a process, and so on. In most cases, the data flows one way. For two-way data flows, two-headed arrows can be used. Diverging and converging flows are also possible, as illustrated in Figure B.5.

Each data flow in a DFD is labeled with a unique name. When the same flow appears on more than one level, the same name is used for each appearance.

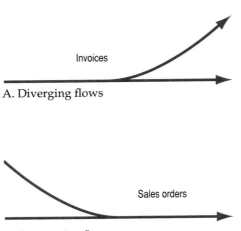

Invoices

A. Diverging flows

Sales orders

B. Converging flows

Figure B.5
Diverging and converging flows.

Data flow arrows can be straight lines, pipe lines, or curved lines. Straight lines are the shortest distance between two points. We use straight lines in Figures 10.9 and 10.10. Pipe lines are straight lines that are always horizontal or vertical to the page and that contain only right-angle turns. We use pipe lines in Figures B.2, B.3, and B.4. Curved lines are also possible, as shown in Figure B.5.

Data Store Symbols   Whereas data flows represent data on the move, data stores represent data maintained in fixed locations. Think of a data store as "data at rest." Examples of data stores are a master inventory file that is kept current and a history file that is held in archival storage.

Data stores are most often depicted with open-ended rectangles such as the ones in Figures B.3 and B.4. However, parallel lines and ovals can also be used, as illustrated in Figure B.6.

Some DFD experts maintain that a data flow name should describe the data, rather than a form or a file. For example, a better name for the Sales Order form would be *Sales order data*. Using names that describe the data makes the DFD more generic, but it also makes data flow naming more difficult. In addition, users are more familiar with form names and file names and will more easily understand the documentation when those names are used. For these reasons, we recommend using form names and file names when they accurately describe the data.

Timecard
history file

A. Open-ended rectangle

Personnel
file

B. Parallel lines

Inventory
master
file

C. Oval

Figure B.6
Data store symbols.

Connector Symbols   You can help the viewer follow the flow of data from one DFD to another by using connector symbols. A **connector symbol** is a circle that contains the number of the process to which the data flow goes or from which the data flow comes. In Figure B.4, the circle labeled 2 connects the Rejected sales order file data flow to Process 2. To follow the flow, you would next turn to the Figure 2 diagram.

Tips for Using DFDs

As you can see from the symbols available for processes, terminators, and data stores, you have much flexibility in how you draw DFDs. However, once you decide on the symbols to use—or more likely, your firm decides for you—you should consider the following tips:

- Begin with a Figure 0 diagram. At that level, you identify the main processes and their linkages to each other and to the environment. When you are sure the Figure 0 diagram represents the basic system structure, complete the documentation by preparing the context diagram and lower-level DFDs. The context diagram is much easier to draw when you have a Figure 0 diagram to serve as a guide.

- Include all the terminators in the context diagram, the Figure 0 diagram, and the lower-level diagrams when they are involved with the data flow. Do not introduce a new terminator on a lower-level DFD.

- To make context diagrams as simple as possible, do not include data stores in them. Also, you might choose not to include data stores in Figure 0 diagrams that contain many processes and terminators.

- When documenting a process with a lower-level DFD, exactly duplicate the data flows previously defined on an upper-level DFD. See Figures B.3 and B.4 for an example.

- When updating the records of a master file with transaction data, show a flow of master file records to be updated leading to the process. Also, show a flow of updated master file records leading to the data store. Figure B.7 is an example.

Putting Data Flow Diagrams in Perspective

Although they are user friendly and do a good job of capturing top-down designs, DFDs were slow to catch on.[2] It took a while for information specialists to get over their infatuation with flowcharting and be receptive to this new tool. Now, DFDs appear to be the main process documentation tool of the future. The DFDs' position in the mainstream of modern systems development is reflected in CASE tools.

---

[2]Jane M. Carey and Raymond McLeod, Jr., "Use of System Development Methodology and Tools," *Journal of Systems Management* 39 (March 1988): 30–35.

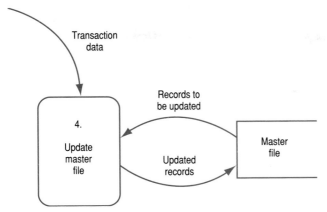

Figure B.7
Updating a master file.

Almost all CASE tools support DFDs. Information specialists, then, should be competent in using DFDs.

Whether users will use DFDs in end-user computing is still unknown. Users typically don't see the need to document their designs as thoroughly as do information specialists. However, if any tool has a chance of being adopted by users, it would appear to be the DFD. The few symbols required and the way that leveled DFDs allow the design to gradually unfold make DFDs especially appealing.

Problems

(Note: Additional DFD problems can be found at the end of Chapter 10.)

1. Draw a Figure 0 diagram of the following procedure. Regard the employees as part of the system's environment. The transmission of data from one location to another without transformation is not considered to be a process.

   A. Employees provide the departmental secretary with their weekly timecards. The departmental secretary prepares a weekly time sheet that lists each employee, the total hours worked, and the hourly rate. The timecards are filed in a Time card history file.

   B. The departmental supervisor signs the time sheet if everything seems to be in order. If any figures are unacceptable, the supervisor returns the sheet to the secretary. The acceptable sheets from all departments are sent in the company mail to the accounting department.

   C. The accounting clerk uses the acceptable time sheets to calculate payroll earnings and writes the earnings on the time sheets.

TM

B

D. Another clerk uses the time sheets to prepare the payroll checks, which are forwarded to the employees. After this operation, the time sheets are filed in an offline Time sheet history file.

2. Draw a context diagram of the system described in Problem 1.

3. Draw a Figure 0 diagram of the following system, used by a bank to open checking accounts. Regard the customer and the new account department as elements in the system's environment. Also, regard each computer program as a separate process.

A. The customer provides the bank with a completed new account application form.

B. A data entry operator enters data from the form into the computer, using a keyboard terminal. Label the output of this process *Entered new account data.*

C. The computer edits the new account data and prepares an error listing of all applications that contain errors. The error listing is sent to the new accounts department. New account data that does not contain errors is written onto a magnetic tape file named New account file.

D. The New account file is input to another computer program that adds the new account data to the Customer master file on magnetic disk and prints the transactions on a transaction listing. The listing is then sent to the new accounts department.

4. Draw a context diagram of the system described in Problem 3.

5. Draw a Figure 0 diagram of the following system, which processes supplier receipts. Regard the suppliers and the purchasing department as part of the system's environment.

A. Supplier shipments include packing lists. A data entry operator keys the packing list data into the computer and then files the packing lists in an offline Packing list history file.

B. The computer program obtains the corresponding records from the Outstanding purchase order file, which is on magnetic disk. The same program prints a received purchases report, which is sent to the purchasing department through the company mail, and writes a Supplier payables file on magnetic disk.

C. The Supplier payables file is input to a program that obtains supplier data from the disk-based Supplier master file and prepares checks that are mailed to the suppliers. The program also creates a Supplier check history file on magnetic tape.

6. Draw a context diagram of the system described in Problem 5.

Systems Methodologies

Development of an information system is an enormous task—so large that it is impossible to do everything at once. The work must be subdivided into separate activities, the activities assigned to persons who have the required skills, and the work spread over time. Even a simple system requires weeks to develop, and most systems require months. Some take years.

This part of the text describes the evolution of an information system. In it, we assume that the system is computer-based. However, the same general process can be followed for systems of all kinds. Also, users who engage in end-user computing can perform all the activities.

This part of the text introduces the basic methodologies used in systems analysis and design. A methodology is simply a recommended way of doing something. Chapter 7 explains the systems approach, a problem-solving process based on the concept of systems. Users can use the systems approach to solve problems facing their units. Information specialists can use it to solve problems of systems development.

Chapter 8 applies the systems approach in the form of a system life cycle. The system life cycle is the traditional systems methodology. In recent years, much effort has been expended to adapt the system life cycle to the needs of modern systems users. This effort paid off by producing two additional methodologies— prototyping and RAD. RAD stands for Rapid Application Development. Chapter 8 explains the three methodologies and provides guidelines that specify when each should be used.

IV—THE COPS LIFE CYCLE

(Use this scenario in solving the HB problem at the end of Chapter 8.)

As Harcourt Brace acquired other publishing companies over the years, it also acquired the software systems. Rather than integrate the new systems into its existing system, Harcourt Brace kept all the systems—old and new— intact. For example, after acquiring Academic Press in 1971, Harcourt Brace continued to use the Academic Press software to process Academic Press transactions. After acquiring CBS in 1987, Harcourt Brace added the CBS Educational Publishing Ordering System (CEPOS), giving itself a total of three distribution systems.

The Birth of the COPS System

It soon became obvious that CEPOS had its limitations. For one thing, it could handle only three distribution centers, and Harcourt Brace had eight. In addition, CBS's program documentation left much to be desired, making it difficult for Harcourt Brace's MIS staff to maintain the software. Because of these shortcomings, top management decided in 1988 to develop a system that could handle *all* Harcourt Brace distribution. The project was named COPS, for Customer Order Processing System. COPS was to handle order entry, inventory, and billing.

The MIS staff at Harcourt Brace had very little experience in developing large systems. Previously, they had concentrated on systems maintenance. The lack of experience and the magnitude of the COPS project prompted the firm's top management to bring in outside help. Andersen Consulting was invited to help plan what the new system would do and how the system would be implemented. The planning lasted about one year, and then Andersen turned the project over to the MIS staff. During a period spanning two years, the MIS staff worked with users to analyze their needs, design the system, develop the needed software, and create the database.

In working with the users, the MIS staff encountered the typical communications difficulties. Generally, the users knew little about computers and the MIS specialists knew little about the applications. To overcome this communications problem, Harcourt Brace management decided to add an applications expert to the MIS staff. Ira Lerner, previously in customer services, was transferred to MIS and made manager of the COPS order entry and billing group. The addition of Ira to the MIS staff proved to be a wise decision, even though the move ran counter to the overall trend in end-user computing. In many firms, information specialists are being transferred *to* user areas to help users implement their own systems.

COPS Cutover

The cutover to COPS originally was scheduled for April 1991, but as that date approached, it became obvious that the new system would not be ready. The cutover was postponed until November, a slack time of the year. Management recognized that the implementation of the new system would place a severe strain on the company and rescheduled the cutover to minimize the disruption.

The COPS Development Team

Bill Presby, now the associate director of sales and marketing systems, manages COPS development, reporting to David Willis, the director of application development support. Refer to Figure HB1.1 to position Bill and his staff in the MIS organizational structure. Bill's COPS development staff is divided into four groups, each with a manager. Three of the groups specialize in COPS subsystems—file maintenance and order entry, order processing and billing, and inventory and depository. The fourth addresses the problems that arise in converting old systems to new.

Personal Profile—Bill Presby, Associate Director

William Presby, shown leading a meeting, is a native of New Bedford, Massachusetts. Bill majored in business administration at Northeastern University and joined Harcourt Brace in 1983. He enjoys bowling and takes advantage of the Orlando climate to pursue his other favorite hobbies of golf, tennis, and fishing. Bill credits his personal success to a God-given desire to serve others. He believes the same attitude is the key to the success of a business. Put simply, a firm must provide a better product or service at a better price than its competitors, and it must maintain a fair return to investors.

CHAPTER 7

The Systems Approach to Problem Solving

Learning Objectives

After studying this chapter, you should:

- know that business problems can be good or bad
- understand that problem solvers must distinguish the symptoms of a problem from its cause, and know how to do this
- understand the difference between problem solving and decision making
- know who solves business problems and know the basic steps that they follow
- understand the systems approach to problem solving

Introduction

Since its beginning, the computer has been regarded as a problem-solving tool. That is one reason the computer has been so popular in business. Businesses have numerous problems to solve, and many of these problems lend themselves to a computer solution.

We ordinarily think of problems as being bad, but problems can also represent opportunities. Business problem solving involves responding to situations that can either benefit or harm the firm.

In responding to these situations, the problem solver must distinguish between the problem's symptoms and its cause. We call this *following the symptom chain.*

After identifying a problem or potential problem, the problems solver begins to make decisions, choosing from among possible avenues that can be pursued. As stated in the previous chapter, the problem solver can apply the computer to decision making. This use of the computer is called a decision support system, or DSS. The DSS has been popular among managers and among nonmanagers for solving problems affecting their units.

This potential contribution of the computer to decision making should also be kept in mind by problem solvers such as information specialists as they develop computer-based systems. The computer can help the information specialists make some of the crucial development decisions.

One of the most widely understood frameworks for problem solving comes from Herbert A. Simon, a management scientist. A framework that is very similar to Simon's is one that is tailored to solving systems problems. The framework is called the systems approach. Managers, nonmanagers, and information specialists can all use the systems approach.

Bad and Good Problems

Although we tend to think of a problem as being bad, it can also be good.

When we hear the term *problem solver,* we tend to think of managers. Nonmanagers also perform this function, as do information specialists. All these business problem solvers have the same motivation; they want to protect their organizations from harm.

When Coca-Cola announced the new formula Coke, many customers complained because they preferred the old formula. This was not the response

Figure 7.1
The decision to market Classic Coke was in response to a bad situation.

Figure 7.2
The decision to establish Federal Express was in response to a good situation.

Coca-Cola management had envisioned. Rather than strengthen Coke's competitive position in relation to Pepsi, the new product announcement had the opposite effect. Coca-Cola management quickly responded by retaining the original formula and calling it Coke Classic. In this case, management acted to minimize the potentially harmful effects of a marketing strategy that had gone wrong.

Business problem solvers do not spend all their time attending to bad situations. They realize they can strengthen their firms as much, or more, by doing things in a new or better way. They take advantage of opportunities for improving the firm.

Federal Express is an example of a firm that seized an opportunity. Its founders saw a need for faster package delivery than was offered by the post office. They formed an overnight delivery service that now covers the world and has been copied by many other organizations—including the post office.

The preceding examples make it clear that problem solvers react to opportunities as well as to threats. For that reason, we define a **problem** as a condition or event that is harmful or potentially harmful to a firm, or beneficial or potentially beneficial to a firm.

Problems Versus Symptoms

To really solve a problem, you must identify the cause. Only by identifying the cause can you ensure that the problem, if bad, does not recur, or if good, does recur. The cause is not always obvious because it is often obscured by one or more symptoms. A **symptom** is a condition that is produced by the problem.

Medical Doctors Follow Symptom Chains

When you have a medical problem, you go to a doctor. You have an ache or a pain that you want relieved. Actually, the ache or pain is not the problem. It is a symptom. The doctor asks questions and conducts tests aimed at clarifying the symptoms and helping pinpoint the cause. In asking questions and giving tests, the doctor follows a symptom chain. A **symptom chain** is a series of symptoms, one causing the next, that leads to the cause of the problem.

Assume you have all the symptoms of a cold—runny nose, watery eyes, and a sore throat. But, you know it isn't a cold because it's mid-July. Besides, the same thing happened last July and the July before that. You go to an eye, ear, nose, and throat doctor. Her initial questions reveal the fact that you always visit your grandmother in July. More questions reveal that your grandmother has a fluffy Persian cat named Homer. The doctor suspects that you may be allergic to Homer and gives you some allergy tests. The tests confirm the doctor's suspicions. The runny nose, watery eyes, and sore throat are the symptoms, and Homer is the problem.

Now, the doctor helps you solve the problem. One solution is to stop visiting your grandmother. Another is to take allergy medicine. You decide to go the medicine route so that you can continue your trips to grandmother's house.

Business Problem Solvers Also Follow Symptom Chains

Business problem solvers approach problems in the same way as doctors. In fact, the business problem solver has the same responsibility as the doctor—to keep the patient in good health. In this case, the patient is the firm.

Take, for example, a manager who comes to work in the morning, turns on the computer, and brings up a display of an income statement. The manager is shocked to see that profits are down. The manager studies the display and sees that revenues are normal but expenses are too high. The manager brings up a detailed breakdown of expenses and sees that the travel expenses are way out of line. Another display of travel expenses reveals that airline expenses have skyrocketed. The manager learns from several sales reps that they had to increase their air travel to cover their territories. When asked why the high air travel expenses had not been incurred in the past, the reps answer that their territories have recently been realigned. The decision is made to realign the territories again, this time considering the impact on travel.

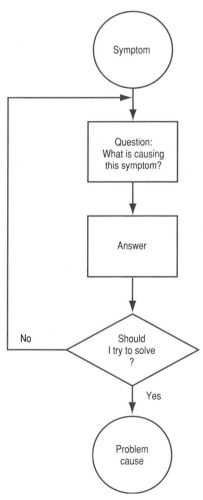

Figure 7.3
The logic of the symptom chain.

In this example, low profits were not the problem. Neither were high total expenses, excessive travel expenses, or high airline expenses. All these things were symptoms of the real problem, which was poor assignment of territories.

Recognizing the End of the Symptom Chain

How do you know when you have reached the end of the symptom chain and identified the cause? The logic of the process is illustrated in Figure 7.3. For each symptom, ask, "What is causing this?" For the answer, ask yourself, "Is this a problem that I should try to solve?" If the answer is no, you have only identified another symptom, and you must repeat the question. When the answer is yes, you have identified the problem cause. The **problem cause** is the condition that is the reason for an exceptionally good or bad influence on the situation under study.

The symptom chain is not a strictly automatic line of questioning. You must practice it over and over with real problems to develop the judgment that is needed.

Problem Solving Versus Decision Making

When you read or hear about the computer being used for problem solving, you also often encounter the term *decision making*. It is easy to get the idea that decision making and problem solving mean the same thing. They do not. **Problem solving** involves all the activities that lead to the solution of a problem. Every problem has a solution. It may not be a perfect solution, but it is the best possible solution under the circumstances.

Many Decisions are Made in Solving One Problem

In solving a single problem, the problem solver must make many decisions. Each **decision** is a selected course of action that leads the problem solver to an effective solution. Listed below are five points at which decisions are made during the problem-solving process:

1. **Separate Symptoms from the Cause** To begin the process, the problem solver must decide whether something is a symptom or a cause. The diamond symbol in Figure 7.3 marks the point where a decision is required for each symptom.

2. **Identify Feasible Solutions** After defining the cause of the problem, the problem solver must identify alternate solutions and decide whether each is feasible. A **feasible solution** is one that fits within the constraints that exist. Limited funds, a tight time schedule, laws, and ethics are just a few of the constraints.

3. **Identify the Best Solution** After identifying the feasible solutions, the problem solver must make decisions about each to identify the best.

4. **Determine the Implementation Strategy** The problem solver makes decisions that determine the best way to implement the solution, including the proper time to put the solution into effect and how the solution will be accomplished.

5. **Determine the Effectiveness of the Solution** Finally, the problem solver must make decisions in evaluating whether the implemented solution is working as it should.

The process of selecting a course of action leading to a problem solution is called **decision making.** The problem solver therefore engages in decision making while solving problems.

Who Solves Problems?

In many cases, the problem solver is a manager. One key to advancing in management is an ability to solve problems. Although no manager has a perfect record in this regard, the road of corporate advancement is paved with effective problem solutions.

Nonmanagers also solve problems. Some nonmanagers are inside the manager's unit, and some are outside.

Problem Solvers Inside the Manager's Unit

The manager's unit contains *specialists* who provide expertise in solving particular problems. Included in this category are information specialists who are assigned to the manager's unit to facilitate end-user computing. Other specialists are marketing researchers in the marketing unit, financial analysts in the financial unit, and industrial engineers in the manufacturing unit.

Other nonmanagers in the manager's unit who solve problems are the *operational employees,* who do the work of the unit. Examples are sales representatives, production workers, warehouse employees, and delivery persons. Finally, the manager's unit contains *administrative employees*—such as secretaries, clerks, and typists—who perform office tasks for the managers, specialists, and operational employees.

Problem Solvers Outside the Manager's Unit

The manager can call on persons outside the unit to help solve problems. Special units in the firm have this support as their main mission. The oldest such unit consists of *accountants* in the accounting department. The unit that has grown the most in recent years consists of *information specialists* in the information services department. Other examples of organizational problem-solving specialists include *economists* and *statisticians*.

Everyone Is a Problem Solver

Chapter 1 made the point that everyone is a systems analyst. Everyone is also a problem solver. As a result, everyone can benefit by learning a process for solving problems.

Using CASE to Define Organizational Dependencies

Upper-CASE tools that support the early stages of the SDLC usually facilitate dependency diagramming. A **dependency diagram** shows which activities are dependent on other activities. Dependencies can apply to functions, processes, or procedures. Therefore, dependency diagrams can illustrate business functions during the planning phase, business processes during the analysis phase, or business procedures during the implementation phase.

Dependencies can be:

- **Time related.** Process B cannot be performed until Process A has been completed.

- **Resource related.** Activity A produces a tangible resource that Activity B uses. For example, Fill-order must be completed before Deliver-order can be performed.

- **Data related.** Activity A creates or modifies some data that Activity B needs. For example, Fill-order cannot occur until Create-account has occurred.

A fourth type of dependency, constraint related, could exist but is not recommended. In a **constraint-related** dependency, an activity is dependent on a constraint set up in another activity. For example, a flag created in Activity A is passed to Activity B. Constraint-related dependencies result in a tight coupling between modules, whereas modules should be as loosely coupled as possible.

The Problem-Solving Process

Many researchers have endeavored to identify the best approach to problem solving—an approach that would be the key to a successful career and a successful organization. The approach that has received the most attention was developed by Herbert A. Simon, a management scientist noted for his contributions to management theory and artificial intelligence.

Simon's Types of Decisions

In a 1977 book, Simon wrote that the decisions made in solving a problem exist on a continuum. At one end of the continuum are **programmed decisions,** which are "repetitive and routine, to the extent that a definite procedure has been worked out for handling them so that they don't have to be treated *de novo* (as new) each time they occur." At the other end of the continuum are **nonprogrammed decisions,** which are "novel, unstructured, and unusually consequential."[1]

The terms *programmed* and *nonprogrammed* referred to a decision's structure or lack of structure. A decision is structured if all its parts are familiar to the problem solver. A decision is unstructured if none of its parts is familiar.

Simon recognized that most decisions lie somewhere in between programmed and nonprogrammed. He concluded that the world is neither black nor white, but "mostly gray."[2]

MIT professors G. Anthony Gorry and Michael S. Scott Morton based their decision support system concept, described in the previous chapter, on Simon's description of decision structure.

Simon's Phases of Decision Making

According to Simon, problem solving involves four phases, with the problem solver making decisions in each phase.

Intelligence Activity   The problem solver searches her or his environment for conditions requiring solutions. In some cases, the problem solver simply observes what is going on in the unit and in the unit's environment. In other cases, the problem solver uses computer-generated outputs, such as the one illustrated in Figure 7.4. This display was produced from a corporate intelligence database called DIALOG. Regardless of how it is performed, the process of learning about a problem or potential problem, is called intelligence activity.

Design Activity   After defining the problem, the problem solver can solve it by inventing, developing, and analyzing possible courses of action. These are the solution alternatives. The problem solver evaluates each alternative to determine its potential for solving the problem. This process is called design activity.

For some problems, the computer can contribute to the design process. Perhaps a mathematical model can simulate the possible outcomes of an alternative, should it be implemented.

[1]Herbert A. Simon, *The New Science of Management Decision*, revised ed. (Englewood Cliffs, NJ: Prentice-Hall, 1977): 46.
[2]Ibid.

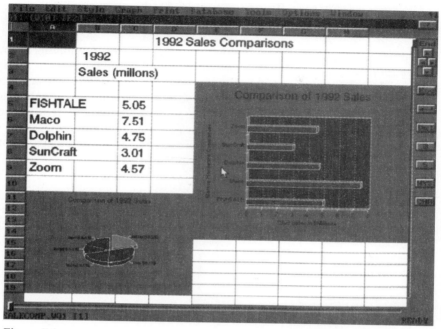

Figure 7.4
Intelligence databases help identify problems to solve.

Choice Activity   The process of choosing the seemingly best alternative is the choice activity. If the problem solver has identified and evaluated the alternatives well, the choice can be straightforward. This is where the problem solver's experience and intuition are put to best use. Even when the computer is applied in the most thorough way, the problem solver ultimately chooses the solution.

Review Activity   Once the solution is implemented and given a chance to work, it is subjected to review activity that assesses its effectiveness. The review accomplishes two objectives. First, it reveals whether the problem has been solved. If performance does not meet expectations, the problem-solving process must be repeated, in part or in whole. Second, the review provides the problem solver with an opportunity to learn from experience. A successful solution is remembered, to be used again in similar circumstances.

Although Simon is not the only person to describe the problem-solving process, all the others have followed the same general approach of defining the problem, considering alternate solutions, selecting the best solution, implementing the solution, and following up. This is an intuitively logical approach. *Any* problem solver can use such an approach in solving *any* type of problem.

The Systems Approach

Information specialists have tailored the general problem-solving approach to the solution of systems problems, and called it the systems approach. The **systems approach** is the recommended procedure for the analysis and design of systems. It consists of defining the problem to be solved by the system, identifying and evaluating the possible system alternatives, selecting and implementing the best alternative, and evaluating the system's performance.

Our description of the systems approach consists of ten steps in three phases. These phases and steps are illustrated in Figure 7.5.

Phase I—Preparation Effort

Before you can solve a systems problem, you should take a **systems view,** also called a **systems orientation.** Your education in systems theory will give you the

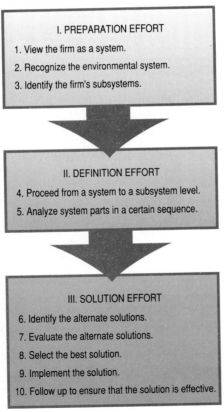

I. PREPARATION EFFORT

1. View the firm as a system.
2. Recognize the environmental system.
3. Identify the firm's subsystems.

II. DEFINITION EFFORT

4. Proceed from a system to a subsystem level.
5. Analyze system parts in a certain sequence.

III. SOLUTION EFFORT

6. Identify the alternate solutions.
7. Evaluate the alternate solutions.
8. Select the best solution.
9. Implement the solution.
10. Follow up to ensure that the solution is effective.

Figure 7.5
The steps of the systems approach.

foundation for a systems orientation, and your career experience will build on that foundation.

When you take a systems view, you are engaging in the first phase of problem solving—the **preparation effort.** This phase consists of three steps.

Step 1—View the Firm as a System   In developing a system to help the firm solve a problem, it is best to view the firm as a system. Such a view is illustrated with the graphic model in Figure 7.6. This is the general systems model that we developed in Chapter 4. The system structure recognizes the importance of a conceptual information system that reflects the status of the physical system. The conceptual system consists of the three rectangles at the top, along with their associated data, information, and decision flows.

When you participate in projects to develop computer-based systems, you should do so with the view of the firm as a system. Your system should help management manage the physical system of the firm.

Step 2—Recognize the Environmental System   By viewing the firm as a system, you acknowledge that the firm exists in an environment. Chapter 5 identified flows of physical and conceptual resources between the firm and eight environmental elements. Those flows are illustrated in Figure 7.7.

The resource flows are critical to the firm. The systems that you develop should facilitate the flows.

Step 3—Identify the Firm's Subsystems   The firm consists of subsidiary systems, or subsystems. The most obvious subsystems are the *functional areas* such as marketing, finance, and human resources, which exist in almost all firms. Less obvious are other subdivisions such as *geographic territories* (for example, domestic operations, Latin America, and the Pacific Rim), *product lines* (Buick, Pontiac, and Oldsmobile), and *customers* (government, health care, and airlines).

Each subsystem is a complete system, and all the systems are arranged in a hierarchy. Because each subsystem is a system, as shown in Figure 7.8, the systems approach can be followed on any organizational level. You can develop a system for a lower-level organizational unit, using the same approach that you use on the top level.

When you begin your career, you can tailor the graphic models to your firm. At that time, evaluate your firm against the general systems model, understand your firm's relationship to its environment, and understand how your firm is organized into subsystems.

Phase II—Definition Effort

After adopting the systems view, you are ready to solve a systems problem. Well, not quite—first you must define the problem. This task of researching a problem is what Simon meant by intelligence activity. We use the term **definition effort** for the phase of *systems analysis* in which you determine (1) what the problem is, (2) where the problem is located, and (3) what caused the problem.

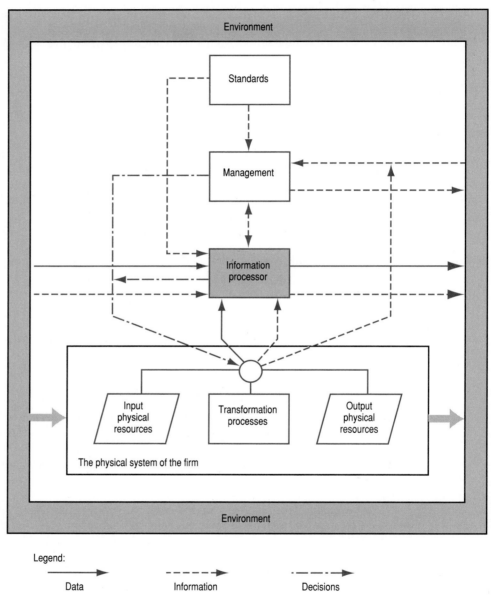

Figure 7.6
The model of the firm as a system.

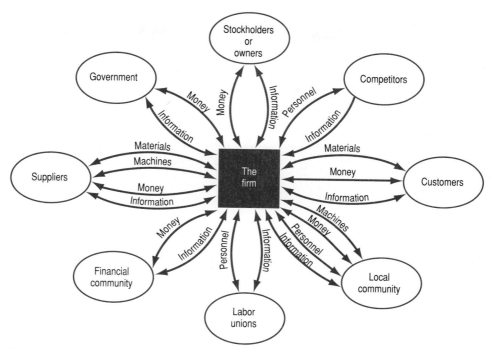

Figure 7.7
The environment of the firm.

Step 4—Proceed from a System Level to a Subsystem Level   In defining the problem, it is always a good idea to start at the top of the organization and work your way down. This method makes sense because all the firm's systems exist to support the entire organization. In addition, the top-down approach gives you the advantage of understanding the big picture before you try to fill in the details.

The top-down approach does *not* mean that you always begin with a study of the entire firm. Instead, it means that you begin with a study of the unit that you are assigned to help. You study that unit as a system in its environment. For example, if you are solving a problem for the sales department, you learn about the elements of the sales system and about its environment.

Once you study the sales department, you focus on the next lower system level, such as a regional sales office. Then, you proceed to a still lower level, such as a branch sales office. Your objective is to identify the organizational level at which the cause of the problem is located. The symptoms might appear on an upper level, but the cause might be on a lower level. For example, a symptom might be low profits, but the cause might be poorly trained sales reps in Los Angeles.

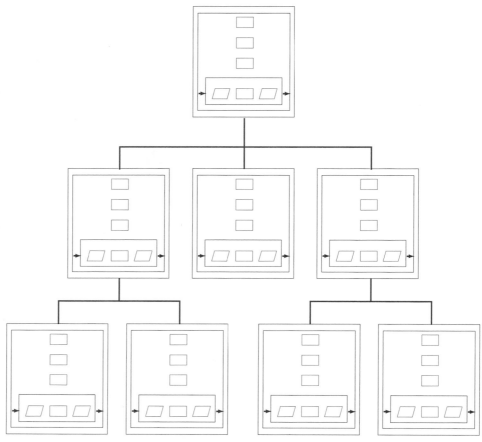

Figure 7.8
The firm's organizational units are subsystems.

Step 5—Analyze System Parts in a Certain Sequence   During your top-down study, you examine the elements *on each level* in the sequence shown in Figure 7.9.

1. **Evaluate the Standards** First, determine whether the system has good standards. If it does not, you cannot engage in problem solving until that deficiency is overcome. Once good standards exist, you can proceed with the element-by-element analysis.

2. **Evaluate the Outputs** Once you are certain the standards are good, determine whether the system's outputs (products or services) meet the standards. If the outputs meet the standards, the problem is not located at this system level. The system is doing what it is supposed to do. If the outputs do

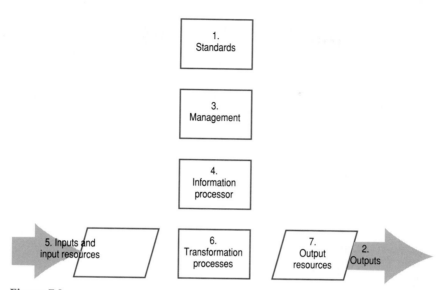

Figure 7.9
The sequence for determining the location of a problem in a system.

not meet the standards, find out why. The answer lies somewhere in the remaining elements of the system.

3. **Evaluate Management** In most cases, management is the reason the system is not producing the desired outputs. Management sets the policies and makes the major decisions. Management may have made the wrong decisions or no decisions at all. However, management is not always to blame. Extenuating circumstances may be involved. For example, the unit might not have enough managers to take care of all their responsibilities. This situation is typical in small, growing firms.

4. **Evaluate the Information Processor** In many cases, lack of information prevents management from performing its functions. If such is the case, then management must be provided with an adequate information processor. It is at this point in the problem-solving sequence that the work of the systems analyst usually ends. The systems analyst is concerned with the conceptual system, and the information processor is the final element in that system. If the problem is *not* a lack of information, then it is somewhere in the physical system. A manager or nonmanager problem solver continues with the analysis of the final three elements—the physical system. A physical systems specialist, such as an industrial engineer, might help.

5. **Evaluate the Inputs and the Input Resources** Poor inputs and input resources might be the reason the outputs don't measure up. The *inputs* are the raw materials that are transformed. The *input resources* are the resources that

receive the inputs into the firm. Input resources include receiving clerks, receiving inspectors, forklifts, and raw materials storage areas.

6. **Evaluate the Transformation Processes** The transformation processes can harbor many possible causes of deficient output. Sloppy workmanship, worn-out machines, and poor working conditions are examples.

7. **Evaluate the Output Resources** The *output resources* that make the transformed products available to the customers can also be the cause of the problem. Examples of output resources are finished goods storage areas, stock pullers, packers, and delivery trucks.

As soon as a problem element is encountered, the search moves to the next lower level system. Figure 7.10 illustrates this top-down process.

Assume that you are solving a marketing problem. The marketing division is the system under study. You examine the elements of the marketing system in the proper sequence until you determine that the information processor is the problem element. Marketing managers are not getting the information they need from their marketing research system.

You turn your attention to the marketing research section, the next lower level, and subject that section to the same element-by-element study. You evaluate its standards, outputs, and management. You find that the marketing research section is not providing the needed information because its management is not skilled in computer use. The problem cause is in the management element of the marketing research system.

Phase III—Solution Effort

You are now ready to solve the problem by developing a new or improved system. The **solution effort** involves all the steps necessary to solve a problem that has been identified and defined. This effort is also called *systems design*. It is the activity that Simon described as design and choice.

Step 6—Identify the Alternate Solutions   Every problem has more than one solution. The first challenge is to identify all feasible alternatives. The more possible solutions you identify, the greater your chance of selecting the one that does the best job.

This is perhaps the most difficult step in the problem-solving process. It is also the one least supported by the computer. You must call upon all your mental resources. Experience plays an especially important role. Solutions that have worked in similar circumstances in the past may work again.

Inexperienced designers can identify creative solutions by applying systems analysis knowledge identified in Chapter 1.

One technique is to apply systems theory and identify possible solutions for each part of the new system—the input, processing, secondary storage, and output elements. Figure 7.11 illustrates that each system element can have multiple solution options, in terms of hardware and how it is used.

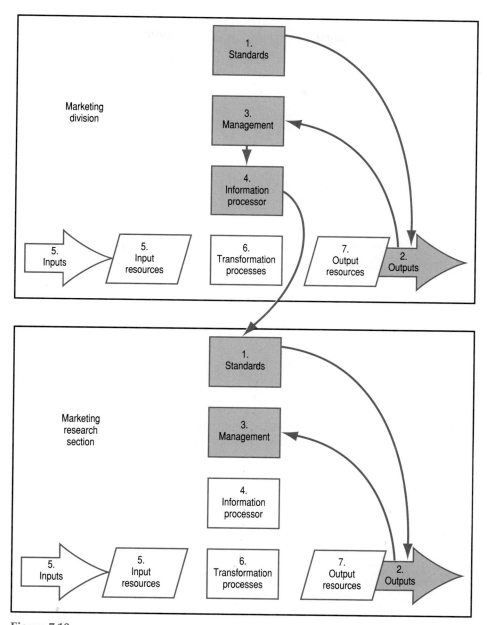

Figure 7.10
The path that is followed in solving the problem in the management element of the marketing research section.

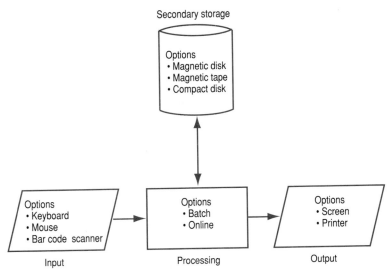

Figure 7.11
Each system element provides solution options.

In a situation such as this, where several options exist for several system elements, the options are *multiplicative.* You calculate the total number of options by multiplying the number of options for each system element. The example in the figure has three input options, two processing options, three storage options, and two output options—a total of 36 options ($3 \times 2 \times 3 \times 2 = 36$). In other words, 36 unique system configurations could be considered in solving the problem.

Some alternatives invariably feature illogical combinations and can be eliminated. For example, online processing would not be performed using magnetic tape storage because the tape does not offer the necessary direct access.

Other alternatives also may not be feasible for one reason or another. For example, some might be too expensive, and some others might take too long. As with the illogical combinations, the infeasible alternatives can be eliminated from consideration.

Ideally, you want to cull down the total number of alternatives to a reasonable number, say, two to five. A host of alternatives makes evaluation difficult.

Step 7—Evaluate the Alternate Solutions   The remaining alternatives are evaluated using certain judgment standards called **evaluation criteria.** For example, each alternative might be evaluated in terms of its cost and user friendliness. Cost and user friendliness are the evaluation criteria. A fair evaluation hinges on the evaluation criteria remaining constant from one alternative to the next.

The evaluation takes three forms:[3]

- analysis—a systematic evaluation of the advantages and disadvantages of each alternative, using quantitative data
- judgment—a subjective evaluation of each alternative by one or more problem solvers
- bargaining—negotiations among several problem solvers

Ideally, the evaluation involves the analysis of quantitative data, such as prices and speeds. In the absence of such data, judgment fills the gap. Bargaining is always a consideration. Bargaining includes company politics, which plays a more important role in systems development than one might expect.

Step 8—Select the Best Solution   The objective of the evaluation is to identify the best solution. Although the selection may consider multiple evaluation criteria, the process boils down to deciding which solution best helps the firm meet its objectives.

Step 9—Implement the Selected Solution   The design that exists on paper is converted to a system that consists of hardware, software, data, and personnel, and is put into use.

Step 10—Follow Up to Ensure that the Solution Is Effective   Some time after cutover to the new system, audits are conducted to ensure that the system is performing as intended. Managers who engage in end-user computing audit their own systems. When information specialists are involved, they conduct the audits. In some situations, internal auditors from the firm's auditing department conduct the audits. The problem is not considered solved until the audits verify that the solution is working.

The systems approach begins in the classroom with preparation effort, includes a definition effort to locate the problem cause, and ends with solution effort that continues until you are assured that you did a good job.

An Example of the Systems Approach

Pretend you have just been hired as a systems analyst by Hardy's Food Market, a Missouri supermarket chain. After your brief orientation period your boss assigns you the task of helping the store managers with an inventory problem. You decide to follow the systems approach.

Step 1—View the Firm as a System

You evaluate your company against the general systems model. You are pleased to learn that Hardy's has short- and long-term objectives for the entire firm and

[3]Based on Henry Mintzberg, "Planning on the Left Side and Managing on the Right." *Harvard Business Review* 54 (July–August 1976): 55–56.

each of its 14 stores. It also has performance standards for all its managers, including the store managers.

Management's task is to ensure that the physical system provides food and other products to the customers when and where they need them. Management uses a computer network to manage the physical system, which consists of the stores, three warehouses, and a fleet of delivery trucks. A mainframe computer is located at headquarters, and minicomputers are located in each warehouse and store.

The input resources include the personnel, facilities, and equipment that are used in providing the stores with their merchandise. The warehouses and the stores' receiving areas are also input resources. The transformation process involves the arrangement and display of the products in the stores to facilitate purchase. The output resources are the checkout registers, checkers, and sackers.

Step 2—Recognize the Environmental System

Hardy's relationship with its suppliers is crucial to its success. Food products, especially perishables, must arrive at the stores in a timely manner.

Hardy's customers are also important. Competition for the consumer's food dollar is keen, and very little distinguishes one supermarket chain from the next. Over the years, Hardy's has been able to operate very profitably, emphasizing quality and service. The resulting financial stability has made it unnecessary to borrow money from financial institutions. All of the store's employees belong to labor unions, so management must consider the influence of the unions when establishing personnel policies. One move that helped union relations was the decision to sell company stock to the employees at a discount. The employees take pride in being owners.

Hardy's stores strive to maintain good relations with their local communities, often participating in charities and fund raisers. For example, Girl Scouts and Boy Scouts are welcome to sell their products near store entrances.

Step 3—Identify the Firm's Subsystems

Hardy's is organized into three main units, each with its own vice-president. The vice-president of finance is responsible for all financial activities, the vice-president of distribution is responsible for the warehouses and the truck fleet, and the vice-president of marketing is responsible for the stores. Managers of the warehouses, the fleet, and the stores report to their vice-presidents.

Steps 4 and 5—Proceed from a System Level to a Subsystem Level, and Analyze System Parts in a Certain Sequence

At the Hardy's headquarters, you review the chain's objectives and standards. Next, you study the recent annual reports, which reveal that last year was the first time in six years that the chain failed to meet its profit objectives. You interview the president and vice-presidents and learn that unusually high costs were incurred due to food spoilage. Often, the stores were overstocked and the perishables spoiled before they could be sold. The problem exists only in the stores. The warehouses have been able to maintain their inventories at the proper levels.

The marketing vice-president complains that the store managers often place emergency orders without knowing that they have received shipments from previous orders. The lack of information results in overstocking. The CIO confirms the story and explains that a delay exists in the entering of the receipt data into the computer. Usually, the delay is only a few hours, but it can be an entire day.

Because the problem exists at the store level, you visit several stores. At each, you verify that performance standards exist and measure the store's sales against the standards. The store managers give the same explanation as did the headquarters executives—that outdated receipt data makes the inventory database worthless. In studying the receiving systems at the stores, you learn that the receiving clerks enter the data into the stores' minicomputers. They must enter the data after 5 P.M. because the minis are used during the day to key in price changes for sale items. Even if an inventory shipment is received at 10 A.M. the inventory records will not be updated until after 5 P.M.

The information that you gather in the stores enables you to state the problem as "The store managers cannot maintain proper inventory levels due to a lack of current information concerning inventory receipts."

Your analysis has enabled you to pinpoint the problem in the information processors used by the stores. A new system must be designed that enables the receipt data to be entered more quickly.

Step 6—Identify the Alternate Solutions

You know that the new system will be computer based because all the chain's systems use computers. You identify the different options that can be considered for each system element.

- **Input Options** The input can be accomplished by keying the data into a terminal or using a bar code scanner. Either piece of equipment could be located in the receiving area and connected to the store's mini. Because the mini's operating system permits multiprogramming, the inventory records could be updated in the daytime while price changes were also being entered.

- **Processing Options** The processing can be either batch or online. If batch processing is performed, the receipt data can be recorded on a secondary storage device and then all inventory records can be updated at one time. If online processing is performed, the inventory records can be updated as the receipt data is entered.

- **Secondary Storage Options** The minicomputer has magnetic disk storage and adequate capacity for added applications. You do not need to make a decision concerning secondary storage.

• **Output Options** Store managers can receive the system's output in displayed, printed, or audio form. The displayed output is fastest, but it does not provide the permanent record that printed output provides. The main advantage of the audio output is that the store managers can use their telephones as terminals.

You now have 12 possible solution combinations: two input options, two processing options, one storage option, and three output options ($2 \times 2 \times 1 \times 3 = 12$). You study the 12 configurations and reduce the number to the four identified in Table 7.1. Two alternatives rely on bar code input, and two rely on terminal input. The bar code systems are distinguished by their output; one uses displays and the other uses hard copy. The two terminal configurations are distinguished the same way. All four alternatives feature online processing using magnetic disk.

Step 7—Evaluate the Alternate Solutions

You confer with store management and headquarters management and decide to base the selection on four factors—cost, speed, flexibility, and ease of operation. Table 7.2 ranks the alternate solutions in terms of the evaluation criteria.

Ranking is only one way to compare the alternatives. You can also use grades such as "Good," "Average," and "Poor," and actual data, such as costs and speeds. You can also assign different weights to the criteria, recognizing that not all criteria have equal importance in the decision.

Step 8—Select the Best Solution

Hardy's management chooses the solution featuring terminal input and displayed output. The main factor influencing the decision is the cost-effectiveness and flexibility of the terminal-oriented configuration.

Table 7.1   The Four Alternate Solutions for Hardy's Food Market

System element	Alternate solutions			
	1	2	3	4
Input	Bar code scanner	Bar code scanner	Terminal	Terminal
Processing	Online	Online	Online	Online
Secondary storage	Magnetic disk	Magnetic disk	Magnetic disk	Magnetic disk
Output	Display	Hardcopy	Display	Audio

Table 7.2   Evaluation of the Alternate Solutions Based on Rank

Evaluation criteria	Alternate solutions			
	1	2	3	4
	Scanner/ display	Scanner/ hardcopy	Terminal/ display	Terminal/ audio
Cost	3	4	1	2
Speed	1	4	2	3
Flexibility	3	4	1	2
Ease of operation	3	4	2	1

Explanation of rankings:
  1 = Excellent
  2 = Good
  3 = Fair
  4 = Poor

Step 9—Implement the Selected Solution

Hardy's buys two terminals for each store. One terminal is installed in the receiving area. It is used to record the receipt data. The other terminal is installed in the store manager's office. It is used to retrieve up-to-date inventory information. Receiving clerks and managers are trained to operate the terminals. The system is implemented.

Step 10—Follow Up to Ensure that the Solution Is Effective

For several weeks after cutover to the new system, you stay in almost daily contact with the store managers and receiving clerks to make certain that they are not having any difficulties. When you are confident of the system performance, you contact the store less often. You report to your boss that everything is going fine. Approximately three months after cutover, Hardy's auditing department assigns an auditor to evaluate the new system's performance. The system passes with flying colors, and the auditor submits a glowing report to top management.

The Significance of the Systems Approach

The Hardy's example shows how a systems developer uses the systems approach as a guideline. In this example, a systems analyst followed the procedure, but a store manager or nonmanager could have taken the same steps. The systems approach provides the basis for all systems analysis and design work, regardless of who does it. Without such a systematic approach, a firm relies on trial and error, increasing the chances of overlooking vital information or a good alternative. The systems approach assures that systems development proceeds in the most efficient and effective way.

Of course, following the systems approach does not guarantee that the systems developer will produce the best system. The systems approach is not a foolproof recipe. Its steps must be taken with much skill, knowledge, and judgment. When these human ingredients exist, however, the systems developer can be confident of doing the best job possible. *The systems approach is the basic systems methodology.* It forms the basis for refinements such as the system life cycle, prototyping, and rapid application development, which we will describe in the next chapter.

Using CASE to Draw a Dependency Diagram

Earlier in the chapter, we recognized that CASE can be used to define organizational dependencies. In drawing dependency diagrams, it is possible to represent such concepts as:

- **Optional.** An activity may or may not be done.
- **Conditional.** An activity is done if a condition exists.
- **Mutually exclusive.** Either one or another activity exists.
- **Cardinal.** A one-to-many relationship exists.
- **Sequential.** Activities are performed in a certain order.
- **Concurrent.** Activities can be performed at the same time.

Figure 7.12 illustrates the structure of a process-based dependency diagram, which can be read in the following manner: Retrieve Student Record. If new student, create new Student Record else check Balance Due. If Balance Due is greater than zero, reject Course Request else check Course Availability. If Course Availability is zero, reject Course Request else accept Course Request and subtract one from Course Availability and add Charge to Student Record.

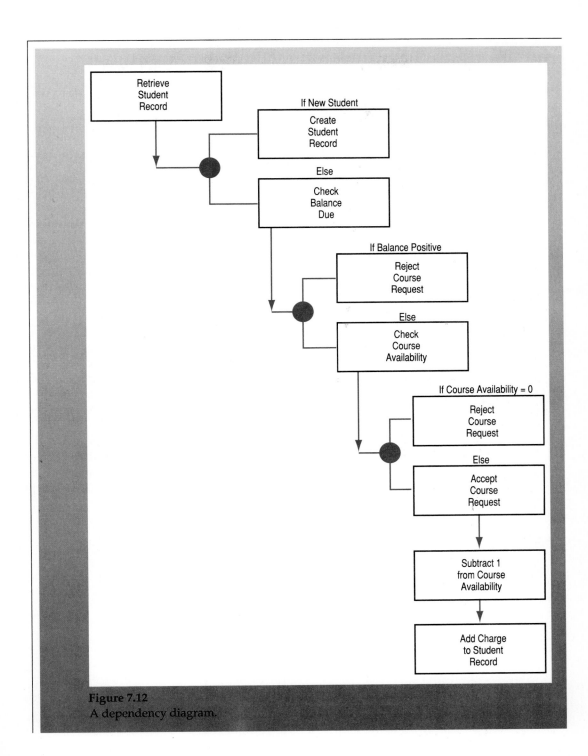

Figure 7.12
A dependency diagram.

Summary

A problem can be something good as well as something bad. Problem solvers usually become aware of a problem by noticing symptoms. They follow a symptom chain to identify the problem cause. It is important to know the cause so that you can ensure that a bad problem does not happen again or that a good problem does happen again. Problems are solved by taking a series of steps, each of which requires one or more decisions.

When we see or hear the term *problem solver,* we normally think of a manager. However, specialists, operational employees, and administrative employees in the manager's unit also solve problems. In addition, problem-solving specialists exist outside the manager's unit. Examples are accountants and information specialists.

Many people have studied the problem-solving process, but none has achieved more fame than Herbert Simon. Simon identified a continuum of decisions, ranging from programmed to nonprogrammed. Programmed decisions are repetitive and well understood. Nonprogrammed decisions are novel and unique. Simon also identified the four phases of decision making used in solving a problem—intelligence, design, choice, and review.

System developers have created an approach that is well suited to solving systems problems. It is called the systems approach. It consists of three phases of effort—preparation, definition, and solution. Preparation effort requires you to view the firm as a system, recognize the environmental system, and identify the firm's subsystems. In applying the definition effort, you proceed from a system level to a subsystem level while analyzing the system parts in a certain sequence. Solution effort demands that you identify the alternate solutions, evaluate them, select the best one, implement it, and follow up.

It is easy to become deceived by the simplicity of the systems approach and fail to appreciate its value. However, it forms the basis for everything that a systems developer does. Managers, nonmanagers, and information specialists can follow the systems approach when developing systems to solve problems of all types. The systems approach is the basic systems methodology.

Key Terms

Problem

Symptom

Problem cause

Problem solving

Decision

Feasible solution

Decision making

Dependency diagram

Evaluation criteria

Key Concepts

A problem as something good or something bad

The symptom chain

Programmed and nonprogrammed decisions

Simon's phases of decision making—intelligence, design, choice, and review activity

The systems approach

Systems view, systems orientation

Applicability of the systems approach to any system level

Questions

1. Why would a manager want to know when something good happens?

2. How does a problem solver know when the end of the symptom chain has been reached?

3. What five types of decisions are made during problem solving?

4. Who are the nonmanagers in the manager's unit who solve problems?

5. Where in the firm would you look to find the information specialists who help the manager solve problems?

6. What are programmed and nonprogrammed decisions? Are there any other kinds? Explain.

7. Simon's concept of programmed and nonprogrammed decisions provided a basis for which of the five CBIS subsystems?

8. When someone has a systems orientation, exactly what does that mean?

9. When do you engage in preparation effort?

10. How is the firm connected to its environment?

11. What is the most obvious way to divide a firm into its major subsystems? What are three other ways?

12. Which phase of systems approach effort is concerned with systems analysis? Systems design?

13. Of the ten steps of the systems approach, which two steps are taken simultaneously?

14. Which system elements receive the attention of the systems analyst when solving problems?

15. Which system element is most often the problem cause? Explain your reasoning.

16. What is the difference between inputs and input resources? Between outputs and output resources?

17. In which of the ten steps of the systems approach does experience play an especially important role?

18. As a rule, how many alternatives should be evaluated? Why this number range?

19. What are three ways to evaluate alternatives?

20. Ideally, what fundamental feature characterizes the best solution?

21. When would a manager have the responsibility of auditing the new system?

Topics for Discussion

1. What are some programmed decisions you make in your daily life? Nonprogrammed decisions? Can you think of any gray areas?

2. How do Simon's four decision phases compare to the ten steps of the systems approach? List both sequences, and draw lines to connect similar steps.

3. Where can politics enter into the systems approach? How can it influence the outcome?

Problems

1. Assume that you are going to buy a new car. Make a list of five evaluation criteria that you would use. Identify three cars that you would like to evaluate. Make a table similar to Table 7.2, and evaluate the three cars. Identify which car you should buy. Briefly explain why.

2. Go to the library and do research on Herbert A. Simon. Write a paper titled "Herbert Simon's Contributions to Problem Solving." Your instructor will specify paper length and format.

Case Problem
Palomar Plastics

Palomar Plastics is located in Logan, Utah. It manufactures the clear plastic covers that protect computer keyboards from dust and spills. Palomar is a classic example of a company that recognized that problems can be good or bad. The booming computer industry created a need for protecting the keyboards, and Palomar seized the opportunity to satisfy the need.

You are Fred Feree, a senior systems analyst at Palomar. Your manager, Walter Clark, has asked you to determine why customer credits are increasing. Each month, more customers are asking for credits against previous purchases. The credits are issued by the accounts receivable section of the accounting department. You visit accounts receivable, talk with the supervisor, Amy Matula, and dig through a file of recent credit records. You learn that the main reason for the credits is poor product quality. The customers often want their money back.

You track down Amos Nash, the manager of quality control, and learn that he is aware of the situation. The problem, as Amos sees it, is low morale among the production workers. The workers just don't seem to try.

You make appointments to talk with several production supervisors. The supervisors reveal that the workers are unhappy because of low pay. Other area companies have increased worker salaries each year to compensate for the rising cost of living, but Palomar has not. Some Palomar workers are making the same money today as they did five years ago.

You suspect that Palomar has had financial difficulties that made it impossible to give raises to the factory workers. You check with Dale Barnett, manager of the accounting department, who shows you the income statements for the past five years. The statements reveal a substantial profit each year.

Why haven't the workers been receiving raises? The best person to answer that question is Ralph Burton, the vice-president of manufacturing. You make an appointment to see him, and get straight to the point. Ralph tells you that he knows of the morale problem but that one of his annual objectives is to keep manufacturing expenses to a minimum. Salaries are one of the largest expense items, so he concentrates his attention there. You ask whether he established the minimum expense objective himself or whether the objective was imposed on him. Ralph says that Palomar president Rebecca Sandoval establishes the annual objectives for each vice-president. The vice-presidents' annual bonuses depend on the degree to which the objectives are met. You ask to see Ralph's list of objectives. He gladly obliges. All the items on the list look reasonable. In fact, one objective is "achieve high-quality production."

You assure Ralph that you will copy him on all of your reports. As you walk back to your office, you say to yourself, "Whew, and to think that all this started with excessive credits."

Assignments

1. Continue to assume that you are Fred, and write a memo to your manager, Walter Clark, apprising him of what you have done. Include a problem statement based on your data. Conclude by identifying the next step, or steps, that you should take to complete your data gathering.

2. Assume that you are Walter Clark. You want to give Fred feedback on how well he has followed the systems approach. List the things he has done that indicate he has followed the approach. List the things that indicate he has not.

Selected Bibliography

Ahn, Taesik, and Grudnitski, Gary. "Conceptual Perspectives on Key Factors in DSS Development: A Systems Approach." *Journal of Management Information Systems* 2 (Summer 1985): 18–32.

Cerveny, Robert P.; Garrity, Edward J.; and Sanders, G. Lawrence. "A Problem-Solving Perspective on Systems Development." *Journal of Management Information Systems* 6 (Spring 1990): 103–122.

Einhorn, Hillel J., and Hogarth, Robin M. "Decision Making: Going Forward in Reverse." *Harvard Business Review* 65 (January–February 1987): 66–70.

Floyd, Barry, and Ronen, Boaz. "Where Best to System Invest." *Datamation* 35 (November 15, 1989): 111ff.

Hogarth, Robin M. *Judgment and Choice.* (New York: John Wiley & Sons, 1980.) 130–154.

Lederer, Albert L., and Smith, George L., Jr. "Individual Differences and Decision-Making Using Various Levels of Aggregation of Information." *Journal of Management Information Systems* 5 (Winter 1988–89): 53–69.

MacCrimmon, Kenneth R., and Wagner, Christian. "The Architecture of an Information System for the Support of Alternative Generation." *Journal of Management Information Systems* 8 (Winter 1991–92): 49–67.

Markus, M. Lynne. "Power, Politics, and MIS Implementation." *Communications of the ACM* 26 (June 1983): 430–444.

Martin, Merle P. "Problem Identification." *Journal of Systems Management* 28 (December 1977): 10–15.

Martin, Merle P. "Problem Identification Indicators." *Journal of Systems Management* 29 (September 1978): 36–39.

Mosard, Gil. "Problem Definition: Tasks and Techniques." *Journal of Systems Management* 34 (June 1983): 16–21.

Sabherwal, Rajiv, and Grover, Varun. "Computer Support for Strategic Decision-Making Processes: Review and Analysis." *Decision Sciences* 20 (Winter 1989): 54–76.

Wedberg, George H. "But First, Understand the Problem." *Journal of Systems Management* 41 (June 1990): 20–28.

Yadav, Surya B. "Classifying an Organization to Identify Its Information Requirements: A Comprehensive Framework." *Journal of Management Information Systems* 2 (Summer 1985): 39–60.

CHAPTER 8

System Life Cycle Methodologies

Learning Objectives

After studying this chapter, you should:

- understand that a computer-based system evolves through a life cycle, and know what happens in each phase of that cycle
- understand the principles of structured systems analysis and design, and know how they affect the system life cycle
- understand prototyping and its effect on the system life cycle
- know how business process redesign is being accomplished with conceptual systems by means of reverse engineering, restructuring, and reengineering
- be familiar with the concepts of information engineering and rapid application development

Introduction

Each of a firm's computer-based systems evolves through a series of phases called the *system life cycle,* or *SLC.* The SLC methodology has provided the basis for computer projects since the beginning of the computer era, but it can be modified to incorporate modern techniques.

The popularity of structured programming triggered the development of a structured approach to systems analysis, design, and implementation. The structured system development methodologies can be integrated into the SLC framework.

Another modification to the SLC is prototyping. Information specialists build a prototype, or model, of a new system to obtain a clearer understanding of the users' information needs. Some prototypes eventually become operational systems. Others serve as blueprints for systems developed the traditional way. Even though prototyping offers some advantages, it also has some drawbacks. Not all systems lend themselves to this approach.

Many of today's firms face the enormous task of maintaining older, poorly designed systems. A solution is to rebuild the systems, using *business process redesign,* or *BPR.* Three BPR techniques—reverse engineering, restructuring, and reengineering—are well suited to conceptual systems.

A new life cycle methodology that lends itself to BPR is called *RAD*, for *rapid application development*.

This chapter lays the important methodological foundation upon which modern systems are analyzed and designed.

The System Life Cycle

The **system life cycle,** or **SLC,** is the process that is followed in developing and using a computer-based system. Many people are involved, and many separate steps must be taken. It is helpful to divide the work into phases devoted to planning, analysis, design, implementation, and use. The first four phases involve systems development and are called the **system development life cycle (SDLC).**

Phase I—Planning

In planning, the user and the systems analyst react to a problem signal and determine whether to develop a computer-based solution. They work under the direction of the MIS steering committee.[1] The eight steps involved are illustrated in Figure 8.1.

Figure 8.1 and the diagrams for the other phases are general models. They show a step-by-step process that fits *most* projects. The process is well suited to large systems, but the particular situation will cause some steps to be reordered or skipped. Also, activity can overlap more than is shown in the diagrams.

Step 1—Recognize the Problem

Of all the life cycle participants, the user is in the best position to recognize a problem because he or she is on the scene every day to spot the problem symptoms. In many cases, the user solves the problem alone. If the task appears too great, the user asks for help. When it looks as if the solution will involve the computer, the user calls information services. Information services management assigns one or more systems analysts to help the user consider a computer-based solution.

Step 2—Define the Problem

The systems analyst meets with the user in the user's office. They follow the symptom chain until they identify and understand the problem cause. This problem definition step includes identifying the cause with a particular system level and element.

---

[1]Chapter 2 described the role of the MIS steering committee. Chapter 1 recognized that database and network specialists can help the systems analyst interact with the user. This assistance can begin during the planning phase.

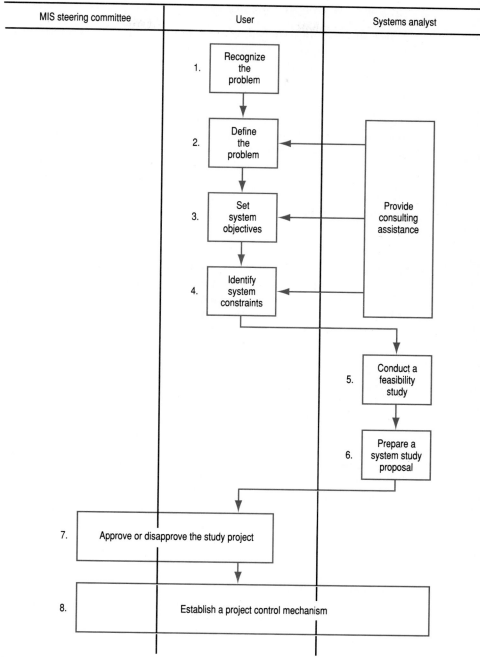

Figure 8.1
The planning phase.

Step 3—Set System Objectives

When the user and analyst agree that a computer-based solution is required, the next step is to specify the objectives. At this early stage, the objectives can only be stated in general terms.

Step 4—Identify System Constraints

The user and systems analyst should understand that certain constraints will limit the scope of the solution. The two types of constraints are internal and environmental. **Internal constraints** are limits that the user or the firm's executives impose on resources. Executives usually impose internal constraints in the form of the firm's operating budget. **Environmental constraints** are limits that environmental elements impose on resources. Each environmental element can shut off certain resources if it so chooses. The new system must fit within all the constraints.

Step 5—Conduct a Feasibility Study

A **feasibility study** is an abbreviated form of systems analysis that determines whether a computer-based solution should be pursued. The analyst interviews key people, in addition to the user, and perhaps examines historical files. The feasibility study usually requires only a few days. It is conducted in the user's area.

Step 6—Prepare a System Study Proposal

If the feasibility study indicates that a computer project is not appropriate, the systems analyst notifies the user and the project ends. If the study indicates that a computer project is appropriate, the analyst writes a report, called a *system study proposal*, recommending that a system study be conducted.

Figure 8.2 presents an outline of such a proposal. The first section summarizes the proposal for the MIS steering committee. The introduction ensures that the readers understand the purpose of the project. Sections 3 and 4 document the findings from planning phase steps 1 through 4. Sections 5 and 6 contain the findings of the feasibility study, recommend that a system study be conducted, and summarize the results that can be expected from the new system. Section 7 offers a general development plan for the remainder of the SDLC. Section 8 is the proposal summary.

At this early stage, the systems analyst can only address the topics in a general way. However, the analyst's estimates, while not perfect, are much better than no information at all. This reasoning typifies how the participants carry out the entire SLC. *All key points are addressed generally at first and in greater detail as more is learned.*

Step 7—Approve or Disapprove the Study Project

After reading the system study proposal, the MIS steering committee reaches a **go–no go decision point**. The committee decides whether to proceed to the next

1. Executive summary
2. Introduction
3. System objectives and constraints
4. Possible system alternatives
5. Recommended system study project
    5.1   Tasks to be performed
    5.2   Human resource requirements
    5.3   Schedule of work
    5.4   Estimated cost
6. Expected impact of the system
    6.1   Impact on the firm's organization structure
    6.2   Impact on the firm's operations
    6.3   Impact on the firm's resources
7. General development plan (analysis, design, and implementation phases)
8. Summary

Figure 8.2
An outline of the system study proposal.

phase or scrap the project. In this way management funds the project phase by phase, an approach that has been termed *creeping commitment*.

Step 8—Establish a Project Control Mechanism

The last step of the planning phase is taken after management approves the study project proposal. Management establishes a control mechanism that will enable it to control the project as long as it continues. A **project control mechanism** is any means of comparing what *is* accomplished with what *should be* accomplished. The mechanism can take the form of a project plan, regular meetings, periodic reports, and graphic techniques, such as bar charts. Some of these mechanisms have been incorporated into computer-based project management systems. With the control mechanism in place, systems analysis can begin.

Phase II—Analysis

The purpose of the analysis phase is to thoroughly understand the user's needs and prepare documentation that can serve as the basis for the system design. The six analysis steps are illustrated in Figure 8.3.

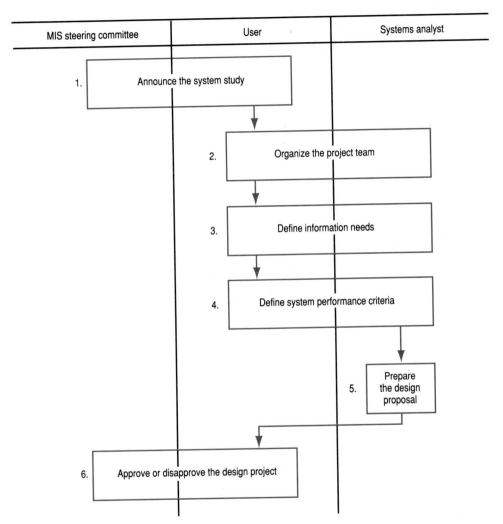

Figure 8.3
The analysis phase.

Step 1—Announce the System Study

Everyone in the user area should be alerted to the system study before the work begins. The users must be aware of the objectives of the new system and how the system will affect their jobs.

The announcement is made by a manager on the highest organizational level affected. For example, the president should announce system studies that affect the entire firm and vice-presidents should announce studies that affect their areas.

Ideally, the announcement combines written and oral media. Newsletter articles, memos to individual employees, and group meetings are all good methods. Some firms use videotapes to ensure communication to everyone, even those in remote locations.

Step 2—Organize the Project Team

The CIO, working with top information services managers, and managers in the user area, determines the composition of the project team. The team leader is designated, and the members who will conduct the analysis are assembled.

Figure 8.4 illustrates the potential makeup of the team. Some members participate throughout the entire SLC. Others participate only when their expertise is needed.

Note that one member is an internal auditor. An internal auditor is a member of the firm's auditing unit who studies conceptual systems for the purpose of ensuring that they function as intended. The internal auditor seeks to confirm **system integrity**—the condition that exists when the conceptual system accurately reflects the physical system. The term *EDP auditor* was once used to describe an internal auditor with computer literacy, but because such literacy is commonplace today, the *EDP auditor* term is no longer in vogue. *EDP* stands for *electronic data processing*. It was used to distinguish computer technology during its early days.

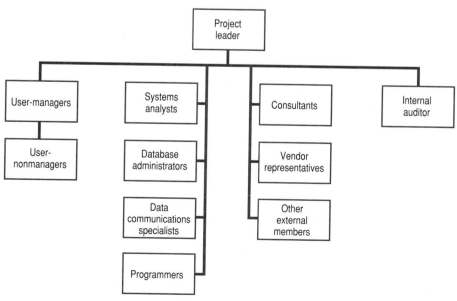

Figure 8.4
A project team.

Early practice was to call in the internal auditor *after* a system was implemented, to seek her or his approval. Disapproval meant repeating some or all development work. It quickly became clear that the internal auditor should be involved as early as possible. This approach enables the auditor to contribute expertise as the design unfolds.

Step 3—Define Information Needs

To the systems analyst, defining the user's information needs is the most important step in the SLC. If the system is to meet the user's information needs, then those needs must be accurately identified. During this step, the systems analyst spends much time in the user area, gathering data by personal interview, survey, observation, and record search.

While defining the user's needs, the analyst makes narrative notes and graphic diagrams. The documentation describes the existing system's processes and the data and consists of tools such as data flow diagrams, flowcharts, and the data dictionary. Traditionally, the documentation has been prepared by hand. Today, many firms use CASE. Some CASE tools can be introduced as early as enterprise planning and modeling.

All the documentation is maintained in a **project dictionary,** a collection of system documentation that the project team members keep throughout the SLC. When the documentation is prepared by hand, the dictionary resides in a hard-copy form. When CASE is used, most of the dictionary can reside in computer storage. The term **repository** describes systems documentation that is in an electronic form.

Step 4—Define System Performance Criteria

Before designing the new system, the systems analyst must understand the level of performance the user demands. The performance specifications are the **performance criteria.** This portion of the SDLC, where the analyst and user define the performance criteria, is often referred to as **requirements analysis** or **requirements specification.**

The performance criteria include dimensions such as speed and accuracy. For example, the user may demand that a new order processing system process each customer order in no more than three hours and fill 98 percent of the orders without error. Note that because these criteria are stated quantitatively it will be easy to evaluate system performance later.

Step 5—Prepare the Design Proposal

Before beginning the design phase, the systems analyst must again justify going ahead. The analyst prepares a *design proposal* that identifies possible system alternatives and recommends that a design project be conducted. The expected impact of the new system is updated with findings from the analysis phase. The general

1. Executive summary
2. Introduction
3. Problem definition
4. System objectives and constraints*
5. Performance criteria
6. Possible system alternatives*
7. Recommended design project
   7.1   Tasks to be performed
   7.2   Human resource requirements
   7.3   Schedule of work
   7.4   Estimated cost
8. Expected impact of the system
   8.1   Impact on the firm's organization structure
   8.2   Impact on the firm's operation
   8.3   Impact on the firm's resources
9. General development plan (design and implementation phases)
10. Summary

*From the system study proposal

Figure 8.5
An outline of the design proposal.

development plan focuses on the design and implementation phases. Figure 8.5 presents a basic design proposal outline.

Step 6—Approve or Disapprove the Design Project

The MIS steering committee and the user decide whether to end the project or proceed to the design phase.

Phase III—Design

The purpose of the design phase is to determine the best way to solve the user's problem. With help, the systems analyst identifies and documents the best configuration of hardware and software in terms of the budget and other constraints. The systems analyst is joined by other information specialists as illustrated in

Figure 8.6. Database and network specialists play especially important roles when the design requires their expertise.

Step 1—Prepare the Detailed System Design

The project team discusses the problem and decides how best to solve it. They document the design, using the same tools used to document the existing system. The documentation is added to the project dictionary.

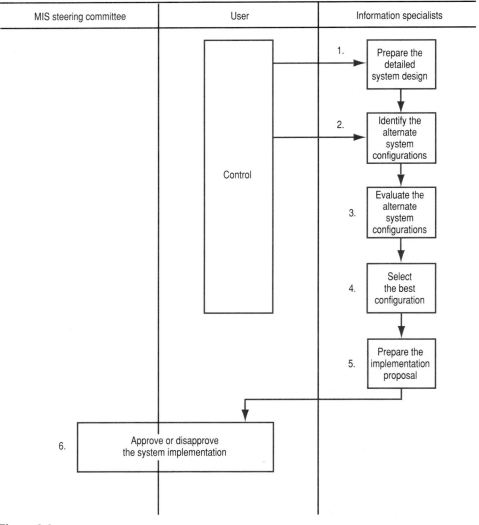

Figure 8.6
The design phase.

Step 2—Identify Alternate System Configurations

The system design prepared in step 1 does not specify the exact configuration of hardware that will be used. For example, the design might specify that a file be maintained in secondary storage but does not specify whether magnetic tape, magnetic disk, or compact disk is to be used.

Several hardware options exist for each system part—input, processing, secondary storage, and output. These options enable the information specialists to identify alternate solutions to the problem.

Step 3—Evaluate Alternate System Configurations

The list of alternate solutions is pared down to those that the information specialists wish to give detailed consideration. These configurations are evaluated, using a set of evaluation criteria. This process was illustrated in Chapter 7 with Tables 7.1 and 7.2.

Step 4—Select the Best Configuration

The information specialists and the user select the configuration that they believe will do the best job of (1) satisfying the user's performance criteria, and (2) enabling the firm to meet its objectives. The solution is tailored to the needs of the user and the firm.

Step 5—Prepare the Implementation Proposal

The information specialists prepare an *implementation proposal* that recommends that the designed system be implemented. Figure 8.7. presents an outline. The proposal summarizes the new system, prescribes a particular implementation project, updates the expected impact on the firm with findings from the design phase, and presents a general plan for the implementation.

Step 6—Approve or Disapprove the System Implementation

The MIS steering committee and the user read the implementation proposal. Then they decide whether to implement the system. The decision to go ahead is an important one because the implementation will be expensive. It will involve many more people and the acquisition of more hardware, software, and data resources.

Before authorizing this expenditure, the MIS committee verifies that (1) the proposal recommends the best design and (2) the proposal recommends the best way to implement that design. When the committee is satisfied that both conditions have been met, implementation can begin.

Phase IV—Implementation

The purpose of the implementation phase is to convert the design documentation into a functioning system. Up to this point, the SLC activity has been sequential.

1. Executive summary
2. Introduction
3. Problem definition
4. System objectives and constraints
5. Performance criteria
6. System design
    6.1   Summary description
    6.2   Performance criteria
    6.3   Equipment configuration
7. The recommended implementation project
    7.1   Tasks to be performed
    7.2   Human resource requirements
    7.3   Schedule of work
    7.4   Estimated cost
8. Expected impact of the system
    8.1   Impact on the firm's organizational structure
    8.2   Impact on the firm's operations
    8.3   Impact on the firm's resources
9. General implementation plan
10. Summary

Figure 8.7
An outline of the implementation proposal.

Now it becomes parallel. The systems analyst and the database and data communications specialists are joined by programmers and computer operations personnel, and possibly by outside specialists such as vendor representatives and building contractors. Many of their jobs can be performed simultaneously. This overlapping is depicted in Figure 8.8 with the two-headed arrows connecting steps 3 through 7.

Step 1—Plan the Implementation

It is now time for a detailed plan devoted exclusively to this particular implementation. Many resources will be involved, and it is crucial that they function together well. The project team prepares a detailed implementation plan that is approved by the MIS steering committee.

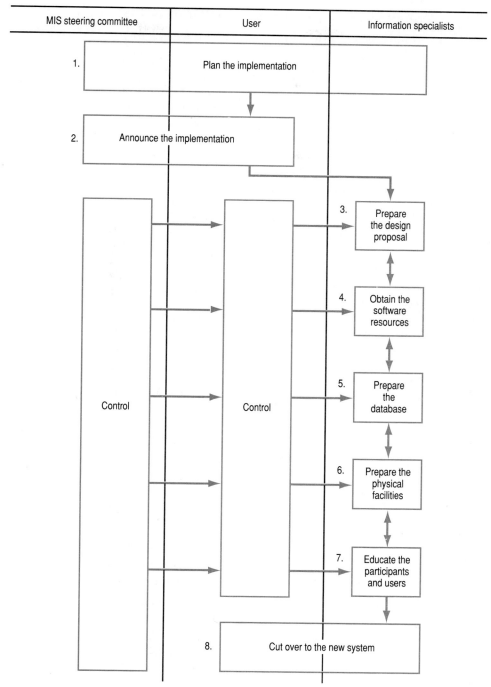

MIS steering committee	User	Information specialists
1.	Plan the implementation	
2.	Announce the implementation	
Control	Control	3. Prepare the design proposal
		4. Obtain the software resources
		5. Prepare the database
		6. Prepare the physical facilities
		7. Educate the participants and users
8.	Cut over to the new system	

Figure 8.8
The implementation phase.

Step 2—Announce the Implementation

Management's decision to implement the system must be passed along to the employees who will be affected. The announcement is made by the same level of management that made the original announcement. This announcement emphasizes the anticipated benefits to the firm and the employees, and management asks the employees for their cooperation.

Step 3—Obtain the Hardware Resources

Before it can obtain the hardware resources, the team must identify the computing equipment that will be required. The new system might use existing hardware, or it might require new hardware. For new hardware, the information specialists prepare a **request for proposal,** commonly known as an **RFP.** The RFP is sent to hardware vendors, asking them to submit bids. Figure 8.9 provides a sample outline. The RFP includes much design and implementation data so that the vendors will clearly understand the demands that will be placed upon them in terms of hardware and support.

Each vendor who wishes to bid responds with a proposal that describes its products and services and argues its case for selection. Figure 8.10 is a sample proposal format. When large orders are involved, vendors supplement the written proposal with formal oral presentations to the MIS steering committee and the project team.

The MIS committee may have the project team and information services management evaluate the vendor proposals. Even in that case, the committee still decides which vendor, or vendors, to use. Once the selection has been made, the

1. Cover letter
2. Introduction
3. System objectives and constraints
4. System design
    4.1   Summary description
    4.2   Performance criteria
    4.3   Equipment configuration
    4.4   Transaction volumes
    4.5   File sizes
5. Installation schedule

Figure 8.9
An outline of a request for proposal.

1. Cover letter
2. Summary of recommendations
3. Introduction
4. Advantages of the proposed equipment
5. Equipment configuration
6. Equipment specifications
    6.1 Performance data
    6.2 Prices
7. Satisfaction of the performance criteria
8. Delivery schedule

Figure 8.10
An outline of a vendor proposal.

firm orders the hardware, scheduled to be delivered and installed in accordance with the implementation plan.

Step 4—Obtain the Software Resources

For about the first 20 years of the computer era, the software acquisition step was called *code the programs.* The availability of prewritten software changed that. Today, the firm faces a **make-or-buy decision** concerning its software—should the firm's programmers produce custom software, or should the firm obtain prewritten packages from software vendors?

When custom software is the choice, the traditional approach is for the programmers to code the programs. The debugged and tested programs are entered in the software library, and program documentation is added to the project dictionary. A more modern approach involves using CASE. Many CASE tools have a code generator that produces computer code directly from the design documentation.

When the choice is to acquire prewritten software, vendor selection can follow the same procedure used for hardware vendors. RFPs for software can be distributed, and software vendors can submit proposals.

Step 5—Prepare the Database

The database administrator, or DBA, prepares the database that the new system will require. The project team includes one or more DBAs. During analysis and design, the DBAs document the data requirements of the new system. During implementation, they use this documentation as the basis for creating the database.

The amount of work necessary depends on the condition of the data required by the new system. The decision logic table in Figure 8.11 illustrates the four possible conditions.[2]

The four conditions are:

- The data has not yet been gathered. Database preparation most likely will be very difficult.
- The data has been gathered but has not been entered into the computer. Database task preparation will be difficult at best.
- The data has been gathered. It is in computer storage, but it is not in the proper format. Database preparation will be moderately difficult.
- The data has been gathered. It is in computer storage in the proper format. Database preparation is skipped.

Because of the wide variation in the amount of time required to prepare the database, it is important to determine the data condition far in advance. Otherwise, the entire project can be held up, waiting for the data.

Step 6—Prepare the Physical Facilities

Perhaps the computer is already installed and the new application will simply be added. In that case, no facility is required and the facilities preparation step is omitted. The same situation usually exists when the computer is a PC. You know

IF data is gathered	N	Y	Y	Y
data is in computer storage	—	N	Y	Y
data is in proper format	—	—	N	Y
THEN preparation is very difficult	X			
preparation is difficult		X		
preparation is moderately difficult			X	
preparation is skipped				X

Figure 8.11
Variations in database preparation difficulty.

[2]A *decision logic table (DLT)* lists the conditions at the top and the actions at the bottom. The columns at the right indicate the combinations of conditions that can apply, identified by Y (does exist), N (does not exist), or a dash (not applicable). Each combination of conditions produces an appropriate action (identified by X). The DLT effectively reflects how combinations of conditions influence an action or decision, and it is especially applicable to the design of decision support systems. DLTs have been around since the beginning of the computer era, but they have never achieved widespread use.

from experience that you simply find a place to set it, connect some cables to it, and plug it in.

However, when the new system requires a complex hardware configuration, the task of preparing the facilities can be a challenging one. When new facilities are required, the operations manager usually is given the responsibility of working with building contractors and hardware vendors in designing and constructing the facilities.

The ordered hardware is scheduled to be delivered shortly after the facilities are ready.

Step 7—Educate the Participants and Users

The information specialists and the users who designed and built the system understand it. Other users, however, must be educated to use the system's output.

Also requiring education are those persons who will provide the input data, enter the data into the computer, operate the equipment, and so on. These people are the system **participants.**

Training sessions for users and participants can be conducted by systems analysts. Analysts are a good choice for this assignment because they have good communications skills and know the system inside and out.

Step 8—Cut Over to the New System

When all implementation work is completed, it is time to put the new system to use. The process of changing from the old system to the new one is called **cutover.** The term **conversion** is also used. The cutover signals the end of the SDLC.

The implementation phase does not include a formal proposal to proceed to the next phase. The MIS steering committee maintains control during implementation, and at any time, can cancel or delay the cutover because of the way the work is progressing. The committee will approve the cutover only when the system is ready for the firm and the firm is ready for the system.

Phase V—Use

The purpose of the use phase is to make the system available to the user. Figure 8.12 illustrates the activities.

Step 1—Use the System

The user uses the system to solve the problem identified in the planning phase. Immediately following cutover, the project team members stay in close contact with the user to ensure that all is going well. Any difficulties that the user encounters are addressed, and remedies are incorporated into the system.

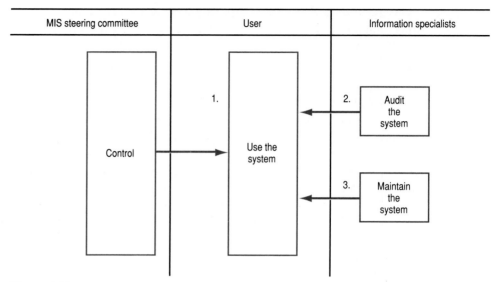

Figure 8.12
The use phase.

Step 2—Audit the System

About 90 days after cutover, when the system has had time to settle down, two audits are conducted. The term **postimplementation review** is used to describe these audits. One audit is performed by the systems analyst. The other is performed by the internal auditor. The audits have separate objectives:

- **Audit to Ensure User Satisfaction** The system analyst's audit uses personal interview or survey to ensure that the user is satisfied with the system. Any indication of dissatisfaction triggers a formal review by information services. Shortcomings are resolved.

- **Audit to Ensure System Integrity** The internal auditor's audit ensures that the system accurately represents the physical system. For example, do the computer's inventory records match the warehouse stock? Any deficiencies are called to the attention of the MIS steering committee in a written report, and the deficiencies are corrected.

These two types of audits are repeated regularly, perhaps once a year, throughout the life of the system. We use the term **annual audit** to describe these ongoing audits.

Step 3—Maintain the System

Throughout the use phase, it is necessary to modify the system so that it continues to meet the user's needs. These modifications are called **systems maintenance.**

Systems maintenance has three purposes:

- To correct errors—Some system flaws and program bugs do not surface until the systems are in use.
- To keep systems current—The environment changes, necessitating changes in the systems. For example, the federal government changes the rules for computing employee social security tax.
- To improve systems—Users see ways to make their systems more effective.

The longer a firm uses its computer, the greater the maintenance workload becomes. In some firms, it has been estimated that the information specialists devote 50 to 90 percent of their time to maintaining existing systems. The term **legacy system** describes such existing systems that drain a firm's information resources. The systems are legacies left to the current staff by previous information specialists and users. A large number of legacy systems makes it virtually impossible for the firm to free itself so that it can expand its scope of computer applications.

Repeating the System Life Cycle

Chapter 1 pointed out the repeating nature of the SLC and illustrated it with Figure 1.6. An existing system is redeveloped by initiating a new SLC.

It becomes necessary to redevelop systems when any of three conditions exists:

- an annual audit reveals deficiences that cannot be corrected by modifications
- management determines that maintenance costs have gotten out of hand
- a change in the firm's physical system requires a change in the corresponding conceptual system

System redevelopment currently is receiving much attention from top management as a means of achieving a competitive advantage. We discuss redevelopment later in the chapter when we address business process redesign.

Putting the System Life Cycle in Perspective

The system life cycle has been around throughout the entire computer era. The first computer was probably installed by following the steps in a trial-and-error manner.

The SLC works because it is so heavily rooted in the systems approach. Like the systems approach, the SLC boils down to defining what you want to do, considering alternate solutions, implementing the one that appears to be best, and following up to make certain that it works. Even though the SLC will continue to be refined, it is difficult to imagine that it will ever be replaced.

Structured Systems Methodologies

When they first followed the classical SLC, information specialists took a bottom-up approach. They created the subsystems of a system, and as the last step, integrated the subsystems to form a system. The main weakness of this approach was delaying the system integration until the final step. Often, the subsystems did not function together as planned. It was like trying to put together a jigsaw puzzle when all of the pieces do not fit.

The solution to this systems problem started in the computer programming area with the discovery of structured programming. A structured program is organized in modules that are arranged in hierarchical levels. The module at the top is called the *driver module*. You code it first, and then you code the subsidiary modules on the next lower level. You proceed in this top-down fashion until all the modules are coded. Structured programming has many advantages, but perhaps the most important is that all of the modules fit.

Structured programming became so popular that systems analysts decided to apply the approach to the development of entire systems. Figure 8.13 is a hierarchy diagram that shows how a system can be viewed as a hierarchy of modules.[3]

The depicted system is a data processing system that is used by a distribution firm—manufacturer, wholesaler, or retailer. The distribution system contains five

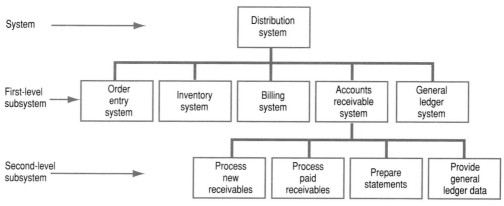

Figure 8.13
A system structure.

[3]A *hierarchy diagram*, also called a *structure chart*, is a systems analysis and design tool that uses rectangles to represent modules. The modules are arranged in a hierarchy to show superior-subordinate relationships. Hierarchy diagrams can be used to document systems, as in Figure 8.13 but they are especially effective for showing program structure. We used a hierarchy diagram in Figure 6.3 to illustrate the different types of software.

first-level subsystems—order entry, inventory, billing, accounts receivable, and general ledger. Each of these systems, in turn, consists of second-level subsystems. The subsystems of the accounts receivable system are shown in the figure. The four other first-level subsystems are subdivided in the same manner.

When systems analysis is performed in a structured way, attention is first focused on the top system level, in this case, the entire distribution system. The systems analyst understands how the overall system works. Next, attention is focused on the first-level subsystems. When they are understood, attention is focused on the second-level subsystems, and so on, down the hierarchy. This top-down sequence is a structured approach to systems analysis. **Structured systems analysis** is the process of studying a system by addressing its subsystems and their interrelationships in a top-down, hierarchical manner.

Systems design can be carried out in the same top-down fashion. **Structured systems design** is the process of designing a new system by approaching the task in a top-down manner, first specifying the characteristics of the entire system and then designing each subsystem to meet the desired characteristics.

Our main concern is with structured systems analysis and design, but the same top-down approach can be applied to the implementation phase. **Structured systems implementation** is the process of implementing a system in a top-down, hierarchical manner. For example, program testing and cutover can be done top-down.

The Influence of the Structured Methodologies on the System Life Cycle

The guideline provided by the SLC prior to the structured methodologies is known as the **classical life cycle approach.** The term **waterfall life cycle** is also used. In this pattern, the phases are religiously executed in sequence, one after the other. Also, the work is done in a bottom-up fashion during each phase after the planning is done. This pattern is shown in Figure 8.14. The numbers and arrows indicate the sequence. Design work begins only when all the analysis is completed. Implementation begins only when all the designing is completed.

When the structured approach to system development is followed, the planning phase is completed, as shown in Figure 8.15, before analysis is begun. The analysis begins at the top level. When the analysis is completed for that level (step 2A), the design work (step 2D) *on that same level* can begin. While the new system is being designed at the top level (step 2D), the analysis of the middle level of the existing system can be performed (step 3A). When the design work is completed at the top level (step 2D), the implementation at that level (step 2I) can begin. The work proceeds in this combined horizontal and vertical pattern until the project is completed.

By overlapping analysis, design, and implementation, more people can work on the project at the same time. The project is completed sooner. Also, the integration of the subsystems is assured because it is built in.

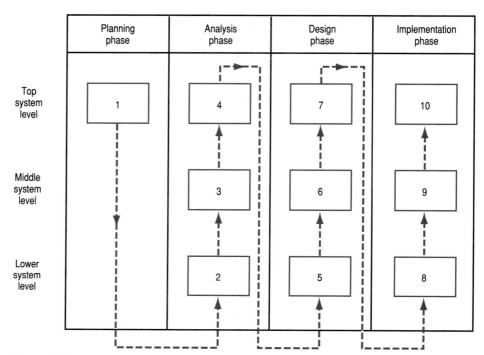

Figure 8.14
The classical system life cycle approach to system development.

Putting the Structured Methodologies in Perspective

The first information specialists performed their work in the manner most convenient to them, without much thought about the future. The result was a hodgepodge of program structures that made inefficient use of the computer and were virtually impossible to maintain. Structured methodologies were devised as a way to lend order to this chaos. Information specialists can now take a team approach to system development. Work goes faster because it is subdivided, and communication among team members is facilitated. The result is better systems, implemented sooner.

Prototyping

Another modern approach, prototyping, has generated even more interest than the structured methodologies.

Communication between the user and the information specialists has always been difficult. The user has had difficulty explaining exactly what the problem is, what information is needed, how the information should be provided, and so on.

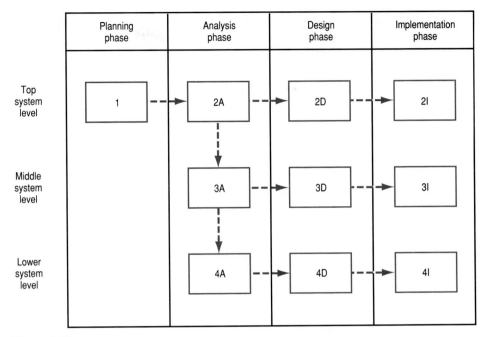

Figure 8.15
The structured approach to system development.

Information specialists decided that the communication problem could be reduced, if not eliminated, using a technique that had been effective in design engineering—prototyping. A **prototype** is a model that contains all the essential elements of the object to be produced. It is used as a pattern. An automobile prototype, for example, enables the design engineer to more effectively communicate the auto's appearance, and perhaps performance, to others.

In the computer field, **prototyping** refers to the process of quickly building a model of the desired software system, to be used primarily as a communication tool to assess and meet the information needs of the user.[4]

Prototyping Ingredients

Prototyping has three key ingredients:

- **Quick Delivery** The information specialist tries to provide the prototype as soon as possible—often the day after the user indicates what is needed.

- **Preview of System Performance** The user can key data into the prototype and view the results on the screen just as with an operational system.

[4]This definition and the following prototyping discussion are based on Jane M. Carey, "Prototyping: Alternative Systems Development Methodology," *Information and Software Technology* 32 (March 1990): 119–126.

- **Easier Communication of Needs** Using the prototype, the user can better see the level of support the operational system will provide, and can more easily make suggestions for improvement.

Each of these three ingredients benefits the user. Prototyping is a user-oriented methodology.

Types of Prototypes

There are two types of prototypes. A **Type I prototype** becomes the operational system after repeated changes based on user feedback. The program flowchart in Figure 8.16 shows the sequence of events. The sequence assumes that the planning and analysis have already been done. In step 4 of the example, the user indicates that the prototype is unacceptable and makes suggestions for improvement.

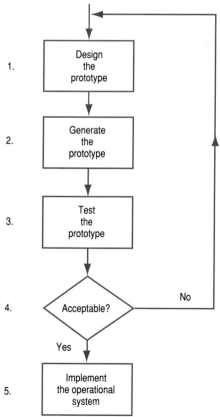

Figure 8.16
The design and implementation of a type I prototype.

Steps 1 through 3 are repeated, and the user's suggestions are incorporated. When the prototype is acceptable, it is implemented as an operational system in step 5.

A **Type II prototype** is a throwaway model that serves as the blueprint for the operational system. Figure 8.17 shows the evolution. Again, planning and analysis are assumed to be done. The prototype must first satisfy the user in step 4, and then the operational system must be accepted in step 7.

How Prototypes are Built

Prototyping would not be possible if not for the software tools that permit the quick delivery. The tools provide two basic prototyping environments, integrated application generators and prototyping toolkits.

Integrated application generators are prewritten software systems capable of producing *all* the desired prototype features—menus, reports, and screens that are tied to a database. Examples of such integrated generators are R:base 5000 and System V for micros and NOMAD2 for mainframes.

Prototyping toolkits consist of a collection of *separate software systems*, each capable of producing a *portion* of the prototype features. Examples of these separate software systems are report generators, screen generators, database management systems, and electronic spreadsheets. Some CASE tools offer one or more of these capabilities.

The two prototyping environments enable information specialists to generate prototypes without going through time-consuming and costly program coding.

Prototyping and the System Life Cycle

The introduction of prototyping caused much debate about whether prototyping would replace the SLC. Now it appears that prototyping can replace most of the SLC for some projects and supplement the SLC for others.

Prototyping as a Life Cycle Replacement   For small projects, a prototype can replace the analysis, design, and implementation phases. Planning and systems maintenance are still required. This developmental pattern is popular with users who develop their own systems, such as electronic spreadsheets.

Prototyping as a Life Cycle Supplement   For large projects, a prototype can be used during analysis, design, and implementation to:

- define information needs (analysis)
- prepare the detailed system design (design)
- evaluate alternate system configurations (design)
- obtain hardware resources (implementation)
- obtain software resources (implementation)

At each point, the prototype serves as a stimulus if the user cannot articulate the needs.

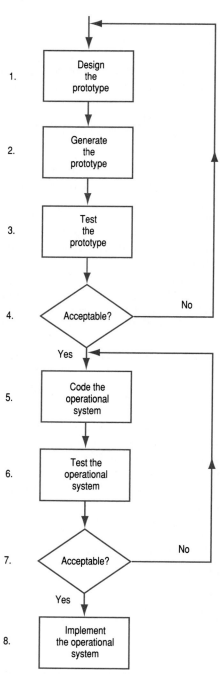

Figure 8.17
The design and implementation of a type II prototype.

Potential Advantages of Prototyping

Prototyping offers the following potential advantages:

- better communication between the user and the information specialists
- easier determination of user needs
- active involvement of the user in system development
- less time and effort spent on systems development, by information specialists and the user alike
- easier implementation because the user knows what to expect

These advantages reduce development costs and ensure that the systems better meet the user's needs.

Potential Disadvantages of Prototyping

It is easy to become so excited about prototyping that you lose sight of its potential disadvantages:

- shortcuts in problem definition, alternative evaluation, and documentation caused by the haste to deliver the prototype
- unrealistic expectations of users inordinately impressed with the prototype
- lack of efficiency of a Type I prototype as compared to a system coded in a programming language
- computer-human interaction that fails to reflect the characteristics of good design, from a behavioral or physical point of view

When prototyping is accomplished by information specialists, the CIO can implement controls that reduce the chances that these disadvantages will occur. The prototyping efforts of users, however, are much more difficult to control.

Applications That Are Good Prospects for Prototyping

Not all applications lend themselves to prototyping. Prototyping works best if:[5]

- a *high risk* is involved (the problem is not well structured, a high rate of change over time is likely, and data requirements are uncertain)
- *user interaction* abounds (the system features online dialog between the user and a microcomputer or a terminal)
- the *users are many* and agreement on design details is difficult to achieve without hands-on experience
- the system must be *delivered quickly* and is expected to have a short life span

[5]David Avison, and David Wilson, "Controls for Effective Prototyping," *Journal of Management Systems* 3 (Number 1, 1991): 45.

- the system is *innovative*, either in the way it solves the problem or in its use of hardware
- the *user's behavior with the system is unpredictable* due to lack of experience

Applications that do not have the preceding characteristics can be developed using the structured approach.

Putting Prototyping in Perspective

A seasoned systems analyst once joked that systems have always been developed as prototypes, but not intentionally. The systems were implemented, the users rejected them, and the systems had to be redesigned. Unfortunately, such a situation occurred too often. Eventually, the information specialists learned the value of involving the user from the beginning. Modern prototyping software tools made possible a developmental strategy that information specialists probably knew all along was the best approach. Prototyping just makes good sense because the system belongs to the user, and the user should, therefore, play a major role in its development.

Business Process Redesign

Chapter 2 described ways that firms seek to gain competitive advantage. After achieving all the benefits possible by fine tuning, many firms are beginning to question the ways they perform their basic processes. In some cases, executives strive to *not* be influenced by the current processes. This leaves them free to define the best procedure and then implement the ideal design. This redevelopment of existing physical and conceptual systems, without being constrained by their current form, is called **business process redesign (BPR)**. The term **business process reengineering** is also popular.

An Example of Business Process Redesign

Possibly the best example of business process redesign is the JIT (just-in-time) manufacturing technique made popular by the Japanese. Traditional manufacturing emphasized mass production. Raw materials were purchased in large quantities, so that the firm could receive quantity discounts, and large numbers of finished goods were produced simultaneously to minimize setup costs. Computers provided management with a conceptual system that was used to manage the physical system.

The logic of the mass-production approach seemed sound, but the Japanese were able to see the benefits of an opposite approach. JIT is characterized by frequent deliveries of small quantities of raw materials and small production lots. Computers share the spotlight with noncomputer techniques. A production worker who is ready for more materials sends a *kanban,* a signal for the preceding work station to release the materials. The kanban can be a card, a flashing light, or even a golf ball rolling through a pipe.

Japan's success with JIT has stimulated firms around the world to take the same approach. The replacement of traditional manufacturing processes with JIT is an example of business process redesign.

BPR and Information Technology

BPR does not always affect information technology (IT). For example, a noncomputer process can be redesigned to function differently but still not involve the computer. However, IT is so tightly intertwined with the firm's activities that a BPR project often results in the need to redevelop computer-based systems. In some cases, a new system will be developed. In other cases, an existing system will be redeveloped. BPR often triggers the redevelopment of legacy systems.

The Three Rs

Three approaches have been devised for the redevelopment of computer-based systems. They are reverse engineering, restructuring, and reengineering—the *three Rs*.

Reverse Engineering   One problem faced by information specialists who redevelop existing systems is understanding what the systems were originally intended to do. This task is made difficult by the lack of good documentation. To solve the problem, the information specialists can use reverse engineering to produce the documentation. **Reverse engineering** is the process of analyzing a system to identify its elements and their interrelationships and to create documentation on a higher level of abstraction than currently exists.[6]

Figure 8.18 shows how you develop the needed documentation in a reverse sequence.[7] The following numbers match those in the figure:

1. You begin in the use phase with the code, and you produce the program documentation in the form of program flowcharts, structured English, and so on.

2. Using the documentation of all the programs from the implementation phase, you produce the system documentation in forms such as system flowcharts, data flow diagrams, and Warnier-Orr diagrams.

3. Using the system documentation from the design phase, you define the user's information needs and performance criteria.

4. Using the information needs and performance criteria from the analysis phase, you define the problem and determine the system objectives and constraints.

[6]The BPR definitions included here are based on Elliot J. Chikofsky and James H. Cross II, "Reverse Engineering and Design Recovery: A Taxomony," *IEEE Software* 7 (January 1990): 13–17.

[7]The diagrams of the three Rs in Figures 8.18, 8.19, and 8.20 are based on Chikofsky and Cross, 14.

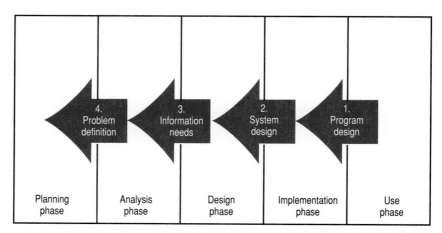

System life cycle

Figure 8.18
Reverse engineering.

Each time you backtrack to an earlier life cycle phase, you produce documentation on a higher system level. However, it is not necessary to backtrack through all the phases.

In reverse engineering you do not change the system's functionality. **Functionality** is the job the system performs. A system redeveloped by reverse engineering performs exactly the same function as the original system. The objective of reverse engineering is to achieve an improved understanding of the system. This understanding can provide the basis for redeveloping the system by means of restructuring or reengineering.

Restructuring   Many existing systems consist of computer code created before structured programming. This code typically is a tangle of unstructured logic and other processing termed *spaghetti code.* Spaghetti code is characteristic of legacy systems, and accounts for much of the difficulty in maintaining them. At some point, it becomes clear that the programs should be redeveloped and coded in a structured form. This conversion of a nonstructured system to a structured one is called **restructuring.**

Restructuring typically begins in the use phase, as shown in Figure 8.19, and can proceed backward through the entire SLC. The restructuring of the entire SLC produces the completely documented, structured system that would have been developed originally had structured methodology been followed.

The curved arrows in the figure illustrate how the restructuring is accomplished in an earlier phase and is then put into operation. For example, to restructure the spaghetti code of the use phase, you repeat the software preparation step of the implementation phase and put the resulting structured code into operation in the use phase. This activity can progress phase-by-phase back

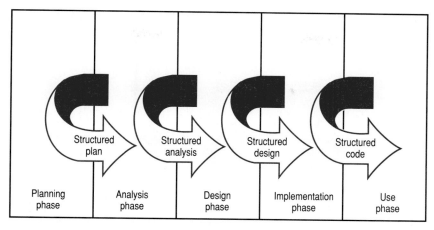

Figure 8.19
Restructuring.

through the life cycle, producing the structured design, structured analysis, and structured plan.

Restructuring updates obsolete systems, but like reverse engineering, it does not change their functionality.

Reengineering   If a system's functionality is to be changed, it is necessary to repeat the SLC. The redesign of a system with the objective of changing its functionality is called **reengineering**. Reengineering consists of the two basic processes illustrated in Figure 8.20. First comes reverse engineering for the purpose of salvaging the existing system's usable components. Next comes **forward engineering,** in which the phases of the SLC are executed in the normal manner. The forward engineering produces an operational system.

The three Rs are three methodologies information specialists can use when involved in BPR projects. Each project makes use of BPR software.

BPR Software

Business process redesign has become so popular during the last few years that software vendors have responded with a wide variety of software that performs all or part of the redesign.

- *Program-analyzing software* examines operational code to determine whether it is a candidate for BPR. Examples are Inspector from KnowledgeWare, and SCAN/COBOL from Computer Data Systems.

- *Reverse engineering software* prepares documentation on a higher level of abstraction. Examples are BACHMAN/Analyst Capture for COBOL from BACHMAN Information Systems, and PacBase/PacReverse from CGI.

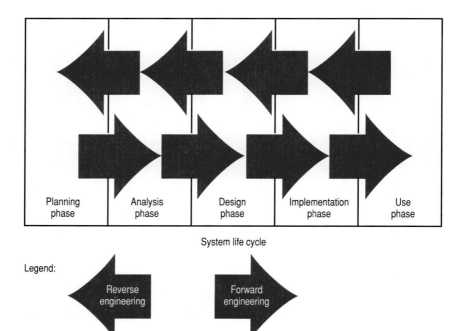

Figure 8.20
Reengineering.

- *Restructuring software* produces structured code and documentation. Examples are Documentation System III from Marble Computer, and InterCASE from InterPort Software.

- *Reengineering software* enables the complete redesign of an existing system. Examples are Application Development Workbench from KnowledgeWare, and Oracle*CASE from Oracle.

The availability of modern software tools makes it possible for firms to pursue a BPR strategy.

Putting Business Process Redesign in Perspective

Many business processes were implemented long before the computer was invented. Innovations in information technology make it possible to perform the processes in a new and better way than the original design allowed. Tough international competition forces firms to consider BPR, but BPR software makes the task feasible. BPR is one of the hottest areas in business computing, and your chances of becoming involved in a BPR project are high.

Rapid Application Development

Legacy systems impose a heavy workload on today's information specialists. However, they are only one force that influences systems work. The other force is end-user computing. The computing climate in many firms today demands quick implementation of a large volume of systems to achieve and maintain a competitive advantage. The human information resources face many constraints, however. Information specialists can implement only so many systems, and users are constrained in what they can do and are willing to do.

James Martin, probably the best-known computer consultant in the world, responded to this climate by devising a new methodology called rapid application development. **Rapid application development, or RAD,** is a developmental life cycle designed to give much faster development and higher-quality results than those achieved with the traditional cycle.[8] RAD is a component in a higher-level methodology called information engineering.

Information Engineering

Information Engineering (IE) is Martin's name for the top-down process of developing computer-based systems that begins with information strategy planning, progresses to business area analysis, and concludes with the design and construction of systems. Figure 8.21 illustrates these IE components. The two faces of the pyramid represent the two basic dimensions of systems—data and activities.

- *Information strategy planning* involves the enterprise planning and strategic planning for information resources described in Chapter 2.

- *Business area analysis* is an analysis of the firm's processes and the data that is needed.

- *Systems design* involves specifying the conceptual systems that will be needed to support the processes of the firm's physical systems, with heavy emphasis on user involvement.

- *Construction* is the conversion of the design into working systems.

The top level of the pyramid forms the organizational setting for systems work. The three lower levels are achieved by following the RAD life cycle.

Phases of the RAD Life Cycle

The four phases of the RAD life cycle are illustrated at the right of the pyramid in Figure 8.21.

- **Requirements Planning Phase** Users work with information specialists to identify business problems to be solved.

[8]The definition of RAD and the basis for the discussion in this section comes from James Martin, *Rapid Application Development* (New York: Macmillan, 1991).

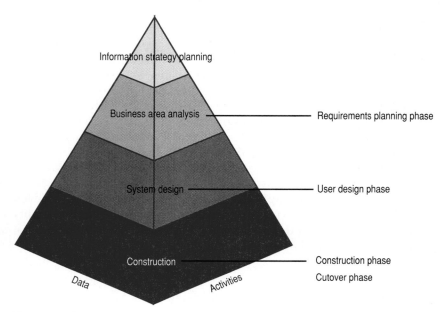

Figure 8.21
The information engineering pyramid.

- **User Design Phase** Users play a big role in the nontechnical design of new systems, assisted by the information specialists.
- **Construction Phase** Information specialists use CASE tools to build prototypes, which are reviewed by users. The reviews trigger refinements. This process is repeated until an acceptable system is produced.
- **Cutover Phase** After thorough planning, the new system is quickly put into use.

The RAD life cycle emphasizes speed. Development time is shortened, but more effort is required of users. Figure 8.22 shows this higher level of effort and how it is shifted forward in the life cycle.

SWAT Teams

An interesting RAD strategy is the use of SWAT teams. SWAT stands for *Skilled With Advanced Tools*. A **SWAT team,** therefore, is a team of information specialists skilled in carrying out a particular aspect of system development. For example, one team might specialize in data modeling, another in system justification, and another in project management. Each team performs only its specialty, moving from project to project.

The SWAT approach recognizes the many specialized skills required to develop modern systems and the difficulty of single teams being expert in them all. SWAT teams are especially skilled in the use of software tools such as CASE.

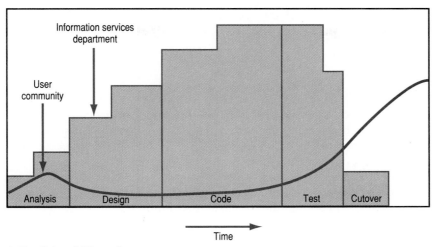

A Traditional life cycle

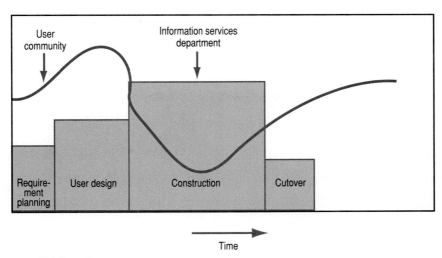

B RAD life cycle

Figure 8.22
A comparison of effort required by the traditional life cycle and a RAD life cycle.

Putting Rapid Application Development in Perspective

RAD was conceived as a solution to the problem of too few information specialists to handle the increasing volume of systems projects. Because the number of specialists is limited, the solution is to use computer-based tools and achieve more user involvement. RAD combines modern tools and end-user computing to form a methodology well suited to the demands of modern systems work.

Putting the System Life Cycle Methodologies in Perspective

Chapter 7 identified the systems approach as the basic systems methodology. In this chapter, we have described methodologies for developing and using computer-based information systems. The life cycle methodologies are the SLC, structured techniques, prototyping, the three Rs, and RAD. All of these methodologies are adaptations of the systems approach.

Experience will teach you when to use a particular methodology and how to tailor each methodology to your particular situation. This and the previous chapter provide a good methodological foundation upon which to build your systems career.

The Impact of CASE on the System Life Cycle

The impact of CASE tools on the SLC is not as straightforward as you might expecrt. In general, once organizational use of CASE tools has matured, the development time is greatly reduced. However, some CASE requirements may actually lengthen certain life cycle phases.

One aspect of development that CASE tools encourage is an iteration of phases rather than a straight, sequential completion of one phase before entering the next. This iterative cycling may increase the length of certain phases, which before CASE may have been shorter. This is particularly true when end users are involved in the iterative cycling in order to enhance requirements definition or design specifications. Upper-CASE tools also require an emphasis on the early phases of development that an organization may have previously neglected. This emphasis will likely result in better systems but might also serve to lengthen the life cycle.

In spite of these influences that add to development time, most of the phases are shortened by the use of CASE tools. Once the analysis and design phases have been completed, the code generation can be done with a touch of a button. The time required for program testing, which is still required to ensure bug-free performance, is also reduced.

Documentation time is reduced as well. In general, documentation is automatically done when the phases of the CASE tool are completed. In addition, the documentation is automatically updated when changes are made to the system. Therefore, documentation is not only easier to complete, it is current with the system functions.

Maintenance is greatly enhanced by the use of a CASE tool. Maintenance is not performed by changing the existing code, as in traditional systems. Rather, it is accomplished by changing the specifications, which then allow code to be regenerated. In this way, maintenance is easy to perform and takes little effort. In addition, the code that is produced is structured, which means that it does not impact existing modules. The modules feature loose coupling and high levels of cohesion within each module. **Cohesion** is achieved when modules perform only single tasks.

Figure 8.23 illustrates the impact of CASE on a life cycle. The ovals represent iteration. You can see that CASE has its greatest positive impact on the duration of the later phases.

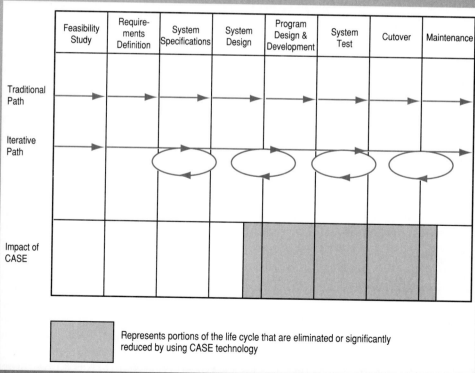

Figure 8.23
CASE can lengthen early life cycle phases and shorten later ones.

Summary

The evolutionary process that an information system follows is called the system life cycle, or SLC. The SLC consists of five phases—planning, analysis, design, implementation, and use. The first four phases constitute the system development life cycle (SDLC). When a system reaches the end of its use phase, it can be scrapped or it can trigger the planning phase of a replacement system.

In the planning phase, the user recognizes the problem and solicits the systems analyst's assistance in defining the problem and exploring the feasibility of a computer-based solution. The MIS steering committee and the user review the systems analyst's proposal, recommending a system study. The proposal gives management an opportunity to make a go-no go decision that relates to only the next phase, an approach called creeping commitment. A go decision causes a project control mechanism to be established that will enable the MIS steering committee, the user, and the project leader to control the development activities.

During the analysis phase, the analyst and user continue to jointly explore the user's information needs. This important step is followed by a statement of the performance criteria, the level of system performance that the user will require. This phase concludes with the preparation of another proposal to proceed, followed by a second go-no go decision.

The design phase begins with the important task of designing the new system. Then, it vividly reflects the underlying influence of the systems approach by identifying and evaluating alternate system configurations, and selecting the best. Again, a proposal determines whether the SLC continues to the implementation phase.

The implementation phase begins with planning how the various tasks will be performed in parallel by the increased number of participants. A key step in this phase involves the acquisition of the equipment that will be used to carry out the design. Hardware vendors respond to RFPs with proposals, and orders are placed with those vendors who make the most attractive offers. If the make-or-buy decision indicates *make*, the choice is whether to develop the code using programming languages or a code generator. If the decision indicates *buy*, the same process that was used for hardware selection can be used to solicit software proposals. The final tasks before cutover include preparing the database, preparing the physical facilities, and educating users and participants.

The use phase consists of use of the system by the user, assisted by information specialists who perform audits and systems maintenance.

Because the SLC is based on the proven logic of the systems approach, the basic tasks have changed little over the years. However, recent refinements have come in the form of structured development methodologies, prototyping, business process redesign, and rapid application development.

The structured methodologies, patterned after structured programming, take a top-down approach that ensures the integration of all of the subsystems. The analysis, design, and implementation tasks are performed in parallel, reducing the length of the project.

Two kinds of prototypes exist. Type I prototypes eventually become operational systems. Type II prototypes are throwaway models that provide the basis for the development of operational systems. Prototyping can replace the analysis, design, and implementation phases for small projects, and can be fitted into the SLC for large projects. On the whole, prototyping is a prescribed system development methodology, but it has its disadvantages that accrue mostly from lack of control. Good prototyping prospects involve high-risk and innovative applications, interaction between the system and many of users, and the need for quick delivery.

The repetition of the SLC can fit within a larger scheme of business process redesign in which the firm seeks to improve the basic ways it does business. When BPR affects information technology, three developmental approaches—reverse engineering, restructuring, and reengineering—can be considered. Only reengineering changes the functionality of the system. The three Rs are made possible by BPR software.

The need to redevelop existing systems and the need to respond to users' needs for new systems have forced information specialists to seek ways to develop systems quickly, with the least effort possible. James Martin devised rapid application development (RAD), a subset of information engineering, to cope with the mounting backlog of systems work. RAD consists of a four-phase life cycle, heavy user involvement, and SWAT teams.

The systems approach, the SLC, structured development, prototyping, the three Rs, and RAD are examples of methodologies. An understanding of these methodologies is essential for anyone who develops computer-based systems.

Key Terms

Internal constraint

Environmental constraint

Feasibility study

Go–no go decision point

Project control mechanism

System integrity

Project dictionary

Repository

Performance criteria

Request for proposal (RFP)

Make-or-buy decision

Cutover, conversion

Postimplementation review

Annual audit
Systems maintenance
Legacy system
Structured systems analysis
Structured systems design
Prototype
Type I prototype
Type II prototype
Integrated application generator
Prototyping toolkit
Business process redesign (BPR)
Functionality
Forward engineering
SWAT team

Key Concepts

The system life cycle (SLC)
The system development life cycle (SDLC)
How management commits resources to the SLC one phase at a time
The repeating nature of the SLC
The bottom-up nature of the classical life cycle approach
The structured approach to systems development
Prototyping
The three Rs as business process redesign methodologies
Information engineering (IE)
Rapid application development (RAD)

Questions

1. Why divide the system life cycle into phases and steps?
2. What is the purpose of the planning phase? Of the analysis phase, the design phase, the implementation phase, and the use phase?
3. Why does an information specialist seldom identify a problem signal?
4. What are the two types of constraints? What do they have in common?
5. How do a feasibility study and a system study differ?

6. Which SLC phases include a formal, written proposal to proceed to the next phase?

7. What is meant by the term *creeping commitment?*

8. What does a project control mechanism compare? Give three examples of such mechanisms.

9. Which SLC phases are introduced with an announcement?

10. Which members of the project team are not members of information services?

11. Where is system documentation maintained?

12. What is a repository?

13. How do information needs and performance criteria differ?

14. What are the two characteristics of the best equipment configuration?

15. What two conditions does the MIS steering committee expect the implementation proposal to meet before they authorize proceeding to the implementation phase?

16. What decision remains if a firm decides to *make* its software?

17. What three factors influence the difficulty of preparing the database?

18. Which information services manager usually sees to it that the physical facilities are ready?

19. Who is a participant?

20. What does cutover signal the end of?

21. How do a postimplementation review and an annual audit differ?

22. What does a systems analyst look for during a postimplementation review? What about an internal auditor?

23. What are the three types of systems maintenance?

24. What is a legacy system?

25. Do you agree that "old systems never die; they are just reborn"? Explain your answer.

26. What is a serious shortcoming of the bottom-up approach to system development?

27. Why can a project be completed sooner when a structured approach, rather than a bottom-up approach, is taken?

28. Which type of prototype does not have to be maintained? Why?

29. Which SLC phases can be replaced on small projects by prototyping?

30. What is the purpose of reverse engineering? Of restructuring? Of reengineering?

31. What does program-analyzing software do?

32. What are the four layers of the information engineering pyramid? On which layer would enterprise planning be performed? On which layer would a code generator be used?

33. How does a SWAT team differ from a project team?

Topics for Discussion

1. Give an example of an environmental constraint that might be imposed by suppliers. By customers. By competitors.

2. Shouldn't the implementation and use phases end with formal proposals?

3. How could an MIS steering committee ensure that prototyping projects do not skip over problem definition, alternative evaluation, and documentation?

4. A set of house blueprints is an example of a methodology, and a carpenter uses various tools in applying that methodology. What are some other examples of everyday methodologies and corresponding tools?

Problems

1. Draw a flowchart segment that shows the logic performed by the MIS steering committee at each go-no go decision point. Give the committee the ability to request more information at each decision point. Use the same symbols as the flowchart in Figure 8.16. The rectangle represents an action. The diamond represents a decision.

2. Assume that you use your PC primarily for word processing and have decided to replace your dot matrix printer with a laser printer. Do library research and identify three manufacturers of laser printers. Prepare a one-page letter that can be sent to each, serving as an RFP. You are interested primarily in the cost of the printer, the cost of replacement toner cartridges, and the terms of the guarantee. Make copies of the three letters, each with the manufacturer's name and address.

3. Go to the library and do research on system methodologies. Write a three-page report summarizing your findings. Cite your references. The references in this chapter might provide some good starting points.

Consulting with

Harcourt Brace

(Use the Harcourt Brace scenario prior to Chapter 7 in solving this problem.)

Pretend you are Bill Presby, the associate director of sales and marketing systems. Mike Byrnes, the CIO, has asked for your recommendations for reducing communications difficulties with users. The addition of specialists such as Ira Lerner to the MIS staff has worked so well that Mike is interested in other options. Prepare a one-page memo stating your recommendations, and briefly explain each.

Case Problem
Splashdown (A)

You have just been hired as a systems analyst by Splashdown, a large water park near NASA, on the outskirts of Houston. On your first day, the systems analyst manager, Mildred Wiggins, explains that an important decision in a system life cycle is whether to use prototyping. Job requests come in from users every day, and the prototyping decision must be made for each request. Mildred thinks that a printed form would make the decision easier. Because you are an expert in prototyping, she asks you to design the form. The form should list the factors that go into the decision and include a five-point scale for each factor. For example, assume that one factor is the extent to which the application involves an online dialog between the user and the computer. The factor and its scale might appear as:

1. The amount of online dialog between the user and the computer is:

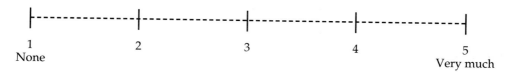

```
1                 2                 3                 4                 5
None                                                           Very much
```

The scale points have values ranging from 1 to 5. The systems analyst who must make the prototyping decision will circle the scale points that best describe a potential application. The scale points for all the factors will be added together. Design the form so that a high total indicates a good prototyping prospect.

After you design the form, Mildred wants you to test it on three prospective applications. In interviewing the people who submitted the job requests, you obtain the following information on the three applications.

Cash flow model—The controller wants to implement an electronic spreadsheet on his micro to show the flow of money into and out of the firm. This is a common application, with all firms doing the job a standard way. Currently, the controller performs the calculations using pencil and paper. However, he is in no big hurry to make the switch. He feels comfortable with the manual system, and nobody else is involved.

Sales commission system—The sales manager wants to implement a system that computes salespersons' commissions according to a new formula he heard about on his last trip to California. The formula takes into account each salesperson's zodiac sign. The new system must be approved by all six marketing managers at headquarters. The mainframe computer program will make the computations in a batch manner, and print the results as a monthly sales commission report. The sales manager is eager to get the system going because he was recently transferred from shipping and has no experience in computing sales commissions. Although commission accounting is usually cut and dried, this approach contains some gray areas that have not yet been explored.

Job openings system—The director of human resources wants to create a system that employees can use from their terminals to retrieve lists of current job openings in the company. The system will be strictly a database retrieval application—no computations. Most of the firm's 1,200 employees use a database in their daily work. The director wants the system in place within 30 days, before negotiations on the new union contract begin.

Assignments

1. Design the form that Mildred has requested. Use your word processor or similar software. Be creative in how you lay out the form. Print a blank copy.

2. Make three additional copies of your form, and complete them using the findings of your data gathering.

3. Prepare a memo to Mildred, advising her of your forms design efforts and your recommendations. Do you recommend prototyping? If so, how strong is your recommendation? Attach the blank form, followed by the three completed forms.

Selected Bibliography

Alavi, Maryam. "An Assessment of the Prototyping Approach to Information Systems Development." *Communications of the ACM* 27 (June 1984): 556–563.

Barlow, John F. "Putting Information Systems Planning Methodologies Into Perspective." *Journal of Systems Management* 41 (July 1990): 6ff.

Bersoff, Edward H., and Davis, Alan M. "Impacts of Life Cycle Models on Software Configuration Management." *Communications of the ACM* 34 (August 1991): 104–118.

Cerveny, Robert P.; Garrity, Edward J.; and Sanders, G. Lawrence. "The Application of Prototyping to Systems Development: A Rationale and Model." *Journal of Management Information Systems* 3 (Fall 1986): 52–62.

El Louadi, Mohamed; Pollalis, Yannis A.; and Teng, James T. C. "Selecting a Systems Development Methodology: A Contingency Framework." *Information Resources Management Journal* 4 (Winter 1991): 11–19.

Ewusi-Mensah, Kweku, and Przasnyski, Zbigniew H. "On Information Systems Project Abandonment: An Exploratory Study of Organizational Practices." *MIS Quarterly* 15 (March 1991): 67–86.

Foss, W. Burry. "Early Wins Are Key to System Success." *Datamation* 36 (January 15, 1990): 79–82.

Gavurin, Stuart L. "Where Does Prototyping Fit In IS Development?" *Journal of Systems Management* 42 (February 1991): 13–17.

Hammer, Michael. "Reengineering Work: Don't Automate, Obliterate." *Harvard Business Review* 68 (July–August 1990): 104–112.

Huff, Sid L. "Information Systems Maintenance." *Business Quarterly* 55 (Autumn 1990): 30–32.

Kerr, James M. "The Information Engineering Paradigm." *Journal of Systems Management* 42 (April 1991): 28ff.

Li, Eldon Y. "Software Testing In a System Development Process: A Life Cycle Perspective." *Journal of Systems Management* 41 (August 1990): 23–31.

Mahmood, Mo Adam. "Information Systems Implementation Success: A Causal Analysis Using the Linear Structural Relations Model." *Information Resources Management Journal* 3 (Fall 1990): 2–14.

Markus, M. Lynne. "Power, Politics, and MIS Implementation." *Communications of the ACM* 26 (June 1983): 430–444.

Mashaw, Bijan, and Casey, Salley D. "An Assessment of the Phases of the System Life Cycle." *Proceedings 1990 Annual Meeting Decision Sciences Institute,* coordinated by Betty Whitten and James Gilbert. (San Diego: 1990.) 1026–1029.

Moad, Jeff. "Maintaining the Competitive Edge." *Datamation* 36 (February 15, 1990): 61ff.

Naumann, Justus D., and Jenkins, A. Milton. "Prototyping: The New Paradigm for Systems Development." *MIS Quarterly* 6 (September 1982): 29–44.

Palvia, Prashant, and Nosek, John T. "An Empirical Evaluation of System Development Methodologies." *Information Resources Management Journal* 3 (Summer 1990): 23–32.

Srinivasan, Ananth, and Kaiser, Kate M. "Relationships between Selected Organizational Factors and Systems Development." *Communications of the ACM* 30 (June 1987): 556–562.

Swift, Michael K. "Prototyping in IS Design and Development." *Journal of Systems Management* 40 (July 1989): 14–20.

Willis, T. Hillman, and Tesch, Debbie B. "An Assessment of Systems Development Methodologies." *Journal of Information Technology Management* 2 (Number 2, 1991): 39-45.

TM
C

The Data Dictionary

In the early years of computing, information specialists designed systems to meet the unique needs of certain organizational units or individuals. If they needed data that was not already in the computer, they created new files. They created these files without paying much attention to the overall needs of the firm. Often, they produced similar or identical files.

Besides being an unnecessary cost, the data redundancy presented problems when information produced from the multiple files did not match. For example, if a sales analysis from marketing's sales data did not match a sales analysis from manufacturing's data, whose data did you believe?

In the early 1970s, this situation stimulated interest in the **database concept,** the idea of a central data repository. Made available to users throughout the firm, a data repository could keep redundancy to a minimum. The concept produced software systems called database management systems (DBMSs) and something called the data dictionary. The DBMS would manage the firm's central data resource. The data dictionary would provide a common basis for all the firm's database development activities.

What Is a Data Dictionary?

A data dictionary is a written description of a firm's computer-based data. It is not the data itself. The first data dictionaries were kept in three-ring binders, with each page describing a data element that was maintained in secondary storage. For example, there might be a page for *Employee number* and another page for *Employee name.*

Information specialists in the information services unit agreed on the characteristics of each data element: the data name to be used in programs, the type of data (alphabetic, numeric, or alphanumeric), the size of the field, the number of decimal positions, and so on. These characteristics were recorded in the data dictionary, which served as a reference guide for all the information specialists as new systems were being developed.[1]

---

[1]Before these data element standards were implemented, the situation could be chaotic. The author recalls one programmer who used names of cigars, such as El Roi Tan and Dutch Master. Another programmer could look at one of his programs and not have the slightest idea of what it did.

TM

C

Data dictionaries became so popular that information specialists soon converted the pages of the three-ring binders to computer storage media. This conversion was facilitated by the development of a **data dictionary system,** or **DDS.** The DDS was software that maintained the data descriptions in secondary storage. Some DDSs were standalone systems. Others were contained in the DBMS.

Today, a firm's data dictionary is maintained in the computer and used by information specialists and users throughout the firm. The data dictionary represents the detailed documentation of the firm's data resource. The data dictionary can supplement the entity-relationship diagram, which provides the summary picture.

The Hierarchical Nature of Data Dictionary Documentation

The data dictionary contains descriptions of the different levels in the data hierarchy. Business data has always been organized in the following hierarchy:

- Files
  - Records in files
    - Data elements in records

For example, a payroll file contains a record for each employee, and each record contains multiple data elements, such as employee number and pay rate. The database comprises all the files.

The data dictionary describes the data in this same hierarchical manner, using a special form called a **data dictionary entry form** for each level. Three types of entry forms exist: the data store dictionary entry, the data structure dictionary entry, and the data element dictionary entry.

Data Store Dictionary Entry

The **data store dictionary entry** describes one or more files that constitute a data store. You can see the relationship between the data dictionary and a DFD. The highest level of data documented by the data dictionary is the data store. The data store dictionary entry form describes a data store in a DFD. Figure C.1 illustrates the data store dictionary entry form.

The *Data Store Name* identifies the data store in the DFD. The *Description* briefly describes the data store. The *Data Structures* area lists the data structures that exist in the store. In most cases, a store contains a single structure. For example, a Payroll master file contains a single type of record—a Payroll master record. The *Volume* area provides an idea of the size of the data store, in terms of the number of records. The *Activity* area explains how often the data in the store is used.

TM
C

DATA STORE DICTIONARY ENTRY

Use: To describe each unique data store on a data flow diagram.

DATA STORE NAME: *Inventory master file*

DESCRIPTION: *The file of records that represents the firm's products*

DATA STRUCTURES: *Inventory master record*

VOLUME (NUMBER OF RECORDS):   *Approximately 20,000*

ACTIVITY: *Approximately 20 percent of the file is active daily.*

Figure C.1
The data store dictionary entry.

Data Structure Dictionary Entry

The **data structure dictionary entry** describes one or more data structures that exist in a data store. Figure C.2 illustrates the data structure dictionary entry form.

DATA STRUCTURE DICTIONARY ENTRY

Use: To describe each unique data structure that exists in the data flows and
                the data stores.

STRUCTURE NAME: *Inventory master record*

DESCRIPTION: *The record that describes an item in the
                      Inventory master file*

DATA ELEMENTS: *Item number*
                      Item class
                      Item description
                      Warehouse location
                      Unit of measure
                      Unit cost
                      Unit price
                      Reorder point
                      Economic order quantity
                      Annual sales units
                      Quantity on hand
                      Quantity on order
                      Backorder quantity

(Continued on Back)

Figure C.2
The data structure dictionary entry.

TM

C

The *Structure Name* is the same as that used in the data store dictionary entry. A brief *Description* of the data structure is provided. The *Data Elements* area lists all the data elements found in the structure. For lengthy structures, the data element listing is continued on the back.

Data Element Dictionary Entry

The **data element dictionary entry** describes one data element that exists in a data structure. Figure C.3 illustrates the data element dictionary entry form.

The *Data Element Name* is the same as that used in the data structure dictionary entry. A brief *Description* of the data element is provided, and the *Type of Data* is specified, along with the *Number of Positions*. For numeric data, the *Number of Decimal Places* is specified.

If a data element is known by several names, or *Aliases,* those names are provided. Examples of data in terms of the *Range of Values*, a *Typical Value,* and *Specific Values* are given when they are appropriate. As an example, a range of values for Employee hourly payroll rate might be $5.00 through $35.00, a typical value of the Number of weekly hours worked might be 40.0, and a specific value for Payroll code might be 3, which designates an hourly employee. The range and the specific values help programmers when they code routines to edit the data for accuracy. Additional editing features are noted under *Other Editing Details.*

These data dictionary entry forms are only one way to document data. Other formats are possible. As an example, a firm might assign certain persons as data managers, with responsibility for certain data elements. This technique is popular with executive information systems. In such a case, the data element dictionary entry would identify the appropriate data manager. In other situations, entries in the entry forms will relate to system security. For example, a data store dictionary entry form can identify who has access to the data.

Data Flow Dictionary Entry

The above three dictionary entries (store, structure, and element) all relate to "data at rest." When the dictionary is used with a DFD, a special entry can be used for the data flows. A sample of a data flow dictionary entry appears in Figure C.4.

The data flow dictionary entry is on the same level in the data hierarchy as the data store dictionary entry. This means that a data flow is documented with three entry forms: (1) a data flow dictionary entry, (2) usually one data structure dictionary entry, and (3) multiple data element dictionary entries. Think of a data flow as a file of data "on the move."

DATA ELEMENT DICTIONARY ENTRY

Use: To describe each data element contained in a data structure.

DATA ELEMENT NAME: *Reorder point*

DESCRIPTION: *The quantity of an item in inventory that triggers a purchase of replenishment stock*

TYPE OF DATA: *Numeric*

NUMBER OF POSITIONS: *7*

NUMBER OF DECIMAL PLACES: *None*

ALIASES: *Reorder quantity*
                        Order point

RANGE OF VALUE: *1 to 1,500,000*

TYPICAL VALUE: *215*

SPECIFIC VALUE: *None*

OTHER EDITING
DETAILS: *Must be a positive number*

TM

C

Figure C.3
The data element dictionary entry.

DATA FLOW DICTIONARY ENTRY

Use: To describe each data flow on a data flow diagram.

SYSTEM NAME: *Distribution system*

DATA FLOW NAME: *Update item records*

DESCRIPTION: *Inventory records that have been updated*
                     to reflect the day's sales activity

FROM: *1. 2. 2 Check the reorder point*

TO: *Inventory file*

DATA STRUCTURES: *Inventory master record*

Figure C.4
The data flow dictionary entry.

Putting the Data Dictionary in Perspective

Although many documentation tools can be used in documenting processes, only one can document data in a detailed way. That is the data dictionary. If a firm wants to document its data, it uses a data dictionary.

The data dictionary can be used alone or with an entity-relationship diagram. Ideally, the entity-relationship diagram is prepared first and the data dictionary provides the detailed support. This is a top-down approach to documenting the firm's data resource.

TM

Problems

(Note: Additional data dictionary problems can be found at the end of Chapter 10.)

1. Use a data store dictionary entry form to document a Customer master file. The data structure is named the Customer master record, and there are approximately 450 records in the file. However, only approximately 20 percent of the file is active.

2. Now use a data structure dictionary entry form to document the Customer master record identified in Problem 1. This record contains the following data elements: Customer number, Customer name, Customer Address, Customer phone number, Territory number, Credit rating, and Customer class.

3. Document the following data elements of the Customer master record with data element dictionary entry forms: Customer number, Customer phone number (including area code), Territory number, Credit rating, and Customer class. The Customer number begins with two alphabetic letters and is followed by six decimal positions. An example of Customer number is AC230711. The firm has 18 sales territories, with numbers ranging from 01 through 99. Customers are classified in terms of three Credit ratings: A = Excellent; B = Good; C = Poor. There are two Customer classes: 1 = High sales history; and 2 = Low sales history.

4. Use a data flow dictionary entry form to document a data flow named Sales order. There is only a single data structure, called the Sales order record. The data flow consists of all data elements necessary to process a customer's sales order: Customer number, Customer order number, Customer order date, Item number, Item quantity, and Item unit price.

5. Use a data structure dictionary entry form to document the Sales order data flow.

6. Use data element dictionary entry forms to document the following data elements in the Sales order record: Customer order date, Item number, Item quantity, and Item unit price. Items are numbered 000001 through 999999. Item quantities can range from 1 through 999. Unit prices can be as low as $0.25 and as high as $12,500.00.

Systems Analysis

An information system evolves in a life cycle. A simple system can require weeks to develop, and projects that span months or even years are not uncommon. And this includes only the time it takes to develop the system. Once it is implemented, the users hope it enjoys a long and happy life.

In this part of the text we begin an expansion of our earlier descriptions of this system life cycle. Although we assume that the system is computer based, the same general process can be followed for systems of all kinds. Another important point is that all the activities can be performed by users when they engage in end-user computing.

This part of the text is devoted to the systems analysis that is so important to systems development. Here is where you will see how the tools of modern systems work are applied in studying and understanding existing systems for the purpose of developing new or improved systems. You will use these tools throughout your career in information systems or management.

Chapter 9 describes the planning that must precede the developmental work and discusses the way management establishes and maintains control over the total effort. Chapter 10 is concerned with systems analysis. The next part of the text, Part Five, will continue the discussion of important life-cycle activities by explaining systems design.

As time goes by, some of the material in this and the next part will no doubt become obsolete as new computer-based methodologies and tools are made available. Your task will be to remain up to date on the newer material. But that is the major challenge of any career.

V—CONTROLLING THE COPS PROJECT

The COPS project is of such importance to Harcourt Brace that it is controlled at both the executive and operational level.

Executive Control

At the executive level, control comes in the form of the COPS review committee, whose members include:

Michael Banks, Vice President and Controller
Mike Byrnes, Vice-President of MIS
John Misiura, Vice-President of Distribution

The composition of this committee reflects the importance of the COPS project to Harcourt Brace. All members are executives, and all will be directly affected by COPS.

The committee plans the work to be done and then reviews the progress. Most of the planning and reviewing occur during meetings held every two or three weeks.

Control Tactics

When the April 1991 cutover date was missed, the committee decided to exert stronger control to ensure that another postponement did not occur. Two tactics were devised. First, it was decided to identify those new system features that, in the minds of the users, were *absolute requirements* in the new system. To make this determination, the MIS staff met with users, and the two parties agreed on a list of 190 items. Some items were minor (such as redesigning screen formats); however, some represented substantial work that would require several months to accomplish.

Once the COPS items list had been prepared, the second tactic was implemented. The committee decided to *freeze* all system changes not on the list. If users identified any additional features they would like in COPS, they would have to wait until the COPS cutover was complete. Features beyond the 190 items would represent the next generation of COPS—COPS2.

Operational Control

Mike Byrnes has the overall responsibility for seeing that COPS is implemented—as designed, and on schedule. Mike looks to his entire staff for support, but primarily relies on Bill Presby, the manager in charge of COPS development. Bill, in turn, relies on the managers of each of the four COPS groups for support.

To stay on top of the situation, Mike uses a software system called NETMAN for project control. NETMAN, a product of Computer Associates of Garden City, New York, produces a series of reports that permit a quick review of the status of the 190 items. Figure HB5.1 is a sample NETMAN printout.

```
COPNETR1
JOB:NETUSRNP                    HARCOURT BRACE JOVANOVICH INC.                    DATE  06/11/93   PAGE   1
SYSIN:NET001DY                  COPS PRIORITY TRACKING REPORT BY USER
                                             COLE

MIS    POINT     USER     CONTROL            DATE        DATE MIS      DATE USER
RESP   CATEGORY  PRIORITY NUMBER             OCCURRED    RESOLVED      CONFIRMED

LERNER COPS2     02       P006253            08/05/92    10/28/92

            SERIES CREDITING
            SERIES ORIGINATED BY A SALESREP MUST HAVE AN INDICATOR WHICH
            WOULD GIVE CREDIT TO THE SALESREP FOR SALES. APPEARS NOT TO BE
            WORKING CORRECTLY. SEE POINT FOR MORE INCLUDING PRINTOUT NEEDS...
========================================================================================
MIS    POINT     USER     CONTROL            DATE        DATE MIS      DATE USER
RESP   CATEGORY  PRIORITY NUMBER             OCCURRED    RESOLVED      CONFIRMED

LERNER COPS2     10       P007730            10/16/92    12/17/92

            CHANGE ORDER OF OUTLET SALES CALC.
            S COLE WOULD LIKE TO CHANGE THE SALES OUTLET CALCULATIONS.  SHE
            WANTS TO MOVE UP THE SALES CREDIT INDICATORS BEFORE CUSTOMER
            DISCOUNT
========================================================================================
MIS    POINT     USER     CONTROL            DATE        DATE MIS      DATE USER
RESP   CATEGORY  PRIORITY NUMBER             OCCURRED    RESOLVED      CONFIRMED

LERNER COPS1     07       P008542            11/19/92    04/22/93      04/22/93

            HOSPITAL ORDERS
            PROBLEM WITH ASSIGNING HOSPITAL ORDERS OF MULTIPLE LINE ITEMS.
            CRITERIA IS FOR THE ENTIRE ORDER TO GO TO OUTLET 03 WHEN THERE IS
            A QUANTITY OF MORE THAN ONE ON ORDER. SOME LINE ITEM WENT TO 02..
========================================================================================
                                    2    TOTAL ACTIVE
                                    1    TOTAL MIS RESOLVED
                                    3    TOTAL USER CONFIRMED
                                         TOTAL POINTS
```

Figure HB5.1
A sample output of the NETMAN project control system.

The report is prepared weekly, but Mike can request a special printing at any time. Mike shares the report information with both the COPS review committee and his staff. The sequences of the information and the report format enable Harcourt Brace management to track COPS progress by user and by the COPS group. NETMAN is a good example of prewritten project management software.

Conduct of a COPS Review Committee Meeting

A COPS review committee meeting begins by first identifying the items scheduled for work since the group last assembled. These items take the highest priority during the meeting. Mike Byrnes then reports on the progress of each item, often using some type of documentation as a visual aid. Sometimes the documentation consists of copies of the NETMAN report. At other times, Mike uses large wall charts.

For each high-priority item, Mike identifies whether it has been completed, is in process, or is awaiting work. Before an item is officially considered complete, the user(s) must acknowledge that the work is satisfactory.

After the high-priority items have been discussed, the meeting is thrown open to the discussion of other issues. When that discussion winds down, the committee identifies the work to be done during the coming week or two. This information usually comes from Michael Banks and John Misiura, who represent the user areas most affected by COPS. Very often, Michael and John will invite personnel from their areas to participate in the meeting. Mike Byrnes does the same, often inviting members of his MIS staff to report on topics that require their expertise. After agreeing on the next set of priority items, principal committee members set the date and time of the next meeting.

Project Planning and Control

Learning Objectives

After studying this chapter, you should:

- Know more about how a user identifies a problem and how the systems analyst works with the user in defining the problem
- Know more about setting system objectives and identifying system constraints
- Understand the multiple dimensions of system and project feasibility
- Be familiar with a process for incorporating ethical considerations into systems designs
- Know how to set the date for cutting over to the new system
- Be aware of some tips to follow in writing systems proposals
- Know more about the control mechanism that management uses to control systems projects

Introduction

A computer-based system project usually begins when a manager recognizes a problem in a user area of the firm. A systems analyst then helps the user define the problem by following a symptom chain until the cause is identified.

Once the user and the analyst agree that a problem indeed exists and that the computer might play a role in its solution, they continue to work together in setting system objectives and identifying constraints.

The systems analyst then conducts a feasibility study to confirm that a computer-based solution is appropriate. In conducting the study, the analyst considers six dimensions of feasibility—technical, economic, noneconomic, legal and ethical, operational, and schedule.

The systems analyst uses the information gained from the user and the feasibility study to prepare a proposal that recommends either a system study or project termination. When the MIS steering committee and user grant authorization to proceed, a mechanism is put in place that aids project planning and control.

This chapter describes the activities performed before systems analysis begins—the activities of the planning phase of the system life cycle.

Using CASE to Plan and Control Projects

I-CASE tools not only support the entire SLC but also provide a project-management capability. I-CASE tools are often referred to as Integrated Project Support Environments, or IPSE.

All projects,. regardless of their intended purpose, have the following characteristics:
- They are unique within an organization and have defined start and end points.
- They have a work scope that can be decomposed into definable tasks.
- They have a budget.
- They require the use of resources.
- They often cross organizational boundaries.

Projects require management of scope, time, human resources, costs, quality, and communications. Over the years, various tools, such as the following, have been developed to help project managers satisfy these requirements:
- Budgets
- Work breakdown structures
- Network diagrams (CPM and PERT) and Gantt charts
- Performance tracking graphics (resource histogram)
- Forecasting, analysis, and corrective action routines

It is important to track actual performance against planned performance so that the organization can learn from doing. This constant striving to improve the final product is a basic characteristic of TQM (total quality management). Future systems projects should benefit from past experience by increased accuracy of estimations and forecasts.

If the CASE tool that a firm is using does not contain project management tools, it is possible to purchase additional software that aids in the management task. This approach loses the feature of integration but is much more productive than foregoing the use of software-based management tools altogether.

The Planning Phase

The purpose of the planning phase is to define the user's problem, decide whether a computer-based solution is feasible, and decide whether to conduct a system study. The planning phase was described in general terms in Chapter 8, and the model is reproduced in Figure 9.1. This chapter discusses each step in greater detail, concentrating on the tools that the user and the systems analyst employ in planning and controlling a computer-based system project.

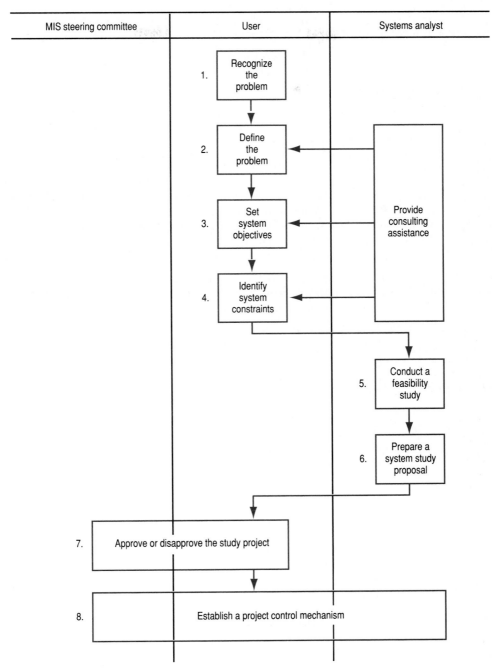

Figure 9.1
The planning phase.

Step 1—Recognize the Problem

Anyone inside or outside the firm can recognize a problem, but a manager in a user area is the most likely to do so. The manager is responsible for everything that happens in his or her unit and is sensitive to all possible influences—whether good or bad.

Problem-Sensing Styles

We typically think of managers as always being alert to problems or potential problems. Yet this is not always the case. Managers differ in how they respond to problem signals. A **problem signal** is an indication that a problem exists or is about to occur. When managers spot one of these indicators, they can react in various ways. The term **problem-sensing style** describes the way a person responds to problem signals. There are three such styles—problem seeker, problem solver, and problem avoider.[1]

- A *problem seeker* is constantly on the alert for problems to solve. This manager assigns a high priority to implementing information systems designed to provide the problem signals.

- A *problem solver* does not back away from problems once they are signaled but makes no special effort to seek them out. This manager implements problem-signaling systems but does not give them a top priority.

- A *problem avoider* makes a special effort to *not* get involved in problem solving, and one way to do so is to not implement problem-signaling systems.

The systems analyst must be aware of these styles because they influence the role of the user in the project. Problem seekers are likely to be enthusiastic and cooperative and, in many cases, to assume a position of project leadership. Problem avoiders, on the other hand, are likely to drag their feet every step of the way.

Step 2—Define the Problem

Although managers devote much time to problem solving, that is only one part of their job. Systems analysts, on the other hand, are professional problem solvers. That is their full-time job. For this reason, managers often call on systems analysts for help when it appears that a problem exists and that its solution could involve the computer.

Distinguishing the Symptoms from the Cause

The systems analyst's first task is to help the manager define the problem. This involves following the symptom chain discussed in Chapter 7 and illustrated

[1]Andrew D. Szilagyi, Jr. *Management and Performance* (Santa Monica, CA: Goodyear, 1981):220–225.

with the flowchart in Figure 7.3. Following the chain leads to the problem cause.

In identifying the problem cause, the analyst and the user should determine whether it originates in the firm's environment or inside the firm. One or more environmental elements can be the cause, or the cause can exist in one of the elements of the firm as a system. The model in Figure 9.2 can be used as a guide in making this determination.

An Information Processor Problem Triggers a System Life Cycle

The systems analyst is most concerned with problems in the conceptual system—the standards, management, and the information processor.

- **Standards Problems** When the standards are inadequate, a project is undertaken to set things straight. The project typically consists of top-level managers

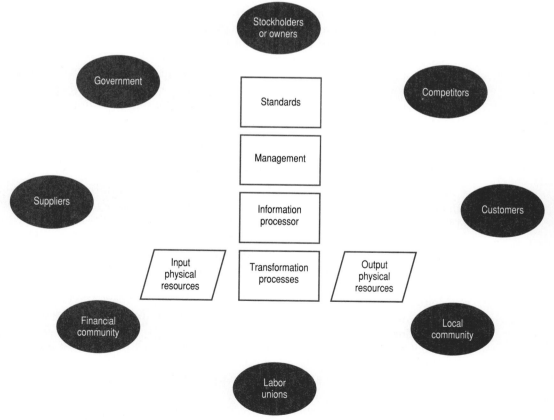

Figure 9.2
The problem cause model.

who study the situation and make recommendations to the executive committee.

- **Management Problems** When management problems exist, they are solved by recruiting new or additional managers, training, reorganization, or similar strategies.

- **Information Processor Problems** When the information processor is not providing the support that it should, a new system is in order. The new system is developed by following the SLC.

This special status of the information processor does not mean that the systems analyst does not contribute to solving problems dealing with the firm's standards or management—just that the solutions to those problems do not require an SLC.

The Problem Statement

The problem definition step concludes with a **problem statement,** one or more sentences that identify both the problem cause and its effect. The problem statement should begin with the words "The problem is," followed by the cause and effect. Table 9.1 gives four examples, for both good and bad problems with both internal and environmental causes.

Note that each problem statement identifies the cause with a particular element in the problem cause model. The system study during the analysis phase may change the problem location, but an effort should be made during the planning phase to define the problem as precisely as possible.

Step 3—Set System Objectives

A particular system might be expected to achieve any number of objectives. However, one guideline that you can use in identifying *types* of objectives is based on the four dimensions of information. Chapter 6 pointed out that information has value because of its (1) accuracy, (2) timeliness, (3) completeness, and (4) relevance. Because the purpose of the system will be to provide problem-solving information, the objectives may be stated in terms of these information dimensions.

The objectives depend on the problem cause. For example, if the cause of the problem is an information system that is not keeping the user current on the activity within the physical system, then the objective (or objectives) would deal with timeliness.

Table 9.2 provides an example of an objective for each information dimension. The objectives are not all stated in quantitative terms. Specific measures are desired, but those can come during the analysis phase, when performance criteria are defined.

The use of guidelines such as the information dimensions is an example of making good use of theory. Rather than wasting time exploring every possible avenue and still running the risk of missing important items, the systems analyst

Table 9.1   Examples of Problem Statements

Origin	Type of Problem	
	Good	Bad
Internal	The problem is that the new computer network makes it possible to distribute decision making through-out the organization to lower management levels.	The problem is that managers in the human resources unit do not have good systems skills, preventing them from assuming positions of leadership in systems development projects.
Environmental	The problem is that the currently low interest rates increase the attractiveness of the long-discussed plant expansion.	The problem is that our biggest competitor's price increase has caused our customers to switch to our products, making it diffi-cult for our order entry system to handle the larger order volume.

uses theory as a basis for a systematic search. This contribution of theory to systems work is an important point to keep in mind as we describe the SLC in this and the next several chapters. You will see many opportunities for applying the systems fundamentals covered earlier in the text.

Step 4—Identify System Constraints

Any number of system constraints can exist and they can originate either within the firm as a system or outside, in the firm's environment.

Internal System Constraints

Internal system constraints are limits on the performance of the system that are imposed by the firm's resources. A good way to come to grips with the internal system constraints is to think in terms of each of the resource types: personnel,

Table 9.2   Information Dimensions Can Provide a Basis for System Objectives

Dimension	Sample System Objective
Accuracy	The new distribution system will substantially reduce customer complaints arising from shipment of the wrong merchandise.
Timeliness	Warehouse managers will be notified not later than 8:30 A.M. of the volume of orders to be filled for the day.
Completeness	The program that determines whether to reject a customer's order because of poor credit will take into account, among other things, the customer's sales history for the past six months.
Relevance	The department head's report of excessive overtime hours for the week will include only those employees who regularly work overtime.

material, machines, money, data and information. The system will most likely be constrained in some way by a limitation on one or more of these resources. Table 9.3 lists examples for each of the six resource types.

Environmental System Constraints

Environmental system constraints are requirements imposed on the system by elements in the firm's environment. These constraints can be seen as restrictions on the basic system elements: input, processing, secondary storage, and output. Figure 9.3 shows these influences.

- **Constraints on Input Data** Environmental elements may be *unable* to provide the system with the needed input data. For example, a supermarket may not store its sales data in the form that a manufacturer can use to track its retail sales. Environmental elements also may *refuse* to provide the data even though they are capable of doing so. A good example is competitors who keep their planned price changes secret.

- **Constraints on Processing** Some environmental elements influence the way the firm processes its data. A classic example is the body of income tax regulations imposed by the government. Payroll programs must compute the tax amounts exactly as prescribed.

Table 9.3   The System Can Be Constrained by Limits on the Firm's Resources

Resource	Sample Constraint
Personnel	The information services unit shall have no more than three full-time employees.
Material	Acquisition of supplies such as paper and diskettes will not exceed $1,000 per month.
Machines	Workstations will be shared by a minimum of two users.
Money	The total annual budget for the computer operation is $375,000.
Data	Salespersons shall not spend more than 30 minutes per day entering sales order data.
Information	Preformatted screens for the executive information system will be updated at least daily.

- **Constraints on Secondary Storage** Environmentally imposed constraints on data storage occur mainly in the form of restrictions on databases. The federal government has passed legislation to protect a person's right to privacy, and these laws restrict the types of data maintained and how that data is used.

- **Constraints on Information Output** Because the firm's data processing system must provide environmental elements with information, those elements frequently specify the form the information must take. One example is when the firm wants to borrow money and the bank requires a financial statement. Another is when the firm wants to transmit its purchase orders electronically and the supplier insists that the data be in a particular format.

In these ways the environmental elements can exert an influence on each element of the firm's conceptual systems.

Putting the Constraints in Perspective

When considering the constraints that might influence the performance of the system, it is best to think in terms of resource flows. All the needed system resources may not be available within the firm. Also, the system may be forced to operate in certain ways because of the environmental influences on system data and information.

All these constraints might not be known during the planning phase, but the analyst and user do their best to estimate them. The estimate provides the initial boundaries within which the system must operate.

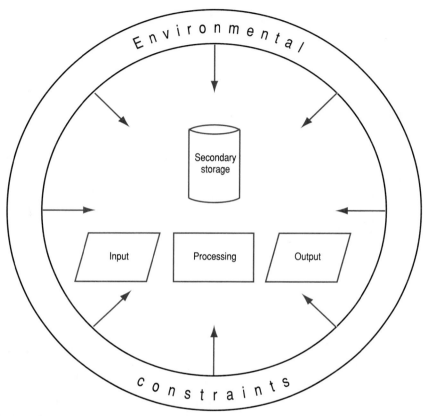

Figure 9.3
Environmental elements can affect any element of the computer-based system.

Step 5—Conduct a Feasibility Study

Thus far, the project has not been costly, requiring only a few hours of the user's and analyst's time. However, if the project continues, many more personnel will become involved, and the cost will rise. Before management commits to a large cash outlay, it wants to be assured that the project is sound. The systems analyst provides this assurance by conducting a feasibility study.

A **feasibility study** is an abbreviated form of systems analysis intended to determine whether a computer-based solution should be pursued. The systems analyst wants to learn only enough about such a solution to justify continuing; this can be accomplished by a few personal interviews and, perhaps, a search through some records. The interviews are limited to persons in the manager's area who are especially knowledgeable about the problem. The records search may involve digging through some of the unit's files, either to gather data that the interviews fail to yield or to substantiate the interview findings.

When conducting the feasibility study, the analyst is concerned with six types of feasibility. You can remember them by using the acronym TENLOS.[2] The letters stand for:

- Technical feasibility
- Economic feasibility
- Noneconomic feasibility
- Legal and ethical feasibility
- Operational feasibility
- Schedule feasibility

Technical Feasibility

During the early years of the computer, managers tended to be skeptical of its abilities to solve their problems. They were concerned about **technical feasibility**—the availability of computing hardware and software to do a particular job. In many cases the managers' concerns were well founded because the technology did not exist.

Hardware vendors were aware of these limitations and gradually closed the gaps in their product lines. Today, technology is available to do practically any job the user wants done. The only real question is whether the user can afford it.

The same applies to the other aspect of technical feasibility—software. Programming can enable the computer to do practically any job, but the question is whether management wants to incur the expense.

Economic Feasibility

The feasibility that has always been of most importance to management has been **economic feasibility**—the ability of a system to pay for itself in monetary terms. Management does not want to spend a dollar unless it receives more than a dollar in return.

The Cost-Benefit Analysis

To provide management with the economic justification it wants, the systems analyst conducts a cost-benefit analysis. A **cost-benefit analysis** is a study of the

[2]This acronym is based on one named TELOS, coined by John G. Burch, Jr., Felix R. Strater, and Gary Grudnitski, and described in their 1983 book, *Information Systems: Theory and Practice,* 3d ed.: John Wiley, 341–342.

costs and benefits that will be incurred in developing and using a computer-based system.

Developmental Costs and Benefits   The **developmental costs** are information resources costs (people, hardware, software, supplies, and facilities) incurred during the SDLC, and these costs can be substantial. **Developmental benefits,** on the other hand, are benefits received by virtue of the development project. These benefits are rare. Examples of developmental benefits include investment tax credits given for investing in new technology and money received through the sale of displaced equipment.

Operational Costs and Benefits   The **operational costs** are the costs of the information resources required to operate the system. These costs are especially visible in the firm's computing center. **Operational benefits** can be in the form of reduced costs or increased revenues made possible through system use. Operational benefits are the reason for developing the system, and the firm hopes that the benefits are substantial.

Before management approves a computer project, it must believe that the total cost of developing and using the new system will be less than the benefits the system provides. Otherwise, management will see no reason to embark on the project.

Figure 9.4
The firm's operational costs are concentrated in the computing center.

Cost Reduction versus Cost Avoidance

Early computer systems reflected a philosophy of **cost reduction,** the assumption that the computer is implemented for the purpose of reducing operating costs—primarily clerical costs. The flaw in this reasoning became obvious to management when it realized that the personnel whose work had been converted to the computer were being reassigned rather than terminated.

Although the computer did not meet expectations in terms of its impact on *current* costs, it did have an effect on *future* costs. By assuming the clerical workload, the computer delayed the need to hire more employees of the same type in the future. Costs were avoided rather than reduced—an economic justification called **cost avoidance.**

Cost avoidance, currently the prevalent view, reflects a more realistic attitude toward the impact of the computer than does cost reduction.

Methods for Conducting the Cost-Benefit Analysis

Three main approaches have been developed for conducting the cost-benefit analysis: break-even analysis, payback analysis, and net present value.

Break-even analysis compares the costs of the existing system with the costs of the new one and determines the point at which the costs are equal—the **break-even point.** Break-even analysis is appropriate when management fears that the firm might not be able to stand the costs of the new system.

Payback analysis compares the costs of the new system to its benefits and identifies the point at which the costs have been recovered—the **payback point.** Payback analysis recognizes that significant costs are incurred in developing a new system and that these costs postpone a realization of a return on the investment.

Net present value (NPV) is a technique appropriate for investments, such as computer projects, that produce benefits far in the future. NPV recognizes that the value of future benefits is less than current value. For example, one dollar five years from now will not be worth as much as one dollar is today. NPV makes it more difficult to economically justify computer projects because it discounts the future benefits.

These three approaches to economic justification are described in greater detail in Technical Module E.

The Current Emphasis on Increasing Revenues

Although management first viewed the computer in terms of its potential impact on costs, recent emphasis has been on revenues. Modern computer-based systems are expected to increase revenues in three basic ways—by reducing inventory investment, by increasing productivity, and by improving decision making. The first two ways can be measured in economic terms.

Reduced Inventory Investments   The single data processing subsystem that has the capability of contributing more than simply reduced costs is inventory. By using scientific inventory methods such as economic order quantities and statistically computed reorder points, a computer-based inventory system can reduce the amount of inventory needed without compromising the service level.

This reduction can exert a big influence on a firm's revenue. Firms typically have much of their assets tied up in their inventories. In fact, an investment of several million dollars is not uncommon. If the computer can reduce the inventory by only a small percentage, a substantial amount of money is freed up for other purposes. For example, if the computer can reduce a $5 million inventory by 5 percent, $250,000 is made available. If the $250,000 is invested in a security that carries an annual interest rate of 12 percent, the result is a $30,000 increase in the firm's revenue.

Inventory is one area of a firm's operations where the economic impact of computer use can be measured in exact, quantitative terms.

Increased Productivity   The manufacturing area provides an opportunity to use the computer to increase productivity. The first computer applications determined raw material requirements, scheduled jobs, tracked jobs, and provided information to manufacturing management. Eventually, the computer was applied in more direct ways. Some of the first minicomputers were used to replace humans in controlling machine tools, such as lathes and drill presses. Computers could do the jobs more accurately and more rapidly than could human operators, thereby increasing productivity. Today, these applications are called CAM, for *computer-aided manufacturing.* Sometimes you hear the term *CAD/CAM.* CAD stands for *computer-aided design,* where a design engineer uses the computer to design the firm's products. Figure 9.5 shows a design engineer using CAD software in the design process.

As with reducing inventory investment, the use of computers to increase productivity can be measured quite accurately in monetary terms.

Improved Decision Making   Today's management tends to regard the firm's data processing system as a given. In many firms it has been in place for a quarter of a century or more, and the bugs have been ironed out. One area of untapped potential for computer use is the area of decision making. Investments in MIS, DSS, OA, and expert systems are being made with the assumption that the information they provide will result in better decisions. The better decisions will, in turn, produce problem solutions that will benefit the firm.

Putting Economic Feasibility in Perspective

Management does not always require absolute economic proof when considering the development of a computer-based system. Management often bases investment in computer projects on judgment and on a hunch that the system will pay its way, even though the benefits cannot be accurately measured.

Figure 9.5
Computer-aided design is an example of using the computer to increase productivity.

What does this mean to the systems analyst? It means that the analyst has to provide management with the support for the project, even though much of that support will not be expressed in monetary terms. In demonstrating the feasibility of the new system, the systems analyst must also recognize the importance of noneconomic feasibility—the *N* of the TENLOS acronym.

Noneconomic Feasibility

Noneconomic feasibility is the justification for computer use that is not expressed in monetary terms. Noneconomic feasibility provides the major justification for the MIS, DSS, office automation, and expert systems of the CBIS. The difficulty of expressing the benefits of these information-producing systems in dollars and cents forces management to use noneconomic measures. Two such measures are the portfolio approach and information economics.

The Portfolio Approach

The **portfolio approach** to justifying a computer-based system holds that evaluation should consider the entire set, or portfolio, of systems.[3] This is the same reasoning a firm's financial analysts use when they determine whether to make an investment, such as the purchase of a stock. The reasoning is that while certain systems may perform better or worse than expected, the overall performance of the portfolio is what counts.

When applying this approach to a computer project, management focuses on how the system produced by that project will fit into the entire portfolio. Management employs a three-pronged strategy to ensure that the portfolio includes the largest proportion of successful systems:[4]

- Keep the number of high-risk projects to a minimum.

- Include several sure winners, even if their anticipated payoffs are not large.

- Scrap a project as soon as the anticipated risk appears to be greater than the anticipated reward.

The use of portfolio analysis relieves top management of the responsibility of justifying economically each and every computer project that comes along.

Information Economics

Information economics recognizes a wide range of potential values that a system can achieve, as well as a wide range of risks that it can incur.[5]

Potential System Values   A proposed computer-based system can provide value to the firm in the following ways:

- By providing information to the elements in the firm's environment

- By supporting the firm's long-term strategy

- By providing the firm with a competitive advantage

- By enabling management to track its critical activities

- By enabling management to respond quickly to competitive actions

- By supporting the firm's strategic plan for information systems

The more of these values a potential system can deliver, the better its chances of being approved for implementation.

---

[3]A description of the portfolio approach first appeared in F. Warren McFarlan, "Portfolio Approach to Information Systems," *Harvard Business Review* 59 (September–October 1981): 142–150.
[4]Lee L. Gremillion and Philip J. Pyburn, "Justifying Decision Support and Office Automation Systems," *Journal of Management Information Systems* 2 (Summer 1985): 15.
[5]Marilyn M. Parker and Robert J. Benson, "Information Economics: An Introduction," *Datamation* 33 (December 1, 1987): 86ff.

Figure 9.6
Hand-held computers provide Avis with system value that leads to competitive advantage.

Potential System Risks   The risks that might prevent a potential system from contributing to a strong portfolio also deserve attention. A proposed system should not:

- Support a risky business strategy
- Depend on unproven capabilities within the firm
- Depend on cooperation from multiple functional areas of the firm when that cooperation might be difficult to achieve
- Be based on poorly defined user needs
- Depend on new and untried technology

The fewer of these risks that characterize a potential system, the better the chances of its development being approved.

Putting Noneconomic Feasibility in Perspective

Management seeks to develop systems that create value while mimimizing risk. The systems analyst provides the information that management needs in order to evaluate a prospective application on these two dimensions. For the systems analyst to satisfy this responsibility, he or she "must really understand what makes the business tick, what factors are most critical to overall success, and which business areas are likely to hold the best opportunities for application of the technology."[6]

Legal and Ethical Feasibility

One of the six dimensions of feasibility pertains to the proposed system's requirement to function within the law and within the limits the firm imposes upon itself for ethical conduct.

Legal Considerations

The fact of the matter is that very few laws have been passed in the United States to regulate computer use. The major U.S. laws include:[7]

Freedom of Information Act (1966)—Citizens and organizations have the right to gain access to federal government records with few exceptions.

Fair Credit Reporting Act (1970)—Credit agencies cannot share information with anyone but customers, must give people access to their data, and must notify people when their data is used in employment or credit checks.

Right to Federal Privacy Act (1978)—Federal government agencies are limited in terms of their searches of the bank records of individuals.

Computer Matching and Privacy Act (1988)—Federal government agencies are restricted in matching computer files in different agencies for the purposes of determining eligibility for federal programs or identifying debtors.

You can see that most of the legislation aimed at computer use is at the federal level. Therefore, the systems analyst who works for the government has a greater responsibility in evaluating the legal feasibility of a new system than does an analyst doing the same type of work in private industry. The federal government analyst should study computer-related legislation and ensure that the new system does not violate its intent.

It is also apparent that the legislation is aimed primarily at credit information. Therefore, the analyst working in private industry runs the greatest risk of violating the law when designing systems that maintain credit and personal information in a database. Two systems that fall into this category are financial and human resource information systems. Such systems should be designed so that the information is always kept current, kept secure, and made available to only those persons authorized to view it.

---

[6]Gremillion and Pyburn, 16.
[7]From Kenneth C. Laudon and Jane Price Laudon, *Business Information Systems: A Problem-Solving Approach* (Fort Worth, TX: The Dryden Press, 1991), 557–560.

Ethical Considerations

Although systems developers have always been aware that their systems must fit within legal frameworks, complying with moral and ethical guidelines is a much more recent concern. *Morals* are generally accepted standards of what is right and what is wrong in terms of conduct or character, and *ethics* are expressions of morals in the form of codes or guidelines.[8] Many professional organizations, such as the Association for Computing Machinery (ACM), have established codes of ethics intended to guide information specialists. The same codes apply to users engaged in end-user computing.

PAPA and a Social Contract for Computer Use   Richard O. Mason, a professor at Southern Methodist University, has coined the acronym **PAPA** to represent the four primary ethical issues that face the information age.[9] The letters stand for **Privacy, Accuracy, Property,** and **Accessibility.** Each ingredient demands that certain questions be answered when developing computer-based systems.

- **Privacy** What information about oneself or one's associations must a person reveal to others, under what conditions, and with what safeguards? What things can people keep to themselves and not be forced to reveal to others?

- **Accuracy** Who is responsible for the authenticity, fidelity, and accuracy of information? Similarly, who is to be held accountable for errors in information, and how is the injured party to be made whole?

- **Property** Who owns information? What are the just and fair prices for its exchange? Who owns the channels, especially the airways, through which information is transmitted? How should access to this scarce resource be allocated?

- **Accessibility** What information does a person or an organization have a right or a privilege to obtain, under what conditions, and with what safeguards?

For computer-based systems to satisfy the PAPA requirement, Professor Mason urges that management enter into a **social contract for computer use** by ensuring that the firm's information systems:

- Do not unduly invade a person's privacy

- Are accurate

- Protect the viability of the fixed conduit resource through which information is transmitted to avoid noise and jamming pollution

- Protect the sanctity of intellectual property

- Are accessible to avoid the indignities of information illiteracy and deprivation

---

[8]Karen A. Forcht, "Assessing the Ethical Standards and Policies in Computer-Based Environments," in *Ethical Issues in Information Systems,* edited by Roy Dejoie, George Fowler, and David Paradice (Boston, MA: Boyd & Fraser, 1991), 57–59.

[9]Richard O. Mason, "Four Ethical Issues of the Information Age," *MIS Quarterly* 10 (March 1986), 4–12.

Mason believes that when these principles are followed, information will flow to create the kind of world in which we wish to live.

Operational Feasibility

Operational feasibility relates to the ability of the people working within a system to do their jobs in a prescribed manner. The basic premise is the fact that no system, regardless of the soundness of its design from a technical standpoint, can succeed if it does not have the support of its personnel resource. Without question, one of the most frequent causes of system failure is a lack of cooperation from the system's users and participants.

Operational feasibility has two dimensions—ability and attitude.

Abilities of Users and Participants

System designs should consider the abilities of users and participants. For example, a report should include terms and figures that the user understands. Similarly, a system that requires factory workers to enter data on the shop floor

Figure 9.7
The abilities of participants to enter data are considered when designing systems.

should not require the workers to perform the same clerical tasks as those performed by office workers.

Attitudes of Users and Participants

Attitudes can influence system success as much as abilities. A user or a participant must *want* the system to succeed. System developers can achieve this positive attitude by paying attention to human factors considerations, which are addressed in Chapter 11.

Schedule Feasibility

The final element in the TENLOS acronym relates to schedule. An important question is whether the system can be put into use by the time that it is needed. This is the system's **schedule feasibility,** and it depends on several considerations that must be resolved prior to cutover.

Constraints on the System Schedule

The date for the cutover can be influenced by both internal and environmental constraints. These are *not* the same constraints described earlier, constraints on the *system.* Here, the concern is constraints on the *development* of the system.

Internal schedule constraints, imposed by the firm's management, require that the system be operational by a certain date. For example, the firm might acquire a smaller company and management decide to integrate that firm's systems into those of the parent firm by September 1. **Environmental schedule constraints,** on the other hand, are imposed by one or more elements in the firm's environment that the system be operational by a certain date. A good example is the requirement to implement, by January 1, a payroll system designed to incorporate a new tax law.

The Procedure for Setting the Cutover Date

In the case of the environmental schedule constraints, the firm has little room for flexibility. The system must be up and running on a certain date, no matter what. For internal schedule constraints, however, management is free to pick the time for the cutover. When making this determination, management can follow one of two strategies:

- Begin with the current date and estimate the development time. The development process is divided into small, manageable subtasks, and best-time estimates are made for each. Then, all the subtask estimates are added to produce the date when all work will be completed and cutover can occur. This is the most realistic way to set the cutover date.

- Begin with the cutover date and establish a development schedule. Management tends to become impatient when it comes to implementing improvements and likes to keep a certain amount of pressure on the development effort. One way to do this is to select a cutover date that appears to management to be reasonable and then work backward from that date, scheduling the various subtasks.

In selecting the desired cutover date, management is especially sensitive to the impact the new system might have on the business. To minimize the impact, management usually times the cutover so that it occurs during a slack period. For example, a university will implement a new registration system during the summer term, when enrollments are down. And, a homebuilder will implement a new job-scheduling system during the winter, when few homes are built because of the weather.

Putting System and Project Feasibility in Perspective

Most of the questions relating to system and project feasibility cannot be answered in a detailed way during the planning phase. The systems analyst can give only estimates, but those estimates are necessary if management is to make a sound decision concerning continuation. Later, as more is learned about the system and its development, management will expect the analyst to update the feasibility evaluations.

Step 6—Prepare a System Study Proposal

When the systems analyst completes the feasibility study, the findings are incorporated into a written report. This report is the system study proposal presented to both the user and the MIS steering committee. (The proposal format is illustrated in Figure 8.2.) In preparing the proposal, the analyst must be aware of three important points:

1. There should be no surprises. The analyst should work closely with the user throughout the planning phase, keeping the user informed of the findings. The analyst must also keep the manager of systems analysis informed, that person should keep the CIO informed, and the CIO should keep the MIS steering committee informed. This communication chain is illustrated in Figure 9.8. When followed during the planning phase, the proposal is merely a formality, explaining facts that everyone already knows. Management uses the proposal as a basis for the decision to proceed with the SLC.

2. Do not sell a bad system. The analyst should not feel an obligation to recommend just *any* type of system. If the data does not support a computer-based

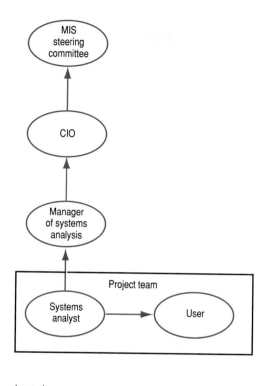

Legend:

Communication of findings information

Figure 9.8
All participants in the planning phase should be kept informed of the findings.

solution, the analyst should recommend that the project be terminated. There will always be other projects.

3. Do sell a good system. On the other hand, if the data supports the need for the system, the systems analyst should make certain that management appreciates the full potential. This assurance is gained by writing a positive, succinct proposal.

Positive language stirs management to action. For example, rather than saying that the new system *can* reduce costs, say that the system *will* reduce costs. *Can* implies uncertainty. *Will* leaves little doubt.

Succinct language does not rely on flowery words that serve only as excess baggage. As an example, rather than saying that the new system will have a "fantastically good impact on customer relations," simply say that it will have a "good" impact. When possible, back up statements with figures, as in "the system will reduce inventory costs by $1 million." However, if statistics are not available, do not attempt to overcome the limitation with verbage. The extra words do more harm than good and will not be received well.

Once the analyst submits the proposal, the decision is in the hands of the user and, primarily, the MIS steering committee.

Step 7—Approve or Disapprove the Study Project

The user and the MIS steering committee read the system study proposal and meet to discuss its outcome. The systems analyst should attend. Perhaps the analyst will be expected to make a formal oral presentation, using visual aids such as transparencies, color slides, or computer-generated presentation graphics.

Management can decide to *go*, to *no go*, or to request additional information. A *no go* decision terminates the project. A request for more information causes the analyst, and perhaps the user, to retrace some of the steps and try again. A *go* decision means that management must prepare to control a development project.

Step 8—Establish a Project Control Mechanism

The system study will be conducted in the form of a project. The firm's management, consisting of the MIS steering committee, information services management, and management in the user area, is responsible for project management. **Project management** consists of all the actions taken to ensure that a project is carried out as planned. Therefore, in order to manage, there must be a plan. This final step of the planning phase is included (1) to establish a project plan, and (2) to establish a control mechanism to ensure that the plan is met.

A **project control mechanism** is any means of comparing what *is* accomplished with what *should be* accomplished in carrying out a project. Ingredients of the control mechanism include the project plan, periodic reports, scheduled meetings, and graphic techniques.

The Project Plan

The project plan in its simplest form consists of time estimates for each step in the remaining development phases—analysis, design, and implementation. The analyst assigns an estimated start date and an estimated completion date to each step. The estimates are approved by information services management, the user, and the MIS steering committee.

Periodic Reports

Once the project team is formed in the analysis phase, the project leader must prepare periodic reports, usually weekly, that keep the MIS steering committee current on the team's activities. The reports should be short and to the point. They should address three topics:

- Progress since the last report
- Any problems that must be resolved
- Tasks to be performed during the coming reporting period

These reports provide a written record of the project team's activity.

Scheduled Meetings

The MIS steering committee meets periodically to monitor project progress. Early in the SDLC, the meetings may be held only once a month. As the cutover date approaches, the meetings occur more frequently, perhaps once a week. During a meeting, the committee chair follows the general outline of the periodic report.

The committee chair invites persons such as members of the project team, consultants, and vendor representatives to attend meetings when their expertise is needed.

Graphic Techniques

The MIS steering committee has many projects to control, and the members' time is precious. The members recognize the value of graphics as a way to make control easier and less time consuming. Two graphic tools that have enjoyed widespread use are Gantt charts and network diagrams.

Gantt Charts   Harry L. Gantt, a management consultant who practiced his trade around 1900, is credited with the idea of using a bar chart to schedule work. The chart, called a **Gantt chart,** is illustrated in Figure 9.9. All activities are listed down the left side, and bars show when the work will be done.

The biggest limitation of the Gantt chart is that it fails to show the interrelationships among activities. One way to show the interrelationships is to arrange the activities in the form of a network diagram.

Network Diagrams   A **network diagram** is a pattern of interconnected arrows that represents work to be done. Figure 9.10 portrays the same project plan as the Gantt chart in a network form.

Perhaps you have heard a network diagram referred to as a PERT chart or CPM diagram. PERT, which stands for Program Evaluation and Review Technique, and CPM, which stands for Critical Path Method, are two styles that the network diagram can take. Details of network diagramming are discussed in Technical Module D.

For years project management graphics were drawn by hand. More recently, special project management software packages have been developed to relieve the analyst of much of the time-consuming artwork. One such system is the Harvard Project Manager, from Software Publishing Corporation. It can produce presentation-quality graphics, such as Gantt and PERT charts, as well as many standard reports. The user can also design customized reports.

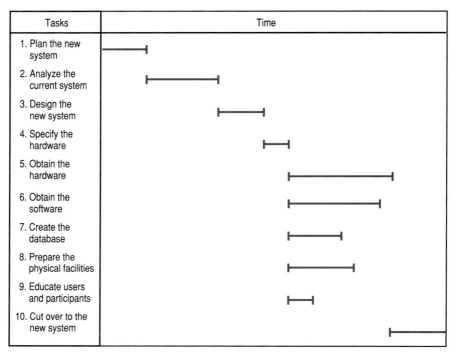

Tasks	Time
1. Plan the new system	
2. Analyze the current system	
3. Design the new system	
4. Specify the hardware	
5. Obtain the hardware	
6. Obtain the software	
7. Create the database	
8. Prepare the physical facilities	
9. Educate users and participants	
10. Cut over to the new system	

Figure 9.9
A Gantt chart.

Putting Project Control in Perspective

Although project management software is important, it is not the key to establishing control over a project. Rather, the key is the attention that management and the systems analyst devote to the planning process. These planners must invest the necessary time in identifying the steps to be done, determining their interrelationships, and estimating their schedules. Once this is done, the software enables the plan to be updated quickly and easily and provides management with vital information in both a graphic and tabular form.

Putting the Planning Phase in Perspective

It would be easy to get the idea that the first step of a computer project is to conduct a system study so that you can understand how the work is currently being done. This approach might have worked fine during the early days of computing, when the new systems were intended to reflect the same functionality as the older systems they were to replace. During those early days, the systems primarily affected persons in the accounting department.

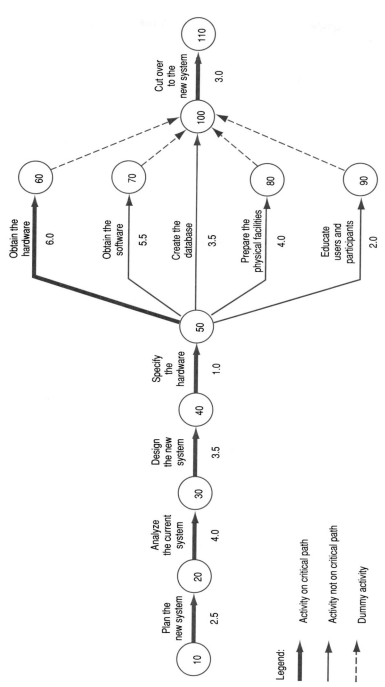

Figure 9.10
A network diagram.

Today's computer-based systems are integrated in all the firm's operations to such an extent that practically everyone in the firm is affected when new systems are developed or existing systems are changed. When systems work has such a widespread effect on the organization, management wants the assurance that the work will go well. This assurance is gained through planning. You have seen that systems planning includes important steps, such as defining objectives, identifying possible constraints, and ensuring that both the project and the system are feasible from various standpoints. If system developers ignore any of these steps by rushing to start the system study, the chances of achieving a good system are diminished. The planning phase is perhaps the most important of the life cycle phases—it gets the project off on the right foot.

Using CASE to Build a Resource Histogram

Figure 9.11 is a diagram called a **resource histogram.** It charts the number of employees needed on each day of a project—in this example, the construction of a swimming pool. You can see that Day 2 requires two employees and Day 17 requires four. All employees

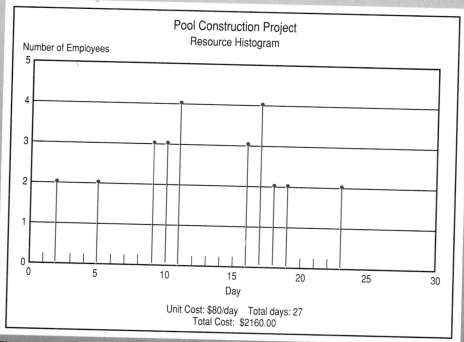

Figure 9.11
A resource histogram.

in this example are paid $80 per day. The total number of person-days is 27, producing a total employee cost of $2160.

The resource histogram can reflect projected costs, actual costs, or both, making it a valuable tool for human resource planning. A resource histogram for a complex computer project would be more difficult to read than the example, but it would still give a visual picture of the distribution of human resources over the project and the bottom-line expenditures.

Summary

The planning phase of the system life cycle consists of eight steps that determine whether a system study should be conducted and, if so, establish a project control mechanism.

Managers in user areas are most likely to recognize a problem; they reflect three problem-sensing styles—problem seeker, problem solver, and problem avoider.

Systems analysts help users define the problem by separating the symptoms from the cause. Problems in the information processor element can trigger an SLC. A problem statement links the cause with an element of the firm or its environment and describes the effect of the problem on the firm.

Many types of objectives can be identified for a new system, but one approach is to base them on the dimensions of information—accuracy, timeliness, completeness, and relevance. Another application of systems theory can help identify system constraints. Internal constraints can be expressed in terms of limits on resources, and environmental constraints can be related to each system element.

There are six types of feasibility—technical, economic, noneconomic, legal and ethical, operational, and schedule. You can use the acronym TENLOS as a memory aid.

Technical feasibility considers whether needed hardware and software are available and affordable. Economic feasibility is concerned with justifying a computer application in monetary terms. In determining economic feasibility, the systems analyst considers various types of costs and benefits, both developmental and operational. Cost objectives originally sought reduction but now stress avoidance. Cost-benefit studies can take the form of break-even analysis, payback analysis, and net present value. Attention is also being given to revenue increase through reduced inventory investments, increased productivity, and improved decision making.

One approach to noneconomic feasibility is to consider the firm's entire portfolio of applications rather than single applications. Another approach is information economics, which recognizes the wide range of values that a new system can generate as well as the risks that it can incur.

Legal feasibility is concerned with laws that deal mainly with computer use by the federal government and credit reporting. Ethical considerations focus on individuals' rights to privacy, accuracy of information maintained about individuals, rights to information property, and accessibility. Firms can channel their energies toward ethical computer use by entering into a social contract.

Operational feasibility relates to the abilities and attitudes of the system users and participants—how well they can support the system. Schedule feasibility considers the time required to implement the system. Constraints can be imposed both externally and internally, and management can schedule forward from the current date or backward from the cutover date.

The systems analyst incorporates the findings of the planning process into a study project proposal, which should not include surprises or sell a bad system, but which should sell a good system. The MIS steering committee uses the proposal as a basis in deciding whether to continue. A *go* decision prompts management to establish a control mechanism consisting of a project plan, periodic reports, scheduled meetings, and graphic techniques.

With the project planning accomplished and the control mechanism in place, the SLC is ready to move to the analysis phase, the subject of the next chapter.

Key Terms

Problem signal

Problem statement

Feasibility study

Cost-benefit analysis

Break-even analysis

Break-even point

Payback analysis

Payback point

Net present value (NPV)

Project management

Project control mechanism

Gantt chart

Network diagram

Key Concepts

Problem-sensing styles

How constraints can be imposed by both the firm and its environment on both system development and use

The six dimensions of feasibility—technical, economic, noneconomic, legal and ethical, operational, and schedule

Cost reduction

Cost avoidance

How management, in its desire to increase profits, has shifted its focus from decreasing costs to increasing revenues

Portfolio approach to system justification

Information economics

Mason's PAPA acronym as a way of cataloging ethical concerns

Questions

1. What are the three problem-sensing styles?
2. Where can a problem cause originate?
3. What is triggered when a problem is identified with the information processor element of the firm as a system?
4. What should the problem statement include?
5. What approach does the text take in setting system objectives?
6. What do environmental system constraints affect?
7. How does the systems analyst gather data during the feasibility study?
8. In today's world, what is the main concern relating to technical feasibility?
9. What does cost-benefit analysis seek to accomplish?
10. Does management approve system projects for the purpose of receiving benefits during system development or during system use?
11. Explain the difference between cost reduction and cost avoidance.
12. What does break-even analysis compare?
13. What does payback analysis compare?
14. Does net present value help or hinder you when seeking to economically justify a new computer system? Explain why.
15. In what areas of a firm's operations can the economic impact of the computer be measured in exact, quantitative terms?
16. What is the current emphasis in computer justification?
17. Distinguish between economic and noneconomic feasibility.
18. What effect, if any, would the portfolio approach to system justification have on the go–no go decision points at the end of each phase of the SDLC?
19. What are the two main concerns of information economics?

20. Which element of the firm's environment has been the focus of most legislation concerning computer use?

21. Name two applications that are exceptionally susceptible to legislation concerning computer use.

22. What do the letters PAPA stand for?

23. Operational feasibility is concerned with two groups of people. Name the groups. It is also concerned with two characteristics of these people. Name the characteristics.

24. Are system constraints the same as schedule constraints? Explain.

25. What are the two basic approaches to setting the cutover date? Which would the MIS steering committee probably prefer? Why? Which would the systems analyst probably prefer? Why?

26. Why prepare a written study proposal if the information in it is common knowledge, as the text states?

27. What are the ingredients of a project control mechanism?

28. What are the three main elements in the project leader's periodic report to the MIS steering committee?

29. What do the bars in a Gantt chart represent?

30. What does a network diagram show that a Gantt chart does not?

Topics for Discussion

1. What types of software, if any, do you think a problem seeker would be most interested in using? What types would a problem solver be interested in using? A problem avoider?

2. The text says that a problem in the information processor triggers an SLC. What, if anything, can the systems analyst do when he or she identifies the problem in the system's standards? In the system's management?

3. If you were a systems analyst, how would you know whether to spend a lot of time conducting a cost-benefit analysis?

4. Information economics identifies five risks that a proposed system can incur. Give an example of each.

5. With whom does management enter into a social contract for computer use? What explicit acts can management perform to ensure that such a contract is upheld?

6. If a systems analyst's feasibility studies constantly indicate that systems projects should *not* be pursued, is it that a bad reflection on the analyst's abilities?

Consulting with

Harcourt Brace

(Use the Harcourt Brace scenario prior to this chapter in solving this problem.)

Although Mike Byrnes is pleased with his use of NETMAN as a project control mechanism, he is continually searching for improvements. He has asked you to conduct a brief review of project management software and identify three systems that appear to warrant a closer look by him. The only requirement is that the software run on a PC. Conduct research in your library to identify the packages, and write a one-page memo that briefly describes each package, provides the name and address of the supplier, and specifies the cost.

Case Problem
Splashdown (B) _____

(For a description of the firm and your previous project as a systems analyst, see the Splashdown (A) case at the end of Chapter 8.)

You did such a good job for your boss, Mildred Wiggins, in designing the prototyping form that she asks you to do another—dealing with system feasibility. She explains that feasibility must be determined for each project and that an orderly approach will ensure that the same criteria are always applied, and might even save some time.

She discusses the subject of feasibility with you, and you have heard the story before—the old TENLOS that your college systems instructor spent so much time explaining. Mildred tells you that Splashdown doesn't regard all of the six feasibility types as having the same amount of influence on the decision to develop a system. She hands you a piece of paper with the following weights written on it:

Technical feasibility	.10
Economic feasibility	.15
Noneconomic feasibility	.20
Legal and ethical feasibility	.05
Operational feasibility	.35
Schedule feasibility	.15
Total weight	1.00

Mildred also wants you to rate a prospective system on each of the feasibility types using the following point values:

Good performance	5 points
Fair performance	3 points
Poor performance	1 point

For example, if a system does a good job of satisfying its economic feasibility, the analyst awards it 5 points. That process is repeated for each feasibility type. Then, each point value is multiplied by its weight, and the weighted values added.

A system will be evaluated as a system development opportunity according to the following total weighted point values:

Good   3.51–5.00

Fair   2.51–3.50

Poor   1.00–2.50

Mildred asks whether you understand your assignment, and you assure her that you do. She then asks you to get on with the form design, and, when you are finished, to try it out on two managers who have requested projects. One is Flora Henry, the director of corporate financial planning, and the other is Emma Ward, the president's executive secretary.

You go back to your office and use up about half a yellow pad, making rough sketches of the form. You want it to be extremely user friendly so that the analyst need only (1) briefly describe the application, (2) enter the six ratings, (3) make the weight and total calculations, and (4) write a brief explanation at the bottom. The form should be self-explanatory—identifying the weights, the meaning of the rating points, and the meaning of the ranges of total points.

You finally come up with a form that you like and decide to test it on Emma. You go to the executive suite, and she explains that the members of the executive committee—the president and four vice-presidents—want to implement an electronic calendaring system, some packaged OA software that enables the secretaries to maintain their executives' appointments calendars in the computer. The executives can use the terminals that are currently located in their offices to check one another's calendars when they want to schedule a meeting. You impress Emma by saying that you know all about electronic calendaring, having just read an article in an office systems magazine that described the many packages available. You ask Emma whether the system can be justified economically, and she replies, "No question about it. When you consider how valuable an executive's time is, if the system saves only a few seconds a day, it's justified." You then ask whether anyone else in the firm will have access to the executives' calendars, and she explains that their secretaries will, plus all managers in the firm, plus quite a few people outside the firm who frequently make appointments.

You note that on your yellow pad and continue the line of questioning: "Will there be any confidential or personal information included in the calendars?" Emma assures you that there will not. Finally, one last question: "When does the system have to be up and running?" Emma smiles, and says "Yesterday." You look puzzled, and she explains "No, I'm just kidding. The executives would like to have it within 90 days." You jot that down, say good-bye, and head for Flora Henry's office.

Flora wants to implement a new competitor forecasting subsystem for the MIS. She wants to be able to use economic indicators to project the annual revenues for all the other water parks in the country. She tells you that it has never been done before, but, if it works, it could give Splashdown a real competitive edge. The system would use existing hardware, and the forecasting program would be written in APL, a powerful yet relatively user-unfriendly language that none of the programmers in information services knows anything about. You ask who the system users would be, and Flora says that they would include all of the forecasting analysts in the financial planning unit, and that they are all checked out on their terminals. You follow up with the big question about economic justification, and Flora says: "Believe it or not, it's expected of us. Eileen [Jacobs, the vice-president of finance, and Flora's boss] told me that the only way we could get the system approved would be if we displaced two of our forecasting analysts." You ask Flora whether that is likely to happen, and she says: "I'm not sure. The analysts have gotten wind of Eileen's requirement, and they're all pretty upset. I think they'll fight it tooth and nail." You write all that down and then explain that you are concerned about not breaking any computer laws or violating anyone's rights. Flora assures you that there is "no sweat." The system will use data available in commercial databases that Splashdown can purchase for a reasonable price. You sense that Flora is getting a little impatient, so you ask your last question concerning the schedule. Flora surprises you by saying that she would like cutover five weeks from now, when the new fiscal year begins. That way, the displaced analysts' salaries will be taken out of her budget.

Having filled the remaining pages of your yellow pad, you decide to call it a day and take a dip in the company pool.

Assignments

1. Design the form that Mildred has requested. Use your word processor or similar software. Print a blank copy.

2. Make two additional copies of your form and complete them using the findings of your data gathering.

3. Prepare a memo to Mildred, advising her of your forms design efforts and your recommendations concerning the two systems. Do you recommend implementation for either or both and, if so, how strong is your recommendation? Attach the blank form, followed by the two completed forms.

Selected Bibliography

Bahl, Harish C., and Dadashzadeh, Mohammad. "A Framework for Improving Effectiveness of MIS Steering Committees." *Information Resources Management Journal* 5 (Summer 1992): 33–44.

Bender, Donald H. "Financial Impact of Information Processing." *Journal of Management Information Systems* 3 (Fall 1986): 22–32.

Christoff, Kurt A. "On Time, Under Budget, Out of Control." *EDGE* 3 (January–February 1990): 37–40.

Doll, William J., and Torkzadeh, Golamreza. "The Relationship of MIS Steering Committees to Size of Firm and Formalization of MIS Planning." *Communications of the ACM* 30 (November 1987): 972–978.

"EDGE's Guide to Project Management Software." *EDGE* 3 (January–February 1990): 40–43.

Gupta, Yash P., and Raghunathan, T. S. "Impact of Information Systems (IS) Steering Committees on IS Planning." *Decision Sciences* 20 (Fall 1989): 777–793.

Henry, Bill. "Measuring IS for Business Value." *Datamation* 36 (April 1, 1990): 89–91.

Hollist, Pen, and Utterback, Jon. "The CASE-Project Management Connection." *System Builder* 4 (April–May 1991): 39–42.

Jenkins, Avery. "Project Planning and Administration." *EDGE* 2 (March–April 1989): 30ff.

Keen, Peter G. W. "Value Analysis: Justifying Decision Support Systems." *MIS Quarterly* 5 (March 1981): 1–15.

Lederer, Albert L.; Mirani, Rajesh; Neo, Boon Siong; Pollard, Carol; Prasad, Jayesh; and Ramamurthy, K. "Information System Cost Estimating: A Management Perspective." *MIS Quarterly* 14 (June 1990): 159–176.

McCusker, Tom. "Tools to Manage Big Projects." *Datamation* 37 (January 15, 1991): 71ff.

Marks, William W. "Successful Project Management Means Business Management." *System Builder* 4 (April–May 1991): 45ff.

Palvia, Shailendra, and Palvia, Prashant. "Timing Strategies for Feasibility Studies in Information Systems Development." *Information Resources Management Journal* 3 (Winter 1990): 15–27.

Rivard, Edward, and Kaiser, Kate. "The Benefit of Quality IS." *Datamation* 35 (January 15, 1989): 53ff.

Roetzheim, William H. "Estimating Project Risk." *Information Executive* 3 (Winter 1990): 47–50.

Shneiderman, Ben. "Socially Responsible Computing I: A Call to Action Following the L.A. Riots." *SIGCHI Bulletin* 24 (July 1992): 14–15.

Shneiderman, Ben. "Socially Responsible Computing II: First Steps on the Path to Positive Contributions." *SIGCHI Bulletin* 24 (July 1992): 16–17.

Stegwee, Robert A., and Van Waes, Ria M. C. "The Development of Information Systems Planning Towards a Mature Management Tool." *Information Resources Management Journal* 3 (Summer 1990): 8–21.

Wedberg, George H. "But First, Understand the Problem." *Journal of Systems Management* 41 (June 1990): 20–28.

Winkler, Connie. "Better Project Management in Uncertain Times." *Datamation* 36 (June 1, 1990): 95–98.

Network Diagrams

A **network diagram** is a pattern of interconnected arrows that represents work to be done. Each arrow is called an **activity.** The activities are performed in a left-to-right sequence, and you cannot begin an activity until all the activities leading to it have been completed. Figure D.1 is a network diagram of the activities performed during a pit stop of an Indianapolis 500 automobile race.

In this example the car enters the pit area and stops. Then both the front end and back end are raised. With the car raised, all four wheels can be replaced. While the wheel work is being done, the driver is given a drink of water and the windshield is cleaned. The car is then lowered, and it leaves the pit area.

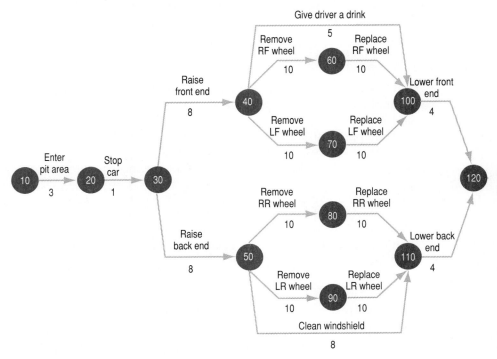

Figure D.1
A network diagram of an Indianapolis 500 pit stop.

TM

D

Activities and Nodes

Each activity in a network diagram is labeled with a brief description, usually consisting of a verb and object, plus an estimate of the amount of time required for completing the activity. In the Indy diagram, the estimates are in seconds.

At the beginning and end of each arrow is a circle, called a **node.** The purpose of a node is to provide a unique identification to each activity. For example, in referring to the first activity in the Indy diagram, you would say "Activity 10 20."

The nodes are numbered from left to right, usually by tens. In assigning the node numbers to an activity, the only rule is to assign a lower number to the left-hand node than to the right-hand node.

Serial Processes

The pattern of the network arrows reveals the manner in which the work is performed. When single arrows are arranged in a series, such as activities 10 20 and 20 30 in the Indy diagram, the work is performed serially—one activity after the other. In a serial arrangement, the same person (or persons) can accomplish all the activities, moving from one to the next. This assumes that the person (or persons) is capable of performing all the activities.

Parallel Processes

When more than a single activity can be performed at a time, this is illustrated by arranging the arrows in parallel, as shown in the Indy diagram between nodes 30 and 120.

The important point to remember about parallel processes is that they require separate resources. For example, if one person is giving the driver a drink, another is removing the right-front wheel, another is removing the left-front wheel, and so on.

Assuming that each activity in the Indy diagram can be performed by a single person, what is the minimum number of pit crew members required?

If you answered six, you are correct. Four for the wheels, one with the water bottle, and one bug removal specialist. These crew members can perform the other activities as well.

Dummy Activities

You will often see a network diagram with activities represented by dotted lines, as in Figure D.2. These are dummy activities. A **dummy activity** does not represent work to be done but is included so that each activity will have a unique pair of node numbers.

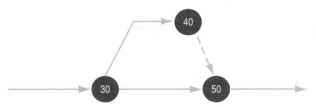

Figure D.2
Dummy activities provide for unique activity identification.

TM

D

The Critical Path

The overall network can represent a maze of multiple paths. All the paths may not require the same amount of time to complete. The most important path is the one that takes the *greatest amount of time*. This path is called the **critical path** because it determines the maximum time of the project.

Figure D.3 is a network diagram of an analysis of a banking operation, with the time expressed in working days. Dummy activities are required to provide activities 60 80, 60 90, 60 100, and 60 110 with unique node numbers. In this example, the critical path is highlighted with a heavy line.

The critical path contains no **slack,** or excess time. Conversely, activities *not* on the critical path *do* contain slack. An activity not on the critical path can be delayed by an amount of time not exceeding the amount of slack for the *path*. Such a delay will not affect the project time. For example, the preparation of the network diagram, activity 40 50, can be delayed as much as six days without affecting the project completion date.

Time Increments

You can use any increment of time—months, weeks, days, hours, or even seconds—to represent activity times. For lengthy computer projects, a time increment in days is usually appropriate. If an activity requires a half day, for example, the increment can be shown as 0.5 days. The diagram should include some type of notation that identifies the time increment.

Early and Late Times

Each activity can be described in terms of early and late times. The **early start time** is the earliest that an activity can begin. The **late start time** is the latest that an activity can begin without affecting the time of the overall project. In the same manner, the **early completion time** is the earliest that an activity can be completed,

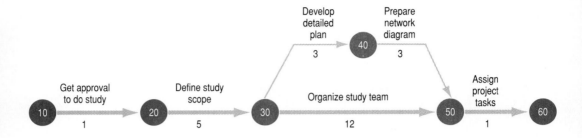

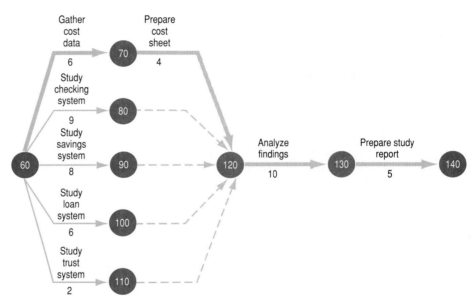

Figure D.3
The processes required to analyze a banking system, expressed in working days.

and the **late completion time** is the latest that an activity can be completed without affecting project time.

The early times are computed by beginning with the *project* early start time and working through the diagram from left to right, adding the activity times. The late times are then computed by beginning with the *project* late completion time and working back through the diagram right to left, subtracting the activity times.

Figure D.4 shows the early and late times for the bank project.

When activities are on the critical path, their early and late times are the same. Activities not on the critical path have a difference, and the difference is the amount of slack on the path.

TM

D

Nodes		Activity	Estimated Time (Days)	Early Start Date	Early Completion Date	Late Start Date	Late Completion Date
10	20	Get approval to do study	1	0	1	0	1
20	30	Define study scope	5	1	6	1	6
30	40	Develop detailed plan	3	6	9	12	15
30	50	Organize study team	12	6	18	6	18
40	50	Prepare network diagram	3	9	12	15	18
50	60	Assign project tasks	1	18	19	18	19
60	70	Gather cost data	6	19	25	19	25
60	80	Study checking system	9	19	28	20	29
60	90	Study savings system	8	19	27	21	29
60	100	Study loan system	6	19	25	23	29
60	110	Study trust system	2	19	21	27	29
70	120	Prepare cost sheet	4	25	29	25	29
120	130	Analyze findings	10	29	39	29	39
130	140	Prepare study report	5	39	44	39	44

Note: The dates are the number of days from the beginning of the project.

Figure D.4
Early and late dates for the banking project.

Estimates of Activity Times

Two techniques have been devised for estimating the activity times: the critical path method and the program evaluation and review technique.

The **critical path method (CPM)** uses a *single* time estimate for each activity. The estimate is the best that can be made using the available information.

The **program evaluation and review technique (PERT)** uses *three* time estimates—an optimistic, a pessimistic, and a most likely. The following often-used formula shows how the most likely estimate has the greatest influence:

$$\text{Activity time} = \frac{T_p + 4T_m + T_o}{6}$$

Where: T_p = Pessimistic time
T_m = Most likely time
T_o = Optimistic time

TM

D

For example, if $T_p = 5.0$, $T_m = 4.0$, and $T_o = 1.2$, the activity time would be:

$$\frac{5.0 + 16.0 + 1.2}{6} = \frac{22.2}{6} = 3.7$$

The appearance of a PERT diagram is the same as that of a CPM diagram. Both reflect single times for each activity. In the case of the PERT diagram, the single times are computed from the three estimates.

Computer-Based Network Diagrams

You can draw network diagrams by hand, or you can use a software package. Both CPM and PERT software packages have been incorporated into **project management systems** that prepare network diagrams and other graphics, and produce reports such as the one listing the dates in Figure D.4. The software performs all the necessary computations, such as determining the early and late times and identifying the critical path.

Putting Network Diagrams in Perspective

Network diagrams are excellent project management tools. They enable the MIS steering committee, information services management, and members of the project team to quickly and easily monitor the status of a development project. By using computer-based project management software, you can prepare high-quality diagrams without investing a lot of time in artwork, and changes can be made quickly and easily as you learn more about the project.

Problems

1. Draw a network diagram of the implementation phase of the SLC illustrated in Figure 8.8. After the hardware resources have been specified in step 3, work on steps 4 through 7 can be performed in parallel. Step 8 can be taken when steps 4 through 7 have been completed.

2. Determine the critical path for your network diagram in problem 1, using the following activity times:

Plan the implementation	3.0 weeks
Announce the implementation	0.2 weeks
Obtain the hardware resources	4.0 weeks
Obtain the software resources	12.0 weeks
Prepare the database	8.0 weeks
Prepare the physical facilities	10.0 weeks
Educate the participants and users	1.6 weeks
Cut over to the new system	3.0 weeks

Write a memo to your instructor, identifying the activities on the critical path and explaining your logic.

VI—ANALYZING THE EXISTING DISTRIBUTION SYSTEM

(Use this scenario in solving the Harcourt Brace problem at the end of Chapter 10.)

As one of the fastest-growing cities in the United States, Orlando has a shortage of good employees. This is especially true of computer specialists, and Harcourt Brace has had its share of high turnover. As soon as MIS employees gain experience, they often receive offers to work for other companies at higher salaries.

Companies combat employee attrition by making the surroundings attractive and the work challenging, and with ongoing recruiting. Companies visit college campuses each year to interview job candidates. Harcourt Brace follows such a practice and maintains a high recruiting visibility on college campuses—especially those in the Southeast.

During a recent campus visit, Harcourt Brace recruiters identified two senior MIS majors, Russ March and Sandi Salinas, as outstanding systems analyst candidates. Both were invited to Orlando for a company visit and were hired after additional interviews.

Today is their first day on the job, and they have just been ushered into Mike Byrnes's office by his secretary.

MIKE BYRNES: Have a seat, Sandi and Russ. I know you're glad that you have a successful college career behind you and that you're now a part of the Harcourt Brace team. It takes a lot of work to earn a degree, and we are just as happy as you that you made it. Now it's our responsibility to get you started on your systems analyst careers. You had your company tour when you were here before. Do you have any questions about what you saw?

RUSS MARCH: No, it was all very thorough and professional. I really liked being shown around the departments and getting to hear what the managers had to say about their areas. Being able to see something, rather than just hear about it, makes it a lot more meaningful.

SANDI SALINAS: Russ and I have talked about it, and the treatment we received at Harcourt Brace was the best of the companies we visited.

MIKE BYRNES: Well, I'm glad to hear that we're doing a good job. The tour is new. I think it's reflective of the entire operation. We try to emphasize quality in everything we do.

Since you were here last, Bill Presby and I have given a lot of thought to your first assignment. We would like to assign you both to the same COPS group. COPS is going to be with us a long time, and it forms the core of our applications software. I've asked Bill to be more specific about what you'll be doing. Let me take you to his office.

(Mike escorts Sandi and Russ to Bill Presby's office. Before leaving, he wishes them luck on their first assignment.)

BILL PRESBY: You remember Ira Lerner from your company visit. He's in charge of the order processing and billing systems. Ira has seven people working for him, but he could use some help. We've had a request from the Bellmawr distribution center to help them better schedule their people to fit the workload each day. As you know, we've frozen all changes to COPS1—that's what we're going to implement in November. But, the request from Bellmawr looks like a good feature for COPS2. Here is the service request form that users fill out for a system enhancement.

(Russ and Sandi study the form, which is reproduced in Figure HB6.1.)

SANDI SALINAS: It sounds like they already know what the problem is—an inability to efficiently schedule their employees.

BILL PRESBY: You're very perceptive. You must have sat through a lot of exciting lectures on problem definition in school. (Everyone has a good laugh, and Bill continues.) But, Bellmawr hasn't explored *why* they're not doing a good scheduling job. And, they haven't given much thought to a solution. They need some professional help in solving a systems problem, and you're going to be that help.

I want you two to go to Bellmawr and perform a systems analysis. Document the existing system. That will be good experience. After you've accomplished that, you can work with the rest of Ira's team to design the new system. But, before we send you off to Bellmawr, you need more details on our customer services department here in Orlando. You need to really understand the interface with Bellmawr. I've asked Sally Wizdo, the customer services VP, to fill you in. Let's go to her office now.

(Bill, Russ, and Sandi go to Sally's office, which looks out onto a large area where the order entry and customer relations clerical staffs are working. Bill returns to his office, and Sally begins her description. As Sally talks, Sandi and Russ take notes on notepads they carry for systems analysis.)

SALLY WIZDO: Our customers place orders two different ways—by mail and by phone. Let's talk about the mail first. It's the larger volume. The mailroom sends all mail orders to the mail coordinators, who open the envelopes and batch the orders in groups of 25 or so. (Sally gives Russ and Sandi an organization chart of customer services. Sally explains that customer entry services is on the left side and points to the mail unit. The chart is shown in Figure HB6.2.)

A seven-digit sequential number is assigned to each batch, and an entry is made in the batch log. It shows the batch number, the date, and the time of day.

Next, each batch is edited by the order editors. They ensure that certain key data elements are present. If the data elements are not there, then the order editors use a CEPOS terminal to obtain the data from the database. Then, they add it to the orders.

SANDI SALINAS: What's a CEPOS terminal?

SALLY WIZDO: CEPOS is the distribution system we obtained from CBS. It's our primary mainframe system. The terminal makes it possible to query the CEPOS database and get information about customers, inventory, and so on.

After the editing, the batches go to the CRT order entry operators, where the data is keyed into the computer. Then, a second group, called reviewers, repeats the

No. 50349	⬤ EDP SERVICE REQUEST	DATE:

System: _COPS_
Requestor: (Print) _Dave Mattson_   _Dave Mattson_ ☐ NEW ☐ CHANGE
Location: _Bellmawr_   Requestors Signature
CO: _____ DEPT: _Dist._ EXT: _____   Departmental Approval/Date

Request & Objective: Improve our employee scheduling program, by linking the warehouse standard performance data with the data from daily orders to be processed.

☐ ADDITIONAL INFORMATION ATTACHED

Anticipated User Savings; MAN DAYS / YEAR _____ YEARLY OPERATING EXPENSES: _____
DESCRIPTION

The system will provide the daily manpower schedule, based on Pack Lists received, and will reduce our supervisor's daily calculations for scheduling.

☐ ADDITIONAL INFORMATION ATTACHED

EDP USE ONLY   **REQUEST REVIEWED**

Request: ☐ ACCEPTED. TARGET COMPLETION DATE: _____
☐ NOT ACCEPTED AT THIS TIME   Manager Signature/Date
☐ REQUIRES FEASIBILITY STUDY TO BE DONE BY: _____
HBJ 429 ISD   Departmental Signature/Date

Figure HB6.1
Service request form.

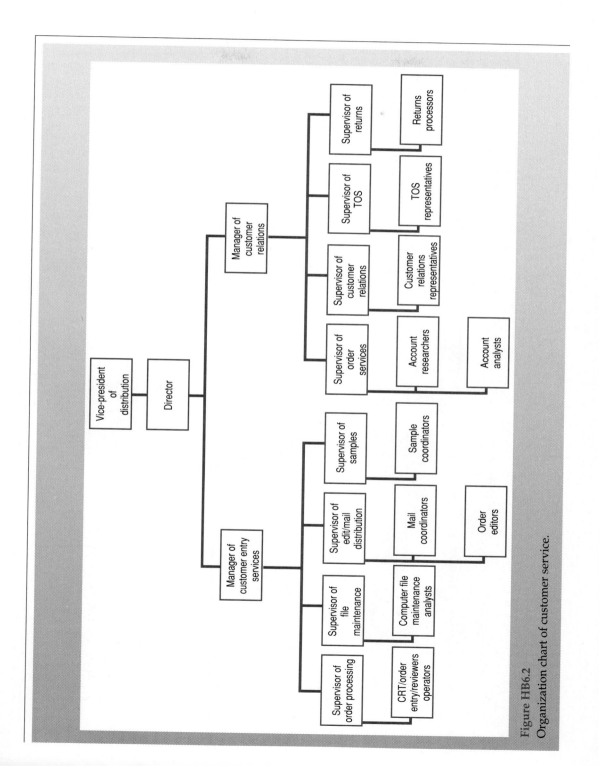

Figure HB6.2
Organization chart of customer service.

data is keyed into the computer. Then, a second group, called reviewers, repeats the keying process—character by character—to verify its accuracy. Inaccurate records are researched by the reviewers and rekeyed. After a successful review, the computer determines which distribution center should fill the order and writes that data on the disk. The order records are batched on the disk by distribution center and can be retrieved by the centers the next morning. The data is printed in the distribution centers in the form of pack lists that the warehouse clerks use when picking the items off the shelves for shipment. But, the warehouse operation is the physical system. What we have here in Orlando is the conceptual system.

RUSS MARCH: What about the telephone orders?

SALLY WIZDO: Good question. This is the point where those orders enter the system. The telephone operators are called TOS reps. TOS stands for *telephone order service.* They are part of the customer relations unit. (See the right side of Figure HB6.2.)

The TOS reps take the phone calls and enter the data into the computer using their CRTs.

Figure HB6.3
An order-entry operator keying order data into the computer.

SALLY WIZDO: Only rush orders. When we get a rush order, the TOS operator fills out a rush order form, and the form is faxed to the appropriate warehouse. For routine telephone orders, no form is filled out. This means that we can't verify the routine TOS orders as we can the mail orders.

RUSS MARCH: Does the lack of a review hurt you? Do you have many errors?

SALLY WIZDO: We used to not have many, but the number has picked up in the last couple of years. I guess it's because of the increase in volume. Our daily order volume increases every year. We might have to consider changing the system to keep the TOS errors to a minimum.

SANDI SALINAS: What happens in the distribution centers when they print out the pack lists?

SALLY WIZDO: I could tell you, but Bill Presby wants you to see for yourselves. I'll leave that part of the system up to the warehouse people.

(Sandi and Russ ask for a guided tour. Sally takes them to the various workstations where they talk with the supervisors, take notes, and obtain sample forms. The next day, Sandi and Russ fly to Philadelphia and rent a car for the short drive to the Bellmawr, New Jersey, distribution center. There, they are greeted by the center manager, Dave Mattson. Dave has a large floor plan of the warehouse on his office wall. He explains the various areas. See Figure HB6.4 for the layout.)

DAVE MATTSON: The books come to the receiving dock from the bindery. Receiving clerks inspect each shipment against the pack list enclosed by the bindery and send the pack list to the receiving office. There, a clerk uses the SLS stock locator system to determine where the shipment should be stored. SLS prints out stock location tags, which are attached to the cartons.

RUSS MARCH: What happens to the pack lists from the bindery?

DAVE MATTSON: Oh, they're filed in a history file. OK?

(Russ nods that he understands, and Dave continues.)

If the front of the warehouse has space for the newly arrived shipment, the cartons are placed on shelves in the elementary, high school, college, or health science areas. That way, the books are easily accessible to the order fillers. If no up-front space exists, the cartons are stored in the reserve area, where they sit on pallets. As many as six pallets can be stacked. The pallets are stored and removed using a forklift with a telescoping boom. Any questions?

RUSS MARCH: What kind of computer do the receiving people use?

DAVE MATTSON: A PC, networked to a network server. We have PCs in receiving, returns, packing, and freight staging. We also have PCs in the elementary, high school, college, and health science areas. All the PCs form a local area network, or LAN, under the control of a network server. The network server stores the SLS software and databases. I guess you studied LANs in college, didn't you?

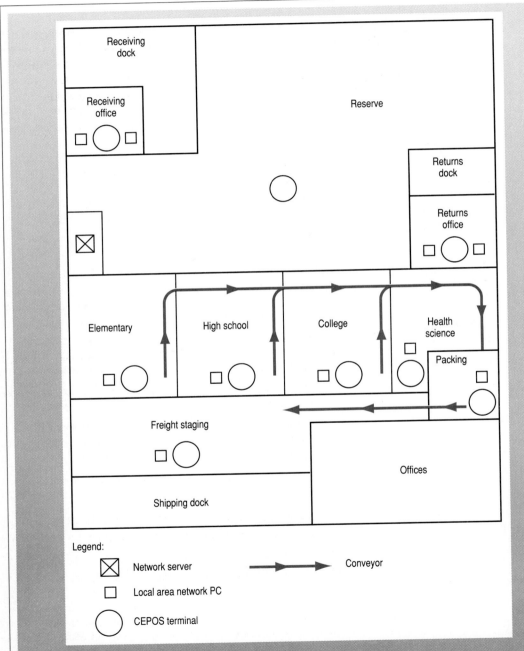

Figure HB6.4
Floor plan of the Bellmawr, New Jersey, distribution center.

(Both Sandi and Russ assure Dave that they have a good understanding of LANs. Russ asks what returns are, and Dave explains that they are books that customers return for some reason. The returns follow the same procedure as the receipts. Together, the returns and receipts represent the flow of books into the warehouse.)

DAVE MATTSON: That's it for receiving. Now, let's look at order filling and shipping. Let me show you the high-speed printer. That's were it begins. (Dave leads Russ and Sandi to a large room in the office area. In one corner is an IBM 6262 1200-line-per-minute printer, a CEPOS terminal, and a fax machine. Three people are sorting through computer printouts. Dave introduces Russ and Sandi to everyone and explains the operation.)

This crew comes to work at 7:00 A.M. and uses the terminal to instruct the computer in Orlando to transmit the pack lists. It takes about an hour to print the lists. The staff here sorts them on the worktables by warehouse area—elementary, high school, college, and health science. The pack lists are in a random sequence when they are printed out. At about 8 o'clock, the supervisors of those four areas come to get their lists. The supervisors take the lists back to their areas, pass them out to their order fillers, and the filling begins.

(Dave gives Sandi and Russ each a copy of the two-part pack list, which is illustrated in Figure HB6.5.)

The items are printed on the pack lists in a sequence that enables the order fillers to pick the items most efficiently. As an item is pulled, the clerk makes a check next

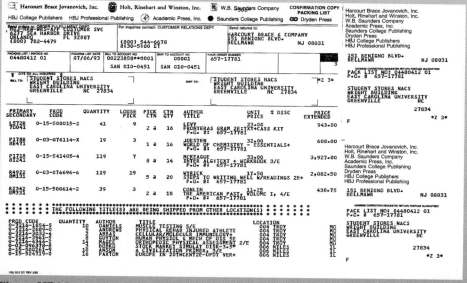

Figure HB6.5
Pack list.

to the item on the pack list. The items are placed in a large plastic box. When the order has been completely picked, the box travels on a conveyor to the packing area. There, the books are placed in shipping cartons, packing material is added, and a copy of the pack list is enclosed. The boxes travel on another conveyor to a scale area where they are weighed and postage is added. At this point, the shipping labels are glued to the boxes.

RUSS MARCH: Where do the shipping labels come from?

DAVE MATTSON: They're on the right side of the pack list. There are two on each of the two parts—making four labels in all. If we need more, we use a photocopier. After being labeled, the boxes are held temporarily in the staging area and then taken to the shipping dock by forklift. UPS makes its pickup at the dock each afternoon.

RUSS MARCH: What happened to the other copy of the pack list?

DAVE MATTSON: After the labels are removed, the originals are sent to the traffic department, where they are faxed to Orlando. There, the invoices are prepared and the inventory records are updated in CEPOS to reflect the items shipped.

SANDI SALINAS: So, what we saw in Orlando in Ms. Wizdo's area is order processing. It feeds the physical distribution system here, which feeds the billing and inventory systems back in Orlando. Correct?

DAVE MATTSON: You've got it.

RUSS MARCH: Does distribution feed both billing and inventory, or only one?

DAVE MATTSON: It feeds billing, and then billing feeds inventory.

RUSS MARCH: What about the faxes? You receive them for rush orders, don't you?

DAVE MATTSON: They come in on the fax machine you saw next to the high-speed printer. They look just like a pack list, except that they have RUSH printed at the top, are hand printed, and, of course, are only a one-part form. One of the staff makes a photocopy, staples the two together, and hand carries the set to the appropriate warehouse area. From that point, the order follows the same route as the regular orders.

SANDI SALINAS: I have one final question. What about the problem? As Mr. Presby described it, you can't efficiently schedule your people. Could you elaborate on that?

DAVE MATTSON: Certainly. Each warehouse area—elementary, high school, college, and health science—has a particular number of employees assigned. On a given day, the workload might be heavy in one area and light in another. The employees are cross trained to work in any area. When the area supervisors pick up their pack lists, they estimate the relative difference in the work loads and decide how many people will be shifted from one area to another for that day. It sounds good, but it's all very imprecise. Sometimes the supervisors guess right, and sometimes they don't.

RUSS MARCH: Do you have any figures on how long it takes to pick a certain volume of books?

DAVE MATTSON: As a matter of fact, we do. A few years ago, some consultants did a time-and-motion study and developed work standards. We can estimate quite accurately, based on the number of books and their locations, how long it should take to pick an order.

RUSS MARCH: Are those figures in the computer?

DAVE MATTSON: They certainly are. They are in a file called the Warehouse Standards file. We use it occasionally to determine the number of people to hire. Any other questions?

(Russ and Sandi indicate that they have run out of questions. They ask Dave to let them observe each warehouse area in operation. On the tour, they take notes and collect forms.)

Conclusion of the Systems Analysis

Sandi and Russ remain at Bellmawr for three more days, interviewing people in the various areas and departments. The analysts ask many questions, observe operations, and fill several notepads with notes. They are not concerned about the neatness of the notes, but want only to capture key points that can be expanded upon later. That opportunity comes when they return to Orlando. They document the processes with a series of data flow diagrams, supported by structured English. They document the data with data dictionary entry forms. All this documentation is placed in a three-ring notebook labeled *Project Dictionary—Bellmawr Employee Scheduling System*. In addition, Russ and Sandi write a system overview that precedes the documentation and explains the system in general terms.

Personal Profile—Ira Lerner, Group Manager (Order Processing and Billing)

Ira Lerner (on the right in the photo on page 397) joined Harcourt Brace in 1970 after studying business at Manhattan Community College, Baruch College, Rutgers University, and the University of Texas at San Antonio. Ira's hobbies include swimming and home repair. He perceives the keys to his success to be the same qualities that mark a good systems analyst. He is open to change and is able to analyze situations and determine the best option possible. As for Harcourt Brace's success, Ira anticipates that the new solid financial base will enable a return to the innovation that marked Harcourt Brace's traditional position of publishing leadership.

CHAPTER 10

Systems Analysis

Learning Objectives

After studying this chapter, you should:

- Be able to recommend to top management the process to follow in announcing the system study
- Know how the composition of the project team evolves during the system life cycle
- Know the techniques that the systems analyst uses to gather data during a system study and know some keys that contribute to their success
- Appreciate why documentation is so important in analyzing and designing business systems
- Have a general understanding of some popular documentation tools
- Understand the strengths and weaknesses of the documentation tools in documenting each of the CBIS subsystems
- Know an approach to defining the performance criteria for the new system
- Have an improved understanding of the project dictionary and its contents at the end of the systems analysis phase
- Understand the role that CASE tools play in system documentation
- Be aware of IBM's AD/Cycle framework for integrating CASE tools from multiple vendors

Introduction

When management authorizes a system study, the systems analyst has the responsibility of gaining an understanding of the existing system. This knowledge will provide the basis for the design of the new system.

The systems analyst has a major responsibility during the analysis phase but does not work alone. A project team is formed; it eventually includes members from the information services unit, the user area, and specialists from both inside and outside the firm.

The systems analyst uses a variety of data-gathering techniques while conducting the system study. As the analyst gathers the data, the findings are recorded, using both narrative and graphic notation. This system documentation enables the project team members to understand the system and to communicate their findings.

Documentation is accomplished using the tools of analysis and design. Traditional tools include flowcharting and record layouts. Currently popular tools include data flow diagrams, structured English, Warnier-Orr diagrams, the data dictionary, and entity-relationship diagrams.

Not all system studies are alike. Many possible influences on a study exist, but a major one is the type of CBIS subsystem being studied. Data processing, MIS, DSS, office automation, and expert system projects all present unique documentation demands.

Once the existing system has been documented, the user and the systems analyst jointly define the performance criteria of the new system. These are the standards that the system must meet in order to satisfy the user.

All documentation produced during the analysis phase is maintained in the project dictionary. Until recently, this documentation was created by hand. With the development of computer-aided software engineering, or CASE, it is now possible for the computer to assume a large portion of the system developers' workload. The popularity of CASE has resulted in a wealth of tools providing various degrees of support during the SLC. IBM has specified a framework called AD/Cycle that allows users to integrate CASE tools from several vendors.

Using CASE for Systems Analysis

The specific steps in system development may vary from firm to firm and from CASE tool to CASE tool, but all SDLCs include three basic steps. The first is analysis, the second is design, and the third is implementation. Of these, the most important is analysis. Without adequate analysis, the other two are meaningless. Mistakes made in analysis are magnified in design and implementation. CASE tools ensure that adequate analysis is accomplished and, therefore, enhance the likelihood that design and implementation will be successful.

Systems analysis can be viewed as the methodological process of understanding the informational and procedural needs and goals of a target system. CASE tools provide a methodology and tools for achieving this understanding. If the CASE tool is an I-CASE tool, the planning will have been completed before the analysis is begun. In systems analysis, the most important aspect is the completion of an **information requirements assessment.** Users and analysts work together to determine which features the target system requires and perhaps an additional list of desired but not mandatory features as well.

One of the techniques for analysis is to model the existing system in order to understand what it does, what it does well, and what it does not do well. In addition, the analysis phase also generates user interface requirements and standards, system performance criteria, target hardware and software environment requirements, and perhaps, most importantly, data requirements.

CASE tool outputs from the analysis phase may include decomposition diagrams, dependency diagrams, decision trees and tables, subschema data models, and data flow diagrams.

The Analysis Phase of the System Life Cycle

The purpose of the analysis phase, illustrated in Figure 10.1, is to define the user's needs in specific terms. The definition consists of two parts—the information the user needs in order to solve the defined problem, and the level of performance the user will expect of the computer-based system that provides the information. Once these needs have been defined, the systems analyst prepares a proposal that recommends either proceeding to the design phase or scrapping the project. Management considers the proposal and makes the go–no go decision.

Step 1—Announce the System Study

Before the system study begins, management announces to the employees who will be affected. The announcement has two important objectives. First, it helps quiet unfounded employee fears, and, second, it contributes to acceptance of the systems analyst by employees in the user area.

Employees are most interested in how a new system will affect them. The new system might change the nature of their work, change their relationships with the people around them, or even cause them to lose their jobs.

Employees should have little to fear from a new system for two reasons. First, management seldom implements a system for the purpose of terminating employees. Second, a system usually takes over the boring, redundant portion of the employees' tasks.

Coming from the information services unit, the systems analyst initially will be regarded as an outsider. The announcement of the study is the first step toward the analyst being regarded by the employees as one of their own. Management endorses the analyst as the person selected to do this important work.

The Content of the Announcement

In making the announcement, management should make four key points:

1. **State why the firm has decided to conduct the study.** Management should identify the problem or potential problem and describe the main features

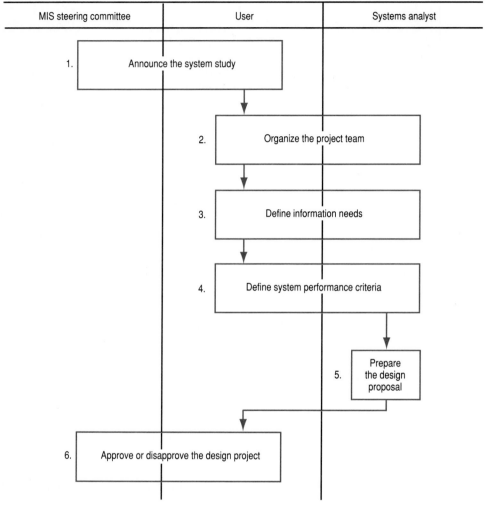

Figure 10.1
The analysis phase.

the new system offers. This material comes from the problem definition and the system objectives established during the planning phase.

2. **Explain the value of the new system to the firm.** Management should explain how the new system will enable the firm to achieve a competitive advantage. By being a stronger competitor in the marketplace, the firm will better meet its customers' needs and will improve its long-term financial stability.

3. **Explain the value of the new system to the employees.** Management should emphasize the new system's potential for making the firm a better employer and for making the employees' work more challenging and interesting.

4. **Ask for employee cooperation.** Management should clarify the relationship between employee cooperation and system success. First, employees must cooperate by providing data during the system study. Later, employees must cooperate as system participants and users.

The announcement gets the project off on the right foot. It makes the point that a problem exists and that management is dedicated to solving it.

Step 2—Organize the Project Team

Although a project team develops the new computer-based system, not all team members participate during the analysis phase. Figure 10.2 is a chart that shows how the members contribute their expertise at certain points during development. During some phases, team members play major roles; during others, they play

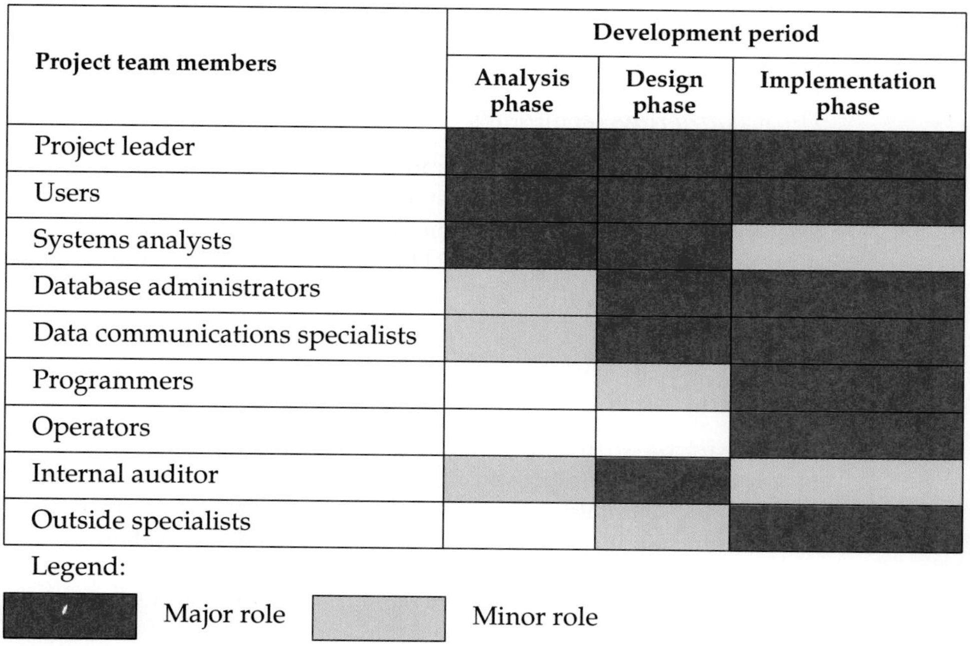

Project team members	Development period		
	Analysis phase	Design phase	Implementation phase
Project leader	Major role	Major role	Major role
Users	Major role	Major role	Major role
Systems analysts	Major role	Major role	Minor role
Database administrators	Minor role	Major role	Major role
Data communications specialists	Minor role	Major role	Major role
Programmers		Minor role	Major role
Operators			Major role
Internal auditor	Minor role	Major role	Minor role
Outside specialists		Minor role	Major role

Legend: ▓ Major role ░ Minor role

Figure 10.2
Project team members participate throughout the development period.

minor roles. As the figure shows, the project leader, users, and systems analysts play major roles during the analysis phase.

Recall from Chapter 5 that the electronic exchange of data by multiple firms is called an interorganizational system (IOS). When the project involves the development of an IOS, the team can include users from outside the firm. For example, a manufacturer's project team can include representatives of suppliers, wholesalers, and retailers.

Step 3— Define Information Needs

The systems analyst determines the information requirements of the new system by studying the existing system. This enables the analyst to identify the weaknesses that the new system must overcome.

How Thorough Should the Study Be?

Some proponents of modern systems development, such as the rapid application development (RAD) methodology explained in Chapter 8, urge that a minimum amount of time be devoted to studying the existing system. They argue that it is pointless to learn everything there is to know about a system that will soon be replaced. You cannot argue with this logic. Therefore, the attitude of the systems analyst toward the system study should be: "I'm going to study the existing system to the extent that is necessary to design the new one—and no more."

Data Gathering Techniques

The systems analyst uses any and all means available for gathering necessary data. A good place for the analyst to start is the documentation of the existing system. These findings can be augmented by in-depth personal interviews, surveys, observation, and record search. The analyst uses the combination of these data-gathering techniques that best fits the problem being solved.

Analysis of Existing System Documentation

It is rare for the systems analyst to develop a brand new system. In most cases the task is to replace or improve a current system, and the chances are good that some form of documentation is present. When the system is a manual one, the documentation may include only written job procedures. When the system is computer-based, the documentation may include nothing more than program listings, or it might include graphic documentation such as flowcharts and print layout forms.

The older the existing system, the less likely the documentation will be a big help to the analyst. One reason is that the system has probably been patched so many times that its documentation presents a poor picture of the current design.

The condition of the existing system documentation gives the analyst a starting point. The analyst supplements this documentation by employing other data-gathering methods.

The In-Depth Personal Interview

An **in-depth personal interview** is a lengthy, face-to-face session involving an interviewer and an interviewee. The systems analyst is the *interviewer,* the person who asks the questions, and the user is the *interviewee,* the person who answers them. The analyst conducts the interview in the setting best suited to gathering the data: the user's office or work area. Here, the user feels the most comfortable and, as the user talks, the analyst takes notes.

In most cases several interviews are required, spread over a period of several days or even weeks. The first meeting is not intended to provide all the data. It is mainly a get-acquainted session. The talk may center on such nonbusiness topics as favorite TV programs, sports, and hobbies. It is important that a bond of friendliness and mutual respect be established between the user and the systems analyst, and the first meeting sets this tone.

Structured and Unstructured Interviews

Once the interviews turn to business, the questioning can take either of two forms—structured or unstructured. In a **structured interview,** the analyst prepares a list of questions ahead of time and makes an effort to follow that sequence. An **unstructured interview,** on the other hand, begins with no formal plan and follows a path determined by the user's answers.

How does the analyst know which type of interview to use? The determining factor is the analyst's knowledge of the problem area. When the knowledge is limited, the unstructured interview is most appropriate. As the user answers each question, the analyst learns more about the problem, and this increased knowledge provides the basis for more questions.

When the analyst is already an expert in the problem area, the structured interview is the way to go. The analyst can draft a list of questions and follow that sequence.

Tips for Conducting In-Depth Personal Interviews

You could devote a lifetime to sharpening your interviewing skills, but here are a few tips that others have found to be especially effective:

- **Make an Appointment** Always make an appointment, either by telephone or in person, with the user. Give the user a brief idea of why you want to conduct the interview. Then, *be on time.*

- **Dress Properly** Dress in a manner that does not distract the user. A good guideline is to dress in such a way that after you leave, the user cannot remember what you were wearing.

- **Use a Question List** A list of the questions you want to ask will ensure that you get the information you need and that you do not waste the user's time or your own. The user will appreciate the fact that you are prepared.

- **Be Flexible** If the user says or does anything to indicate that your line of questioning is leading nowhere or going in the wrong direction, you should revise your questions. This applies even to a structured interview. Break off the planned line of questioning and see where the new path will lead. You might obtain information you never knew existed. Also, keep in mind that a person can communicate in ways other than words. Facial expressions and gestures—called *body language*—often reveal a person's inner feelings.

- **Use a Tape Recorder Whenever Possible** When you make the appointment for the interview, find out whether the user objects to being recorded. Usually, there are no objections—especially at upper management levels, where there is little fear of being quoted. Assure the user you will not make the tape available to anyone outside the project team without first obtaining permission. By taping the interview, you will not have to worry about taking notes and will have an accurate record not only of *what* is said but of *how* it is said.

The in-depth personal interview is the best way for the the analyst and the user to jointly define the user's information needs.

When to Use an In-Depth Personal Interview

A major drawback to the in-depth personal interview is that it is the most time-consuming data-gathering method. For that reason, the analyst uses the technique only with key people in the user's area—the manager and employees active in the problem area.

The Survey

Where the in-depth personal interview asks different questions of a small number of persons, a **survey** asks the *same* questions of a *large* number of persons. The person asking the survey questions (the systems analyst) is the *surveyor*, and the person answering them (the user) is the *respondent*.

The survey is increasing in importance as a data-gathering method because of the trend toward interorganizational and distributed systems. Surveys provide the systems analyst with a way to efficiently and inexpensively gather data from users and participants hundreds or thousands of miles away.

Chances are you have been a respondent in a survey, answering questions about where you like to shop or which political candidates you prefer. The surveyor probably called you on the phone or perhaps interviewed you in a mall. These are telephone and personal surveys, respectively. There is a third type—a mail survey.

Personal Survey

The *personal survey* is one conducted by personally interviewing the respondents. Because more respondents are involved than in an in-depth interview, the length of the personal survey interview is usually much shorter.

The setting for the personal survey interview is the same as for the in-depth personal interview—the user's office or work area. Questions are printed on a form called a **questionnaire.** The analyst reads the questions and writes the user's answers in the appropriate spaces on the questionnnaire.

Objective and Subjective Questions   Although the personal survey interview is always structured, the questions can be either objective or subjective.

An **objective question** is one that can be answered only in certain, predefined ways. As a student, you are familiar with this format. It is used for true/false and multiple-choice questions.

Figure 10.3
The systems analyst conducts in-depth personal interviews and personal surveys in the user's office or work area.

Rather than ask a true/false question, the analyst asks a question such as: "Do you think that the problem is the lack of user training?" The respondent replies "Yes," "No," or "I'm not sure." (Don't you wish you could reply "Not sure" to a true/false question?)

Multiple-choice questions are difficult to ask in a personal interview because the respondent usually likes to view all the choices before answering. This difficulty can be overcome by making a card containing the list of possible answers and showing it to the user as the question is asked.

The main advantage of objective questions is that they make it easy to conduct a computer analysis of the responses. The main limitation is that the format often forces users to respond with answers they would not ordinarily give, creating mistaken impressions about the nature of the problem.

This limitation is overcome by another type of personal survey interview question, called the **subjective question,** also called an **open-ended question** because it permits the respondents to answer in any way they like. For example, the analyst asks: "How do you feel about the existing system?" The questionnaire contains several blank lines where the analyst records the answer. The main advantage of the subjective question is its flexibility. The main disadvantage is the difficulty of analysis because of hard-to-classify answers.

Tips for Designing Personal Survey Questionnaires   Most of the in-depth personal interview tips also apply to the personal survey interview. You should make an appointment, dress properly, and be alert to the user. In addition, you should keep certain tips in mind when designing questionnaires:

- **Keep the Questionnaire Short** By designing the questionnaire so that the questions can be answered in no more than, say, 20 minutes, you conserve not only the user's time, but yours as well. Plus, you make certain that the data analysis will not become too time consuming.

- **Ask the Easy Questions First** Begin with questions that are easy to answer. If you begin with the more difficult ones, the user may decide not to participate, and you could be left with a blank questionnaire. The likelihood of the user not cooperating lessens once the questioning begins.

- **Understand the Importance of Sequence** The answer to one question can influence the answer to another. For example, if you ask the user to list the weaknesses of the existing system and then ask how the user feels about the existing system, you can imagine what the response will be. Ask for general feelings first, and then follow up with specifics.

- **Validate the Responses** You are never certain whether the user is giving answers that reflect the true situation. You should conduct **data validation** by comparing the gathered data with other, related data. For example, if you ask an executive how many telephone calls he receives each day and he replies "About forty," validate the response by asking his secretary to keep a log for several weeks. People are not always completely accurate when it comes to their perceptions. It is *always* a good idea to validate responses. This applies to *any* data-gathering technique.

The personal survey is the best way to directly gather data from a large number of people.

When to Use a Personal Survey   The personal survey is second only to the in-depth personal interview in effectiveness. Once the analyst has used the in-depth personal interview for key people in the user area, he or she will use the personal survey for those others who can be met with personally.

Telephone Survey

The same general survey rules apply when the analyst conducts a *telephone survey*. The main difference is that the analyst does not have as much control. A telephone respondent can more easily terminate the session than in a personal interview.

Tips for Conducting Telephone Surveys   The same tips used in designing personal survey questionnaires also apply to telephone surveys. Keeping the questionnaire short is even more critical. Telephone questionnaires that keep the respondent on the line for more than five or ten minutes should be used only in special circumstances. In addition, you should:

- **Consider Using a Professional Researcher** Conducting a good telephone interview is an art, and is best performed by someone with experience. Marketing research firms can be hired to conduct large-scale telephone surveys. And even though the researcher is a professional, do not assume that he or she will perform exactly as desired. Monitor the researcher's calls until you are sure that the procedure is being followed and that the researcher is conveying the desired image. Professional telephone researchers expect this practice.

- **Design the Questionnaire for Easy Administration** Make certain the questionnaire identifies the question sequence to be followed. For example, notes such as "If YES, go to question 23" should be incorporated.

The attitude of the telephone respondent to the project is greatly influenced by the way the researcher conducts the interview. Do not overlook the potential advantage of the telephone interview to create a positive attitude for the project among the respondents.

When to Use a Telephone Survey   The telephone interview provides a good way to establish a two-way dialogue with persons in remote user areas. It ranks second only to the two forms of personal interviews in effectiveness.

Mail Survey

When the cost of a telephone survey is too great, a *mail survey* might be a good alternative.

Tips for Designing Mail Survey Questionnaires   The main problem with the mail survey is the potentially low response rate. In some commercial mail surveys, only 5 percent or so of the respondents return completed questionnaires. To keep the response rate as high as possible, consider the following tips:

- **Keep the Questionnaire Short** Questionnaire length is extremely critical for mail surveys. Ideally, the questionnaire should be no longer than a single page. If necessary, additional questions can be printed on the back. If still more space is needed, a four-page, folded questionnaire printed on a single sheet is the next best choice. Also, use small type to pack more questions into a limited space.

- **Personalize the Mailing** Enclose the questionnaire in an envelope addressed to a specific individual, not a job title. Include a short cover memo from your boss, the manager of systems analysis or, even better, the CIO, explaining the purpose of the survey.

- **Use the Telephone for Follow-up** If you have not received the completed questionnaire after two weeks, follow up with a telephone call. Offer to send a second questionnaire if one is needed.

In addition to these techniques for achieving a higher response rate, keep the following suggestions in mind for increasing the effectiveness of the questionnaire.

- **Consider Using Scales** Ratings scales offer a good way for users to express their opinions. One type of scale, called a **Likert scale,** consists of several points on a continuum labeled with descriptors. The scale follows a statement. For example:

    The system provides me with valuable problem-solving information.

| Strongly agree | Agree | Indifferent | Disagree | Strongly disagree |

Sometimes you will want to assign weights to the various responses. For example, a "Strongly agree" response might be worth 5 points, an "Agree" worth 4, and so on. The weights facilitate quantitative analysis.

- **Facilitate Computer Input** Optical character recognition (OCR) is the best way to enter mail survey data. The data entry is faster and more accurate than keyboarding. The questionnaire in Figure 10.4 provides a good example of an OCR form. The respondent uses a pencil or pen to enter letters and numbers in the boxes and to darken the ovals. When OCR is not available, design the form to facilitate keyed entry. Key entry is covered in the next chapter, when we address designing the new system inputs.

- **Provide for Unanticipated Answers** Leave a response position for answers you cannot anticipate. Question 1 in Figure 10.4 is a good example. One response position is labeled "Other," with a blank line following.

WARRANTY REGISTRATION FORM   Please return within 2 weeks of purchase or gift.

DO NOT WRITE IN THIS AREA

Completing this information will register your skate warranty. Completion of questions will qualify you to receive a free gift, compliments of Rollerblade.

Marking Instructions
- Use a No. 2 Pencil or, Blue or Black Ink Pen Only
- Fill the Oval Completely
- Make No Stray Marks

Right	Wrong

Mailing Instructions
- When completed return by mail, folded and unsealed
  - Fold Only Where Indicated
  - DO NOT seal the form with tape, glue or staples

○ Mr.   ○ Mrs.   ○ Ms.

Last Name   **First Name**   **Initial**   **Warranty Number**

Street   **Home Telephone No.**

City   **Owner's Signature**

State   **Zip Code**   **Date of Purchase**   **Name of Store where purchased**

Month Day Year

Printed in U.S.A.   Mark Reflex ® by NCS MP83387:321

◄ FOLD HERE

To help us understand our customers and their lifestyles, please answer the following questions.

1. Where did you purchase your Rollerblade skates:
- ○ Received as a gift
- ○ Department Store
- ○ Sportings Goods Store
- ○ Surf Shop
- ○ Bike Shop
- ○ Ski Shop
- ○ Hockey Shop
- ○ Discount Store
- ○ Other: _____

2. Skate model purchased:
- ○ Racerblade 908
- ○ Macroblade 608
- ○ Lightning 608
- ○ Zetra 303
- ○ Zetra 100
- ○ BladeRunner

3. Skate size purchased:
- ○ 1
- ○ 2
- ○ 3
- ○ 3½
- ○ 4
- ○ 4½
- ○ 5
- ○ 5½
- ○ 6
- ○ 6½
- ○ 7
- ○ 7½
- ○ 8
- ○ 8½
- ○ 9
- ○ 9½
- ○ 10
- ○ 10½
- ○ 11
- ○ 11½
- ○ 12
- ○ 12½
- ○ 13
- ○ 14

4. Primary reason for choosing this skate model? (Mark only one)
- ○ Product features
- ○ Rollerblade reputation
- ○ Quality
- ○ Value/Cost
- ○ Color/Appearance
- ○ Other

5. Which of the following factors influenced your purchase?

Advertising	Media Coverage	
○ Newspaper	○ Newspaper	○ Rollerblade Demo Van
○ Magazine	○ Magazine	○ Family/Friend
○ TV	○ TV	○ Other:
○ Outdoor		

6. Is this your first purchase of In-Line skates?
- ○ Yes   ○ No ⟹ If No, how many pair have you previously purchased for yourself?

○ One   ○ Two   ○ Three   ○ Four or more

7. Other members of my family who own Rollerblade skates?
- ○ Only Myself
- ○ Spouse
- ○ Mother
- ○ Father
- ○ Son
- ○ Daughter

8. Are you considering purchasing Rollerblade accessories?
- ○ BLADEGEAR
- ○ Wrist Guards
- ○ Elbow Guards
- ○ Equipment Bag
- ○ Knee Pads
- ○ Gloves
- ○ Helmet
- ○ Other: _____

9. Indicate your current In-Line skating ability:
- ○ Beginner
- ○ Intermediate
- ○ Advanced
- ○ Expert

10. Indicate your primary usage of In-Line Skates: (Mark All That Apply)
- ○ Recreational/Social
- ○ Roller hockey
- ○ In-Line Racing
- ○ Alpine Ski Training
- ○ Nordic Ski Training
- ○ Personal Transportation
- ○ Aerobic/Fitness
- ○ Cross-Training
- ○ Training for Cycling
- ○ Figure Skating/Dancing
- ○ Other:

11. Approximately how many hours per week do you skate?
- ○ Less than 1 hour
- ○ 1 - 5
- ○ 6 - 10
- ○ 11 - 15
- ○ More than 16

12. How many people do you usually skate with?
- ○ Skate Alone
- ○ 1 Other
- ○ 2 - 3 Others
- ○ 4 or More

◄ FOLD HERE

13. What additional physical activities do you regularly participate in?

PRIMARY ACTIVITY (Mark Only One)
- ○ Ice Skating/Figure
- ○ Ice Skating/Speed
- ○ Ice Hockey
- ○ Bicycling
- ○ Running
- ○ Fitness Walking
- ○ Aerobics
- ○ Swimming
- ○ Triathalons
- ○ Alpine Skiing
- ○ Nordic Skiing
- ○ Snowboarding
- ○ Skateboarding
- ○ Surfing/Sailboarding
- ○ Racquet Sports
- ○ Golf

SECONDARY ACTIVITY (Mark Only One)
- ○ Ice Skating/Figure
- ○ Ice Skating/Speed
- ○ Ice Hockey
- ○ Bicycling
- ○ Running
- ○ Fitness Walking
- ○ Aerobics
- ○ Swimming
- ○ Triathalons
- ○ Alpine Skiing
- ○ Nordic Skiing
- ○ Snowboarding
- ○ Skateboarding
- ○ Surfing/Sailboarding
- ○ Racquet Sports
- ○ Golf

⊑ **PLEASE CONTINUE ON THE REVERSE SIDE** ⟹

Figure 10.4
A questionnaire designed for optical character recognition.

These tips not only produce a higher response rate than otherwise might be expected but also provide better data.

When to Use a Mail Survey   If you cannot ask questions of the respondent either in person or over the telephone, the next best choice is the mail survey. It is relatively inexpensive, and much valuable data can be gathered. In fact, you might be able to obtain data by mail that you could not gather in person. People are often more open in written responses than in the ones they give in person or over the telephone. This is especially true when the mail respondent can remain anonymous. Consider the mail survey as a way to gather data on sensitive issues.

Putting the Survey Data Gathering Techniques in Perspective

The survey is the best way to gather data from a large number of respondents. Responses to the objective questions can be entered into the computer and analyzed using special survey software or statistical packages. Answers to subjective questions can also be keyed in, sorted by question, and printed for visual analysis.

Because of the structured nature of the questionnaire, it is difficult to probe for answers as to *why* a particular situation exists. A good strategy is to use a mail survey to get a general overview and then follow up with personal or telephone surveys to focus on specifics.

Observation

The earliest professional systems analysts, industrial engineers (IEs), pioneered data gathering by means of observation. **Observation** is viewing an activity as it occurs. For example, IEs often use stopwatches to time how long it takes workers to perform operations.

The toolkit of the systems analyst does not include a stopwatch, but observation is a good way to gather data. Usually, the analyst will conduct personal interviews and then ask permission to observe the activity in the user area.

Tips for Gathering Data by Observation

When using observation, you should:

- **Be Inconspicuous** The secret to obtaining good data is to do so without influencing the work process. When employees know they are being observed they tend to work faster, do more important jobs, demonstrate greater interest, and so on. Position yourself outside the main work area; do not engage in conversation, do not move around, and do not take notes. After a while the employees will forget you are there. As soon as you get back to your desk, write down what you saw.

- **Observe a Representative Work Flow** Stay long enough to observe periods of both heavy and light activity to acquire a sense of the fluctuation in workload.

The entire time you are in the user area, you should be alert to what is going on. Follow the above tips when conducting observation formally.

When to Use Observation

Use observation when you have gathered all the data you can by in-depth personal interview and survey, yet believe there is more to learn. This is usually the case, if only for the reason that interview findings need to be validated.

In addition, observation provides a better understanding of the setting in which the employees do their work than does any other data-gathering technique. The setting often explains why things happen as they do.

Also, observation increases the opportunity for the analyst to be accepted by workers. For this reason, *it is always a good idea for the analyst to work in the user area rather than in his or her office.* With the analyst on site, users develop an appreciation for the amount of time and effort that goes into the study and are more likely to respect the analyst's recommendations.

Record Search

It is often necessary to search through the user's files to gain an accurate picture of the problem. This activity is called **record search** and is without a doubt the most unglamorous part of analysis work.

For an example of the need to do record search, assume that a unit has had difficulty keeping up with the daily volume of sales orders. On certain days the volume is too great to handle. On other days there is much idle time. A record search can be conducted of the file of sales orders for the past several months and the number of orders for each day tabulated. This data provides a clear picture of the fluctuation in transaction volume that the new system will be expected to handle.

Tips for Gathering Data by Record Search

When you see the need to conduct a record search, make an effort to:

- **Use Part-Time Help** Record search is extremely labor intensive. Consider using part-time help, perhaps college students. The cost is lower than if the analyst does the work, and the analyst is freed to perform tasks that better utilize his or her knowledge and skills.
- **Design a Special Data-Gathering Form** Findings are normally written down as the records are examined. Rather than using scraps of paper or the backs of envelopes, design a recording form tailored to the specific type of data being gathered. In the above example of determining the daily volume of sales orders, a form such as the one in Figure 10.5 could be designed.

DAILY SALES ORDER VOLUME									
Date			Day of week	Volume	Date			Day of week	Volume
Yr	Mo	Da			Yr	Mo	Da		
9 3	0 1	0 4	M O N	155					
		0 5	T U E	78					
		0 6	W E D	82					
		0 7	T H U	111					
		0 8	F R I	93					
		1 1	M O N	165					
		1 2	T U E	89					

Figure 10.5
A special form designed to facilitate record search.

A unit's records provide an accurate view of the past. An understanding of this past is helpful when designing a system for the future.

When to Use Record Search

Use record search when interviews, surveys, and observation cannot provide all the data, or when there is a need to validate findings.

Putting Data Gathering in Perspective

The point in the analysis phase where you identify the user's information needs is where you make or break the project. There is probably no greater challenge in business than being asked by a user to solve a problem that he or she has not been able to solve. The methodology of the systems approach provides the framework for you to meet this challenge, and the data-gathering techniques are applied within this framework. These data-gathering techniques are tools of systems analysis and design—just as important as, or more important than, the tools for documenting systems.

The Role of Documentation

As the systems analyst learns about the existing system, she or he creates a written record of the findings, called **system documentation,** which describes two features of the system: (1) the data that flows through the system, and (2) the processes that transform the data.

The documentation can take several forms, depending on the tools used. A **documentation tool** is a graphic or narrative technique used to describe a system.

One important point must be made concerning the documentation tools: *All documentation tools can be used in both systems analysis and design.* During analysis the tools document the existing system. During design the tools help design the new system.

The Objectives of Systems Documentation

Systems documentation accomplishes two main objectives. First, it enables project team members to understand the system. Second, it enables team members to communicate the details of the system.

Understanding   Documentation enables the information specialists to identify the parts of the system and see how they fit together. The processes usually are performed in a set sequence, and the data is used by each process—as input, output, or storage.

This ability of the documentation to aid understanding is valuable during both system development and use. Because some business systems remain in use for years, it is common for one group of analysts to create a system and another group to maintain it. The group charged with the maintenance may be unfamiliar with the system and depend entirely on the documentation for understanding.

Communication   We have seen that systems are analyzed and designed by more than one person—the team approach. Team members must communicate with one another as they perform their specialized tasks. Team members also need to communicate with system participants and users. The documentation produces formats that are recognized by systems professionals around the world. The documentation is a *universal language* of systems analysis and design.

Traditional Documentation Tools

During the punched-card era and the early years of computing, systems analysts documented systems processes with flowcharts and data with record layouts. Although many firms have switched to structured tools, many existing business systems are documented in the traditional way.

Flowcharts

There are two types of flowcharts—system and program. Figure 10.6 illustrates a system flowchart. A **system flowchart** is a diagram that uses symbols to show

how the major processes and data of a system are interconnected. The processes can be performed manually, with the aid of keydriven machines such as calculators or typewriters, and with computers. Computer processes are illustrated with rectangles, each representing a single program. The system in Figure 10.6 consists of three computer programs. Each processing step is numbered.

System flowcharts are usually drawn by systems analysts. The responsibility for documenting programs in the system falls on the shoulders of the programmers, who may use program flowcharts. A **program flowchart** is a diagram that uses interconnected symbols to show the detailed processes performed by a single computer program. Figure 10.7 illustrates a program flowchart of the program in step 4 of the system flowchart in Figure 10.6. This program flowchart conforms to the rules of structured programming. The driver module is at the left, and four subsidiary modules are at the right.

One feature of flowcharting that distinguishes it from other tools is that flowcharts document the system in either a summary or a detail fashion, the system flowchart providing the summary and the program flowchart the detail.

An inability to document in intermediate degrees of detail when using flowcharts creates two problems for the systems analyst. First, it makes it difficult to take a top-down, structured approach because the analyst cannot gradually progress to the detail. Second, it makes it difficult to communicate designs to users. A user might understand the system flowchart but get lost in the detail of the program flowchart. Most program flowcharts are considerably more complex than the one shown in the figure.

Record Layouts

Early systems analysts documented the records of the data files with record layout forms such as the one pictured in Figure 10.8. A **record layout** is a form that shows the fields of data composing a record. The form typically includes multiple layers, each with a capacity of 100 characters. One or more layers are used to identify the locations of the data fields in a record.

Record layouts and the flowcharts enabled early systems analysts and programmers to document their systems. Record layouts are, for all purposes, obsolete. However, flowcharting remains a valuable documentation tool. For that reason, a detailed description of flowcharting is included in Technical Module G.

Structured Documentation Tools

Structured programming became so popular that systems analysts applied the same approach in documenting complete systems. The structured analysis and design tools that have stimulated the most interest are data flow diagrams, structured English, Warnier-Orr diagrams, the data dictionary, and entity-relationship diagrams.

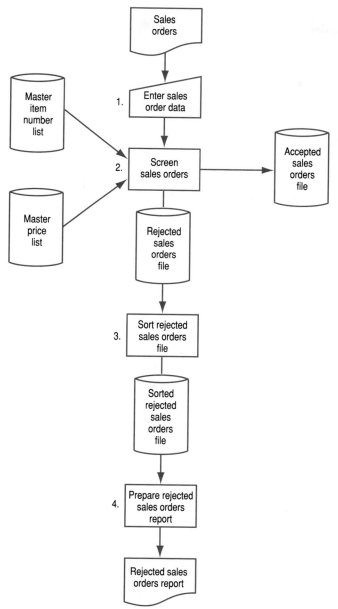

Figure 10.6
A system flowchart.

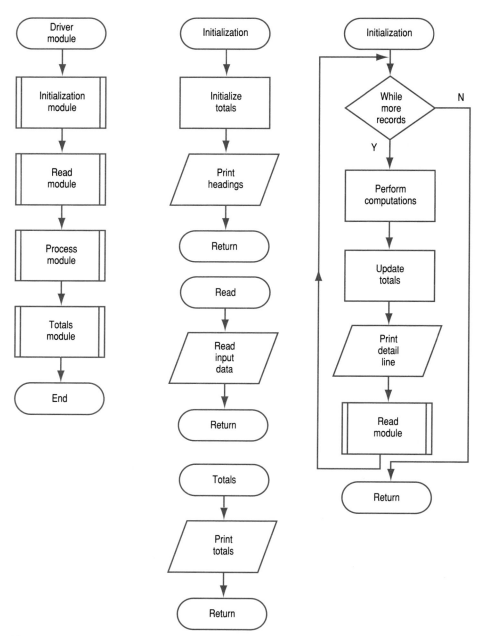

Figure 10.7
A program flowchart.

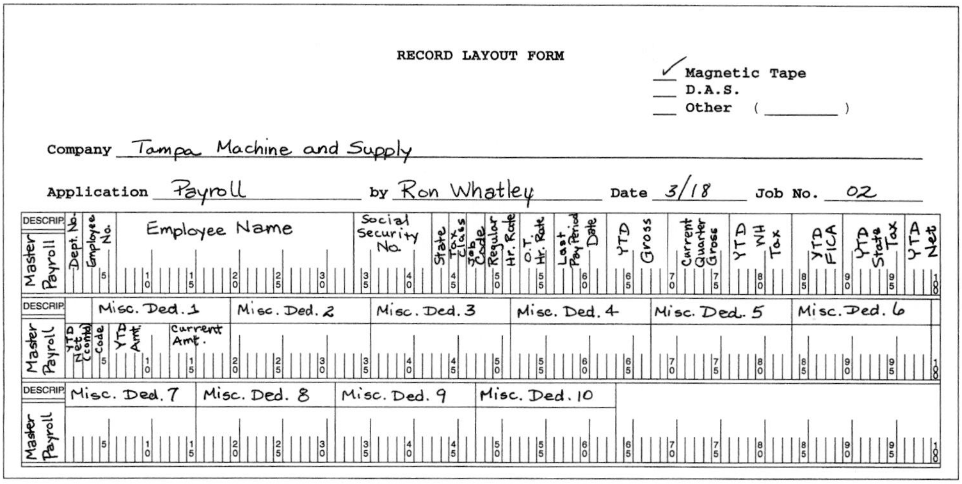
RECORD LAYOUT FORM

✓ Magnetic Tape
__ D.A.S.
__ Other (________)

Company Tampa Machine and Supply

Application Payroll by Ron Whatley Date 3/18 Job No. 02

DESCRIP: Master Payroll — Dept. No. | Employee No. | Employee Name | Social Security No. | State Tax Class | Job Code | Regular Hr. Rate | O.T. Hr. Rate | Last Pay Period Date | YTD Gross | Current Quarter Gross | YTD WH Tax | YTD FICA | YTD State Tax | YTD Net

DESCRIP: Master Payroll — YTD Net (cont'd) | Code | YTD Amt. | Current Amt. | Misc. Ded. 1 | Misc. Ded. 2 | Misc. Ded. 3 | Misc. Ded. 4 | Misc. Ded. 5 | Misc. Ded. 6

DESCRIP: Master Payroll — Misc. Ded. 7 | Misc. Ded. 8 | Misc. Ded. 9 | Misc. Ded. 10

Figure 10.8
A record layout form.

Data Flow Diagrams

A **data flow diagram (DFD)** documents the processes of a system with circles or rectangles with rounded corners and documents the flow of data with arrows. Stationary files of data can be represented with open-ended rectangles, and the elements in the system's environment represented with squares. Note that this is the environment of the system, not the firm.

Unlike flowcharts, DFDs document a system in various levels of detail. The DFD that summarizes the system, illustrated in Figure 10.9, is called a **context diagram.** The single process symbol in the center represents the entire system. The flows of data from and to the system environment are represented with arrows.

We can explode the single process symbol of the context diagram to show the major processes. This produces the **Figure 0 diagram** in Figure 10.10. It is called a Figure 0 diagram because the DFD on the next higher level, the context diagram, does not contain a numbered process. A DFD name refers to the number of its higher-level process.

Each process in the Figure 0 diagram can then be documented with its own DFD. This top-down documentation continues until only the detailed processes remain.

DFDs are a natural way to document a system. The symbols and arrows are much like the notes you sketch on a yellow pad as you interview a user. However, DFDs are best suited to document systems on the higher hierarchical levels. They must be supplemented with other tools for the details, tools such as program flowcharts or structured English. The details of data flow diagramming are explained in Technical Module B.

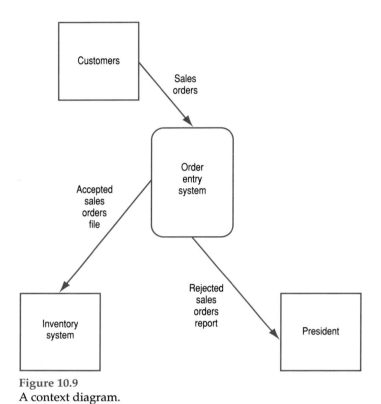

Figure 10.9
A context diagram.

Structured English

Structured English, a narrative in a shorthand form, looks like computer code but is not. Earlier versions of structured English were called **pseudocode.**

An example of structured English appears in Figure 10.11. This structured English explains the logic and arithmetic performed in step 1 of the Figure 0 diagram in Figure 10.10.

This structured English example reflects the format of a structured program. The driver module is bounded by the words START and STOP. Just below STOP, the names of the three subsidiary modules are aligned on the left margin, and the processes within each module are indented to show hierarchy. Names of data elements, such as CUSTOMER.NUMBER, are capitalized, as are all words typical of computer syntax, such as INPUT. This is only one type of format, which can vary from one firm to the next.

A systems analyst can use a combination of DFDs and structured English to document all a system's processes. The DFDs handle the documentation on the higher levels, and the structured English takes care of the lowest level. Technical Module H describes structured English in greater detail.

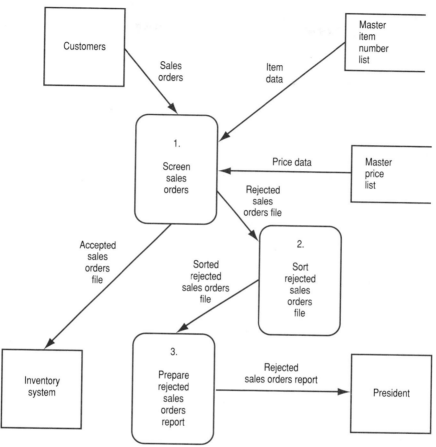

Figure 10.10
A Figure 0 diagram.

Warnier-Orr Diagrams

A structured tool capable of documenting a system at all levels is Warnier-Orr. In a **Warnier-Orr diagram,** pictured in Figure 10.12, the processes are described with a narrative arranged in brackets. The system is identified at the left, the leftmost bracket enclosing the major processes. Each of these processes can then be documented in lower levels of detail, using additional brackets and moving from left to right. At the lowest level of detail, the descriptions can be similar to structured English or computer code.

Because of space limitations, it is usually necessary to draw several Warnier-Orr diagrams to document an entire system from top to bottom. For that reason, the diagram in the figure does not include all the possible details.

```
Screen Sales Orders
START
Perform Enter Sales Order Data
Perform Edit Sales Order Data
Perform Compute Order Amount
STOP
Enter Sales Order Data
  INPUT SALES.ORDER Data
Edit Sales Order Data
  Edit CUSTOMER.NUMBER by ensuring that it is a positive
    numeric field
  IF Edit Is Failed
    THEN WRITE REJECTED.SALES.ORDER Record
  END IF
  DO for Each Item
    EDIT ITEM.NUMBER to ensure that it matches a number in
      the MASTER.ITEM.NUMBER.LIST
    IF Edit Is Failed
      THEN WRITE REJECTED.SALES.ORDER Record
    END IF
    EDIT UNIT.PRICE to ensure that it matches the price in
      the MASTER.PRICE.LIST
    IF Edit Is Failed
      THEN WRITE REJECTED.SALES.ORDER Record
    END IF
    EDIT ITEM.QUANTITY to ensure that it is a positive
      numeric field
    IF Edit Is Failed
      THEN WRITE REJECTED.SALES.ORDER Record
    END IF
  END DO
Compute Order Amount
  IF no edit errors
    THEN DO for Each Item
        COMPUTE ITEM.AMOUNT = ITEM.QUANTITY * UNIT.PRICE
        COMPUTE ORDER.AMOUNT = ORDER.AMOUNT sum
    END DO
    WRITE ACCEPTED.SALES.ORDER Record
  END IF
```

Figure 10.11
Structured English.

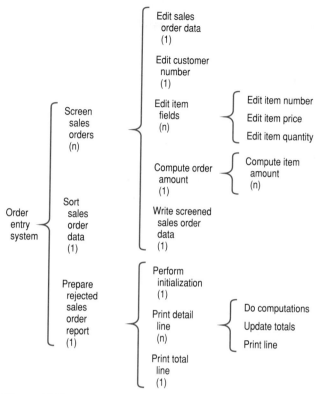

Figure 10.12
A Warnier-Orr diagram.

Although Warnier-Orr is an extremely effective way to document processes, it is not supported by most CASE tools. For this reason, the future of Warnier-Orr is uncertain.

The Data Dictionary

DFDs, structured English, and Warnier-Orr diagrams are structured approaches to documenting *processes*. A structured approach to documenting *data* is the **data dictionary,** a detailed description of all the data used in a system. The descriptions can be stored in the computer using database management system (DBMS) or data dictionary system (DDS) software, or they can be maintained in hardcopy form. Three dictionary forms can document a file in varying degrees of detail.[1]

[1]The dictionary forms presented in this chapter are based on formats developed by James Senn and documented in his *Analysis and Design of Information Systems* (New York: McGraw-Hill, 1984): 125–134.

DATA STORE DICTIONARY ENTRY

Use: To describe each unique data store on a data flow diagram.

DATA STORE NAME: *Master price list file*

DESCRIPTION: *A file that identifies the retail sales price of each item in the product line*

DATA STRUCTURES: *Master price record*

VOLUME (NUMBER OF RECORDS): *Approximately 15,000*

ACTIVITY: *Approximately 15 percent of the records are active each day.*

Figure 10.13
A data store dictionary entry.

Data Store Dictionary Entry   The *data store dictionary entry* describes the contents of a data store in a DFD, or the contents of a file in a system flowchart or Warnier-Orr diagram. Figure 10.13 provides an example. The entry form identifies the data store name and includes a brief description. A data store contains one or more data structures. Think of a structure as a record format. In this example, the Master Price List file contains a single record format, the Master Price record. The rest of the form describes the store in terms of size and use.

Data Structure Dictionary Entry   The *data structure dictionary entry* describes each data structure. Each data element in the structure is listed as shown in Figure 10.14.

Data Element Dictionary Entry   The **data element dictionary entry** describes a single data element in a structure. All pertinent data about each element is in-

```
┌─────────────────────────────────────────────────────────────┐
│                DATA STRUCTURE DICTIONARY ENTRY                │
│                                                               │
│  Use:      To describe each unique data structure that exists │
│            in data flows and data stores.                     │
│                                                               │
│  STRUCTURE NAME:  Master price record                         │
│                                                               │
│  DESCRIPTION:     A record that contains the unit sales price  │
│                   for each item                               │
│                                                               │
│                                                               │
│  DATA ELEMENTS:   Item number                                 │
│                   Item class                                  │
│                   Item description                            │
│                   Unit of measure                            │
│                   Unit sales price                            │
│                                                               │
│                                                               │
│                                                               │
│                                                               │
│                                                               │
│                                                               │
│                                                               │
│                     (Continued on Back)                       │
└─────────────────────────────────────────────────────────────┘
```

Figure 10.14
A data structure dictionary entry.

cluded, such as the type, field size, examples of contents, and so on. Figure 10.15 provides an illustration.

You can see the top-down nature of the data dictionary. The data store describes the big accumulation of data, such as a file. Each store contains structures, and each structure contains data elements.

Entity-Relationship Diagrams

An **entity-relationship diagram (ERD)** identifies the entities that are described with data, as well as the relationships among entities. The entities are often repre-

DATA ELEMENT DICTIONARY ENTRY

Use: To describe each data element contained in a data structure.

DATA ELEMENT NAME: *Unit sales price*

DESCRIPTION: *The retail sales price for one unit of an item
                          in the product line*

TYPE OF DATA: *Numeric*

NUMBER OF POSITIONS: *7*

NUMBER OF DECIMAL PLACES: *2*

ALIASES: *Sales price*
                          Price

RANGE OF VALUES: *00000.50 – 08999.99*

TYPICAL VALUE: *00037.95*

SPECIFIC VALUES: *None*

OTHER EDITING
DETAILS: *Nonnegative*

Figure 10.15
A data element dictionary entry.

sented with rectangles, and the relationships with diamonds. When used in enterprise modeling, the ERD represents all the firm's data. When used on a lower level it can represent only a system's data.

An ERD begins with a rough sketch, which is then iteratively refined. Figure 10.16 shows a rough ERD for a system that processes customer sales orders.

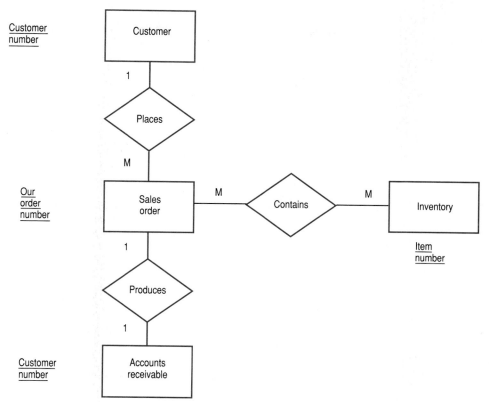

Figure 10.16
A rough entity-relationship diagram.

Letters and numbers are included next to the rectangles to show how many entities are involved in a relationship. A *1* means that the entity occurs only once in the relationship. An *M* means "many." The example in the figure shows that a single customer places many sales orders, the many sales orders contain many inventory items, and each sales order produces a single accounts receivable.

The underlined entries next to each entity are the entity keys. A **key** is a data element that identifies the entity. The term **identifier** is also used. For example, the customer number identifies all data describing one occurrence of the customer entity.

The ERD and the data dictionary can be used together. The ERD provides a summary picture of the relationship among entities, and the data dictionary provides a detailed description of each entity's contents.

The ERD is the newest of the data documentation tools and is currently very popular. However, the ERD does more than simply document data. It provides the basis for data analysis. During enterprise modeling the executives, assisted by information specialists, can use the ERD to conceptualize data requirements necessary for the firm to meet its strategic objectives. During systems analysis or systems design the users, systems analysts, and DBAs can use the ERD to conceptualize data that the system under study will require.

The ERD provides a picture of the data resource, which shows the major groupings of data and how they are related. The ERD enables system developers to determine needed data elements, their attributes and dependencies, and the logic of how certain data elements are derived. The ERD is a powerful data design tool that can be the starting point for developing all the firm's conceptual systems.

A Summary of the Documentation Tools

We have seen that some of the documentation tools are used to document processes and some to document data. The tools also vary in terms of how they reflect the level of detail. Figure 10.17 positions the tools to show their relationships.

Each vertical column represents combinations of tools that are normally used together. The shaded areas represent voids where certain levels of documentation are missing. For example, when using flowcharting the system flowchart provides top-level summary documentation and the program flowchart provides the detail, but there is no support in between. Complete support can be achieved by using DFDs in conjunction with structured English, or by using Warnier-Orr diagrams. On the data side, the top-level view of ERDs is recognized, supplemented by a top-down description provided by the data dictionary.

The task of systems analysts and other information specialists is to use a combination of tools that provides all the needed documentation for both processes and data.

Physical and Logical Models

Effort is expended in the analysis phase to thoroughly understand the existing system. The assumption is that the user's information needs can best be satisfied by a new or improved system if the existing system is understood. The new or improved system will be designed to overcome the limitations of the existing system.

The systems analyst uses the documentation tools to model the existing system in the form of diagrams and descriptions. The tools are used to produce two types of models—physical and logical. A **physical model** documents how data flows and processes are implemented using particular technology. For example, a physical model reveals that a keyboard terminal is used to enter order data, that a master file resides on a direct-access storage device, and so on. A **logical model,** on the other hand, documents the data flow and processes without being concerned with the technology that is used. For example, a logical model reveals that order data is entered into the system and that a master file is used. A logical model is concerned with neither the media nor the machines.

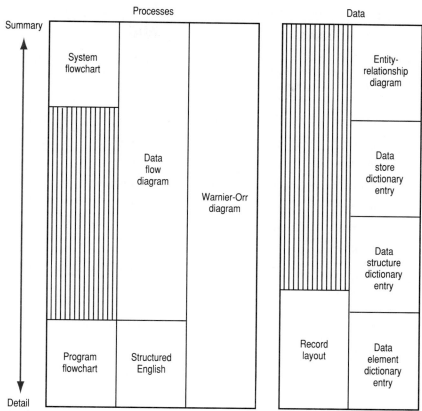

Figure 10.17

A comparison of the documentation tools.

As the systems analyst performs the analysis, the physical model is usually the most obvious. It is easy to see the technology and how it is used. In preparing the *physical model of the existing system,* the best documentation tool is the system flowchart. The flowchart symbols reveal the technology. Understanding the physical model enables the analyst to probe deeper and determine the underlying logic, producing a *logical model of the existing system.* The documentation tools that are best suited for logical models are DFDs, Warnier-Orr diagrams, and structured English.

Unique Documentation Requirements of the CBIS Subsystems

None of the documentation tools is aimed at particular CBIS subsystems. However, characteristics of those subsystems make certain tools more effective than others. Table 10.1 shows the suitability of each tool for each subsytem.

Table 10.1   Suitability of the Documentation Tools for Each CBIS Subsystem

	CBIS Subsystem				
Tools	Data Processing System	Management Information System	Decision Support Systems	Office Automation Systems	Expert Systems
Flowcharts	Excellent	Good	Poor	Not used	Poor
Data flow diagrams	Good	Good	Excellent	Not used	Excellent
Warnier-Orr diagrams	Good	Good	Excellent	Not used	Excellent
Structured English	Good	Good	Excellent	Not used	Excellent
Entity-relationship diagrams	Excellent	Excellent	Excellent	Not used	Excellent
Data dictionary	Excellent	Excellent	Excellent	Not used	Excellent

Flowcharts are excellent for documenting the data processing system because of its sequential nature. Flowcharts are also good for documenting the MIS but are generally poor for DSSs and expert systems.

As you can see from the table, none of the tools is used to document office automation systems. OA requires no custom software and is used to communicate information rather than process it, so OA systems are not documented as are other CBIS subsystems. The documentation, when present, is provided by software vendors.

DFDs are excellent for documenting DSSs and expert systems because the graphic style is user friendly, facilitating user involvement. DFDs are good for documenting data processing and MIS designs, and when firms engage in restructuring they often convert the documentation from flowcharts to DFDs.

The Warnier-Orr diagram is excellent for documenting the complex logic of DSSs and expert systems in a user-friendly manner. Also, like the DFDs, Warnier-Orr documentation is good for data processing and management information systems.

Structured English is of greatest value in documenting the detailed logic of DSSs and expert systems. It is also an alternative to program flowcharts when documenting the detailed logic of the data processing system and the MIS.

All data should be documented using ERDs.

The use of the data dictionary is identical to that of ERDs.

In summary, when documenting processes, flowcharting is best suited to the data processing system. DFDs, Warnier-Orr, and structured English are perfect for DSSs and expert systems. All the organization's data should be documented with ERDs and the data dictionary.

Step 4—Define System Performance Criteria

When the existing system has been documented in step 3 of the analysis phase, the next step is to define the system performance criteria. These criteria represent the standards that the new system must meet in order to satisfy the user.

As you might expect, the systems analyst can approach this task in a number of ways. For the exercise to be carried out in an orderly manner, it helps to draw on theory. Such an approach was taken in the planning phase by using information dimensions as the basis for the system objectives. We also addressed the subject of feasibility in terms of the TENLOS acronym. We can use both constructs as a basis for defining the performance criteria.

Use Dimensions of Information

For each dimension of information, the criterion must be stated in measurable terms. It is not sufficient to say that the new system "must process sales orders in an accurate manner." That might have been fine for the system objectives but is unacceptable for the performance criteria. The analyst continues to question the user until they *both* agree to a criterion, such as "must fill 98 percent of the sales orders with the proper items."

Use Types of Feasibility

Two of the feasibility types provide especially good approaches to defining performance criteria: economic and noneconomic feasibility.

Economic Performance Criteria   When management looks to the new system as a means of reducing costs, the amount of the reduction should be stated in specific terms, such as "Reduce monthly clerical costs in the order department to $12,000." When pursuing a cost avoidance strategy, the performance criterion might be "Maintain the number of clerical personnel in the billing department at 14 for a period of two years after cutover."

Noneconomic Performance Criteria   Noneconomic performance criteria are more difficult to quantify. One approach is for the analyst and the user to design a questionnaire during the analysis phase that the user can complete during the postimplementation review and subsequent annual audits. The questionnaire will enable the user to subjectively evaluate the system on its noneconomic dimensions.

The questionnaire can contain scales such as the Likert scale described earlier in the chapter. Each scale represents a single noneconomic evaluation criterion. Weights can be assigned to each scale, and a total score can be computed to reflect each user's perceptions. Such a quantitative approach is more appropriate when there are many users and a composite evaluation is needed.

Step 5—Prepare the Design Proposal

The systems analyst documents the findings of the analysis phase in a design proposal. Such a proposal was illustrated in Figure 8.5.

Step 6—Approve or Disapprove the Design Project

The MIS steering committee and the user review the design proposal and decide whether to continue with the project. A *go* decision shifts the emphasis from analysis to design.

The Project Dictionary

We defined the *project dictionary* in Chapter 8 as the collection of system documentation that the project team members create throughout the SLC. When management authorizes the design phase, the members of the project team ensure that the project dictionary contains all the documentation that has been prepared up to this point. The contents should include items such as those listed in Figure 10.18. Several sections are taken directly from the design proposal. The remainder consists of findings from the system study. The dictionary's appendix contains working papers such as the data-recording forms used while conducting record search.

The Role of CASE in System Development

Both information specialists and users can take advantage of CASE tools in system development.

We recognized in Chapter 1 that four categories have been established: upper CASE, middle CASE, lower CASE, and integrated CASE.[2] An **upper CASE tool** is used prior to the system life cycle for enterprise modeling and strategic planning for information resources, and during the planning phase of the SLC. Examples of upper CASE tools are Develop-Mate from IBM, PC Prism from Index Technology, and IEW/Planning Workstation from KnowledgeWare. A **middle**

[2]This classification is from Michael Lucas Gibson, "The CASE Philosophy," *BYTE* 14 (April 1989): 209ff.

1. Problem definition*
2. System objectives and constraints*
3. Performance criteria*
4. Possible system alternatives*
5. Documentation of the current system
    5.1   Process documentation (some combination of the following)
        5.1.1   System flowchart
        5.1.2   Program flowcharts
        5.1.3   Data flow diagrams
        5.1.4   Hierarchy diagrams
        5.1.5   Warnier-Orr diagrams
        5.1.6   Structured English (process dictionary entries)
        5.1.7   Other process documentation tools
    5.2   Data documentation
        5.2.1   Entity-relationship diagram
        5.2.2   Data dictionary entries
6. General development plan*
7. Appendix
    7.1   Transcriptions of interview tapes
    7.2   Working papers

*From the design proposal

Figure 10.18
The contents of the project dictionary at the end of the analysis phase.

CASE tool is used during the analysis and design phases in documenting the existing and new systems. Examples are Excelerator from Index Technology, Visible Analyst from Visible Systems, and Analyst/Designer Toolkit from Yourdon, Inc. A **lower CASE tool** is used during the implementation and use phases to help the programmer develop, test, and maintain program code. Examples are Telon from Pansophic Systems and Excelerator for Design Recovery from Index Technology.

A firm can use a combination of upper, middle, and lower CASE tools to provide the coverage it needs, or it can use an **integrated CASE tool (I-CASE)** that supports the firm's strategic planning and all phases of the SLC. Popular examples of integrated CASE tools are Information Engineering Facility (IEF), from Texas Instruments, and IN-CASE from Electronic Data Systems.

A CASE Master Plan—IBM's AD/Cycle

More than 50 software and hardware vendors are currently marketing CASE products.[3] Many firms use several products from several vendors to achieve the variety of support they need. IBM recognized the chaos that could result when firms tried to integrate incompatible CASE products, and so it created AD/Cycle

[3]Carma McClure, "The CASE Experience," *BYTE* 14 (April 1989): 246.

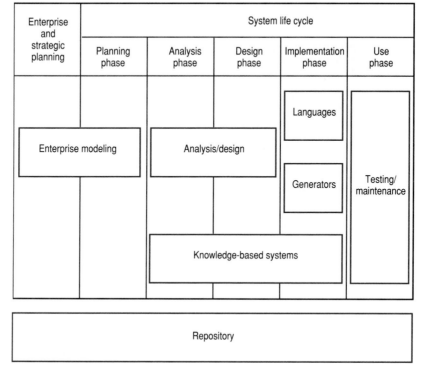

Figure 10.19
IBM's AD/Cycle positions CASE tools for top-down application development.

as a solution. **AD/Cycle** is a set of application development standards all vendors can follow in developing CASE products so that the products can be used in combination. (The letters *AD* stand for application development and are pronounced as in 1776 A.D., not *Ad* as in a TV ad.)

Central to AD/Cycle is a special type of database called the repository. The **repository** is the storage reservoir containing the documentation and code produced by CASE tools during the SLC. The repository serves as the foundation, as shown in Figure 10.19.[4]

The repository provides for the following types of CASE tools:

● *Enterprise modeling*—software designed for use during enterprise planning and strategic planning for information resources, which precedes the SLC, and during the planning phase.

[4]This diagram is based on one copyrighted by International Business Machines Corporation. The IBM diagram appeared in Ralph Carlyle, "Is Your Data Ready For the Repository?" *Datamation* 36 (January 1, 1990): 46 ff.

- *Analysis/design*—software intended for use during the analysis and design phases.

- *Languages*—programming languages such as PL/1, C, and COBOL that can be integrated with middle CASE tools, and dynamic debuggers that can be tightly coupled with the languages' edit/compile facilities.

- *Generators*—application generators such as IBM's Cross System Product (CSP), which generates code from documentation created by the analysis/design software. The user also can integrate custom-written code with the generated code.

- *Knowledge-based systems*—expert systems shells such as Knowledge Engineering Environment (KEE) from IntelliCorp, Inc., and Knowledge Tool (KT) from IBM.

- *Testing/maintenance*—tools designed to enhance software testing and maintenance.

The enterprise modeling software is an example of upper CASE, and the analysis/design software is an example of middle CASE. The languages and generators are examples of lower CASE. The knowledge-based systems consist of middle and lower CASE. The testing/maintenance software is an example of the business process redesign (BPR) tools described in Chapter 8.

AD/Cycle is a step in the right direction, although it can be expensive. One expert estimates that it will cost a minimum of $10 million to build a major repository-based application, and the cost could be as high as $50 million.[5]

Even so, AD/Cycle is an effort to provide a life cycle methodology geared to the use of CASE tools. Although it will take time for AD/Cycle to become a reality for most firms, it is definitely a significant methodology for the future and one with which information specialists should be familiar.

Using CASE to Define Data in the Data Dictionary

Figure 10.20 is a data store dictionary entry for the Rag City Accounts Receivable Master file. CASE can generate such entries as a byproduct of defining entities in an entity-relationship diagram or data stores in a DFD. The description explains that the Accounts Receivable Master file contains all the data about accounts receivables for all customers of Rag City. The file has no known aliases, and it is composed of seven data elements including the customer ID and customer address. No notes are mentioned, and seven DFD processes are identified as locations where the Accounts Receivable Master file is involved.

[5]Carlyle, p. 46ff.

```
┌─────────────────────────────────────────────────────────────────────┐
│ PROJECT:   RAG CITY                                                   │
├─────────────────────────────────────────────────────────────────────┤
│ LABEL:    (A1) ACCOUNTS RECEIVABLE MASTER FILE                        │
├─────────────────────────────────────────────────────────────────────┤
│ ENTRY TYPE:   DATA STORE                                              │
├─────────────────────────────────────────────────────────────────────┤
│ DESCRIPTION:   THE A/R MASTER FILE HOLDS ALL THE                      │
│                INFORMATION ABOUT ACCOUNTS RECEIVABLES                 │
├─────────────────────────────────────────────────────────────────────┤
│ ALIAS:   NONE                                                         │
├─────────────────────────────────────────────────────────────────────┤
│ COMPOSITION:   A/R MASTER FILE =   CUSTOMERID                         │
│                                +   CUST_ADDRESS                       │
│                                +   CUST_REP                           │
│                                +   CREDIT RATING                      │
│                                +   FINANCE CHARGE                     │
│                                +   BALANCE DUE                        │
│                                +   DUE DATE                           │
├─────────────────────────────────────────────────────────────────────┤
│ NOTES:                                                                │
├─────────────────────────────────────────────────────────────────────┤
│ LOCATIONS:   DFD 1.1, DFD 1.1.3, DFD 1.2, DFD 1.2.3, DFD 2.1.         │
│ DFD 2.1.3, and DFD 3.1                                                │
└─────────────────────────────────────────────────────────────────────┘
```

Figure 10.20
A data dictionary entry for Rag City.

Summary

The first step of the analysis phase is management's announcement of the upcoming system study. This announcement is intended to quiet employee fears and foster acceptance of the systems analyst. Management states why the firm has decided to conduct a study, explains how the new system will help the firm and the employees, and asks for cooperation.

The project team is formed before the study begins. Initial members include the project leader, one or more systems analysts, an internal auditor, possibly database and data communications specialists, and representatives from the user area.

The step of defining the user's information needs is accomplished by analyzing existing system documentation, conducting in-depth personal interviews and surveys, observing, and searching through records. The surveys can be conducted in person, over the telephone, and through the mail.

Documentation is important in the development and use of business systems because of their typically long life cycles. The documentation exists in a narrative and graphic form, and describes both data and processes. The documentation enables team members to understand the problem and to communicate among themselves and with others.

Traditional documentation tools include flowcharts and record layouts. Although still used, they have been increasingly replaced by more modern tools that permit a top-down, structured approach. Structured tools include data flow diagrams, structured English, Warnier-Orr diagrams, the data dictionary, and entity-relationship diagrams.

The CBIS subsystems exert an influence on the documentation tools used. Flowcharts are best for data processing systems, but they can also be used for MISs. DFDs, Warnier-Orr, and structured English are excellent for DSSs and expert systems. ERDs and the data dictionary should be used with all CBIS subsystem types except office automation, which does not lend itself to documentation.

Before recommending that the design phase be conducted, the systems analyst works with the user to define the performance criteria of the new system. A systematic approach considers the dimensions of information and types of feasibility—economic and noneconomic. The economic criteria are easily quantified, but the noneconomic criteria are best measured subjectively, using rating scales.

At the end of analysis, the project dictionary includes selections from the design proposal as well as study findings.

The biggest documentation innovation in recent years has been CASE. Upper CASE tools are used during strategic planning and the planning phase. Middle CASE tools are used during analysis and design. Lower CASE tools are used during implementation and use. Integrated CASE tools span strategic planning and the SLC. CASE tools relieve the analyst and other information specialists of much work and are being used to produce an increasing proportion of application software.

Recognizing that users can have difficulty trying to integrate the CASE tools from several vendors, IBM announced its application development strategy called AD/Cycle. AD/Cycle is a framework within which various categories of CASE software can be integrated throughout the SLC. If interest in AD/Cycle continues, its price may eventually come within reach of smaller firms.

Key Terms

In-depth personal interview

Structured interview

Unstructured interview

Survey

Questionnaire

Objective question

Subjective question, open-ended question

Data validation

Likert scale

Observation

Record search

System documentation

Documentation tool

System flowchart

Program flowchart

Record layout

Data flow diagram (DFD)

Structured English, pseudocode

Warnier-Orr diagram

Data dictionary

Entity-relationship diagram (ERD)

Physical model

Logical model

AD/Cycle

Repository

Key Concepts

The fluid composition of the project team, emphasizing varying knowledge and skills as the SLC progresses

The complementary nature of the data gathering tools in gathering the wide range of data needed to analyze a system

How the structured documentation tools explode processes and data structures to show lower levels of detail

How the project dictionary provides a written record of the SLC as it unfolds

The classification of CASE tools based on how they fit in system development

AD/Cycle

Questions

1. What are the two objectives of the system study announcement?
2. What four points does management make in the system study announcement?
3. Which project team members play major roles during the analysis phase?
4. What are the two most effective data-gathering techniques?
5. What distinguishes an in-depth personal interview from a personal survey?
6. Which data-gathering techniques involve an interviewee? A respondent?
7. Distinguish between a structured and an unstructured interview.

8. Which data-gathering techniques use a questionnaire?

9. Distinguish between an objective and a subjective question.

10. Which data-gathering technique might be the best for gathering sensitive data? Explain your answer.

11. Which data-gathering techniques could involve retaining the services of specialists, perhaps part time?

12. Which kind of survey uses a Likert scale?

13. Which two features of a system are documented?

14. What are the two objectives of system documentation?

15. What are the two types of flowcharts?

16. Is flowcharting a good top-down tool? Why or why not?

17. What does a record layout tell about a system's data?

18. What is the name of the top-level DFD? The one on the next lower level?

19. Name two tools that can be used to document the details of the lowest-level DFD processes.

20. Which documentation tool displays a system's hierarchical structure in a left-to-right manner?

21. Explain why the data dictionary is a top-down tool.

22. How are the ERD and the data dictionary used in combination?

23. What is a good tool to use with DFDs to document a system's processes?

24. Which tool documents processes at all levels?

25. Which CBIS subsystem normally is not documented?

26. Which mix of tools would you use to thoroughly document a DSS?

27. Which two types of feasibility provide the best basis for defining performance criteria?

28. Would the forms that you use while conducting a record search be included in the project dictionary? If so, where?

29. Identify the SLC phase or phases where you would use an upper CASE tool. A middle CASE tool. A lower CASE tool. An I-CASE tool.

30. What is the biggest current drawback to AD/Cycle?

Topics for Discussion

1. Which data-gathering technique would most require validation? Would any require none?

2. Why not dispense with the study of the current system and just ask the users their information needs?

3. Why should a systems analyst learn how to flowchart?

4. Keeping in mind that a methodology is a recommended way of doing something, what is AD/Cycle recommending? To whom?

5. Assume that you are a systems analyst and that the user with whom you have been working during the analysis phase tells you: "Why don't you go ahead and figure out the performance criteria by yourself?" What should you do?

Problems

Technology Modules B (Data Flow Diagrams), C (The Data Dictionary), and H (Structured English) provide the necessary techniques for working the following problems.

1. Prepare a Figure 0 diagram of the following procedure:

    A. A clerk opens the mail and pulls out all customer payments. Each payment consists of a payment document and a check. The payment documents are sent to the data entry department. The checks are sent to the cashier's office. All other mail is placed in a Suspense file.

    B. A data entry operator keys the payment document data into a keyboard terminal. After the keying operation, the payment documents are filed in the Payment History file.

    C. The computer sorts the payment records into customer number sequence and writes them onto magnetic tape as a Payments file.

    D. The Payments file is used to update the Accounts Receivable Master file (on magnetic disk). At the same time, a Payments report is printed by the printer. The Payments report is used by the supervisor of the accounts receivable section.

2. Repeat problem 1, except draw a context diagram.

3. Prepare a Figure 0 diagram of the following procedure:

    A. Sales orders are received from customers. A clerk in the order department batches the orders in groups of 25 and uses a desk calculator to accumulate batch totals for each batch. The totals include (1) number of orders and (2) total order amount. The desk calculator prints the totals on a paper strip that is held for use later in the procedure.

    B. The batches of sales orders are given to a data entry operator, who keys the data into the computer. After the keying operation, the sales order batches are filed in the Sales Order History file.

    C. The computer generates the same totals as the order department clerk and prints the totals on a Batch Totals report. During the same operation, the computer writes the sales order data onto a Suspense file that resides on magnetic disk.

D. A control clerk compares the batch totals computed by the order department clerk from step A with those computed by the computer in step C.

E. No action is taken when the totals balance. However, when they do not balance, the control clerk identifies the cause of the imbalance and makes any necessary corrections to the error batches, which are removed from the Sales Order History file. For example, numbers are rewritten to make them more legible.

F. The corrected error batches are sent to the data entry operator for reentry in step B. The desk calculator strips and the Batch Totals report are filed in a Batch Totals History file.

4. Repeat problem 3, except draw a context diagram.

5. Prepare a Figure 0 diagram of the following procedure. Consider the work of the employees in creating their timecards to be part of the payroll system—not the environment. Therefore, the payroll system will not be triggered by an environmental input.

Each day, each employee writes the number of hours worked on her or his timecard. At the end of the week, each employee totals the hours worked , using a pocket calculator, and writes the total at the bottom of the card. The cards are given to the supervisor, who uses a desk calculator to verify the total figure on each card. If a card contains an error, the supervisor corrects it. All the cards for each department for each week are forwarded to the computer department, where the data is keyed into a keyboard terminal. After the keying operation, the timecards are placed in a Timecard History file. The computer writes the data on a Weekly Payroll Transactions file on magnetic tape. The computer, using a program named Update Payroll Master File, then reads from the transactions file and obtains the appropriate master record from the Employee Master file, on magnetic disk. The computer computes the regular, overtime, and total earnings, using data from both the transaction and master files. The master record is updated to reflect the new year-to-date totals. During the same operation, a summary of the payroll transactions are written on a magnetic tape as a Payroll Summary file. Using a program named Print Weekly Payroll Summary Report, the computer then reads from the Payroll Summary file and prints a Weekly Payroll Summary report for the manager of the payroll department.

6. Repeat problem 5, except draw a context diagram.

7. Use structured English to document the Update Payroll Master File program described in problem 5.

8. Document the timecard in problem 5 with a data flow dictionary entry, a data structure dictionary entry, and three data element dictionary entries. Define the record structure as you see fit, and select the three data elements for detailed documentation.

9. Repeat problem 8, except document the Employee Master file. Use data store, data structure, and data element dictionary forms.

Consulting with

Harcourt Brace

(Use the Harcourt Brace scenario prior to this chapter in solving this problem.)

Assume that you are Russ or Sandi. Based on what you have learned during your systems analysis, identify any areas that seem to be prime targets for improvement. Write a one-page memo to Ira Lerner, identifying the area or areas. You need not describe how the improvement will be achieved. Simply describe the need for each area.

As a separate exercise, complete the following documentation of the current system:

1. After reading Technical Module B, draw a set of data flow diagrams to document the order processing system in Orlando. Draw a context diagram and a Figure 0 diagram. The Figure 0 diagram should contain only two processes: process mail orders and process telephone orders. Draw a Figure 1 diagram of the procedure for processing mail orders.

2. After reading Technical Module H, use structured English to document the processes performed by a TOS rep.

3. After reading Technical Module C, prepare data store, data structure, and data element dictionary entry forms for the batch log.

Case Problem
Midcontinent Industries

Midcontinent Industries manufactures garden tools. Of the firm's 425 employees, 275 to 350 are assigned to the manufacturing division. The number depends on the season. The largest production volumes come in the fall. Although Midcontinent has a computer, few of the systems are completely computerized. The hourly payroll system is a good example of an integration of computer and manual processing.

The Hourly Payroll System

All manufacturing employees, except managers, are paid on an hourly rate. Each is given a timecard each week. On arriving and departing, they insert their time-cards in timeclocks installed at the plant entrances. The timeclocks print the date and the time of day on the cards. The time of day is based on a 24-hour clock; for example, 1:30 P.M. is printed as 1330.

When leaving for the weekend, the employees compute the hours worked that week. Using pocket calculators, they subtract the time in from the time out and then subtract another 1½ hours for the lunch hour and two 15-minute coffee breaks. The remainder is the number of hours worked that day. Finally, the employees add the week's daily hours to arrive at the weekly hours. They sign their timecards and give them to their departmental secretaries.

Departmental Secretaries Check the Timecards

The departmental secretaries check the timecards against their employee lists to ensure that they have a card from each employee. If a timecard is missing, the secretary locates the employee and obtains the card. If a card cannot be found, the secretary fills out a new card for 40 hours, and the employee is paid for that amount of time.

Next, the secretary scans each card to ensure that all fields have been filled in for daily and weekly hours. When necessary, the secretary fills in the missing data.

When the secretaries finish their operation, they give the verified timecards to their department managers.

Department Managers Approve Overtime

The department managers scan the cards, looking for those with overtime. Employees with 40 or more hours are eligible for overtime pay—1½ times the hourly rate for all hours worked over 40. If the manager approves an employee's overtime, the manager signs the card. When the manager cannot recall a need for the overtime, the manager writes a 40 above the employee's figure and signs the card. An employee with disapproved overtime is paid for 40 hours. If the employee later shows that the overtime was actually worked, the overtime pay is computed by a separate system.

After signing the overtime cards, the department managers send all the cards to the computer department through the company mail.

The Computer Performs Weekly Payroll Processing

A data entry employee in the computer department keys the data from each time-card into a keyboard terminal and then files the timecards in a Timecard History file. A computer program writes the entered data onto a Weekly Timecard file, on magnetic disk. A sort program sorts the Weekly Timecard file by employee number (minor sort key) within department number (major key). A third program reads the Sorted Weekly Timecard file and prints a Weekly Payroll Report. This report is filed in a Weekly Payroll History file, which provides a printed record if anyone wishes to audit the payroll activity.

The Sorted Weekly Timecard file is also used to update the Employee Payroll Master file, on magnetic disk. This file maintenance program is the most complex of the system. It computes the weekly regular and overtime earnings, the current and year-to-date gross earnings, social security tax, income tax, and net earnings amounts. The appropriate fields are updated in the employee master records, and an output file is written on magnetic disk that contains the most important computations for each employee. This output file, called the Weekly Payroll file, is used to print the payroll checks, which are sent in the company mail to the employees.

The Weekly Payroll files are maintained for four weeks and are then used to prepare a monthly report. First, the weekly files are merged, using a merge program. Then, the Merged Weekly Payroll file is used to print the monthly payroll report, which is sent to the president in the company mail.

Tasks

Note: Technical Modules B, C, and H provide the detailed understanding of DFDs, the data dictionary, and structured English that you will need to do these tasks.

1. Draw a set of data flow diagrams that documents both the weekly and monthly hourly payroll procedures. Use the following guidelines:

    A.   Draw a context diagram and a Figure 0 diagram. The context diagram should reflect that the daily hourly payroll system provides the timecards and that the employees receive the checks. In drawing the Figure 0 diagram, regard the weekly computer processing as Process 1 and the monthly processing as Process 2. Connect the weekly and monthly processes with a data store.

    B.   Draw a Figure 1 diagram of the processing performed by the weekly system. Assume that the processing begins with the computation of daily and weekly hours by the employees. Show the *weekly* work performed by the (1) employees, (2) secretaries, (3) managers, and (4) computer. Document the processes of the employees, secretaries, and managers with single symbols. Do not include the detailed payroll computations in documenting the processes of the file maintenance program. Also, do not include the special payroll system that corrects errors when employees are paid for only 40 hours although they worked overtime.

    C.   Draw a Figure 2 diagram of the processing performed by the monthly system.

2. Use structured English to document the processes required to update the Employee Payroll Master file. Include reading the record to be updated, per-

forming all necessary computations, and writing the updated record back to the file.

3. Fill out a data flow dictionary entry for the Sorted Weekly Timecard file.

4. Fill out a data store dictionary entry for the Employee Payroll Master file.

5. Fill out a data structure dictionary entry for the Employee Payroll Master record.

6. Fill out three data element dictionary entries for (1) employee name, (2) employee number, and (3) year-to-date gross earnings.

Selected Bibliography

Appleton, Daniel S. "The Modern Data Dictionary." *Datamation* 33 (March 1, 1987): 66–68.

Choobineh, Joobin; Mannino, Michael V.; and Tseng, Veronica P. "A Form-Based Approach for Database Analysis and Design." *Communications of the ACM* 35 (February 1992): 108–120.

Forte, Gene, and Norman, Ronald J. "A Self-Assessment by the Software Engineering Community." *Communications of the ACM* 35 (April 1992): 28–32.

Fosdick, Howard. "Ten Steps to AD/Cycle." *Datamation* 36 (December 1, 1990): 59ff.

Foss, W. Burry. "Early Wins Are Key to System Success." *Datamation* 36 (January 15, 1990): 79–82.

Gorman, Kevin, and Choobineh, Joobin. "The Object-Oriented Entity-Relationship Model." *Journal of Management Information Systems* 7 (Winter 1990–91): 41–65.

Hamilton, Robert, and Hamilton, Dennis. "On-Line Documentation Delivers." *Datamation* 36 (July 1, 1990): 45ff.

Hollist, Pen, and Utterback, Jon. "The CASE-Project Management Connection." *System Builder* 4 (April–May 1991): 39–42.

Holton, John B. "Data, Process and Logic (DPL) Flow Diagramming: An Integrated Approach to CIS Development." *Information Executive* 3 (Winter 1990): 57–63.

Huff, Clifford C. "Elements of a Realistic CASE Tool Adoption Budget." *Communications of the ACM* 35 (April 1992): 45–54.

Hughes, Cary T., and Clark, Jon D. "The Stages of CASE Usage." *Datamation* 36 (February 1, 1990): 41–44.

Kievit, KarenAnn. "Software Review: EXCELERATOR 1.9." *Journal of Management Systems* 3 (Number 1, 1991): 83–90.

Kuehn, Ralph R., and Fleck, Robert A., Jr. "Data Flow Diagrams for Managerial Problem Analysis." *Information Executive* 3 (Winter 1990): 11–15.

Lindholm, Elizabeth. "A World of CASE Tools." *Datamation* 38 (March 1, 1992): 75ff.

Loh, Marcus, and Nelson, R. Ryan. "Reaping CASE Harvests." *Datamation* 35 (July 1, 1989): 31ff.

Merlin, Vaughan, and Boone, Gregory. "The Ins and Outs of AD/Cycle." *Datamation* 36 (March 1, 1990): 59ff.

Moran, Robert. "The Case Against CASE." *InformationWeek* (February 17, 1992): 28ff.

Norman, Ronald J., and Nunamaker, Jay F., Jr. "CASE Productivity Perceptions of Software Engineering Professionals." *Communications of the ACM* 32 (September 1989): 1102–1108.

Orr, Ken; Gane, Chris; Yourdon, Edward; Chen, Peter P.; and Constantine, Larry L. "Methodology: The Experts Speak." *BYTE* 14 (April 1989): 221ff.

Saunders, Paul R. "Effective Interviewing Tips for Information Systems Professionals." *Journal of Systems Management* 42 (March 1991): 28–31.

Schussel, George. "The Promise and the Reality of AD/Cycle." *Datamation* 36 (September 15, 1990): 69ff.

Shemer, Itzhak. "Systems Analysis: A Systemic Analysis of a Conceptual Model." *Communications of the ACM* 30 (June 1987): 506–512.

Sullivan-Trainor, Michael L. "TI's IEF Scores High for Integration, Benefits Delivery." *ComputerWorld* 25 (April 22, 1991): 72–73.

Vessey, Iris; Jarvenpaa, Sirkka L.; and Tractinsky, Noam. "Evaluation of Vendor Products: CASE Tools as Methodology Companions." *Communications of the ACM* 35 (April 1992): 90–105.

Warnier, Jean-Dominique. *Logical Construction of Systems* (New York: Van Nostrand Reinhold, 1981).

Yourdon, Edward. "What Ever Happened to Structured Analysis?" *Datamation* 32 (June 1, 1986): 133ff.

Economic Justification

The systems analyst has the responsibility of showing management that money spent on a computer-based system is not only a good investment but as good as, or better than, other investments management could make. Management seeks economic justification.

Cost-related Justification Strategies

During early years of computing, management sought to justify expenditures by means of a cost reduction strategy. That is, it was expected that the costs related to the computer system would be less than those of the old systems. When it became evident that the costs of the displaced resources were seldom reduced, but merely shifted to other areas, management turned to a cost avoidance strategy. This strategy, popular today, reflects an intent to postpone future costs, such as those of increasing personnel, by implementing computer-based systems.

The Difficulty of Economic Justification

Economic justification of data processing systems is usually easier to achieve than that of information-oriented CBIS subsystems, such as MIS, DSS, OA, and expert systems. It is difficult to economically justify information-oriented systems because of the difficulty of placing a monetary value on information. The data processing system that has realized the most success in terms of economic justification is inventory. It is relatively easy to prove that inventory investment is reduced by using a computer-based system. Money previously invested in inventory can then be invested in more profitable ways.

The Responsibility of the Analyst for Economic Justification

The systems analyst must be alert to management's need to justify the computer or computer project and, when management insists on economic justification, the systems analyst must make an effort to provide that justification.

In no case should the analyst indicate that economic justification exists when, in fact, it does not. It is *not* the responsibility of the analyst to prove the justification.

Rather, the analyst is responsible for conducting the analysis and reporting the results to management. Then, it is management's responsibility to decide whether to proceed with the project. In this setting, as in all others dealing with systems development, the analyst recommends, and the manager decides.

Economic Justification Methods

Several methods have been devised to provide the basis for establishing some, or all, of the economic justification. The most popular methods include break-even analysis, payback analysis, and net present value.

Break-even Analysis

Although a new system may not be justfied by cost savings or reduction, management still wants to know how the cost of the new system compares with the cost of the existing one. In most cases, the new system will have lower operational costs because new technology processes larger volumes, with greater accuracy, than older technologies. However, high developmental costs can delay the realization of the operational benefits from the new system. **Break-even analysis** compares the monthly costs of the existing and new systems, and identifies the point when the costs are equal.

Break-even analysis is illustrated with the graph in Figure E.1. The graph contains two cost lines, one for the existing system and one for the new. The cost of using the existing system rises each year as it handles increasing business volume. The cost of the new system is very high in the beginning, reflecting developmental costs, but eventually falls and then has a smaller annual rate of increase than the existing system.

When you want to conduct a break-even analysis, you assemble the cost data on the two systems as shown in Table E.1 The point where the cost of the existing system is equal to that of the new system is the **break-even point.** The time prior to the break-even point is the **investment period** of the new system, and the time after is the **return period.** The firm invests in the new system so that it may enjoy the return.

The strengths of break-even analysis are its ability to compare the costs of the existing and new systems, and its inclusion of developmental costs. The main weakness is that it considers only costs and ignores the benefits. Its popularity is based, in part, on the fact that it is generally easier to compute costs than benefits. Therefore, break-even analysis is appropriate for all of the CBIS subsystems.

Payback Analysis

Once management is convinced that the new system compares favorably to the existing one in terms of costs, the next question is how long it will take for the new system to pay for itself. Payback analysis can provide the answer. **Payback**

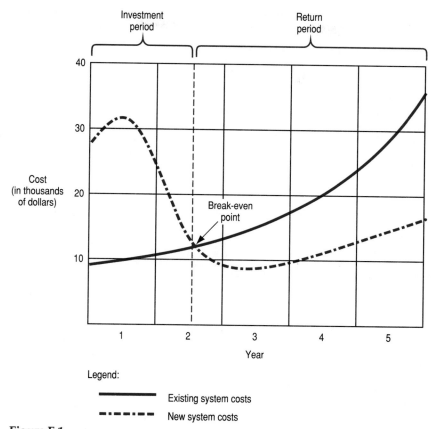

Figure E.1
Break-even analysis.

analysis determines how long it will take for the cumulative benefits of the new system to equal its cumulative costs. The point at which the new system benefits and costs are equal is the **payback point.** The length of time required to reach the payback point is the **payback period.**

The payback concept is illustrated with the graph in Figure E.2 and the data in Table E.2. The computation of the payback period uses the formula:

$$PBP = Y + \frac{C}{V}$$

Where: PBP = Payback period
Y = The last year that the cumulative benefits of the new system were negative

Table E.1   Break-Even Analysis Compares the Costs of the Existing and New Systems

Year	Month	Existing System Costs	New System Costs	Difference
1	1	8	26	+18
1	2	8	27	+19
1	3	9	30	+21
1	4	8	31	+23
1	5	8	31	+23
1	6	9	32	+23
1	7	9	31	+22
1	8	9	30	+21
1	9	9	29	+20
1	10	9	28	+19
1	11	9	27	+18
1	12	10	24	+14
2	1	10	23	+13
2	2	11	21	+10
2	3	11	19	+8
2	4	11	18	+7
2	5	12	16	+4
2	6	12	13	+1
2	**7**	**12**	**12**	**+0**
2	8	12	11	−1
2	9	13	10	−3
2	10	13	9	−4
2	11	13	9	−4
2	12	13	8	−5

C = The absolute value of the cumulative benefits of the new system
for the last year that they were negative

V = The absolute value of the cumulative benefits of the new system
for (1) the last year that they were negative, and (2) the first year
that they are positive

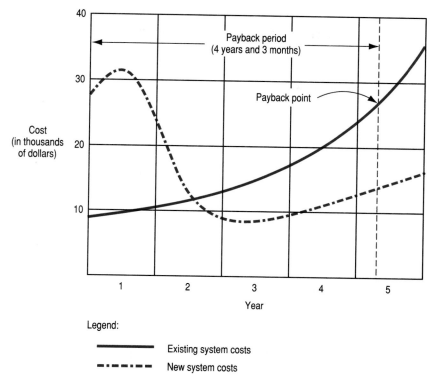

Figure E.2
Payback analysis.

Using the data in the table, the payback period is computed as:

$$PBP = 4 + \frac{4}{4 + 12} = 4 + \frac{4}{16} = 4.25 \text{ years, or 4 years and 3 months}$$

Payback analysis is a more robust method of economic justification than break-even analysis, because payback analysis not only incorporates the costs of the existing and new systems but also projects the point at which the new system development costs will be recovered. The payback point identifies to management when the new system will begin to have a more positive effect on the firm's cash flow than will a continued use of the existing system.

However, payback analysis is not without its limitations. Note from Table E.2 that the new system benefits are calculated by subtracting the new system costs from the existing system costs. This is a rather narrow view of benefits in that it ignores the value of the new system's output. For example, the value of a marketing information system is not based on its lower costs than a previous system but, rather, by the problem-solving information that it can provide.

Table E.2   Payback Analysis Is Concerned with the Cumulative Benefits of the New System

	1	2	Year 3	4	5
Existing system costs	9	12	15	22	29
New system costs	30	13	8	11	13
New system benefits	−21	−1	+7	+11	+16
Cumulative benefits	−21	−22	−15	−4	+12

Note: Annual costs and benefits in thousands of dollars.

Another limitation of payback analysis is the fact that it fails to recognize the time value of money. This limitation can be overcome with another cost-benefit method called net present value.

Net Present Value

If you invest $100 for five years at an interest rate of 12 percent, compounded annually, the investment will increase in value as shown in Table E.3.

Another way to look at the interest influence is to take a value, such as $100, at some time in the future, say five years, and compute the amount necessary to invest now in order to have the $100. The computation is made using the formula:

$$\text{Discount percent} = \frac{1}{(1 + I)^n}$$

Where: I = Annual interest rate
n = Number of years in the future

The computations are made for each of the five n values at a given interest rate. For an interest rate of 12 percent, the discount percentages are:

1 year = 0.8929
2 years = 0.7972
3 years = 0.7118
4 years = 0.6355
5 years = 0.5674

The percentages are multiplied times $100.00, and the results are those illustrated in Table E.4.

Table E.3   Interest Increases the Future Value of an Investment

Amount invested: $100.00
Interest rate: 12 percent
Term: 5 years

Year	Investment value at end of year
1	$112.00
2	125.44
3	140.49
4	157.35
5	176.23

The table shows that the present value of $100 one year from now is $89.29, for two years from now it is $79.72, and so on. The concept is called the **present value of money,** and it recognizes that future money has less value than existing money.

This same approach can be used to determine the **net present value (NPV)** of a new system by computing the present value of both the benefits and the costs. This computation is shown in Table E.5. The data is the same as that used for payback analysis in Table E.2. The only difference is that Table E.5 discounts the

Table E.4   Interest Decreases the Present Value of a Future Amount

Present Value	Number of Years into the Future				
	1	2	3	4	5
$89.29 ←	$100.00				
$79.72 ←		$100.00			
$71.18 ←			$100.00		
$63.55 ←				$100.00	
$56.74 ←					$100.00

Note: Present value of $100.00 invested at a rate of 12 percent.

Table E.5   The Net Present Value of the New System

	Year				
	1	2	3	4	5
Existing system costs	9	12	15	22	29
New system costs	30	13	8	11	3
New system benefits	−21	−1	+7	+11	+16
Discount percent	0.893	0.797	0.712	0.636	0.567
Discounted benefits	−18.8	−.8	+5.0	+7.0	+9.1
Cumulative discounted benefits	−18.8	−19.6	−14.6	−7.6	+1.5

Discounted PBP $= 4 + \dfrac{7.6}{9.1} = 4.83$ years, or 4 years and 10 months

Note: Annual costs in thousands of dollars; annual interest = 0.12.

future benefits computed by payback analysis. Whereas payback analysis looks at new system benefits in terms of current dollars, NPV discounts those dollars to reflect the fact that they are future money.

You can see that the process of discounting the benefits produces two dramatic changes. First, it reduces the cumulative benefits for the five-year period from the $12,000 in Table E.2 to the $1,500 in Table E.5. Second, it prolongs the payback period from the four years and three months computed earlier to four years and ten months.

NPV makes economic justification more difficult, but management might insist on its use. This is especially true when management is considering several alternate ways of investing in the future. For example, the firm might be able to realize a greater profit by investing in a certificate of deposit (CD) than in a computer project. Management could invest $851.10 in a CD today at an interest rate of 12 percent to produce the same $1,500 value five years from now that is produced by the system project.[1]

When you consider the amount of effort required by the system project, its substantial development expense, and the often traumatic effect that implementation can have on the firm, the CD represents a very attractive alternative. However, management still might decide to go with the system project, recognizing that it will produce noneconomic value, such as providing the firm with a greater ability to achieve competitive advantage.

[1]The present value of the CD is computed by multiplying its value five years from now times the discount percent for the 12 percent interest rate: $1,500 × 0.5674 = $851.10.

Putting Economic Justification in Perspective

Management will not invest in a computer or a computer project without assurance that doing so will benefit the firm. As managers become more computer- and information-literate they recognize that computer system benefits cannot be measured only in dollars and cents. Like so many decisions made by top management, the one to invest in the computer can be a largely subjective one, based on faith. Management has faith that the information specialists will be able to develop and implement a good system.

Before management gives the go-ahead based on faith, however, they might insist on an economic analysis. In this case the systems analyst must be prepared to make the economic justification calculations that we have described. As with the other tools in the analyst's toolkit, the ability to use the economic justification tools are the mark of a professional systems analyst.

Problems

1. Use payback analysis to compute the payback period for the new system, using the data below. Write a one-page memo to your instructor, explaining your findings.

	Year				
	1	2	3	4	5
Existing system costs	12	15	19	25	32
New system costs	27	11	12	15	19

Note: Annual costs in thousands of dollars.

2. Use the figures in problem 1 to compute the net present value of the investment. Assume an annual interest rate of 10 percent. Use the formula for calculating the discount percent given earlier. What effect does the discounting of the benefits have on the payback period? Write a one-page memo to your instructor, explaining your findings.

Systems Design

At this point in the system development life cycle, we have completed the systems analysis. We have studied and documented the existing system in the process of understanding and defining the user's information needs. The task now is to design a new or improved system that meets those needs.

Here, we devote three chapters to the important topic of design. Chapter 11 describes the activities performed during the design phase of the SDLC and addresses some important design issues. Chapter 12 explains how controls are built into systems, and Chapter 13 focuses on the important topic of system security.

The design phase, like the analysis phase, provides the systems analyst with an opportunity to use all the tools for documenting both processes and data. In using the tools, the analyst applies a combination of creativity and systems theory, mixed with a good knowledge of information technology. Today's design efforts also pay particular attention to human factors considerations and apply new approaches such as object-oriented design and use of CASE tools. Chapter 11 is devoted to the challenging process of systems design.

Information accuracy has been emphasized since the beginning of the computer era. The first systems were used for accounting processes, and in accounting, accuracy has always been of primary concern. Although management is responsible for achieving accuracy in its information systems—an objective attained by identifying risks and implementing controls—the actual work is usually performed not by management, but by specialists. Chapter 12 describes how information specialists and internal auditors, working with management, identify the potential risks in computer-based systems and incorporate necessary controls.

Concern for security is more recent. During the first years of computing, you could walk down the sidewalk of any large city and view a firms' computing installation through large windows. Firms used their computers as showcases—as indications of their progressive nature. A combination of natural disasters, such as earthquakes and floods, plus intentional damage during the social *unrest of the 1960s and 1970s, caused management to rethink the subject of security. Today, systems are located behind locked doors, inaccessible not only to the general public but also to many employees. Chapter 13 explains how to achieve security in systems designs.*

After reading and studying Part 5, you will be in a position to design computer-based systems that meet users' needs and feature sound controls and tight security.

VII—DESIGNING THE EMPLOYEE SCHEDULING SYSTEM

(Use this scenario in solving the Harcourt Brace problem at the end of Chapter 11.)

Russ and Sandi returned to Orlando after completing their systems analysis of the Bellmawr operation. Ira called together the order processing and billing group the morning after Russ and Sandi's arrival, and that was the first chance for them to meet all of the team members. In addition to Ira, the team included senior systems analyst Anne Hogan, systems analyst Sue Shine, database specialist Frank Greco, network specialist Phyllis Steinham, programmers Suzanne McNulla and Lloyd Hamm, and internal auditor Pam Robinson. The addition of Russ and Sandi brought the team size to ten, the largest of the Harcourt Brace teams.

Russ and Sandi were given the title junior systems analyst, one they would hold until they demonstrated proficiency in the use of the main documentation tools and successfully developed a complete system. At that point, they would be promoted to systems analysts. They hope to then become senior systems analysts, and eventually systems managers. Ultimately, both would like to become a CIO, or become a manager or executive in a user area. With systems backgrounds, they have many options.

Ira's meeting was devoted to a status report on the COPS order processing and billing systems. Of the 63 order processing and billing points identified as prerequisite to November cutover, 35 had been completed, and 20 were in process. Ira and the team members were confident they would be ready well in advance of the July 31 deadline that Mike Brynes had imposed. Ira was so confident of the COPS schedule that he asked Sandi and Russ to make their project dictionary available to the team. By sharing this information, the entire team could engage in the system design and overlap that effort with their COPS activities. That was one thing that neither Sandi nor Russ had anticipated—that an information specialist often works on several systems at the same time.

Russ and Sandi first presented their project dictionary to Ira, who read it and told them it was very good work. He suggested only a few minor changes. Those changes were made, and then the notebook was circulated among the team. Only one team member, Frank Greco, had any suggestions, and they were easily incorporated into the documentation. When everyone had read the project dictionary, Ira called a meeting to explore the design of the new system.

IRA LERNER: Sandi and Russ, we've all had a chance to look over your documentation, and agree that it's very professional—a good base for our system design. Let's begin as we always do by coming up with a good statement of the problem.

[Ira asks the group members for their ideas, and everyone talks back and forth for almost 20 minutes before there is general agreement.]

IRA LERNER: I can live with that. [Ira writes the statement on the chalkboard: "The problem is that Bellmawr warehouse area supervisors cannot accurately reassign personnel to match fluctuating daily work loads due to a lack of good information." Ira underlines the word *information*, emphasizing it is the basis for redesigning the system.]

ANNE HOGAN: Sandi, we all have copies of your DFDs. Why don't you start in the beginning and explain the system. When you reach a point where we think some improvement can be made, we can discuss it. OK?

SANDI SALINAS: That's fine. [Sandi begins with the receipt of mail orders and describes that portion of the system. When she gets to the TOS operation, everyone begins shifting in their seats as if they can hardly wait to spring into action.]

PAM ROBINSON: That lack of written document for the regular TOS orders bugs me. I can't believe the system was designed that way. There's no audit trail. If something gets fouled up farther down the line, there's no way to go back and reconstruct the activity.

ANNE HOGAN: Good point, Pam. But, you have to remember that the system has worked well until recently. It was a design geared to low volumes.

PAM ROBINSON: I recognize that, but when you design a system, you should always build in good controls.

ANNE HOGAN: That's certainly true. Is everyone in agreement that an order form should be completed for nonrush TOS orders? [Everyone responds "Yes."] Not one dissenting vote—that's good. Russ, I want you to work with Sandi and come up with a TOS order form. While you're at it, you might want to reconsider the form used for faxing the rush orders to the distribution centers. Perhaps one form could be used for both purposes. A good approach might be to begin with the fax form and use it as a basis for the design.

RUSS MARCH: We'll do it.

[Anne Hogan asks the team if any further discussion of the TOS order entry portion of the system is necessary. Nobody has anything more to say, so Anne asks Sandi to continue.]

SANDI SALINAS: The starting point for everything in Bellmawr is the printout of the pack lists on the high-speed printer. Keep in mind that the problem, as we have defined it, deals with an inability to reschedule employees based on the relative volumes of these forms for the warehouse areas. That being the case, the pack lists seem to be the key to the problem solution. Would you agree?

IRA LERNER: Most definitely. You know, the physical portion of the Bellmawr operation has been studied and restudied so many times it's not funny. There's always a consulting firm in there trying to make things run smoother. Bellmawr is our best example of warehouse automation. So I'd be surprised if we came up with anything that would change the actual filling of the order. Sandi, what do you think?

SANDI SALINAS: We didn't find anything in the physical side of the system that appeared to need change. Our solution has got to be focused on the conceptual side—more specifically, on the pack lists.

ANNE HOGAN: That's probably true, Sandi, but you should continue with your explanation of the system. That way, we will all have the entire picture. We can then come back to the pack lists and discuss them.

[Sandi continues, ending with the faxing of the completed pack lists to Orlando to trigger the billing operation.]

RUSS MARCH: Sandi and I did raise our eyebrows a little when we first heard about the faxing operation. That seemed to be a crude way to respond to Orlando. We saw a need for an electronic transmission directly into the mainframe. But then we learned that COPS will take care of that. You're going to locate a data entry group in Bellmawr to key confirmation data directly into the Orlando mainframe, aren't you?

IRA LERNER: That's right. That will be a part of the new system. We'll have data entry units in all the distribution centers. It's part of our long-term plan to decentralize our information resources. There's no need to send data to Orlando so that we can key it in. It might as well be keyed in on site.

[The discussion returns to the pack lists, and the team considers various ways to provide the area supervisors with the information they need.]

RUSS MARCH: When I asked Dave Mattson about the availability of warehouse order-filling standards, I thought they might offer a way to solve the problem. They're already in the computer. If somehow we could apply them to each order, the computer could estimate the filling time. Then, the computer could add up the times for each area. The supervisors would then have an exact knowledge of what the daily workloads would be.

LLOYD HAMM: We could even go one step farther. The computer could determine how many order fillers would be needed for each area.

[Everyone agrees that Russ and Lloyd have hit on a good approach to solving the problem. The team tries to come up with other ideas, but cannot. The order-filling standards seem to form the basis for the solution.]

SUE SHINE: Well, how do we get this information to the supervisors? Do we continue to have them come to the computer room and get the printouts? We could transmit the output from the standards computations along with the pack lists.

[The team explores the output portion of the system, and decides that one alternative would be transmission of a special report prior to the pack lists. The report would contain the results of the standards computations, showing the number of order fillers by area. Sandi sketches the report on the chalkboard. That format appears in Figure HB7.1. Sandi uses X's and O's to represent where digits will appear.]

```
                ORDER-FILLING PERSONNEL REALLOCATION FOR DATE

STAFFING BEFORE REALLOCATION

                         STANDARD          AVAILABLE
AREA                      HOURS             HOURS            VARIANCE

ELEMENTARY                 XX                XX                -XX
HIGH SCHOOL                40                56               +16
COLLEGE                    XX                XX               +XX
HEALTH/SCIENCE             XX                XX                -XX

TOTAL                     XXX               XXX              +XXX

RECOMMENDED REALLOCATION (NUMBER OF HOURS/NUMBER OF EMPLOYEES)

FROM                 TO->       EL        HI        COL       H/SC

ELEMENTARY                     00/00     00/00     00/00     00/00
HIGH SCHOOL                    16/02     00/00     00/00     00/00
COLLEGE                        00/00     00/00     00/00     00/00
HEALTH/SCIENCE                 00/00     00/00     00/00     00/00

STAFFING AFTER REALLOCATION

                         STANDARD          AVAILABLE
AREA                      HOURS             HOURS            VARIANCE

ELEMENTARY                 XX                XX               +XX
HIGH SCHOOL                40                40                +0
COLLEGE                    XX                XX               +XX
HEALTH/SCIENCE             XX                XX               +XX

TOTAL                     XXX               XXX              +XXX
```

Figure HB 7.1
A single report of daily order-filling workload by warehouse area.

SANDI SALINAS:　Another approach would be to make the report available to the supervisors in a displayed form. There's a CEPOS terminal available to each area. They will be COPS terminals after the cutover. The supervisors could use their terminals to bring up the reports first thing in the morning.

RUSS MARCH:　We could even tailor a report to each area. Each supervisor could obtain a report for his or her area. It might look like this. [Russ sketches another output on the chalkboard. This appears in Figure HB7.2.]

```
ORDER-FILLING PERSONNEL REALLOCATION HIGH SCHOOL AREA FOR DATE

          STANDARD HOURS:              40 HOURS
          AVAILABLE HOURS:             XX HOURS
          VARIANCE:                   +XX HOURS

     STATUS OF OTHER AREAS BEFORE REALLOCATION:

          ELEMENTARY                  -XX HOURS
          COLLEGE                     +XX HOURS
          HEALTH/SCIENCE              -XX HOURS

          RECOMMENDED REALLOCATION:

     NUMBER OF EMPLOYEES:              XX
          TO:                      AREA NAME

     STATUS OF ALL AREAS AFTER REALLOCATION

          ELEMENTARY                  +XX HOURS
          HIGH SCHOOL                 +XX HOURS
          COLLEGE                     +XX HOURS
          HEALTH/SCIENCE              +XX HOURS
```

Figure HB 7.2
A report of daily order-filling workload tailored to a particular warehouse area.

ANNE HOGAN: I like the idea of the reports for each area, but the supervisors must meet and discuss who will be reassigned for the day. It seems like the neatest design would be one where all the supervisors received the transmission in a group. Then they could get on with the reassignments.

PHYLLIS STEINHAM: That's where I come in. As a network specialist, I have a good idea of the plan for terminals in each of the locations. I know that there's been a lot of discussion about putting COPS terminals in all the Harcourt Brace conference rooms. Then the rooms could be used in a GDSS, or group decision support system, fashion. This sounds like a good way to justify such a setup in Bellmawr.

[The team continues the discussion and concludes that they have identified four feasible solutions. Sandi writes them on the chalkboard. They appear as Table HB7.1.]

Table HB7.1 **Alternate Solutions**

Solutions

1. Print a single supervisor report in the Bellmawr high-speed printer room.
2. Display a single supervisor report on terminals in the warehouse areas.
3. Display tailored supervisor reports on terminals in the warehouse areas.
4. Display tailored supervisor reports on a terminal located in the Bellmawr conference room.

ANNE HOGAN: We seem to have done a thorough job on the outputs. Let's turn our attention to the necessary processing. Russ, what is needed there?

RUSS MARCH: Not much really. Most everything is already being done in CEPOS or will be done in COPS. As I understand it, COPS will sort the pack list records into warehouse area sequence prior to transmission to Bellmawr. That will eliminate all of the manual sorting. Each area's pack lists will come out together, and somebody, either the supervisor or someone else, will pick them up and take them back to the area. So, the addition of the sort is the only change in the computer processing.

Everything else revolves around the order filling standards. Once all the pack list data has been written on the disk for the day, a program can use that file and the standards file to compute the order-filling times. Those times can be written on a file that will be used to print the report—or reports, as the case may be.

SUZANNE McNULLA: The program to compute the order-filling times seems to offer a good challenge. I would look forward to coding it.

ANNE HOGAN: Well Suzanne, you may get your wish. Russ, as you described the processing, you also described the input data. The two files—pack list and standards—give us everything we need. We don't have to worry about creating a new file. That's good.

IRA LERNER: Before we start celebrating, we had better look at the standards file and see what's there. We might have to make some changes so that we can use it. That file is vital to the solution. Frank, you're the database specialist. Take a look at the standards file and work with Sue and Suzanne to explore what has to be done.

Conclusion of the System Design

The general feeling among the team is that they have taken a good first cut at designing a new system. The next step is to evaluate the alternate solutions. Since the alternatives deal with the format of the output, Russ and Sandi return to Bellmawr and solicit the users' opinions.

To help Dave Mattson and the area supervisors evaluate the output alternatives, Sandi and Russ use a CASE tool to prepare prototypes of the displays. Sandi and Russ

Table HB7.2 **Contents of the Project Dictionary at the Completion of the System Design**

   I. System overview
  II. Existing system procedure
      A. Data flow diagrams
      B. Structured English
      C. System flowchart
 III. Existing system data
      A. Data dictionary entry forms
  IV. New system procedure
      A. Data flow diagrams
      B. Structured English
      C. Action diagrams
      D. System flowchart
   V. New system data
      A. Entity-relationship diagram
      B. Data dictionary entry forms
  VI. New system outputs
      A. Printer spacing charts
      B. Screen layout forms

learned how to use CASE in college. The supervisors meet in the Bellmawr conference room and a PC is used to simulate the display of the reports.

The discussion with the Bellmawr managers is followed by another team meeting in Orlando where Frank Greco and his group report no difficulties in using the standards file. The Bellmawr managers' preferences are considered, and the decision is made to go with alternative 4—the conference room setup with a report tailored to each area. This decision is approved by Dave Mattson, Ira Lerner, Bill Presby, and Mike Byrnes.

The final step in the systems design is to complete the documentation. Sandi and Russ add new sections to the project dictionary that document both the processes and the data of the new system. This is no small task. The determination of the logic to be employed by the daily warehouse scheduling program poses the most difficult challenge. Anne Hogan helps out, as does an industrial engineer from a consulting firm that developed the order-filling standards. After several weeks, the documentation is done. Ira insists that a variety of documentation tools be used so as to provide the junior systems analysts with as much experience as possible. Table HB7.2 shows the composition of the dictionary at the completion of the design.

Personal Profile—Anne Hogan, Senior Systems Analyst

Anne M. Hogan is a native of Baltimore who now calls Orlando home. She attended Essex Community College, Johns Hopkins University, and Rollins College, and earned an Associate of Arts degree in General Studies. She joined Harcourt Brace in 1987 with

some programming experience gained in previous jobs. She enjoys running and read-ing, and credits her managers with her success. According to Anne, "Managers decide who gets what project and how much responsibility. They can develop your technical skills and confidence." Overall, she finds the people at Harcourt Brace easy to work with—an ideal situation for someone doing systems work.

CHAPTER 11

Systems Design

Learning Objectives

After studying this chapter, you should:

- Know how the main approaches to the design of computer-based systems have evolved and which ones are currently receiving the most attention
- Know what joint application design is, and how it fits into the systems design process
- Understand one approach to identifying alternate system configurations
- Know how to make a structured evaluation of alternate system configurations
- Know some techniques for designing source documents and reports, and making systems more user friendly
- Be aware of the potential impact that object oriented analysis and design can have on systems development
- Recognize the importance of incorporating human factors considerations into the entire system life cycle

Introduction

The designs of early data processing and management information systems featured a bottom-up approach emphasizing processes, but current thinking favors a top-down approach featuring a central database.

The design phase of the SLC begins with the preparation of a detailed system design. The project team can inititate this effort informally by engaging in brainstorming or formally by engaging in joint application design.

The next step is to identify alternate system configurations. One approach is to consider options for each system element, beginning with the output. Attention then turns to storage and input considerations.

Alternate configurations are then evaluated, and the best one recommended to the MIS steering committee. The committee then decides whether to proceed with the implementation phase.

All areas of systems design are important, but the systems analyst must pay special attention to computer input and output. User friendliness combined with good source document and report design can contribute to user satisfaction.

Although the fundamentals of good system design have evolved over more than a quarter of a century, a new approach is exerting an influence for change. This influence is object-oriented development, a product of the current popularity of object-oriented programming.

Regardless of the approach, behavioral factors should be incorporated into system designs. However, this effort is not limited to the design phase and can span the entire system life cycle.

The Design Task

The design task is essentially one of modeling the new system. We saw in Chapter 10 that during systems analysis, the systems analyst models the existing system, first with a physical model and then with a logical model. The analyst has the same responsibility during the systems design phase, only this time the models relate to the system being designed. Also, during design, the logical model is normally prepared first, followed by the physical model.

Basically, two features of the new system must be modeled—its data and its processes. The analyst prepares a **data model,** which documents the data to be used by the new system, and a **process model,** which documents the processes. The data model usually takes the form of a *logical model,* depicting the logical relationships that exist among the data elements rather than the technology used to house the data. The logical data modeling tools include entity-relationship diagrams and the data dictionary. The process model takes the form of a logical model when documented with tools such as DFDs and structured English. It takes the form of a *physical model* when documented with system flowcharts.

The Evolution in Approaches to Systems Design

During the relatively brief history of the computer, four basic approaches have been followed in designing computer-based systems. The approaches have emphasized document preparation, problem solving, the database, and, most recently, the enterprise data model.

The Document Preparation Approach   Early *data processing systems* were developed in two phases. First, each system—inventory and billing, for example—was designed with emphasis on the processes that transform input into output. Then, all systems were integrated by linking the inputs and outputs. The designs reflected an output emphasis on accounting documents such as income statements, invoices, and payroll checks. Because the systems were designed separately, each one tended to be an island, with its own data files.

The Organizational Problem-Solving Approach   In the mid-1960s, firms began to implement their *management information systems* organization-wide, but there was no grand plan for integrating the separate systems. As a result, the development process continued the same bottom-up emphasis on processes, and separate files, often incompatible, were typical.

The Organizational Database Approach   During the early 1970s, two developments had a dramatic effect on systems design. One was the appearance of *database management system (DBMS)* software that permitted the creation and maintenance of a database that could be shared by users throughout the firm. The DBMS signaled the beginning of a shift in emphasis from processes to data.

The other influence was a refocusing of system design on specific problems of individual managers—the *decision support system (DSS)* approach. The key feature of the DSS era has been an organizational database that applies a top-down constraint on the design of separate systems. The database provides a much-needed vehicle for the integration of the systems.

The Enterprise Data Model Approach   An emerging approach to systems design shifts emphasis from the users' data resource to the organization's. The idea is simple: If all the firm's data is stored in a database, there is no limit to the information support the firm's computer-based systems can provide.

Enterprise modeling represents a step beyond providing a data resource to meet individual problem solvers' needs by also meeting the needs of the overall organization.

Putting Contemporary Systems Design in Perspective

Most firms today follow the systems design approach of the DSS era and make available a database to meet the problem-solving needs of their managers. Relatively few firms engage in enterprise modeling. Both approaches are top-down strategies for applying the computer as a problem-solving tool and represent the mainstream design philosophies of the 1990s. The trend is definitely toward enterprise modeling. Large firms are setting the trend, but whether smaller firms will follow remains to be seen.

Using CASE for Systems Design

When the analysis phase of the SDLC has been completed, using CASE, the next step is to begin the design phase. Design involves modeling the target system, and the model can be detailed or general. The design phase includes several important steps, which usually are not accomplished in a sequential order but rather by means of a great deal of iteration, until the users and the project team are satisfied with the final product.

All systems should adhere to proven guidelines, which are intended to make the systems easier to develop and maintain. Maintenance can represent as much as 50 percent of system costs. It is important to make the maintenance task as easy to accomplish as possible so that enhancements and adjustments can be made in a timely manner with as little effort as possible.

The project team should observe the following guidelines as they use CASE to develop systems that will be easy to use and maintain.

- Information systems are for users, not for the project team. In all decisions, the needs and interests of users should be foremost. Users should be involved in the design process and in all other portions of the SDLC.
- Systems should be divided into cohesive, loosely coupled programs or modules. Cohesion refers to how closely the activities within a module are related to one another. Loose coupling means that a module should not be dependent on other modules.
- Information systems are dynamic, not static. Therefore, it is important to design an information system that has room to grow and change. Storage allocation and flexibility, enhancement support, and clear documentation techniques that change as the system changes all contribute to ongoing system maintenance and enhancement.

The project team has various documents, information, and directives from the CASE tool, users, and the organization that must be considered in constructing the new sytem. Using hierarchy charts, entity-relationship diagrams, and other tools and information, the project team constructs a logical design of the new system.

The logical design can take the form of a leveled set of proposed DFDs. These DFDs are used to communicate the proposed system to users, who must approve the design. Then, the DFDs, along with other process and modeling documentation such as hierarchy charts, ERDs, and a complete data dictionary, are made available to those persons who will have responsibility to implement the system. CASE products support the completion of all these design outputs.

The Design Phase of the System Life Cycle

Regardless of which design strategy is employed, the design phase of the system life cycle is carried out by taking the steps illustrated in Figure 11.1.

Step 1—Prepare the Detailed System Design

At this point in the SLC, each team member has likely identified one or more ways to solve the defined problem, ideas that originated in the feasibility study and began to take shape during the system study. The first step of the design phase provides an opportunity for team members to communicate their ideas. This can be accomplished in various ways. Traditionally an informal approach of brainstorming was considered the best method. Currently, a more formal approach, called joint application design (JAD), is receiving attention.

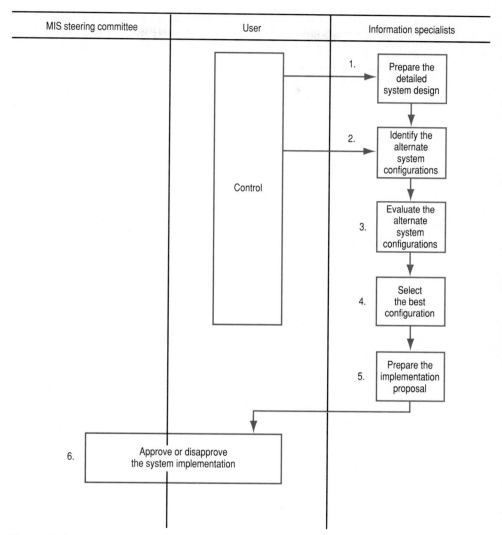

Figure 11.1
The design phase.

Brainstorming

Brainstorming is simply the communication that takes place between two or more people who assemble in a group and follow no specific agenda in discussing a particular topic.

When applied to systems design, brainstorming is conducted by members of the project team, and the topic is the new system. Team members identify the different approaches to solving the problem, cull those that have no potential, and discuss the advantages and disadvantages of the others. Once they agree on which overall approach is best, they discuss the details of the selected design, addressing such topics as user interfaces, basic system logic, data needs, and operating procedures.

The brainstorming session might include several meetings over a period of several days or weeks. The end product is a consensus on the system design that will best address the user's problem.

Joint Application Design

Communication among team members can be made more structured by joint application design. **Joint application design (JAD)** is a formal approach to systems design in a group decision support system setting.[1] JAD consists of a several-day session attended by project team members and three other types of participants: a **session leader** who creates a climate that enables the group to reach a consensus; one or two **scribes** who maintain a record of everything that is said and publish a written report; and an **executive sponsor,** a top-level manager from the user area, who kicks off the session and is present for the conclusion.

The Three Phases of JAD

JAD is performed in three phases. First, planning must be done. Then, the session is held. Finally, the results are documented.

Planning for the Session   The project team prepares for the JAD session by identifying the other participants, selecting the site, making certain that all the required equipment is available, and establishing an agenda.

Conducting the Session   The key to a good JAD is the session leader. This person ensures that all participants contribute their ideas, that the session is not monopolized by the most vocal participants, that the discussion stays on track, and that disagreements are settled. It is not necessary that the session leader be an expert on the application area. Rather, the leader must be skilled in conducting a group session in a stimulating manner.

Documenting the Results   When the scribes deliver the detailed written record of the session, the task of the project team is to prepare a **specification document,** which summarizes the agreed-upon system design. Figure 11.2 identifies sections that might be included in such a document.

Note in the figure that the basis for sections 1 through 4, and also 12, can come during the planning phase. The basis for the remainder can come during the analysis phase. Notice the specific nature of sections 7 through 11. Such attention to detail is necessary regardless of how this step of the design phase is carried out. An informal brainstorming session should produce the same degree of detail.

---

[1]This description is based on Per O. Flaatten, Donald J. McCubbrey, P. Declan O'Riordan, and Keith Burgess, *Foundations of Business Systems*, 2d ed. (Fort Worth, TX: The Dryden Press, 1992): 210–218.

1. Management objectives
2. Scope and limits
3. Business questions the system will answer
4. Information required to answer the questions
5. Relationship with other systems
6. Issues
7. Data element specifications
8. Screen layouts and report layouts
9. Menus
10. Processing rules
11. Operating procedures
12. Performance and operational requirements

Figure 11.2
An outline of a specification document.

Placement of JAD in the System Life Cycle   We have included JAD in the design phase, but it can come earlier. JAD can be used by the firm's top executives during enterprise planning and then throughout each subsequent SLC. During the planning phase, JAD can help define the problem and identify system objectives and constraints. When used in the analysis phase, JAD can facilitate the definition of the user's information needs. During the implementation phase, JAD can help the project team determine the best approach to cutover, and during the use phase, it can provide the basis for selecting the best redevelopment strategy. A firm that is totally committed to JAD will likely use it during each life cycle phase.

Step 2—Identify the Alternate System Configurations

The next step is to identify the system configurations that can accomplish the agreed-upon design. There are probably an infinite number of approaches, and the one that is taken is influenced by the particular situation. However, a process that appears to be especially logical since it is geared to meeting the user's needs is illustrated in Figure 11.3.

The systems analyst and user first determine whether information is to be provided by report generators, mathematical models, or office automation communications. This choice then requires that attention be given to output devices, secondary storage media and organization, and input considerations.

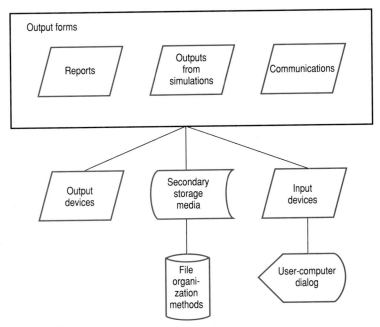

Figure 11.3
The identification of alternate system configurations can begin with basic forms of output.

The design therefore evolves in a reverse sequence, beginning with outputs and ending with inputs. The sections below describe in more detail each design option that influences the configuration of the new system.

Basic Forms of Information Output

In Chapter 6, it was noted that users obtain information from the computer in three basic ways: (1) reports, (2) outputs from simulations, and (3) communications.

Reports   Reports can be prepared periodically or in response to a special situation or request. Both periodic and special reports can be printed or displayed. For example, a user can receive a periodic report in a hardcopy form identifying some type of exceptional activity that requires further study. The user can then use a database query language to display a special report that sheds light on the exception.

Outputs from Simulations   When employing a mathematical model, the user may try many decisions, looking for the best result. In this setting, high speed in displaying the output is ideal. Once the user identifies the decision to be implemented, a hardcopy output can be produced.

Communications   Some office automation (OA) applications provide the user with the choice of printed or displayed output. Electronic mail and electronic calendars are examples. Other OA applications are dedicated to particular output media. Word processing almost invariably produces hardcopy, whereas voice mail is by definition auditory. The dedicated-media applications simplify the project team's output decision. When the user specifies an application such as word processing or voice mail, the output medium is automatically identified.

Putting the Output Forms in Perspective   The project team should pay particular attention to the output forms since they are the end product of system use, and since the user's perception of the system will be influenced primarily by the output.

The design of system output is a point in the development process where prototyping is especially effective. The information specialist can provide the user with prototypes of different forms of output, enabling the user to identify the one that is preferred.

Output Devices

The basic form of output will have a direct influence on the output device used. The following output devices, which can be included in the system configuration, offer alternate ways of producing computer output.

- **Display units** Online keyboard terminals are often called **CRT terminals** because their output device uses a *cathode ray tube (CRT)*. The term **VDT,** for **visual display terminal,** has the same meaning. These display units are also a part of microcomputer configurations.

- **Printers** The first computer printers were called **line printers** because they were so fast they appeared to print an entire line at one time. Line printers, still used in mainframe and mini configurations, can attain speeds as high as 4,000 lines per minute. Even faster output can be achieved by **page printers,** which print in the 15,000 to 20,000 lines-per-minute range. The dot-matrix printers so popular with micros are called **character printers** and print 30 to 350 characters per second.

    Line printers and dot matrix printers are examples of **impact printers**—a metallic object strikes an ink ribbon in the same manner as a typewriter. Impact printers can make multiple copies, but they are relatively slow. Faster speeds are achieved by **nonimpact printers** that apply ink to the paper in much the same manner as a copying machine. Page printers, which use laser and ink-jet technology, are examples of nonimpact printers. Nonimpact printers can produce only single copies.

- **Plotters** Until character printers with a graphic capability came along, the only way to obtain attractive printed graphic output was with a special device called a plotter. Plotters are still used when special graphic needs exist, such as large-sized graphs, heavy volumes, or exceptionally high quality.

Figure 11.4
A plotter.

Plotters, such as the one illustrated in Figure 11.4, are typically attached on-line to mainframes and minicomputers or operated offline from tape or disk input.

- **Computer Output Microform Units** The microform output produced by computers is called **COM,** for **computer output microform.** Microform usually exists in a roll called **microfilm** or in a sheet called **microfiche.**

- **Audio Output Units** Push-button telephones can serve as output devices for both prerecorded messages and voice mail. The message that says "I'm sorry, the number you dialed is no longer in service; the new number is . . ." is assembled from a prerecorded vocabulary in a special type of storage called an audio response unit. The voice mail message is converted from analog to digital form for storage in secondary storage and then converted to analog form for playback. A special unit performs the conversion.

Printed Versus Displayed Output   The trend is away from printed output, called **hardcopy,** to displayed output. Users like the displays because of their speed and ability to produce vivid colors. The only real disadvantage of displayed output is the absence of a paper record.

Printed output is used for lengthy documents, documents that are sent through the mail or viewed by several people simultaneously, such as during a conference, and for historical storage. Printed output is also used for turnaround documents. A **turnaround document** is a computer-prepared form that is mailed to a user, who then returns it so that it may facilitate data entry. The paper forms you include with your credit card payments are examples of turnaround documents. The forms are read optically to identify you and the amount of your payments—all without the need for keyboard entry.

Putting the Output Options in Perspective   Ideally, the basic form of output should influence the choice of output device. If a user prefers a report in hardcopy form, then a printer is required. Likewise, if the application is electronic calendaring, then a CRT is used. However, the user's preferences might be constrained by the existing computer configuration. The task of the systems analyst is to design the system so that the outputs best satisfy the user, given the existing constraints.

Secondary Storage Options

Two main decisions must be made concerning the secondary storage. First, the storage medium must be selected. Then, in the case of direct access storage, the file organization method must be determined.

Secondary Storage Media   There are two basic types of secondary storage—sequential and direct. **Sequential storage** requires you to access records one after the other. Magnetic tape is the best example of a sequential storage medium. **Direct access storage** lets you access a record quickly, without sequentially searching for it. Magnetic disks and diskettes are examples of direct storage media. A unit that has a direct storage capability is called a **DASD,** for **direct access storage device.**

When presented with the task of selecting the secondary storage medium the analyst will invariably select DASD because of its capability for providing a current database. A sequential storage medium such as magnetic tape is useful when batch processing is adequate. Compact disks are expected to be the sequential storage medium of the future.

File Organization Methods   There are three basic methods for organizing data in secondary storage—sequential, indexed sequential, and direct. All three methods can be used with a DASD, but only sequential organization is available with sequential storage.

- **Sequential File Organization** In **sequential file organization,** data records are positioned one after the other in the storage medium, in a particular sequence. The sequence is determined by one or more data elements called *keys*. For example, the key of an inventory file is usually item number, and the records are arranged in item number sequence.

An advantage of sequential file organization is the efficiency with which the secondary storage is used. Very little space is wasted because multiple records are typically blocked together, reducing the number of gaps that separate the records.

The big disadvantage of sequential organization is its inability to provide specified records quickly for processing. The program must engage in a time-consuming search, beginning with the first record in the file and proceeding until the needed record is found.

- **Indexed Sequential Access Method** The efficiency of sequential storage can be achieved along with quick access by providing an index. The index sends the access mechanism of the DASD to the appropriate data record in a sequentially arranged file. This technique is **indexed sequential access method,** better known as **ISAM.** Figure 11.5 shows that the index contains the keys and the DASD addresses of the first records of record groups. In this example, assume

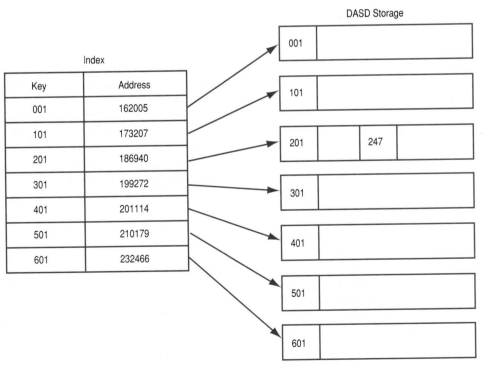

Figure 11.5
The indexed sequential file organization includes an index that provides for direct access to records.

that a file has been divided into seven subsets of approximately 100 records each. When a program must access a certain record in the file, such as record 247, it examines the keys in the index. The program determines that the record is in the third subset since its beginning key is greater than 201 and less than 301. The program directs the access mechanism to location 186940, where record 201 is stored, and then searches each record in the subset until record 247 is reached. Record 247 is read into primary storage and processed by the program.

The main disadvantage of indexed sequential organization is that an index search must be conducted for each record retrieved.

- **Direct File Organization** The need for a separate search can be eliminated by using direct file organization. In **direct file organization** the records are located randomly in the DASD using an address computed from the record key. For example, arithmetic operations are performed on a customer number such as 5398712 to produce a storage address that specifies that the record is located on DASD unit 3, disk surface 8, track 139, in position 4.

A big limitation of direct file organization is that the addressing algorithm may compute the same address for more than one key, producing what are called **synonyms.** The synonyms must be stored in an overflow area where retrieval can be inefficient. Another limitation is the fact that large gaps of unused storage space can exist throughout the randomly located records.

Putting the Secondary Storage Options in Perspective   The application to be performed by the computer determines the type of file organization needed. However, unlike the output options, the storage considerations are usually transparent to the user. In other words, they are of no concern to the user, who is interested only in what the system does, not in how it does it. The systems analyst selects the storage media and determines the file organization method based on the user's needs.

Sequential storage organization is decreasing in popularity and is applicable only to certain data processing systems, such as payroll, which can be processed on a cyclical basis, such as daily or weekly. Information-oriented CBIS subsystems such as MIS, DSS, and expert systems demand DASD, using indexed sequential or direct organization.

Input Options

The data input option is influenced by the volume of data, who is responsible for the process, and the location at which the input occurs.

- **Online Keydriven Units** When used in relation to hardware, the term **online** means connected to the computer. Examples of online keydriven units are keyboard terminals and microcomputers. Although most data input is accomplished by means of the keyboard, many users prefer a **pointing device,** which enables the user to enact certain operations by simply pointing to an area on the

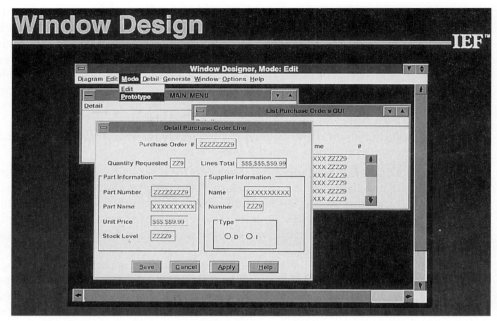

Figure 11.6
Screen designs can combine keyboard and mouse input.

screen. Examples of pointing devices are the mouse, the track ball, touch-screen capability, and hand-held remote control devices similar to those used to change television channels.

The screen illustrated in Figure 11.6 represents a currently popular design. This screen is produced by the Texas Instruments Information Engineering Facility CASE tool, better known as IEF. The user can use the keyboard to enter data into the boxes in the upper portion of the window and use the mouse to select options represented by the buttons at the bottom.

- **Magnetic Ink Character Recognition (MICR)** The specially shaped characters at the bottom of checks are printed in magnetic ink, and MICR input devices read the data into the computer. MICR was conceived as a way to relieve the input bottleneck in banks. The term **input bottleneck** describes the inability of keydriven data entry devices to keep pace with the faster speeds of computer processing and output.

- **Optical Character Recognition (OCR)** OCR devices read data printed in ordinary ink. Some devices, popularly called **scanners,** read machine-printed characters; some read handprinted data; some read barcodes; and some read pencil marks, such as those you make on true-false and multiple-choice exams. The scanners at supermarket checkout counters are examples of OCR, as are the cash registers in department stores that feature reading wands. OCR is an effective way to read large volumes of input data.

Both MICR and OCR represent an input strategy called **source data automation.** The input data is automated so that it enters the computer without keyboarding.

- **Special-Purpose Terminals** Some terminals permit data entry without the need to press keys in the ordinary way. McDonald's, for example, uses terminals that represent each of its products with a single key; you order a large fries, and the clerk enters the transaction with a single keystroke. Other special terminals are ATMs (automatic teller machines) and scanners that read data from the magnetized strip on the back of credit cards.

- **Voice Input** You can also enter data into the computer by speaking commands and data into a microphone. Hardware vendors continue to work on this potentially revolutionary form of input.

- **Push-Button Telephone** The input device that exists in the largest numbers is the push-button telephone. This method of data input is popular with persons who must communicate with the computer while away from their offices. For example, a sales representative who is calling on a customer can use the customer's phone to enter data, such as a request for an inventory status report, and hear the audio response in the telephone's earpiece.

Putting the Input Options in Perspective   In selecting input devices, the systems analyst is influenced by two factors—the requirements of the application and the capabilities of the persons who must operate the devices. For example, data processing systems feature large volumes of data, which can be entered by means of MICR or OCR or by data entry operators using keyboard units. The operators have good typing skills; therefore, the keydriven units represent a practical approach. On the other hand, managers enter very little data into their information systems and usually do not wish to spend their time typing. Their attitude explains the popularity of pointing devices.

User-Computer Dialog

When online keydriven units are selected as the input option, the next step is to determine the type of user–computer dialog. The basic choice is whether the dialog will be directed by the computer program or by the user.

Program-Directed Dialogs   A **program-directed dialog** is one in which the computer provides the user with a screen display that specifies the nature of the user's response. The three most common forms are menus, form filling, and prompting.

A **menu** is a list of choices available to the user, which is displayed on the screen. Popular software packages, such as Lotus 1-2-3 and Microsoft Windows, provide examples of good menu techniques.

Form filling requires the user to enter data into specific areas on the screen, much as one would fill out a paper form. Figure 11.7 shows an input screen that utilizes form filling.

Prompting is accomplished by the computer's telling the user what to do next or by the computer's asking the user a question. An example is "Please enter file name," displayed when you indicate that you want to write a file onto disk.

Operator-Directed Dialogs In an **operator-directed dialog** the user determines the sequence of the processes. This can be accomplished by use of a command language or by direct manipulation.

Many software systems, such as Lotus and dBASE, have a **command language** that causes the system to perform certain operations. For example, a dBASE user enters a command to sort data into a descending sequence.

Direct manipulation is accomplished when the user moves the cursor to a particular location on the screen and causes something to happen. In using a CASE

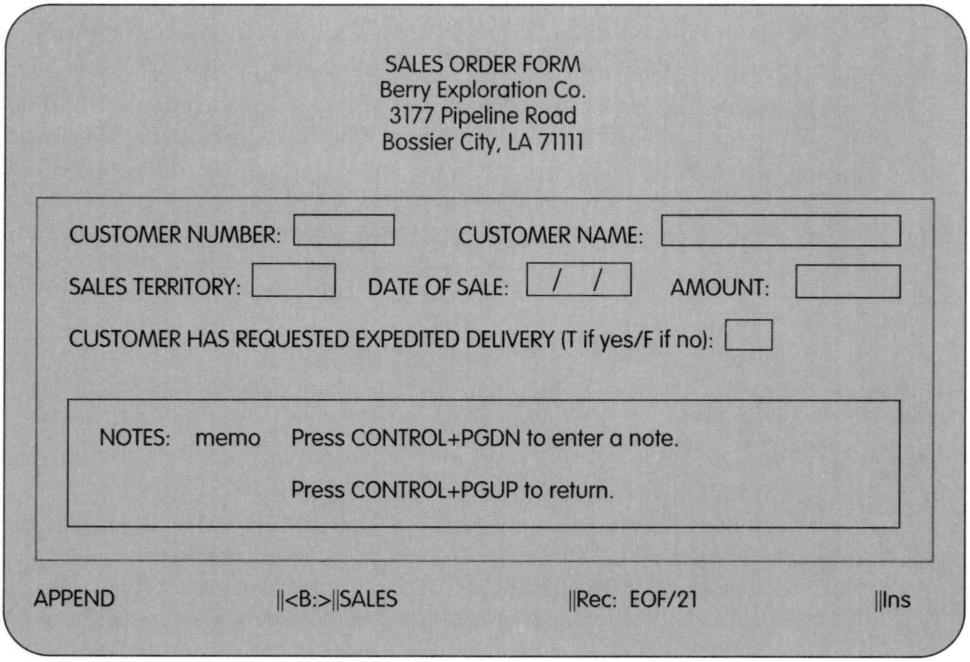

SALES ORDER FORM
Berry Exploration Co.
3177 Pipeline Road
Bossier City, LA 71111

CUSTOMER NUMBER: ______ CUSTOMER NAME: ______

SALES TERRITORY: ______ DATE OF SALE: / / AMOUNT: ______

CUSTOMER HAS REQUESTED EXPEDITED DELIVERY (T if yes/F if no): ___

NOTES: memo Press CONTROL+PGDN to enter a note.

Press CONTROL+PGUP to return.

APPEND ||<B:>||SALES ||Rec: EOF/21 ||Ins

Figure 11.7
The form filling technique.

tool such as Excelerator, for example, the user uses a mouse to position the cursor on the screen and draw a DFD process symbol.

Putting the User-Computer Dialog in Perspective   Users vary in their preferences for particular types of dialog. An expert user may prefer a command language because of the power and speed it provides, while a novice may prefer to use menus. How does the systems analyst handle this variability? One approach is to use prototyping to tailor system dialog to specific users. Another is to design systems so that users can select the dialog type they prefer.

Putting the Identification of Alternate System Configurations in Perspective

The determination of alternate system configurations is a good example of a semistructured problem. Some portions, such as the performance characteristics of various pieces of computing hardware, can be identified and evaluated in specific terms. Other portions, such as user preferences for particular dialogs, must be handled subjectively. The user and the other members of the project team work together to identify system configurations that should be considered.

Step 3—Evaluate the Alternate System Configurations

Once the alternate system configurations have been identified, they can be evaluated. First the evaluation criteria are identified, and then the ability of each configuration to satisfy the criteria is considered.

Specification of Evaluation Criteria

The project team members identify the items that will measure the ability of each alternative to meet the system objectives. These are the **evaluation criteria.**

Although a relationship exists between the evaluation criteria and the performance criteria established in the planning phase, the two are not the same. Performance criteria are what it will take to satisfy the user and will be applied once the system becomes operational. Evaluation criteria, on the other hand, are the measures used during the design phase to select the system that has the best chance of satisfying the performance criteria. In short, the evaluation criteria are a step toward achieving the performance criteria.

All team members are likely to contribute their own evaluation criteria. For example, in a particular project the systems analyst might place a high value on efficiency, the user might emphasize a quick retrieval of current information, and the DBA might stress a disaster recovery capability.

Accomplishment of the Evaluation

Each person involved in the evaluation then rates each alternative on each criterion. This can be accomplished in a completely subjective way, with each person simply rank ordering the alternatives. This technique was described in Chapter 7 and illustrated in Table 7.2 when we explained the systems approach.

Another technique is to quantify the evaluation as illustrated in Figure 11.8. In this example, each criterion is weighted based on its relative importance, and rated on a scale ranging from "very good" to "very poor." A rating of "very good" receives 10 points, "good" receives 8, and so on. The ratings of all the persons evaluating the particular system configuration are averaged, and a total score is computed.

The letter P in the figure stands for probability. Probability—the likelihood of something happening—can be viewed as a percentage. For example, the illustrated configuration has a 20 percent chance of performing "very good" in terms of its input efficiency. The letters EV stand for expected value. The expected value is computed by multiplying the probability by the rating points. For example, the EV for "very good" input efficiency is 2.0 (.20 times 10).

All configurations are rated in the manner illustrated in the figure, and the configuration with the highest total weighted score is the prime candidate for selection.

Putting the Quantitative Evaluation of Alternatives in Perspective

All project teams do not follow a quantitative process such as the one illustrated in the figure. However, the approach has value because it forces the participants to agree on the criteria and to consider the impact of each on the selection. The quantitative approach lends a certain amount of discipline to a process that can be very subjective.

Step 4—Select the Best Configuration

With the evaluation of each alternative completed, the next step is to select the best. This is the point in the process where the cohesiveness of the project team is put to the test since participants tend to support the alternatives that will best benefit their units.

However, the team is motivated to reach agreement because it knows that the choice should *not* be left to the MIS steering committee. The committee has neither the time nor the tools for making such a choice. The project team is in the best position to select the system configuration for implementation.

Step 5—Prepare the Implementation Proposal

The format of the implementation proposal was illustrated in Figure 8.7. The system design is first described in a summary, and then the performance criteria are listed, along with a brief explanation of how the recommended design satisfies each. Finally, the equipment configuration is identified and explained. The re-

Rating

Criteria	Weight	Very good (10) P	EV	Good (8) P	EV	Average (6) P	EV	Poor (4) P	EV	Very poor (2) P	EV	Total points (total EV)	Total weighted points (weight × total EV)
Input efficiency	.20	.2	2.0	.6	4.8	.2	1.2	.0	0	.0	0	8.0	1.60
Responsiveness	.35	.2	2.0	.7	5.6	.1	.6	.0	0	.0	0	8.2	2.87
Currency of information	.25	0	0	.2	1.6	.2	1.2	.3	1.2	.3	.6	4.6	1.15
Disaster recovery capability	.20	0	0	.7	5.6	.3	1.8	.0	0	.0	0	7.4	1.48
Total points												28.2	7.10
Maximum points												40.0	10.00

Legend: P = probability EV = expected value

Figure 11.8
A quantitative approach to the evaluation of a system alternative.

mainder of the proposal is devoted to a general description of how the system can be implemented and the results that can be expected when the system becomes operational.

Step 6—Approve or Disapprove the System Implementation

The chance of the MIS steering committee disapproving the implementation at this late stage in the SLC is small. However, the committee might ask the project team to rethink certain aspects of the design. This is especially likely when the committee feels that the recommended design is too expensive or fails to support the strategic plan for information resources.

Complete the Documentation

The project team has carefully documented the new system design as it evolved. However, additional work may be required to ensure that the documentation tells the whole story to the persons who must achieve the physical design. The key implementation players, such as programmers and operations personnel, should review the documentation and identify areas where further work is needed. When that work is accomplished, the new system documentation is added to the project dictionary. Then, the implementation phase can begin.

Designing Computer Input and Output

The task of the system developers is to achieve good design throughout the system. All system elements are important, but the input and output have the greatest influence on user satisfaction. In recognizing this fact we address three topics related to the design of input and output—source document design, user friendliness, and report design.

Source Document Design

A **source document** is a paper form that provides the input data to a system. It is usually a printed form, such as a timecard or a sales order, and is usually filled out either by hand or with a device such as a typewriter. A data entry operator reads the data from the source document and keys it into the computer. Sometimes, the source document is read magnetically or optically.

Tips for Designing Source Documents   The source documents you design will have a greater chance of contributing to the efficiency and accuracy of system input if you follow these tips:

1. Provide for a natural forms completion pattern. When filling out a form, persons usually enter data in a left-to-right and top-to-bottom sequence, and that format should be reflected in the form design. Note that this sequence is typical in Western cultures but may not be in others.

2. Make the fields the right size. The areas reserved for various data elements should be large enough to contain those elements. If the form is to be completed on a typewriter, the vertical and horizontal registration of the typewriter should be taken into account. Fields to be completed by hand should identify the character spacing, ideally with a box for each letter, as in the upper portion of Figure 11.9.

3. Use carbon sets to provide multiple copies. When more than a single copy is needed, use a **snap-apart form** that contains multiple copies attached to a binding stub at the top or side. Each copy should be in a special color, with its use, such as "Customer copy," identified. The sets can use interleaved carbon paper, or the forms can be printed on **NCR paper.** NCR stands for "no carbon required" and was invented by the National Cash Register Company.

    Regardless of the type of paper used, reserve the top copies of the forms set for the most important uses, such as customer copies and warehouse picking tickets. This design ensures that the most important copies will be the most legible. For sets with many copies, use thin paper to improve the legibility of the bottom copies.

4. Design mail documents to fit envelopes. How many times have you tried to enclose a turnaround document in an envelope that is too small? Poor system design! When source documents must be mailed, they should be designed to fit in the envelopes when properly folded, and the firm's name and address should show through the window.

5. Test forms usage before printing. Give a sampling of users an opportunity to use the form before printing large quantities. Use desktop publishing or word processing software to produce forms prototypes on laser or ink jet printers. Such prototypes can approach the appearance of forms printed by forms companies. As the users go through the motions of using the prototype, they often see the opportunity for improvements. Continue prototyping until the form is acceptable to the users.

Rely on the local sales representatives of forms companies for help in designing source documents and other printed forms. The reps are both knowledgeable and cooperative, and they can provide sample forms and perhaps even forms design handbooks.

User Friendliness

A system is **user friendly** when users and participants can *learn* and *use* the system easily and efficiently. The systems analyst can build user friendliness into a system by achieving consistency, simplicity, complete help support, ease of error correction, and judicious use of screen features.

- **Consistency** Things should happen the same way every time. For example, all systems should use the same log-on procedure, certain types of material should

1. Your First Name   M.I.   Last Name

Street   Apt. No.

City   State   Zip

Telephone

2a. Your age _____ yrs.

2. Date of Purchase: Month   Day   Year

2b. Edition of the game you purchased ☐ Standard
☐ Deluxe
☐ Deluxe Travel

2c. Price paid: $ _____

3. Was this product a gift or a self purchase?
☐ Self purchase
☐ Surprise gift
☐ Gift that was requested by the user

4. What is the age and sex of the person who bought this game?

Sex   **Age**
☐ Male   _____ Years
☐ Female

5. Where was this game purchased?
☐ Toy Store   ☐ Mail Order
☐ Department Store   ☐ General Merchandise
☐ Discount Store   (Sear's, Penney's, etc.)
☐ Catalog Showroom   ☐ Received as a gift
☐ Drug Store   ☐ Other (Specify)
☐ Bookstore   _____

6. This game was purchased for:
☐ Children (up to 12 years)
☐ Teens (13-17)
☐ Adults

☐ Male
☐ Female
☐ Both Male and Female

7. Which of the following influenced the decision to buy this game?
☐ Played with it before
☐ Reputation
☐ Family member's request
☐ Friend's recommendation
☐ Magazine ad
☐ Newspaper ad
☐ Price
☐ Store display
☐ TV commercial
☐ TV show/movie
☐ Newspaper article
☐ Magazine article
☐ Replacement
☐ Gift
☐ Played in family when growing up
☐ Others: _____

8. Did you plan to buy this game before you entered the store?
☐ Yes
☐ No

9. How often do you play Scrabble* Brand Crossword Game?
☐ Once a week
☐ Once a month
☐ Once every 3 months
☐ Once every 6 months
☐ Once a year
☐ Less than once a year

10. Who will this game be played with most often? (CHECK ONLY ONE BOX)
☐ Family of adults over 18 years
☐ Adult friends (non-family)
☐ Children (up to 12 years)
☐ Teens (13 to 17 years)
☐ Family members including teens
☐ Family members including children

11. How much do you enjoy playing Scrabble* Brand Crossword Game?
☐ Very much
☐ Pretty much
☐ Somewhat
☐ Not too much
☐ Not at all

12. Check which edition of Scrabble* Brand Crossword Game you own. Also, which editions would you consider buying in the future.

	Own	Consider Buying
Standard Edition	☐	☐
Deluxe	☐	☐
Deluxe Travel	☐	☐
Scrabble* for Juniors	☐	☐
Gameboy Super Scrabble*	☐	☐

13. What other games have you played with your peers in the past year?
_____
_____
_____
_____

14. What other games have you purchased in the past year?
_____
_____
_____

15. How often do you or family members play board games?
☐ Once a week
☐ Once a month
☐ Once every 3 months
☐ Once every 6 months
☐ Once a year
☐ Less than once a year

16. When are you most likely to watch T.V.? Between . . .
☐ 7 a.m. - 10 a.m.   ☐ 7 p.m. - 8 p.m.
☐ 10 a.m. - 4 p.m.   ☐ 8 p.m. - 11 p.m.
☐ 4 p.m. - 6 p.m.   ☐ 11 p.m. - 11:30 p.m.
☐ 6 p.m. - 7 p.m.   ☐ 11:30 - on

16a. What T.V. shows do you most often watch?
_____
_____
_____

17. Are you . . .
☐ A student   ☐ Employed part-time
☐ Employed full-time ☐ Homemaker
☐ Retired   ☐ Other:
_____

18. Your total annual household income is:
☐ Under $15,000
☐ $15,000 - $24,999
☐ $25,000 - $34,999
☐ $35,000 - $44,999
☐ $45,000 - $54,999
☐ $55,000 or over

19. Are you a member of the National Scrabble* Association?
Yes ☐   No ☐

20. Do you own the official Scrabble* Player's Dictionary?
☐ Yes ☐ No

Figure 11.9
A source document designed to facilitate handprinted recording.

always appear in the same areas of the screen, and the same keys on the keyboard should always be used for the same purposes. Such consistency can be seen in popular prewritten software where the F1 key is typically used for help, menu bars are usually positioned at the top of the screen, and menu choices can be selected by using arrow keys, typing the first letter, pressing Enter, or pointing and clicking.

- **Simplicity** The system should be simple to use. When you can barely fit all the input onto a single screen, consider using multiple screens. Figure 11.10 is an example of a mathematical model input screen that tries to accomplish too much. Such a cluttered display makes it difficult to learn and use the model.

  Figure 11.11 shows how the model input can be redesigned so that it is displayed on three screens.

- **Complete Help Support** Users should be able to obtain help when they need it. There are basically two kinds of help: context sensitive and functional.

  Context-sensitive help answers the user's questions concerning particular features of the software being used. For example, the user might need to enter

CASH FLOW MODEL

BEGINNING BALANCE [$ 100000]   DESIRED CASH LEVEL [$ 150000]
AMOUNT OF NOTE [$ 75000]   BAD DEBT PERCENTAGE [6%]
INTEREST RATE ON LOANS [12%]   INTEREST RATE ON INVESTMENTS [8%]

RECEIVABLES SCHEDULE FOR MONTH   1 [35%]   2 [40%]   3 [25%]   TOTAL 100%
PAYABLES SCHEDULE FOR MONTH   1 [70%]   2 [20%]   3 [10%]   TOTAL 100%

NOTE PAYMENT SCHEDULE

2 [$ 0]   3 [$ 0]   4 [$ 0]   5 [$ 25000]
6 [$ 25000]   7 [$ 25000]   8 [$ 0]   9 [$ 0]
10 [$ 0]   11 [$ 0]   12 [$ 0]   13 [$ 0]
                   NOTE BALANCE $0

SALES FORECAST
2 [7000]   3 [8000]   4 [9000]   5 [12000]
6 [15000]   7 [22000]   8 [28000]   9 [22000]
10 [12000]   11 [9000]   12 [8000]   13 [8000]

PRESS <ALT><H> FOR DATA ITEM HELP, <?> FOR DATA ENTRY SCREEN OPTIONS HELP,
<END> TO RUN SIMULATION, OR <ESC> TO END PROGRAM

Figure 11.10
Cluttered input can overwhelm the user.

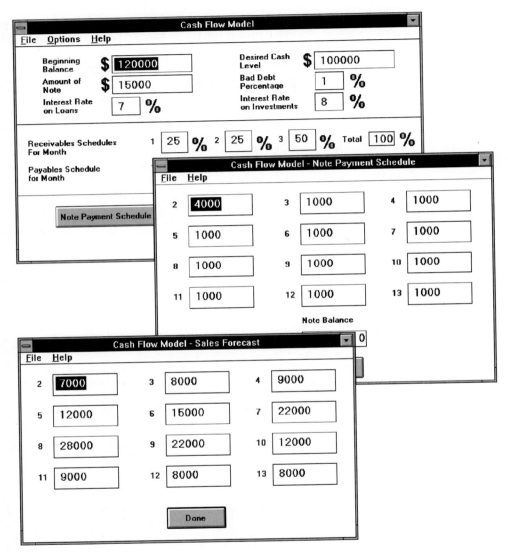

Figure 11.11
One way to achieve uncluttered input is to use multiple screens.

data into a field but not understand the field label. The user can use the mouse to click on the field and obtain a help message such as the one pictured in Figure 11.12.

Functional help provides instructions about how to perform particular operations, such as saving a file, loading a file, printing, and so on.

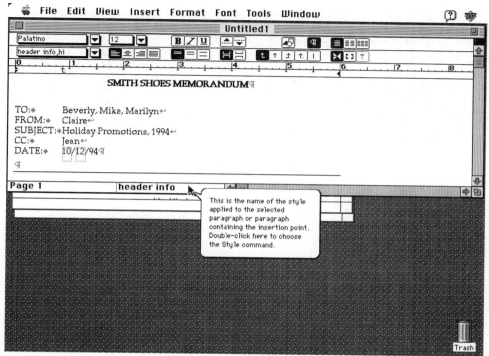

Figure 11.12
Context-sensitive help sheds light on particular model features.

- **Ease of Error Correction** Make it easy for users to correct their errors. Many software systems provide the Escape key as a way for the user to "back out" of error conditions. Also, when the software detects a user mistake, the error message should be stated respectfully and should provide an explanation of the error so that the user can correct it. Figure 11.13 shows how a window can be positioned on the screen in such a way that the user can see both the error and the error message.

- **Judicious Use of Screen Features** Computers provide a variety of features that can attract the user's attention—color, blinking cursors, reverse background, audible beeps, and so on. If these features are overused, they lose their effectiveness. An example of judicious use of color can be found in executive information systems. The EISs typically use **stop-light colors,** where red means that something has gotten out of control, yellow means that it might get out of control, and green means that everything is okay. The colors guide the executives through the information displays.

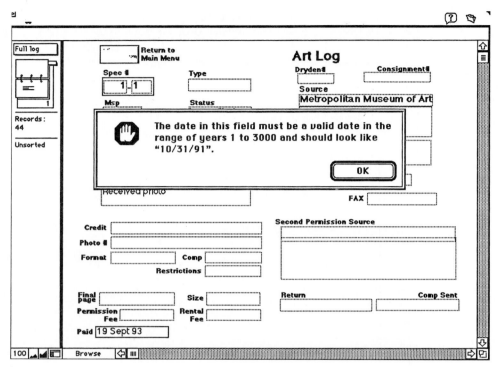

Figure 11.13
Location and content of error messages can facilitate correction.

These are just a few of the considerations that can be incorporated into systems to increase the efficiency and effectiveness of the computer-human interface.

Report Design

In Chapter 6 the ability to design periodic reports was cited as a key to the analysis and design of MISs and DSSs. Both periodic and special reports can be prepared using a tabular or graphic format. A **tabular report** displays its contents in rows and columns, whereas a **graphic report** uses bars, lines, dots, pie slices, and other shapes. Tabular and graphic reports can be either printed or displayed.

Tips for Designing Tabular Reports   Because tabular reports were the standard form for business information long before computers, a number of conventions have been handed down over the years, and users expect to find certain things in certain places.

Tabular reports include three main areas, as shown in Figure 11.14. The **headings area** is at the top and identifies the report and its contents. The **body** is in the

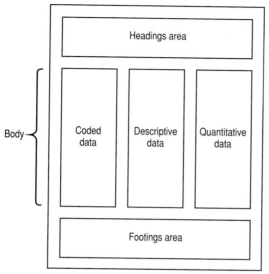

Figure 11.14
Users expect to see certain information in certain
report areas.

middle and contains the rows and columns of data. The **footings area** is at the bottom and includes page and final totals. Within the body, it is common practice to locate *coded data* such as dates and item numbers on the left-hand side, *descriptive data* such as names in the center, and *quantitative data* such as quantities and amounts on the right-hand side.

Follow these tips to make your tabular reports more effective.

1. Use a **layout form** that provides for both horizontal and vertical spacing. The form can be displayed on the screen or printed on paper. Figure 11.15 is an example.

2. Provide complete identifying information. Unless there are reasons to the contrary, you should include the following information in the headings area: report title, time period covered by the report, report preparation date and time, and column headings. EISs frequently identify the name and telephone extension of the person in the firm who is an expert on the report data so that the executive can follow up if necessary. This should be of value to lower-level users as well.

3. Incorporate management by exception. This can be accomplished in three ways:

   • Only print the report when an exception occurs. A good example is a report showing employees who worked overtime the previous week.

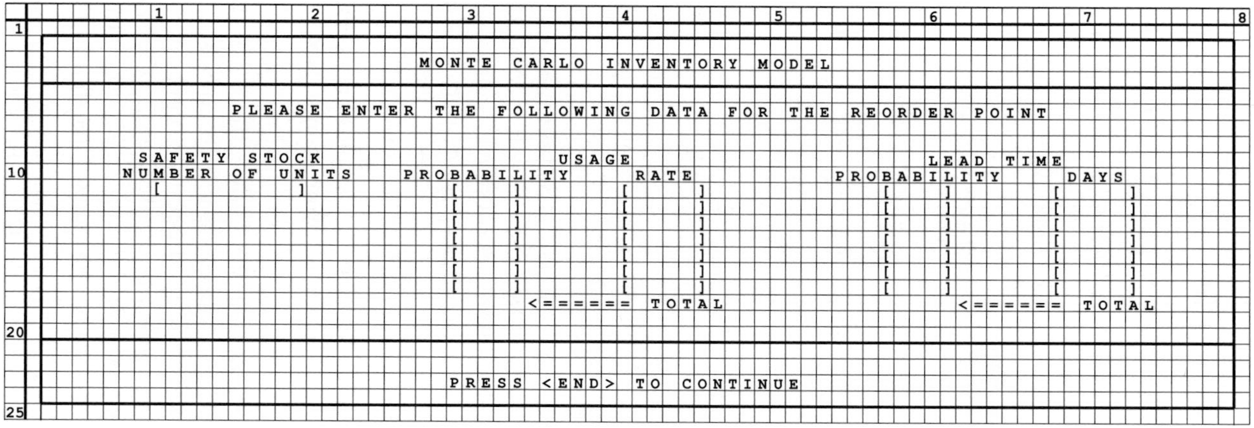

MONTE CARLO INVENTORY MODEL

PLEASE ENTER THE FOLLOWING DATA FOR THE REORDER POINT MODEL

SAFETY STOCK
NUMBER OF UNITS [__]

PROBABILITY USAGE RATE

PROBABILITY [__] [__] [__] [__] [__] [__] [__] TOTAL
[__] = [__] = [__] = [__] < [__] = [__] = TOTAL

LEAD TIME
PROBABILITY DAYS

PROBABILITY [__] [__] [__] [__] [__] [__] [__] TOTAL
[__] = [__] = [__] = [__] < [__] = [__] = TOTAL

PRESS <END> TO CONTINUE

PRESS <END> TO CONTINUE

Figure 11.15
A report layout form.

- Print exceptions in special columns. In Chapter 3, we used Figure 3.12 to illustrate an aged accounts receivable report. Columns are used in this report to show the age of each receivable—30 days, 60 days, 90 days, and so on.

- Use variance columns to highlight exceptions. A standard report format shows actual performance, expected performance, and the variance in separate columns. The user scans the variance column to see exceptions that demand follow-up action. We illustrated this in Figure 6.10 when we explained how executives use the drill down technique.

Tips for Designing Graphic Reports   The past decade has seen a surge in activity related to computer graphics; however, much remains to be learned about their effectiveness in problem solving. When designing reports for users who prefer graphics, consider the following tips.[2]

1. Use line and bar charts to summarize data. Line charts effectively summarize complex material necessary for quick decisions, and bar charts facilitate understanding and readability.

2. Use line and bar charts to show trends over time. Use groups of lines and bars, as shown on the left-hand side of Figure 11.16, rather than single lines and bars.

3. Use grouped bar charts to illustrate parts of a whole. The human eye performs most accurately when reading grouped bars with a common baseline. Neither segmented bar charts nor pie charts offer this.

4. Use grouped line and bar charts to illustrate patterns. Segmented charts are not as effective for showing patterns, such as the change in annual sales.

5. Use horizontal bars to show relationships. Users tend to overestimate the length of vertical bars.

6. Use single-variable bar charts to illustrate data points. Research does not support the idea that graphics are more effective than tabular data in portraying single values. Include the actual data values at the end of the bars for added precision.

These guidelines apply to users in general. Use prototypes to identify the types of graphs that specific users prefer.

[2]Taken from Sirkka Jarvenpaa and Gary W. Dickson, "Graphics and Managerial Decision Making: Research-Based Guidelines." *Communications of the ACM* 31 (June 1988): 764–774.

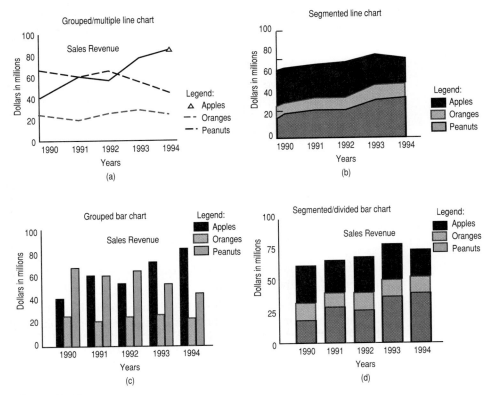

Figure 11.16
Sample graphic report formats.

Taking an Object-Oriented Approach to Systems Design

During the past few years a new systems methodology—object-oriented develop-ment—has gained a relatively large and enthusiastic following. Object-oriented development was triggered by object-oriented programming.

The origin of object-oriented programming can be traced to a modeling lan-guage called SIMULA developed in the late 1960s. The idea was refined by re-searchers at the Xerox Palo Alto Research Center (PARC), and the end result was the Smalltalk language. More recent developments have come in the form of the C++ language and a software engineering tool named Eiffel.

Object-Oriented Systems Development

An **object-oriented system** is one that places primary importance on the objects that compose the system and their relationships. An **object** is an element that ex-

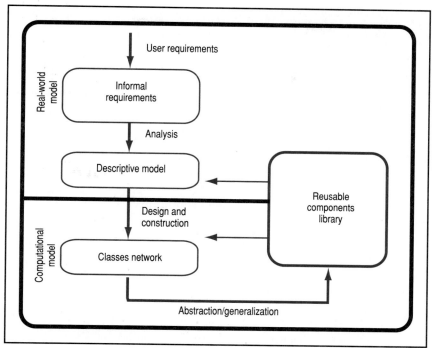

Figure 11.17
The object-oriented software life cycle.

hibits behavior when subjected to some type of stimulus, such as a transaction or a message. In a written description of a system, the objects are generally represented by the nouns.

An object-oriented system is developed by following a process that reflects a life cycle such as the one pictured in Figure 11.17.[3]

The supporters of object-oriented development cite as a major advantage its ability to accurately model the real world. This modeling begins with the systems analyst and user jointly defining the user's requirements, which are then incorporated into a descriptive model. The model provides the basis for design and construction of a computational model consisting of a network of classes. Both the descriptive model and the classes network make use of a reusable components library. One of the advantages of the object-oriented approach is the reusable nature of its modules.

[3]This model and the discussion in this section are taken from Jean-Marc Nerson, "Applying Object-oriented Analysis and Design," *Communications of the ACM* 35 (September 1992): 63–74.

Table 11.1   A Cluster Chart

Class	Definition
CLIENT	Car renter, individual or corporate customer
CONTRACT	Rental terms with payment conditions
RENTAL	Rental information completed when taking out and returning a vehicle
VEHICLE	Automobile selected from the rental fleet
MODEL	Description of selected features
RATE	Pricing conditions

Source: Jean-Marc Nerson. "Applying Object-oriented Analysis and Design," *Communications of the ACM* 35 (September 1992): 64.

Better Object Notation

One approach to documenting the object-oriented design is called **BON,** for **Better Object Notation.** BON consists of a series of steps.

1. Define the system boundary. This is the starting point for any design approach. Decide what is included in the system and what constitutes its environment.

2. Identify the classes. For example, if developing a reservation and billing system for a car rental company, identify the classes of objects as the customer, contract, rental information, vehicle, model, and rate.

3. Group classes into clusters. A **cluster chart** lists the classes of objects, along with brief definitions, as shown in Table 11.1. Each class is defined in terms of the services it provides, its information and constraints, and its relationships to other classes.

4. Define classes. For each class in the cluster chart a **class chart** is completed, which provides English-language descriptions of questions, commands, and constraints. *Questions* are information that other classes might require of this class. *Commands* are services that other classes can require. *Constraints* are forms of knowledge that the class must maintain. Table 11.2 is an example of a class chart for the vehicle class.

5. Produce class descriptions. Class charts are used to produce class descriptions, as shown in Figure 11.18. The legend in the shaded area shows some of the graphical notations used in BON.

Table 11.2   A Class Chart

Vehicle		
Cluster name: *RENTAL PROPERTIES*		
TYPE OF OBJECT: Automobile selected from the rental fleet		Behaves like: *RENTED ITEM*
Questions	Commands	Constraints
Model of car *License plate* *Availiability* *Departing location* *Returning location*	*Check mileage* *Refill gas tank* *Change oil*	*Departing and* * returning locations* * are the same.*

Source: Jean-Marc Nerson. "Applying Object-oriented Analysis and Design," *Communications of the ACM* 35 (September 1992): 67.

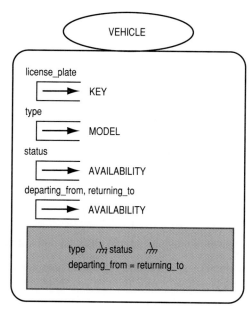

Figure 11.18
A class description.

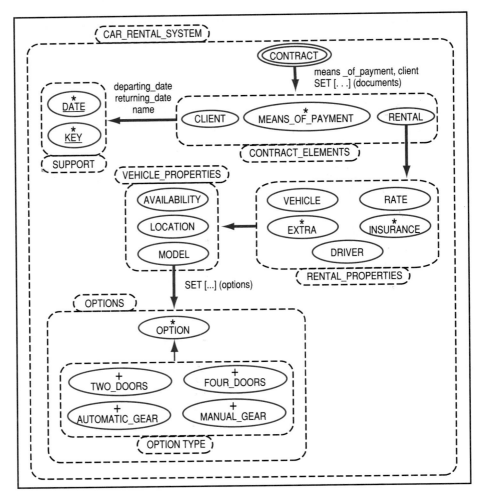

Figure 11.19
A static model.

6. Document the system with static and dynamic models. Class descriptions are used to prepare a static model of the system, as shown in Figure 11.19. Clusters are represented by rounded corner rectangles. Classes are represented by names in ovals. Relationships are illustrated with arrows.

   A dynamic model is produced for each system scenario. A scenario where the client rents a two-door vehicle with manual transmission is shown in Figure 11.20.

   Both the static and dynamic models enable the system to be documented in terms of its structure and processes.

BON is only one way to document an object-oriented design. It is presented here to illustrate the special nature of the object-oriented development methodology and tool set.

Putting Object-Oriented Development in Perspective

Although the object-oriented approach to development is recognized as an innovative way to build systems, it is based on many of the systems fundamentals described in Chapter 3. The auto rental system described here is an open system existing in an environment. The system boundary separates the system from its environment. The system is described in terms of the static relationships that exist among its elements and by its behavior in response to stimuli from its environment.

The object-oriented approach is gaining a tremendous amount of support. No doubt it is a development methodology for the future. For the time being it is primarily used at the workstation level, but it is expected to migrate to larger systems configurations. It has suffered from a lack of CASE tool support, but that shortcoming seems to be lessening with the appearance of systems such as Eiffel.

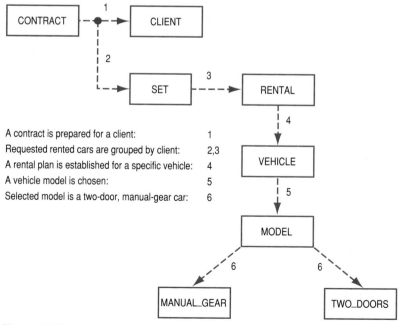

Figure 11.20
A dynamic model.

Although object-oriented development is just now emerging as a viable methodology, chances are good that you will enounter it in the future as you pursue a career as either user or information specialist.

Incorporating Human Factors Considerations into Systems Development

Although methodologies such as the traditional SLC and the object-oriented software cycle define the necessary steps in systems development, the steps cannot simply be followed automatically. **Human factors considerations**—influences that have the potential to affect the system participants and users—must be taken into account. Although a large number of such considerations exist, two stand out: fear of the computer, and computer–human interaction.

Fear of the Computer

During the early years of computer use in business, economic justification for the system was often based on clerical expense reduction. You can imagine the shock when clerical employees heard of a firm's plans to implement a computer. Fearing the worst, the employees' natural response was to refuse to cooperate with the implementation effort or, worse still, sabotage it in some way.

When systems analysts and management recognized the damaging effect that fear could have on a computer project, they sought a solution: communication. The rationale was that if the employees understood why the firm had decided to install a computer and what effect it would have, there would be little or nothing to fear. In most cases management had no intention of terminating the displaced employees. Instead, the plan was to transfer them to other jobs.

Although the fear of job loss is no longer a major concern, there can be a fear of job change. Employees can fear that their work will be altered in such a way that it will no longer be interesting and challenging. Also, employees can fear that their personal relationships with co-workers may be changed or even ended. Because fear can still be a factor in systems projects, management continues to announce its computer-related intentions to employees during the SLC. We have included these announcements at the beginning of the analysis and implementation phases.

Computer–Human Interaction

Management can show an even greater concern for human factors considerations by incorporating special activities in the SLC.[4] The activities, called **user factor stages,** can be added at certain points in each of the five phases, as shown in Figure 11.21. The first three stages in the right-hand column represent *additional steps* that are inserted in the SLC as shown. The last three represent *elaborations of existing steps.*

[4]This description is based on Marilyn M. Mantei and Toby J. Teorey, "Incorporating Behavioral Techniques Into the Systems Development Life Cycle," *MIS Quarterly* 13 (September 1989): 257–274.

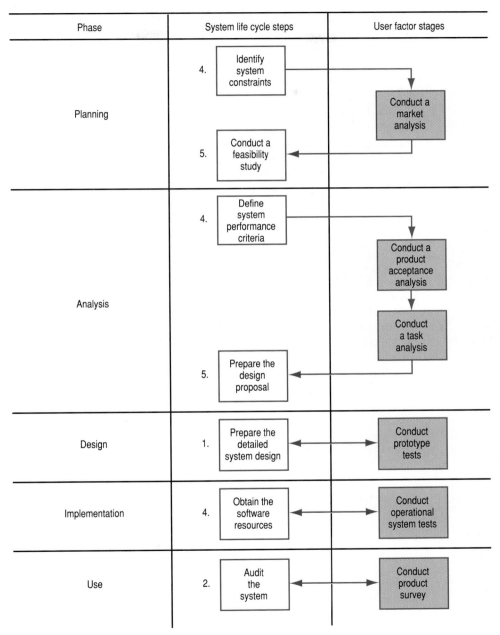

Figure 11.21
Building human factors considerations into the system.

Conduct a Market Analysis   Just as marketers conduct marketing research to determine consumer needs for a product, the systems analyst should examine the users' perceptions and feelings about the computer-related tasks they perform. This **market analysis** is meant to identify general features that a new system should possess and the political problems that might be encountered in its development. This special attention comes before the feasibility study.

Conduct a Product Acceptance Analysis   Mockups of the new system are presented to users, who meet in groups to react to the new design. The mockups are not computer-based prototypes but, rather, videotapes of the system in operation, story boards, or paper forms. The mockups concentrate on user interface.

Conduct a Task Analysis   A **task analysis** is a study of the user's feelings about a current task so that the new system can be designed to accommodate those feelings. One approach is to ask the user to describe his or her thoughts as the task is performed.

Both the product acceptance analysis and task analysis are conducted prior to writing the design proposal.

Conduct Prototype Tests   A prototype tests the user's reaction to a proposed system design. This is the typical way to use a prototype in the design phase and is a part of preparing the detailed system design.

Conduct Operational System Tests   When the operational system is not identical to the prototype, special tests are conducted to measure the user's learning time and performance time. This testing is a part of obtaining the software resources.

Conduct a Product Survey   This is the postimplementation review and annual audit conducted by the systems analyst to determine the extent of user satisfaction with the system.

The user factor stages are refinements to the SLC that pay special attention to the user's needs, with emphasis on the computer–human interface. This interface is especially important in achieving user satisfaction with the system.[5]

Some Design Suggestions

Each new hardware and software development brings with it new challenges in systems design. It is impossible for one individual to keep up with all design developments. The snowballing effect of technological innovations explains the trend toward database and data communications specialization.

Four suggestions appear to be especially appropriate to the systems analyst who is faced with the challenge of remaining current on systems design.

[5]The Association for Computing Machinery, ACM, has a special interest group, SIGCHI, that specializes in the computer–human interaction. It publishes a newsletter and sponsors conferences that provide potentially valuable information to the systems analyst. Special student memberships are available. Ask your instructor whether your college has a student chapter. For more information, contact ACM Headquarters, 11 West 42nd Street, New York, NY 10036.

- Seek the help of specialists, both inside and outside the firm, when designing large, innovative, and especially important systems.
- Use prototyping as a means of improving communications with the user.
- Take advantage of CASE tools, which not only remove much of the drudgery from design and documentation but also can do a better job.
- Be humble. You do not know everything and can learn much from others—especially users.

By taking advantage of existing human and software resources, you can design systems that effectively meet the needs of the users and your organization.

Using CASE to Design a Report

Figure 11.22 is a mockup of a missing item report for Rag City. The report contains a title, subtitle, date field, six columns including vendor, SKU (stockkeeping units), description, quantity ordered, quantity received, and value, as well as page and report totals. During the design phase, as the project team seeks to design a system to meet user needs, an ability to produce report prototypes is of great value. Most CASE tools provide report and screen design capabilities.

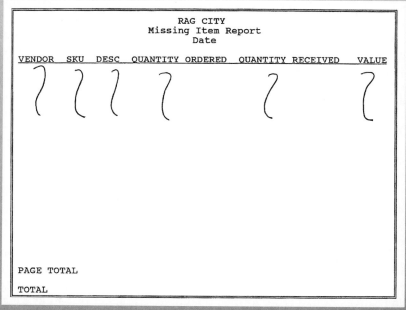

Figure 11.22
Report prototype for Rag City.

Summary

The design task is to model the new system with data and process models.

Four approaches to systems design have characterized the development of business systems. The first, data processing systems, emphasized processing and the document output of lower-level systems, which were separately designed. Next, management information systems emphasized the processes that produce information output but achieved little in the way of systems integration. Integration was finally achieved with decision support systems in the form of a central database that supported the separate DSSs. The approach currently stimulating interest is the idea of a database that captures all the computer data in the enterprise and serves as the basis for all systems development.

The systems design phase of the SLC begins with a detailed system design developed by project team members through such communications techniques as brainstorming or joint application design. JAD is a structured approach that requires planning the session, conducting the session, and documenting the results. JAD sessions can come during enterprise planning or at any point in the SLC.

With the detailed design accomplished, attention turns to identifying and evaluating alternate system configurations. One approach is based on the system elements. The systems analyst first determines whether the user's needs will be satisfied by reports, mathematical modeling, or office automation communications. Then attention is focused on output devices, storage media and file organization methods, and, finally, input devices and user–computer dialog.

Users prefer the speed of displayed output and the permanent record provided by printers and plotters. Microform and audio output are used in special circumstances.

Sequential storage media have almost completely given way to DASDs because of the inability of the sequential media to provide a current database. If you have a sequential storage device, you must process data sequentially. If you have a DASD, you can choose between sequential, indexed sequential, and direct.

MICR and OCR were intended to ease the input bottleneck caused by keyed input. Users can also enter data by means of special-purpose terminals, voice input, and push-button telephone.

Program-directed dialogs include menus, form-filling screen designs, and prompting. Operator-directed dialogs include command language and direct manipulation. It is possible to design a system so that the user can select the type of dialog.

Evaluation of the alternate configurations begins with agreeing on the evaluation criteria and then measuring each alternative in terms of the criteria. The evaluation can proceed in an informal, subjective manner by rank ordering the alternatives or in a more formal, quantitative manner by rating weighted criteria to produce a score for each alternative.

Selection of the best alternative is complicated by the semistructured nature of the problem and the tendency of participants to look out for their own interests.

When these differences are reconciled, the team recommends a system configuration to the MIS steering committee. Using the material provided by the team, the committee makes the important decision of whether to proceed with implementation. A positive decision triggers finalization of the design documentation.

Modern systems design efforts place heavy emphasis on input and output. Attention is paid to source document design, user friendliness, and report design.

Source document design considers the way that persons fill out forms and their reaction to prototypes, as well as issues such as field size, multiple copies, and mailing.

The area in the system design where user friendliness is most critical is computer-human interaction. System features that contribute to user friendliness include consistency, simplicity, complete help support, ease of error correction, and judicious use of screen features.

When designing tabular reports, begin with a layout form. Include enough information to make the report self-explanatory, and incorporate management by exception by preparing the report only when exceptions occur, placing exceptions in special columns, or using variance columns.

When designing graphic reports, keep in mind that research findings suggest that line and bar charts are superior to segmented bar and pie charts. Also, use groupings to show parts of a whole and patterns, and position bars horizontally with accompanying data values at the end. However, do not lose sight of the unique preferences of the user.

Object-oriented programming spawned an object-oriented development methodology. One approach to documenting object-oriented development is Better Object Notation, or BON. BON begins with a definition of the system boundary, followed by a grouping of the system objects into classes. The classes are next grouped into clusters and documented with a cluster chart. Each class is defined in a class chart that addresses questions, commands, and constraints. Each class is then described using the BON notation. The class charts and descriptions enable the developers to build static and dynamic models. Most current interest in object-oriented development is centered on workstation environments, but is expected to progress to larger-scale systems with the availability of CASE tool support.

In an effort to incorporate human factors considerations into the systems development process, management has traditionally made announcements to employees at the beginning of the analysis and implementation phases. Management concern can go farther by including special user-oriented actions at five points in the SLC. A market analysis can be added to the planning phase, and product acceptance analysis and task analysis can be added to the analysis phase. During the design phase, prototype tests can be incorporated in the preparation of the detailed system design, and during the implementation phase, operational system tests can be a part of obtaining the software resources. During the use phase, a product survey can be included in the system audit.

As you carry out your design responsibilities, take advantage of specialists, use prototyping to improve communications, and make use of CASE tools.

At the conclusion of the design phase, the main work of the systems analyst is completed. Some involvement will be necessary during the remaining phases, but it will not measure up to the work that has been accomplished in terms of importance. The systems analyst and the user have taken problem symptoms and transformed them into a concise problem definition and have created the logical design of a computer-based system intended to help the user solve the problem. With these important steps taken, the design can now be implemented.

Key Terms

Data model

Process model

Brainstorming

Joint application design (JAD)

Specification document

Impact printer

Nonimpact printer

Hardcopy

Turnaround document

Sequential file organization

Indexed sequential file organization

Direct file organization

Program-directed dialog

Operator-directed dialog

Source document

Context-sensitive help

Functional help

Tabular report

Graphic report

Object-oriented system

User factor stage

Market analysis

Task analysis

Key Concepts

The evolution in systems design strategies that has progressed from a bottom-up emphasis on processing and output to a top-down emphasis on data

An approach to the definition of system alternatives that begins with user outputs and considers other system elements

The relationship between secondary storage media and file organization methods

Source data automation as a way to combat the input bottleneck

The relationship between performance criteria and evaluation criteria

User friendliness

Report design as a means to achieve management by exception

Object-oriented system development

Human factors considerations

Questions

1. Is a data model usually logical or physical? What about a process model?
2. What software innovation provided the necessary integration between a firm's computer-based systems?
3. What is the output of enterprise modeling? Which analysis and design tool is best suited to document this output?
4. When is JAD performed?
5. What are the three basic ways of producing output information on the computer?
6. Which type of printer is required to produce multiple-copy forms?
7. Name two ways to produce hardcopy graphic output.
8. What device serves as an audio output unit?
9. Why would a turnaround document be used?
10. What are the two basic types of secondary storage? Name the file organization methods that each can use.
11. What role do record keys play in sequential file organization? What is their role in indexed sequential and direct file organization?
12. What is the main drawback of indexed sequential file organization? Name two limitations of direct file organization.
13. Name four examples of pointing devices.
14. What is meant by the input bottleneck? What input strategy addresses it?
15. Distinguish between MICR and OCR.
16. Name three ways to achieve program-directed dialog.

17. Generally speaking, the user enters either commands or data into a keyboard. Which of the program-directed dialogs is best suited to commands? To data?

18. Why go through a quantitative evaluation process when the task of selecting a system configuration is not completely structured?

19. Name five features of a system that make it user friendly.

20. How do EISs use the stop-light colors?

21. What are the three main areas of a tabular report?

22. Would you put the following in the left-hand, middle, or right-hand area of a report body? Employee number. Unit price. Quantity ordered. Customer name. Product class. Age. Territory number.

23. Name three ways to incorporate management by exception into reports.

24. How can you make bar charts more effective in illustrating data points?

25. How are the objects of a system represented in a written description, such as a narrative procedure?

26. What is described in a class chart?

27. What are the two types of models that are developed in following the object-oriented software life cycle?

28. What traditional strategy has management used to combat employee fears of the computer?

29. What do you attempt to learn in a market analysis that you do not expect from a feasibility or system study?

30. Which would be the most effective way to conduct a task analysis—in-depth interview, survey, or observation? Explain your answer.

Topics for Discussion

1. What could joint application design offer that brainstorming could not?

2. Pretend a large corporation has scheduled a JAD session to determine the future direction its computing resources will take. All the top executives will attend. Who would you pick for the session leader—the president, the CIO, or an outside consultant? Support your answer.

3. What determines whether a manager uses displayed or printed special reports?

4. What is the relationship between evaluation criteria and performance criteria?

5. Suppose you are a member of a project team and you want to bias the evaluation of configurations to favor your unit. How would you go about doing it?

Problems

Use the problems at the end of Chapter 10 in preparing the following documentation. Technology Modules A (Entity-Relationship Diagrams) and G (Flowcharts) provide the necessary techniques.

1. Prepare a system flowcart to document the procedure described in problem 1.
2. Prepare a system flowchart for problem 3.
3. Prepare a system flowchart for problem 5.
4. Prepare a normalized entity-relationship diagram to illustrate the data described in problem 5. List the data elements next to the entities. Underline the identifiers in the lists.

Consulting with

Harcourt Brace

(Use the Harcourt Brace scenario prior to this chapter in solving this problem.)

Continue to pretend that you are Sandi or Russ. Prepare the following documentation:

1. After reading Technical Module G (Flowcharts), document the *existing* order processing and distribution system, using a system flowchart. Remember that you are to document the *conceptual* system, not the *physical* system. This flowchart will reflect the same procedure as the DFDs you prepared in the previous Harcourt Brace installment.

2. Document the *new* order processing and distribution system using a system flowchart. The flowchart should reflect the changes the group suggests.

3. After reading Technical Module A (Entity-Relationship Diagrams), document the data of the new order processing and distribution system using an entity-relationship diagram.

4. Use prototyping software to prepare a prototype of the supervisor report illustrated in Figure HB7.1. If such software is not available, use a layout chart.

5. Prepare a prototype of the report illustrated in Figure HB7.2. Use a layout chart or prototyping tool.

Case Problem
Midcontinent Industries

Use the Midcontinent Industries case at the end of Chapter 10 in performing the following documentation.

Tasks

Technical Module G provides the detailed understanding of flowcharts that you will need to perform these tasks.

1. Draw a system flowchart of the weekly procedure. Use the following guidelines:

   A. Include only a single vertical flow on each page. Use offpage connectors to connect the page flows.

   B. Begin the system with the calculation by the employees of the daily and weekly hours.

   C. Do not include the detailed processing performed by the secretaries or department managers. Show a single processing step for each.

   D. Do not include processes for the employees giving the timecards to the secretaries, the secretaries giving the cards to the managers, the managers mailing the cards to the computer department, and mailing the payroll checks to the employees.

   E. Do not include the special processing when unapproved overtime must be reconciled by the special system.

2. Draw a system flowchart of the monthly procedure. Use the following guidelines:

   A. Show separate disk symbols for each of the four weekly payroll files. Add a suffix for each file—1, 2, 3, and 4.

   B. Begin the processes with step 1.

   C. Do not include a process for mailing the monthly report to the president.

Selected Bibliography

Bailin, Sidney C. "An Object-Oriented Requirements Specification Method." *Communications of the ACM* 32 (May 1989): 608–623.

Benbasat, Izak; Dexter, Albert S.; and Todd, Peter. "An Experimental Program Investigating Color-Enhanced and Graphical Information Presentation: An Integration of the Findings." *Communications of the ACM* 29 (November 1986): 1094–1105.

DeSanctis, Gerardine. "Computer Graphics As Decision Aids: Directions for Research." *Decision Sciences* 15 (Fall 1984): 463–487.

Dos Santos, Brian L., and Bariff, Martin L. "A Study of User Interface Aids for Model-Oriented Decision Support Systems." *Management Science* 34 (April 1988): 461–468.

Gorman, Kevin, and Choobineh, Joobin. "The Object-Oriented Entity–Relationship Model." *Journal of Management Information Systems* 7 (Winter 1990–91): 41–65.

Henderson, John C. "Managing the IS Design Environment." *Sloan School of Management Working Paper No. 1887–97.* (Cambridge, MA:Massachusetts Institute of Technology, May 1987.)

Hix, Deborah, and Schulman, Robert S. "Human-Computer Interface Development Tools: A Methodology for their Evaluation." *Communications of the ACM* 34 (March 1991): 74–87.

Hoadley, Ellen D. "Investigating the Effects of Color." *Communications of the ACM* 33 (February 1990): 120–125.

Ives, Blake. "Graphical User Interfaces for Business Information Systems." *MIS Quarterly* (Special Issue 1982): 15–47.

Jain, Hemant K., and Bu-Hulaiga, Mohammed I. "E-R Approach to Distributed Heterogeneous Database Systems for Integrated Manufacturing." *Information Resources Management Journal* 3 (Winter 1990): 29–40.

Klein, Gary, and Beck, Philip O. "A Decision Aid for Selecting Among Information System Alternatives." *MIS Quarterly* 11 (June 1987): 177–185.

Liberatore, Matthew J.; Titus, George J.; and Dixon, Paul W. "The Effects of Display Formats on Information Systems Design." *Journal of Management Information Systems* 5 (Winter 1988–89): 85–99.

McLeod, Poppy Lauretta. "Are Human-Factors People Really So Different? Comparisons of Interpersonal Behavior and Implications for Design Teams." *Journal of Management Information Systems* 9 (Summer 1992): 113–132.

Peters, Lawrence. *Advanced Structured Analysis and Design.* (Englewood Cliffs, NJ: Prentice–Hall, 1987.)

Ricciuti, Mike. "Database Vendors Make Their CASE." *Datamation* 38 (March 1, 1992): 59–60.

Romei, Lura K. "Making Sure Form Follows Function." *Modern Office Technology* 36 (October 1991): 67–69.

Ronen, Boaz; Palley, Michael A.; and Lucas, Henry C., Jr. "Spreadsheet Analysis and Design." *Communications of the ACM* 32 (January 1989): 84–93.

Sauter, Vicki L., and Schofer, Joseph L. "Evolutionary Development of Decision Support Systems: Important Issues for Early Phases of Design." *Journal of Management Information Systems* 4 (Spring 1988): 77–92.

Snodgrass, Coral R., and Szewczak, Edward J. "A Societal Culture Perspective on Systems Analysis and Design." *Journal of Management Information Systems* 3 (Number 1, 1991): 69–78.

Starke, Frederick A., and Ferratt, Thomas W. "Dealing With Organizational Politics." *Information Executive* 3 (Winter 1990): 6–8.

Stevens, W. P.; Myers, G. J.; and Constantine, L. L. "Structured Design." *IBM Systems Journal* 2 (Number 2, 1974): 115–139.

Te'eni, Dov. "Determinants and Consequences of Perceived Complexity in Human-Computer Interaction." *Decision Sciences* 20 (Winter 1989): 166–181.

Teorey, Toby J.; Wei, Guangping; Bolton, Deborah L.; and Koenig, John A. "ER Model Clustering as an Aid for User Communication and Documentation in Database Design." *Communications of the ACM* 32 (August 1989): 975–987.

Umanath, Narayan S., and Scamell, Richard W. "An Experimental Evaluation of the Impact of Data Display Format on Recall Performance." *Communications of the ACM* 31 (May 1988): 562–570.

Warnier, Jean–Dominique. *Logical Construction of Systems.* (New York: Van Nostrand Reinhold, 1981.)

Yoon, Youngohc, and Guimaraes, Tor. "Developing Knowledge–Based Systems: An Object-Oriented Organizational Approach." *Information Resources Management Journal* 5 (Summer 1992): 15–32.

Yourdon, Edward, and Constantine, Larry L. *Structured Design: Fundamentals of a Discipline of Computer Program and Systems Design.* (Englewood Cliffs, NJ: Prentice-Hall, 1979.)

TM

F

Designing Graphical User Interfaces

Users can interface, or interact, with the computer either directly or indirectly. For the first ten years or so of the computer era, all application designs featured an indirect interface, with information specialists serving as intermediaries. The users created source documents, and data entry personnel in information services produced computer-readable input media. The data entry personnel used offline key-driven devices such as keypunch machines, key-to-tape machines, and key-to-disk machines. After the media were produced, they were then read by computer input devices. The hardcopy computer output was delivered to the users, often in the company mail.

The mid-1960s saw the introduction of online keydriven terminals, which popularized computer timesharing. The online terminals made it possible for users to bypass the data entry intermediaries, enter data directly into the computer, and receive the outputs in a displayed or printed form. The popularity of this direct user interface approach received a big boost in the 1980s when the microcomputer stimulated distributed processing, decision support systems, and end-user computing. Although data entry personnel in information services and user areas still account for a large volume of computer input, the recent focus in systems design has been on direct user interfaces. The popularity of this approach is expected to continue to increase in the future.

The Graphical User Interface

The direct interaction between the user and the computer is called the **human-computer interface (HCI)** or the **user interface (UI).** The first forms of the HCI featured a dialog consisting entirely of alphanumeric data in the form of commands, prompts and their responses, form filling, and menu selection. The ability of microcomputers and terminals to display data in a graphic form popularized what has become known as the graphical user interface.

A **graphical user interface,** or **GUI,** is a means for the user to interact with the computer through the use of typography, symbols, color, and other static and dynamic graphics to convey facts, concepts, and emotions.[1] The underlying logic of

---

[1]This definition paraphrases one in Aaron Marcus, "Designing Graphical User Interfaces: Part I," *UnixWorld* 7 (August 1990): 107.

F

TM

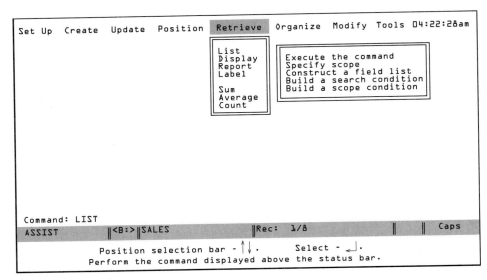

Figure F.1
Menus and windows produced by dBase III Plus.

the GUI is that graphics offer improvements over alphanumeric data by making it easier for the user to learn to use a system, improving system usability, and producing a more favorable user perception. This technical module addresses the special considerations that influence GUI design.

The Influence of Hardware and Software Vendors on GUI Design

Graphical user interfaces are a product of the microcomputer era, and both hardware and software vendors played key roles in the evolution. Several software products during the 1980s, such as the VisiCalc electronic spreadsheet from VisiCorp, the 1-2-3 electronic spreadsheet from Lotus Development Corporation, and the dBASE database management system from Ashton-Tate pioneered features that are used in today's GUIs, including menu bars, pull-down menus, and tiled windows. Figure F.1 illustrates how dBASE III Plus utilizes these features.

The **menu bar** comprises the list of choices across the top of the screen, ranging from *Set Up* to *Tools*. Users make selections by moving the cursor or highlihgt bar to the appropriate choice and pressing Enter, or by typing the first letter of the choice. When a choice is selected, a subsidiary menu appears below to provide the options for that particular choice. The subsidiary menu is called a **pull-down menu.** In the example, the user has selected *Retrieve* from the menu bar, and the pull-down menu prompts the user to specify the particular type of retrieval—a *List, Display, Report,* and so on. When the user makes a selection, such as *List,* a second pull-down menu appears, enabling the user to specify the next action, such as *Execute the command.* These two pull-down menus are examples of tiled windows. A **tiled window** is one that occupies a particular space on the screen into which no other window can infringe.

The menu capabilities of the microcomputer-based spreadsheets and DBMSs provided the foundation upon which Apple built its user interface. Apple's contributions to the GUI have come in the use of the mouse as a pointing device, icons to represent data and processes, means of window manipulation, and buttons.

The Mouse

Users of the Apple Macintosh use the mouse to select certain objects on the screen by positioning the pointer on the object and then performing clicking, pressing, and dragging operations. **Clicking** involves quickly pressing and releasing a button on the mouse to select an object. When an object is selected it is highlighted, or darkened. **Pressing** involves holding down the mouse button and is used to move the pointer to choices in a menu and then select a highlighted choice by releasing the mouse button. **Dragging** involves selecting an object by pressing down on the mouse button and then moving the mouse to reposition the object on the screen.

Icons

Apple also popularized the use of **icons,** or small, pictorial representations of objects or features. Figure F.2 shows the Macintosh icons used to illustrate floppy disks, data folders, programs, documents, and trash.

Floppy disk   Data folder

Program   Document   Trash

Figure F.2
Some Macintosh icons.

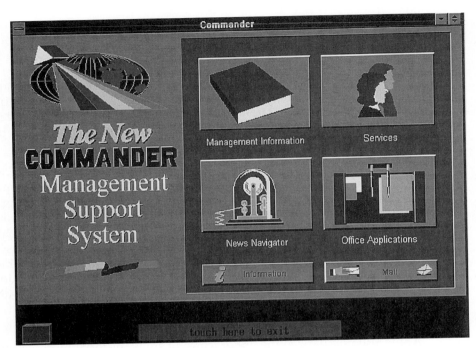

Figure F.3
Icons identify the main subsystems of the Commander EIS from Comshare.

Icons are key ingredients in many modern computer applications. Executive information systems, or EISs, make extensive use of icons. The Commander EIS from Comshare uses icons in its opening menu, as shown in Figure F.3, to identify the major system components.

Window Manipulation Operations

The Macintosh makes extensive use of windows, using overlapped as well as tiled arrangements. An **overlapped window** is one that partially obscures another window that was previously displayed. When several windows are overlapped, the one on top is the **active window**—the one used to perform operations. Figure F.4 shows a Macintosh screen display with overlapped windows. Each window has a name that is indicated in a **title bar** at the top.

Various operations can be performed on windows by using the mouse. You can:

- **Change window size** by pointing to the **size box** in the lower-right corner, as shown in Figure F.5, and dragging the box to any location. The window changes size as the box is dragged.

- **Show all window contents** by pointing to the **zoom box** on the right end of the title bar and clicking on the mouse. You return the window to its previ-

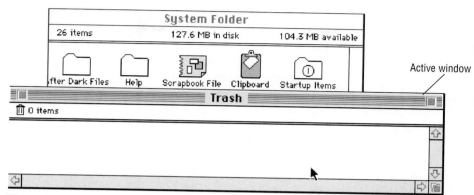

Figure F.4
Overlapped windows.

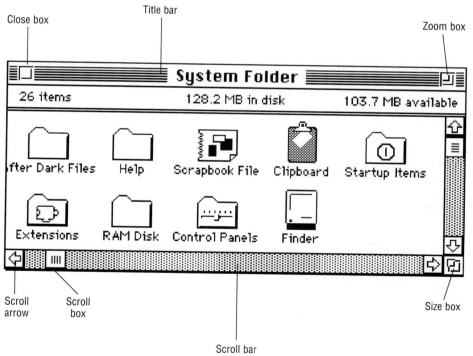

Figure F.5
Boxes facilitate window manipulation.

ous shape by clicking on the zoom box again. Mac window sizes are created in a category, called *resources*. The size of a window at full zoom size is specified by its specification. A person changes the size of a window when the program is running by using the *change window size* box. Clicking the zoom box restores the window to the dimensions originally specified in its resource data. Hopefully, icons will then be visible. If not, use of scroll bars will reveal them.

- **Scroll window contents** by pressing the arrows at either end of a **scroll bar.** Scroll bars are located both at the bottom of the window and on its right side. Each bar contains a pair of scroll arrows, with a scroll box in between. Pressing the appropriate **scroll arrow** causes the window to display contents at its top, bottom, right and left sides. Then, the window can be moved over its contents in the indicated direction by dragging the appropriate **scroll box.**

- **Close a window** by clicking on the **close box** at the left end of the title bar.

- **Move a window** by pointing to any location on the title bar (other than the close or zoom boxes) and dragging.

In addition to these window controls, Apple has utilized two types of boxes that have become popular GUI features—dialog boxes and alert boxes.

A **dialog box** is a window that contains one or more messages from the computer to the user, often requesting a user response. Figure F.6 shows a dialog box used for specifying details of a printing operation.

An **alert box,** pictured in Figure F.7, warns the user of a problem or potential problem, often making use of icons and sound.

Buttons

A **button** is a graphic representation of a symbol that is "pressed" by clicking on the mouse to trigger a particular operation. The button is usually a square or

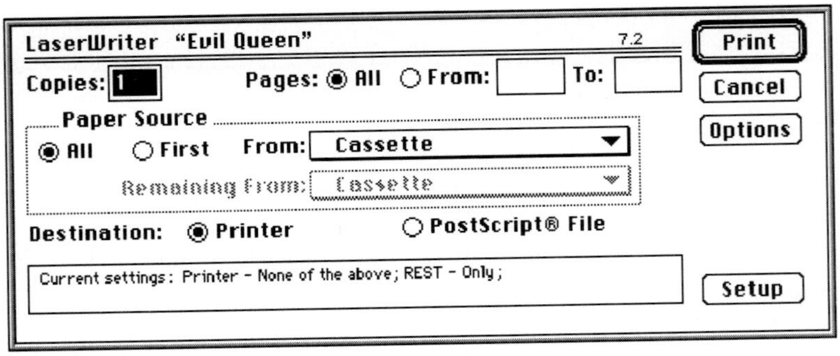

Figure F.6
A dialog box.

Figure F.7
An alert box.

rectangle and is often shaded to give a three-dimensional appearance. The button is labeled with wording or an icon. Figures F.3, F.6, and F.7 illustrate the use of buttons. Often, a group of buttons will be used together and only one can be selected at a time. The term **radio buttons** describes this arrangement.

These basic GUI techniques can also be found in software systems from other vendors, such as NextStep, OPEN LOOK, OSF/Motif, Microsoft Windows, and OS/2. The best way to obtain a first-hand understanding of good GUI techniques is to use systems such as these.

You incorporate GUIs into your own designs by using GUI design toolkits and user-interface management systems. **Design toolkits** enable the creation of the desired screen appearance and capability by direct manipulation, rather than by programming. **User-interface management systems (UIMSs)** build on the toolkits by providing additional functionality such as error messages and "undo" operations.[2]

GUIs as a Means of Achieving Usability

Software developers incorporate GUI capabilities in their designs to achieve usability. **Usability,** according to John S. Hoffman of Skill Dynamics™, an IBM company, is a concept that describes those product attributes that enable users to quickly, efficiently, and effectively use the product to accomplish *their* real work in a way that meets or exceeds *their* needs and expectations.[3]

The key to the GUI's contribution to usability is the favorable perception that the user forms for the software, based on the interface. This favorable perception enables the GUI to achieve an observable improvement in productivity. The

[2]For an example of GUI design software, see Jeff A. Johnson, Bonnie A. Nardi, Craig L. Zarmer, and James R. Miller, "ACE: Building Interactive Graphical Applications," *Communications of the ACM* 36 (April 1993): 40–55.
[3]John S. Hoffman, *Principles of Human-Computer Interface Design,* unpublished manuscript, 1992, 1–5.

ability of a GUI to achieve this productivity depends on its metaphor. A **metaphor** is an invisible web of terms and associations that underlies the way we speak and think about a concept.[4]

A good example of a metaphor is the electronic spreadsheet with its columns, rows, cells, labels, cell contents, and so on. The electronic spreadsheet has the same appearance to the user as a ledger sheet used by accountants; it is a natural way to array business data. When electronic spreadsheets were first introduced, business users were attracted to them because they provided a familiar metaphor. When such a real-life metaphor is used, the meaningful words and symbols are readily processed.

In a similar fashion, the use of icons to represent documents, folders, and floppy diskettes offer an effective metaphor because they have the same appearance as objects typically found on one's desk. Achievement of such a **desktop metaphor** is a goal of systems aimed at microcomputer users.

When building a GUI, it is a good idea to focus on one or two metaphors and then add other functionality. You should study the users and understand what mental models they employ in their everyday work. Then tap into those models by building in metaphors that both reflect and extend them.

GUI Requirements

One of the leading authorities on GUI is Aaron Marcus, a principle of Aaron Marcus and Associates, a Berkeley, California, consulting firm. According to Marcus, a GUI design must satisfy several basic requirements.[5] The GUI must provide for:

- A comprehensible mental image, or metaphor
- An appropriate organization of data, functions, tasks, and roles
- An efficient way to navigate among these data, functions, tasks, and roles
- A quality appearance, known as *the look*
- Effective interaction sequencing, known as *the feel*

A GUI design that satisfies these requirements will provide an effective means of communication for every kind of computer application. This is possible because graphic design relies on established design principles.

---

[4]Thomas D. Erickson, "Working with Interface Metaphors," in Brenda Laurel (ed), *The Art of Human-Computer Interface Design* (Reading, MS: Addison-Wesley), 1990, 66.
[5]This material on GUI requirements and design principles in the next section draws heavily from the three-part series titled "Designing Graphical User Interfaces," by Aaron Marcus, which appeared in the August, September, and October 1990 issues of *UnixWorld*.

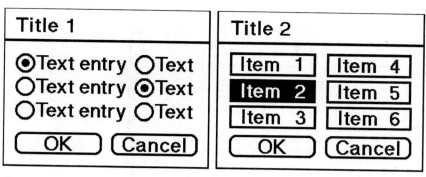

Figure F.8
Internal consistency.

TM
F

GUI Design Principles

According to Marcus, the three main principles of GUI design are organization, economy, and communication.

Organization

Designers achieve organization in their interfaces by means of several techniques—consistency, screen layout, relationships, and navigability.

Consistency is achieved by maximizing the regularity of the location and appearance of all components. There are two kinds of consistency—internal and external. **Internal consistency** deals with the regularity within a single system, and two examples appear in Figure F.8. In both examples, like items have the same general appearance. **External consistency,** on the other hand, deals with regularity across systems.

Screen layout can contribute to organization by using horizontal and vertical grids to provide the framework, as illustrated in Figure F.9. The grid concept applies to entire screens, windows, buttons, and icons.

Relationships are established by grouping like items. This grouping can be enhanced by using features such as background color. Figure F.10 shows how the example on the right does a better job of defining relationships than the one on the left.

Navigability deals with the ability of an interface to focus the user's attention on the appropriate material and to lead the user through the material in the proper manner. Figure F.11 illustrates a poor example on the left and a good example on the right. The good example uses title bars to identify major areas and bullets for subsidiary items within each area.

TM
F

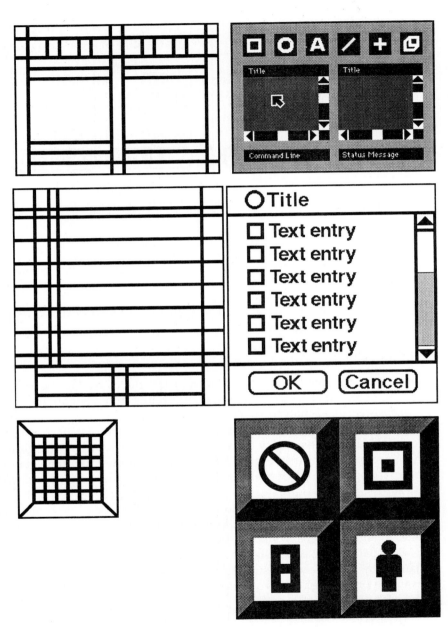

Figure F.9
Grids provide the framework for organization.

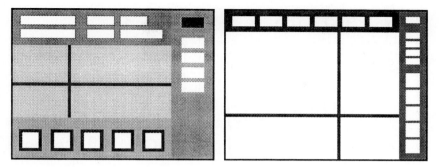

TM

F

Figure F.10
Relationships should be clear, consistent, appropiate, and strong.

Economy

The second GUI design principle is economy. The main idea is to not overuse the various graphical features that are available. Economy can be achieved by means of simplicity, clarity, distinctiveness, and emphasis.

Simplicity minimizes the work the user must expend to understand a display. Although both windows in Figure F.12 contain the same material, the one on the right has the simplest appearance. In this example, simplicity is achieved by means of format. It can also be achieved by including only essential elements.

Clarity means minimizing the opportunity for ambiguity. To the novice user, the right-hand icon in Figure F.13 offers a higher probability of being perceived as a zoom operation than the icon on the left.

Figure F.11
Organization should assist the user in navigating through the display.

TM
F

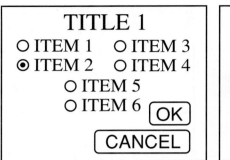
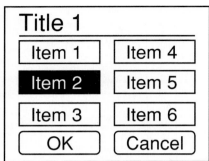

Figure F.12
Simple displays are easier to understand.

Distinctiveness enables the user to separate elements on the screen into logical groupings. In striving to achieve distinctiveness, as much harm can be done by too much as too little. Figure F.14 illustrates both extremes. On the left, there is not enough distinctiveness to help the user focus on the important elements. On the right, the use of too much distinctiveness, called the **Las Vegas approach,** is equally ineffective.

Emphasis is the final means of achieving economy. The designer should make it easy for the user to pick out important elements by keeping them to a minimum. Techniques include minimizing clutter and adhering to vertical and horizontal grid.

Figure F.13
Graphics should be carefully chosen to minimize ambiguity.

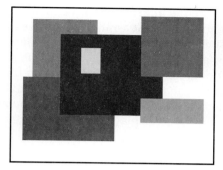 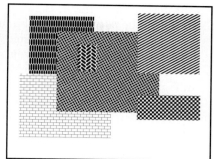

TM

F

Figure F.14
Strive to achieve the proper balance in distinctiveness.

Communication

The third design principle is communication, which is achieved through a balanced offering of legibility, readability, typography, symbolism, multiple views, and color. We discuss color in the next major section.

Legibility can be achieved by using a combination of characters and graphics that show up well. The emphasis is on communication rather than appearance. In applying this design principle, the designer should pay particular attention to the environment in which the GUI will be used. If the room is dark, a brightly lighted screen may produce too much glare. If the room is bright, dark backgrounds may introduce unwanted reflections.

Readability makes the display easy to interpret and understand. As a general rule, lines of alphabetic material should be left-justified, whereas numbers should be right-justified or justified on the decimal point. Attempts to make the display more attractive, such as by centering lines of text, can actually diminish readability.

Typography deals with the manner in which textual material is displayed. Marcus suggests that a maximum of three typefaces in a maximum of three point sizes be used, that lines of text contain no more than 60 characters, and that combinations of uppercase and lowercase be used. Two typography techniques to avoid are the use of uppercase only, and the use of justified right-hand margins with fixed-width fonts. Either can slow reading speed by as much as 12 percent.

Symbolism relates to the use of graphics that convey messages and information in the desired way. This is a complex principle to apply and can benefit from an iterative process that takes advantage of user feedback in refining the graphics until the desired effect is achieved. In some cases, users enjoy the iterative process; in others, they fail to understand why the design was not right the first time.

TM

F

Multiple views enable the user to see the information in various ways. Multiple forms can be used, such as tabular, graphic, and narrative. When we discussed the CBIS in Chapter 6, we illustrated such a screen in Figure 6.12. Information can also be displayed in multiple levels of abstraction, as when an executive uses drill-down to view successively greater degrees of detail. We illustrated this technique in Figure 6.10. The GUI should enable the user to select the desired view. For example, a typical EIS technique is to first display informaiton in a tabular form and then let the user transform it into a graphic display by pressing a key or clicking the mouse.

These design principles provide guidelines to both systems analysts and users as they engage in interface design. The information services organization should incorporate these guidelines in standards manuals, style guides, templates, and clip art so that they can be more easily applied. Figure F.15 is an example of GUI design guidelines followed by a project team in developing a system for an automobile dealer.

The Use of Color in GUI Design

Color is an important means of achieving communication in a GUI design for two reasons. First, the option of using color is available to practically everyone by virtue of the widespread use of color monitors and the increasing use of color plotters and printers. Second, color can offer a powerful communication capability *when used properly.*

Objectives in Using Color

In designing graphical user interfaces that use color, two particular objectives should be kept in mind.[6] First, color can be used to impart information. Second, the interfaces should be designed so that the user can select the desired colors.

Advantages of Using Color

Marcus recognizes that use of color has both its good and bad points.[7] In terms of the advantages, he believes that color can be used in GUIs to:

- Call the user's attention to important material
- Enable the user to organize material into hierarchies or structures
- Portray objects in a more natural manner
- Give graphics a dynamic dimension across both time and space

[6]Gitta Salomon, "New Uses of Color," in Brenda Laurel (ed), *The Art of Human-Computer Interface Design* (Reading, MS: Addison-Wesley), 1990, 271.
[7]Aaron Marcus, "Designing Graphical User Interfaces: Part III," *UnixWorld* 7 (October 1990): 136.

- The program has a graphical user interface in a multitasking environment.

- To simplify the interface to the task of data collection, all panels are full-screen with warning dialogs or help fields.

- Only a few actions are possible in each panel.

- Each action is initiated by clicking a button or by pressing a keystroke combination.

- There is only one entry and one exit to each panel.

- These entries and exits are consistent and uniquely symbolized.

- There is limited movement between panels to allow quick data entry.

Figure F.15
GUI guidelines developed by a project team.

- Contribute to an accurate interpretation
- Provide an additional coding capability
- Make information more believable and appealing

These advantages serve to improve the user's perception of both the information that the system provides and the system itself.

Disadvantages of Using Color

In being alert to the potential disadvantages of using color, Marcus cautions the designer to be aware that it can:

- Increase system cost in the form of more expensive hardware
- Fail to accomplish its objectives with color-deficient users, a situation found in approximately 8 percent of Caucasian males and 0.5 percent of females
- Cause visual discomfort and afterimages when certain combinations are used improperly

TM
F

- Achieve other than the desired effects when not used in accordance with users' particular cultures

These disadvantages can be minimized or avoided by adhering to color design principles.

Color Design Principles

The Marcus principles of organization, economy, and communication can be applied specifically to color.[8]

- **Color Organization** You can use color to organize material by assigning certain color to particular groups and adhering to a consistent use of color from screen to screen and from system to system. The use of similar background colors is especially effective in achieving organization. By applying the same color schemes to hardcopy material such as user manuals and system documentation, the advantages of organization can be expanded to include user training as well as systems maintenance and redesign.

- **Color Economy** A maximum of from three to seven colors should be used in those situations where the user is to remember the meaning of each color. A good design technique is to first design the screen for black and white, and then add color. Another technique is to use shape in conjunction with color as a means of accommodating all users, including those with deficient color vision.

- **Color Communication** Viewers do not have the same sensitivity to colors in all areas of their vision. Typically, viewers are more sensitive to colors in the center of their vision than on the periphery. Keeping this in mind, use bright colors such as red and green in the center of the screen display and colors such as blue, black, white, and yellow near the edges. When users are older, make an effort to use brighter colors and to minimize use of blues, which are difficult to discern. Also pay attention to the environment in which the color will be viewed. When the room is dark, use white, yellow, or red for text, lines, and shapes, and use blue, green, or dark gray for the background. In a bright environment, use blue or black text, lines, and shapes on light backgrounds of magenta, blue, white, or light yellow. Finally, be aware that certain color combinations can cause irritation and produce afterimages. Particularly avoid strong combinations of red and blue, red and green, blue and yellow, and green and blue.

Good use of color is difficult to achieve for several reasons. We have seen how the environment is a factor. The space in which the color is used also has an influence. The size, shape, and location of objects influence how their color is perceived. You should also take into account many physiological and cultural

[8]Ibid, pp. 136–138.

TM

F

influences. In coping with these influences, designers should use experimentation to determine good color sets and then make small refinements to learn the effects that certain color changes produce. In this way, designers can harness the power of this potentially powerful communication tool.

The User Interface Design Process

User interface designs of all types, nongraphic as well as GUI, evolve by following methodologies described elsewhere in the text, such as the system development life cycle, prototyping, rapid application development, and business process redesign. However, particular attention has been paid to the user interface design process, and it is appropriate to recognize that effort here. John Hoffman has done an especially thorough job and suggests that user interface design evolve in ten steps.[9] Figure F.16 shows these steps in relation to the system life cycle phases described in Chapter 8.

1. **Identify User Needs** Learn about users by means of interviews and surveys. Learn who the users are in terms of age, gender, educational background, experience, physical handicaps, and so on. Also determine their expectations of the new system.

2. **Perform a Job Analysis** Understand the user's job by breaking it down into its component parts and producing a task list. This effort lays the foundation for system acceptance by users, who perceive the system as being tailored to their needs.

3. **Select Users' Work Tasks to Be Supported** Include in the system boundary only those tasks that are critical to the job and can benefit from the system. Keep in mind cost versus benefit and adherence to the organization's strategic objectives.

4. **Perform a Task Analysis** A task analysis is a subdivision of each task into its elementary steps that recognizes the knowledge and skills required for each step. Hoffman views task analysis as the *skeleton* for interface design—the architectural base around which the interface will be designed.

5. **Determine the Overall Interface Style** Select from the use of menus, form filling, commands, direct manipulation, or some combination by recognizing the objectives, advantages, and disadvantages of each. Although a wealth of suggestions apply here, some general design advice includes:

   • Provide flexibility.

   • Provide the user with the ability to "undo" errors.

[9]Hoffman, pp. 4-5 through 4-28.

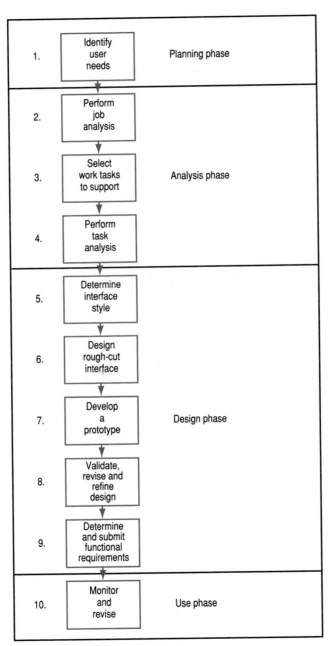

Figure F.16
The ten steps of interface design and their relationship to
the system life cycle.

- Provide user feedback.
- Allow users to be in control.
- Provide optional help.
- Use defaults for common or likely outcomes.

6. **Design a Rough-Cut Interface** Position the major elements of the interface by paying attention to structure rather than appearance. Consider the use of a storyboard, or hardcopy, view of the interface as an alternative to one produced with software. Make certain that you adhere to the design principles and have the layout checked by a subject-matter expert.

7. **Develop a Prototype** Refine the rough-cut interface by putting it in a form so that it can be tried out by one or more users in a walk-through or simulation. Include all essential elements, but not necessarily in a polished form.

8. **Validate, Revise, and Refine the Initial Design** Use walk-throughs, simulations, and tests with persons who either will use the interface or are representative of users. Be alert to responses and incorporate improvements in new prototypes. Continue this process until objectives have been met or resources have been exhausted.

9. **Determine and Submit Functional Requirements** Write functional specifications that will serve as the blueprints for building the software. The specifications should incorporate all screens and explain how they interact.

10. **Monitor Throughout the System Life Cycle and Revise** Be alert to changes in users identified in step 1. Be open-minded to user problems and complaints. Remember that a positive perception of the system by the user is the key to usability. This perception is also influenced by the manner in which you respond to user needs over the life of the system.

These recommended steps emphasize the importance of understanding user needs at the task level and of taking a deliberate approach to interface design in the form of a rough-cut interface, prototypes, and attention to refinement and monitoring. We can see that the SLC methodology provides a general guideline that can be tailored to the specific needs of a certain aspect of system development—in this case, interface design.

Future Developments in Interface Design

Future improvements in interface designs can be expected, due mainly to improvements in hardware. These improvements are classified by

TM

F

the acronym **SILK,** which stands for speech, image, language, and knowledge.[10]

- **Speech** Future interfaces will have an improved speech capability, enabling improved speech input and output.

- **Image** Systems will be able to manipulate images faster, perhaps faster than textual information.

- **Language** Programs will be able to read, write, file, and even translate natural language text.

- **Knowledge** Interfaces will take advantage of artificial intelligence to make systems more tolerant of user error and ambiguity.

In addition to improving existing technologies, future systems can be expected to expand into the area of three-dimensional images and virtual realities. John Walker of Autodesk uses the term **cyberspace** to describe a three-dimensional domain in which cybernetic feedback and control occur.[11] The system provides users with the feeling of being inside a particular world rather than simply viewing an image. Virtual-reality displays can respond in real time to user eye, head, and body movements by employing stereo goggles, DataGloves, or DataSuits.

In conjunction with these dramatic improvements, systems designers can expect to become more involved with producing multimedia documentation and tailoring interfaces to group settings. We can expect interface design to continue to be one of the most exciting and challenging parts of systems work.

Problems

1. Design a GUI that will enable a manager to use the drill-down technique to prepare the three reports illustrated in Figure 6.10. Use an opening screen that will enable the user to specify the operation to be performed. Then, use one or more screens to present the information. Use the storyboard approach. First, use pencil to sketch the layout for each screen on posterboard. Apply as many of the design principles as you can. When you are satisfied with the layouts, use a black marker to redraw the pencil lines.

2. Now add color to your storyboards, applying the color prinicples. Assume that the user is a fifty-year-old executive who has a dark office and does not have color-deficient vision.

3. Make an oral presentation to your class, explaining the screens and how they are used. In the process, explain how you applied the GUI design prin-

[10]Aaron Marcus and Andries van Dam, "User-Interface Developments for the Nineties," *Computer* 24 (September 1991): 53.

[11]John Walker, "Through the Looking Glass," in Brenda Laurel (ed), *The Art of Human-Computer Interface Design* (Reading, MS: Addison-Wesley), 1990, 444.

ciples. Obtain feedback from the class concerning ways to improve your designs.

4. Take the class feedback into account and revise your screens. If you have design software available, use it. Otherwise, prepare a new set of storyboards.

5. Present your revised GUI to the class.

TM

F

Flowcharts

Flowcharts are the oldest of the tools used for documenting processes. They are also the most standardized, their symbols having been specified by both national and international standards organizations. Flowcharts are therefore a common language for information specialists around the world.

Flowcharts are not as appropriate for documenting modern, online, communications-oriented systems as they were for the earlier batch designs. However, flowcharts still provide some features that ensure their continued use. They can play a role in new system development and also serve as a platform for systems maintenance and business process redesign.

Basic Types of Flowcharts

There are two basic types of flowcharts—system and program.

System Flowcharts

A **system flowchart** is a diagram consisting of symbols representing the data files and processes that comprise a system. Figure G.1 is a flowchart of a system that processes sales orders. This particular system consists of four processes, which are numbered.

Program Flowcharts

A **program flowchart** is a diagram consisting of symbols that represent the processes performed by a single program. Figure G.2 is a flowchart of a program that prints a report. A program flowchart similar to this, although perhaps more detailed, would be used to document the fourth process of the sales order system—the step that prepares the Rejected Sales Order report.

This program flowchart documents a structured program. The driver module is on the left side, and four subsidiary modules are in the center and on the right side. Each module is bounded by ovals called **terminal symbols.** The rectangles with the double bars at each end are **predefined process symbols** that represent the subsidiary modules. The parallelograms are **input** and **output symbols** that represent program statements that cause data to be read and written. The rectan-

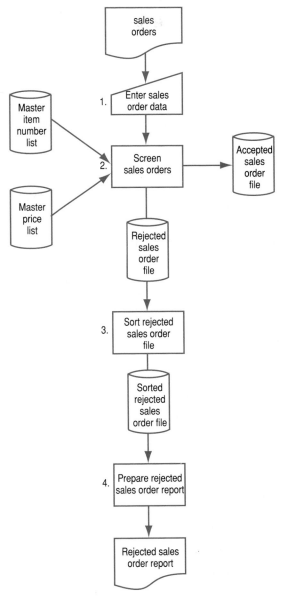

Figure G.1
A system flowchart.

TM
G

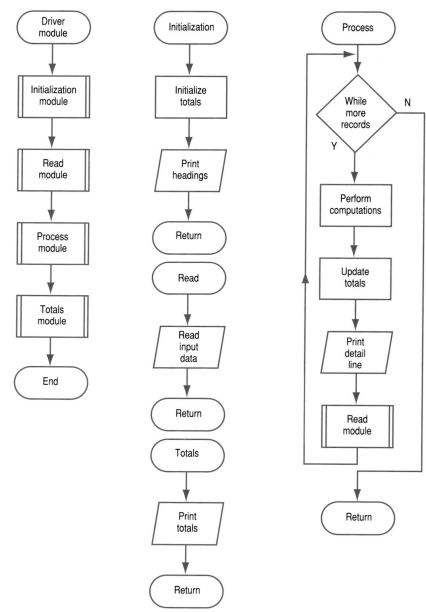

Figure G.2
A program flowchart.

gles are **process symbols** that represent statements that accomplish arithmetic and data movement operations. The diamond is a **logic symbol** that represents control statements that enable the program to choose between alternate paths based on conditions that are encountered.

The terminal, predefined process, input and output, and logic symbols are used only in program flowcharts. They are not used in system flowcharts.

As a general rule, systems analysts prepare system flowcharts, and programmers prepare program flowcharts. Because we are more concerned with the systems analyst, we will not elaborate on program flowcharts here.

System Flowchart Symbols

System flowcharts show the flow of data through the system, from beginning to end. Three sets of symbols are used—one for the processes, another for the data, and a third to establish flow connections. Each symbol contains a brief label.

Process Symbols

There are four different kinds of processes—manual, offline keydriven, online keydriven, and computer. These symbols are illustrated in Figure G.3.

A **manual process** is one performed without the aid of a labor-saving device. The process can be physical, such as opening the mail, or it can be mental, such as checking a document for errors.

An **offline keydriven process** is one performed with the aid of a keydriven device, such as a typewriter or a pocket calculator, that is *not* connected directly to a computer.

An **online keydriven process** is one performed by operating a keydriven device, such as a computer keyboard or a keyboard terminal, that *is* connected to a computer.

A **computer process** is any process performed by a computer, from a program containing only a handful of instructions to one containing thousands.

Data Symbols

Each process transforms input data into output data. Figure G.4 shows five symbols that represent the most popular data forms. Because these symbols were designated some time ago, they do not fit some currently popular media, such as the telephone, microform, and diskettes. In the absence of company guidelines, the systems analyst must improvise when documenting such media.

A **document** is anything printed on paper, including handwritten notes, typed forms such as letters, preprinted forms such as timecards, and all types of computer printouts.

A **display** is anything displayed on a computer screen, ranging from a single data element to a lengthy report.

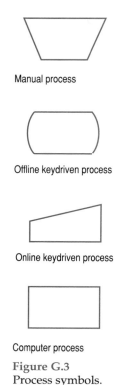

Manual process

Offline keydriven process

Online keydriven process

Computer process

Figure G.3
Process symbols.

Offline storage is any accumulation of documents or data that is maintained apart from the computer. Generally, offline storage refers to material stored in a filing cabinet.

Online storage is secondary storage, both sequential and direct access, that is connected to the computer. Sequential storage is usually *magnetic tape,* which is represented by a circle with a tail. Direct access storage is represented by an upright cylinder.

Connector Symbols

The process and data symbols are connected by lines to show the procedural flow. Two special symbols are used to connect flows from different parts of the system. An **onpage connector symbol** is a small circle containing a letter or number that can be used to connect one point on a page to another point on *the same page.* Figure G.5A illustrates the use of onpage connectors.

An **offpage connector symbol** is a symbol shaped like home plate on a baseball diamond and is used to connect a point on one page to another point *on another page.* Figure G.5B is an example. The point of the offpage connector symbol is always aimed *down;* it is not an arrow.

A   A document

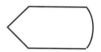

B   A display

C   Offline storage

D   Online magnetic tape storage

E   Online direct access storage

Figure G.4
Data symbols.

Flowchart Layout

The flow of processes in a flowchart should be from top to bottom and from left to right. You will not always be able to achieve this pattern because of the particular arrangement of symbols, but it should be your goal.

In determining the flowchart layout, you should not try to put too much on a single page. It is much better to use multiple pages than to overwhelm the viewer with a complex and confusing layout.

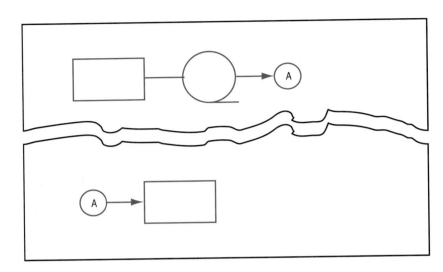

A   Onpage connector

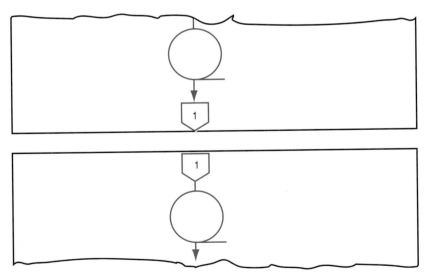

B   Offpage connector

Figure G.5
Connector symbols.

Sample System Flowcharts

One feature of a system flowchart not offered by other documentation tools is the way it shows how computing technology is used. You can glance at a system flowchart and tell whether it documents a manual system, a magnetic tape-oriented system, a magnetic disk-oriented system, and so on.

A Magnetic Disk-Oriented System

Computer configurations today feature magnetic disk storage. Practically all microcomputers reflect this architecture. We illustrate a magnetic disk-oriented order entry system with the system flowchart in Figure G.1. In that system, sales order data is entered into the computer in step 1. In step 2, the sales order data is screened by comparing it with a Master Item Number list and a Master Price list. Data that compares favorably is written on the Accepted Sales Order file, and data that contains errors or possible errors is written on the Rejected Sales Order file. The Rejected Sales Order file is then sorted, in step 3, into a sequence that is used in step 4 to print the Rejected Sales Order report. All four processes involve the computer.

A Manual System

Even though the computer has been applied to practically all business systems, many manual processes remain. A flowchart of a manual order entry system is illustrated in Figure G.6. This system is one that you might find in a very small firm. The editing in step 1 is done manually, and the calculations in step 2 and the typing in step 3 are accomplished with the aid of offline keydriven machines. In a manual system such as this, all files are offline. The Rejected Sales Order file is an example.

A System Using Both Magnetic Tape and Disk

The system flowchart in Figure G.7 gives you an idea of how both magnetic tape and disk can be incorporated into a system design. Here, magnetic tape is used for a history file—the Daily Claims History file. Tape is ideal for this use because of its ability to store large quantities of data at low cost. Magnetic disk is used for the more active data—the Current Claims file. As claims are being processed, the direct access ability of the magnetic disk makes it possible to retrieve information when it is needed.

This flowchart describes a system that might be used by an insurance company to process claims. In step 1, someone opens the mail and separates the claim forms from the other types of mail. Note that a diamond is not used to show the separation of the two paths, as is the practice in a program flowchart.

TM

G

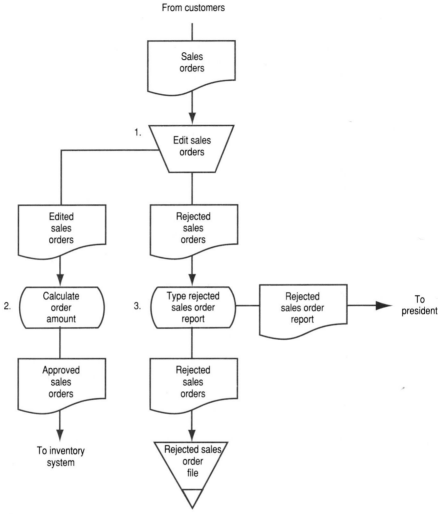

Figure G.6
A system flowchart of a manual order entry system.

In step 2, a calculator is used to create batch totals for control purposes. These totals, created prior to computer entry, are called **external totals**—they are external to the computer.

In step 3, a data entry operator keys the claims data into a computer keyboard or keyboard terminal, and then files the claim forms in the offline Claims History file.

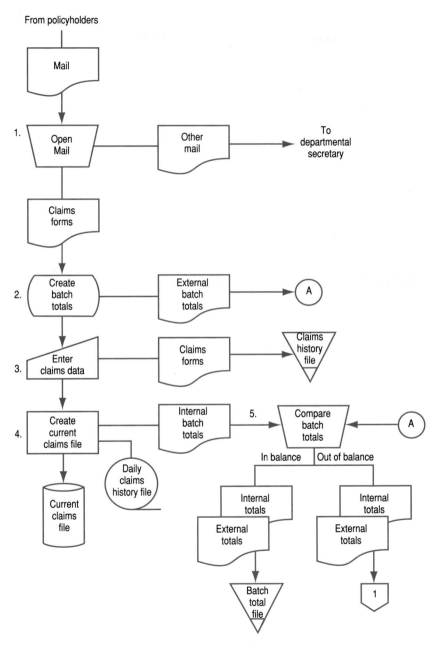

Figure G.7
A system flowchart of a claims processing system.

TM

G

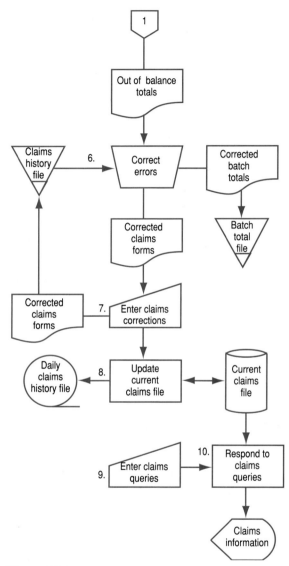

Figure G.7
continued.

Once the data is in the computer, a program is used in step 4 to create another set of batch totals and write the claims data on a Current Claims file. The program also writes the Daily Claims History file to be used as an audit trail. The batch totals created in this step are called **internal totals** because they are produced from data in the computer's storage.

In step 5, the external and internal totals are compared. When an in-balance condition exists, the total forms are filed in the Batch Total file. Out-of-balance totals are handled in step 6. Errors are corrected, using the contents of the Claims History file. The corrected batch total forms are then filed in the Batch Total file.

Steps 7 and 8 show how the Current Claims file is updated with the corrected claims data. The program also adds records to the Daily Claims History file.

In step 9, users query the Current Claims file for information that is displayed on their screens in step 10.

Tips for Drawing System Flowcharts

Your system flowcharts will do a better job of communicating a system design when you follow these tips. Refer to Figure G.7 for examples.

- Number each process step.

- Label each process step with a verb and object. Use specific names. Do not say "Enter data." Rather, say "Enter claims data."

- Label each data symbol with the name of the data—a specific file name, report name, display name, and so on.

- Use arrowheads only on lines that enter a process symbol, *not* on lines that enter a data symbol. See lines entering steps 1 through 4. The only exception is when the lines enter the *last data symbol* in a flow. See the arrowhead entering the "Claims information" symbol at the end of the flowchart. The data symbols should appear to be overlaid on top of the arrows that are connecting the process symbols.

- Add notes to identify source and disposition of data, as done at two points in step 1.

- When the same data passes through multiple steps, it is not necessary to repeat the data symbol for each step. See steps 2 and 3. Both of these steps use the Claims forms as input.

- Use a two-headed arrow to show that data is retrieved from a DASD file and then returned to the same file. See step 8. This is how you show file maintenance for a DASD file, where the master file records are updated with transaction data.

- Always use the online keydriven symbol and the rectangle together for a data entry or database querying operation, as shown in steps 3 and 4, and 9 and 10. In this arrangement, there is no need for an intervening data symbol.

- Remember to always file away source documents after they have been used for data entry. Step 3 is an example.

- Use only a single rectangle for a program, regardless of how many different processes the program performs. Step 4 is a good example. Even though three

TM

G

separate outputs are produced (internal batch totals, the Current Claims file, and the Daily Claims History file), only a single rectangle is used. A good rule to remember is "One program: one rectangle."

- When a logical process separates a single data flow into two or more paths, label either the *documents* or the *arrows*. Step 1 illustrates document labeling (Claims forms and Other mail), and step 5 illustrates arrow labeling (In-balance and Out-of-balance).

- Overlap document symbols when multiple *types* of documents are kept together, as shown for the output of step 5.

- When using both onpage and offpage connectors, use letters for labeling one type and numbers for labeling the other. It does not matter which type of label is used with which type of symbol; however, be consistent within the same flowchart.

- When using offpage connectors, repeat the file symbol on the additional page. Note that the symbol "Out-of-balance totals" at the top of the second page identifies the input to step 6.

Obviously there is a lot to keep in mind when creating a flowchart. Flowcharting is the most difficult of the documentation tools to learn to use properly.

Computer-aided Flowcharting Tools

Many CASE tools provide a flowcharting capability. Excelerator, from Index Technology, is an example. In addition, a number of software systems, such as MacFlow from Mainstay, are especially well suited to flowcharts. It is also possible to use graphics packages, such as MacDraw II from Claris, to prepare flowcharts that are not only attractive but easy to maintain. The software tools eliminate much of the time-consuming artwork that goes into developing good flowcharts, making them an appealing documentation tool.

Putting Flowcharts in Perspective

Although some documentation tools are potentially useful for users as well as for information specialists, such is not the case for flowcharts. Users perceive flowcharts as being too "unfriendly" for end-user computing. Therefore, flowcharting is a tool of the information specialist.

Because flowcharting has been around longer than the other documentation tools, many information specialists believe that it has outlived its usefulness. To a certain extent this is true. However, many systems in use today are documented entirely with flowcharts. If the systems analyst is to maintain these systems and perhaps replace them with reengineered systems, it is necessary to understand flowcharting.

System flowcharts are good for showing how computer technology is applied. This feature has real value in visualizing how the logical design is reflected in the physical design. A good design strategy is to use structured tools such as data flow diagrams, structured English, and action diagrams as logical models, and then use system flowcharts as physical models. System flowcharts are good supplements to the other tools and help round out the documentation package.

TM

G

Problems

1. Draw a system flowchart of the following manual procedure:

   A. The departmental secretary receives timecards from all employees in the department each week. The secretary uses the cards to prepare a time sheet, which lists each employee, the total hours worked, and the hourly rate.

   B. The departmental supervisor audits the time sheet and signs it if it is acceptable. If unacceptable, the sheet is returned to the secretary. Identify the two output paths by labeling the documents.

   C. A clerk in the accounting department uses all the acceptable time sheets to calculate payroll earnings. The earnings figures are calculated on a desk calculator and are written on the time sheets.

   D. The time sheets are used to prepare the payroll checks, which are typed. The payroll checks are forwarded to the employees. After this operation, the time sheets are filed in an offline Time Sheet History file.

2. Draw a system flowchart of the following computer-based system used by a bank to open new checking accounts:

   A. The customer completes a New Account Application form. There is no input to this process step, only output.

   B. A data entry operator enters data from the New Account Application form into the computer, using a keyboard terminal.

   C. The computer edits the new account data and prepares an error listing of all applications that contain errors. The error listing is sent to the new accounts department for correction. New account data that does not contain errors is written onto a magnetic tape file named New Account file.

   D. The New Account file is input to another computer program that adds the new account data to the Customer Master file, on magnetic disk, and prints the transactions on a Transaction Listing. Because data is going only *to* the magnetic disk, use only a one-headed arrow.

3. When the firm receives shipments from suppliers, packing lists are enclosed in the cartons. A data entry operator keys the packing list data into the computer and then files the packing lists in an offline Packing List History file. The computer program obtains the corresponding records from the Outstanding Purchase Order file, on magnetic disk. The same program prints a report

TM

G

named Received Purchases, which is sent to the purchasing department through the company mail, and writes a Supplier Payables file on magnetic disk. The Supplier Payables file is input to another program that obtains supplier data from the disk-based Supplier Master file and prepares checks that are mailed to the suppliers. The same program that prints the checks also creates a Supplier Check History file on magnetic tape.

VIII—Safeguarding the Information Resources

(Use this scenario in solving the Harcourt Brace problem at the end of Chapter 13.)

In addition to its fame as a tourist center, Orlando is also known as the "lightning capital of the world." Frequent lightning storms cause blackouts and brownouts in the municipal electrical system—a problem that can put a computer center out of action. Most companies cannot afford for their computers to be down for any length of time, and so they take precautions to keep the disruptions to a minimum.

Harcourt Brace, recognizing the value of its information resources, has taken steps to protect them. In fact, Harcourt Brace probably has one of the best-secured computer operations in the country. This did not come about by accident. Mike Byrnes has made computer security a top priority and doesn't miss an opportunity to voice his dedication to it. The easiest way to get Mike into a conversation is to bring up computer security, as Russ March found out at his first company picnic.

Mike Byrnes: Hey, Russ. How's it going? Ira has been keeping me posted on the progress that you and Sandi are making on the Bellmawr project. It sounds like you are really on to something.

Russ March: That's right. Things are going great. Sandi and I are learning a lot, and the people on the team and in Bellmawr are great. How are things with you?

Mike Byrnes: Couldn't be better. Except for the glitches that we ran into last week when we went through a dry run on our contingency plan, we really haven't had anything out of the ordinary. In this business, that's a real blessing.

Russ March: We touched on contingency planning in college, but I don't remember much about it. Is that a big thing with Harcourt Brace?

Mike Byrnes: It certainly is. We put a lot of emphasis on system security, and the contingency plan is just one part of it. We follow the plan when something goes wrong, such as a storm that knocks out our system. I think we've got one of the best security setups in the country. If you'd like to learn more, why don't you drop by sometime? Sandi might be interested as well, since we didn't cover the security issue during the orientation review.

[Russ and Sandi take Mike up on his invitation and make an appointment. After exchanging greetings, Mike leads the two analysts through the maze of hallways to the room where the computer equipment is located. The door is locked, and Mike takes a plastic card from his pocket.]

Mike Byrnes: As you know, we have security guards for the building 24 hours a day. Plus, access to the data center is restricted to only those people with these cards. It's called a card key entry system. [Mike inserts the card in a reader next to the door and opens the door.]

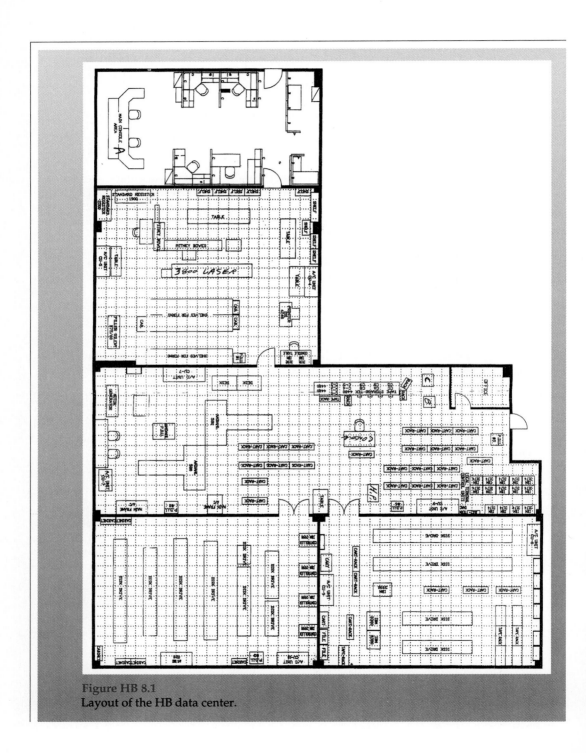

Figure HB 8.1
Layout of the HB data center.

This is the print room. Actually, there are four rooms in the data center—one for the printing equipment, the biggest room for the Amdahl and IBM CPUs, one for the disk drives, and one for the tape drives. All four have 2 feet of crawl space beneath the floor for the cables, and, naturally, all four have special air-conditioning and humidity controls. Each room also has its own PDU.

RUSS MARCH: What's a PDU?

MIKE BYRNES: It stands for power distribution unit. It's sort of like a buffer. Rather than take electricity directly from the municipal service, we run it through a bank of batteries in the basement, which are called a UPS—for uninterruptable power supply. If the municipal power goes out, we run on the batteries. The batteries are continually charged by the municipal power. We have two diesel generators in the basement that can keep the batteries charged if municipal power doesn't kick back in right away.

SANDI SALINAS: That's impressive. I'm glad you're showing us all this. What are those little boxes on the ceiling—smoke detectors?

MIKE BYRNES: That's right. We also have them under the floor. They set off a fire suppression system in case of smoke or fire.

[Russ, Sandi, and Mike watch the print room employees operate the Pitney-Bowes inserting machine. Mike explains that the machine is used to insert statements into envelopes and add postage. "It'll handle about 7,500 statements an hour," he adds as they go into the CPU room.]

RUSS MARCH: Look at that console. It looks like NASA Mission Control. I've never seen a console with eight CRTs. What do you need so many for?

MIKE BYRNES: Two are for the CPU, and they provide the main interface for the operators. The others are for certain system software. One is for TSO, another is for CICS, one is for our network control program called NETVIEW, and so on.[1] Maybe we got a little carried away, by they do provide the operators with a complete picture of what's going on.

[The three walk past the Amdahl and IBM CPUs, which give no hint as to the millions of processes that are going on inside each second. Next is the disk room, and it, too, offers no indication of the real level of activity; there's not an operator in sight. Mike explains how some disk drives are backups in case others go down. "We have 148.7 gigabytes of data on disk. A gigabyte is a billion bytes," he says. Finally, they go to the tape room. There's more activity there, but the most impressive sight is the large number of tape cartridges in racks.]

MIKE BYRNES: We have about 14,000 cartridges, and there are about 5,000 of the older reels over in the corner. The cartridges are a lot faster than the reels. Also, a single cartridge contains about 20 percent more data than a reel—in about one-fourth the space. That's why the cartridges are so popular for archival storage.

[1]TSO and CICS are products of IBM. NETVIEW is a product of Computer Associates.

Figure HB 8.2
The HB data center console.

Well, that's the computer tour.

RUSS MARCH: And, I guess that's it for computer security. Right?

MIKE BYRNES: Wrong. Computer room security is only one part of it. There's also data security and software security. Come on back to my office and I'll tell you about those.

[The three return to Mike's office.]

Actually, the computer room security is the most visible, but we have to secure the data and the software, too. As you know, someone can illegally tap into our files without getting into the computer room.

We have a formal procedure for making changes to software and for implementing new systems or system modules. We have what we call a move request form that is submitted by the programmer when a new program is to be implemented. Suzanne filled out one on your scheduling system. The form must be signed by the group manager. In the case of your project, Ira signed it. This prevents someone from illegally implementing a program. We have similar controls on changes to programs. We keep a backup of the actual changes as a safety precaution, and we also maintain an audit trail as a means of following up on changes—who made them, when, and so on.

You're probably familiar with data security since you've had a database course in college, right?

SANDI SALINAS: That's right. I remember my professor saying that DBMS protects a database like a series of chainlink fences. If you want to get to the data, you must go through the fences. The first one is the user password, the second is a directory of files that the user can access, and the third specifies what kind of access is permitted—read only, read and update, and so forth.

MIKE BYRNES: That's an interesting comparison, and it's exactly what we have here. We have also instituted formal procedures for issuing passwords and password changes. Users must change their passwords every 30 days. If they don't, we cancel them. Also, when someone quits, we're notified so that the employee's password can be revoked. You have to have formal procedures.

RUSS MARCH: What about backing up the database?

MIKE BYRNES: We maintain copies of all disk files on tape cartridges, and the cartridges are stored off-site, in a fireproof vault.

RUSS MARCH: It sounds like you're prepared for anything.

MIKE BYRNES: Almost. And, if anything bad does happen, our contingency plan is put into action. But, we don't wait for an emergency. We test the plan periodically to make sure that it works. This is what happened last week. Mark Arak, whom you've met, is manager of contingency planning. It's his responsibility to come up with the plan and to subject it to tests, just like you would a computer program. We have an agreement with Comdisco Corporation, in Rosemont, Illinois, to use their computer if ours goes down for any length of time.

SANDI SALINAS: I'm really impressed. It's good to know that when our system is implemented it can continue to function, even in the case of some type of disaster. It looks like that is one part of systems work that we won't be concerned with now.

MIKE BYRNES: You're right as far as the present time is concerned. But it will be more of a concern as you are promoted up through management ranks. Then, you will be expected to play a more important role in the decisions that we constantly make concerning security. Until then, you can do like Greyhound, and "leave the driving to us." [Everyone laughs, and Sandi and Russ say good-bye and return to their offices, where they get back to the reality of the new Bellmawr system.]

Personal Profile—Mark Arak, Manager of Contingency Planning

Mark Arak went from his hometown of Hoboken, New Jersey, to Jersey City State College, where he earned a Bachelor of Arts degree in economics. He credits his 20 years of experience in many facets of information systems as the key to his career success. For relaxation, he enjoys cooking, golf, and tennis. Mark believes that Harcourt Brace's future sucess will hinge on its ability to manage successfully, using the financial freedom gained from the acquisition by General Cinema in 1991.

CHAPTER 12

Systems Controls

Learning Objectives

After studying this chapter, you should:

- Understand how management responds to business risks by establishing controls
- Know the components of a firm's system of internal control
- Be aware of the main computer-related risks and know a useful technique for systematically identifying appropriate controls
- Understand how controls can be built into the systems development process, systems design, and day-to-day operation of the computing facility
- Appreciate the different control challenges posed by each CBIS subsystem

Introduction

From the time an entrepreneur decides to form a business enterprise, he or she faces risks. Management has responsibility for recognizing risks, implementing appropriate controls, and following up to ensure that the controls are working.

A firm's internal control system consists of a control environment, plus control procedures, which include general controls that apply to all systems as well as application controls that apply to specific systems.

The three major risks that threaten computer-based systems are unauthorized tampering, loss of system integrity, and disruption. The controls implemented to mitigate these risks ensure information integrity, security, and compliance.

Each system element is subject to specific risks. One or more controls should be put in place for each risk. This direct relationship between risks and controls can be pictured in the form of risk and control matrices.

Computer controls are applied to dynamic processes in three areas: systems development, systems design, and systems operation. Systems development controls begin with the firm's commitment to IRM and strategic planning for information resources, and conclude with the life cycles of individual systems. Systems design controls are incorporated into each of the system elements. Systems operation controls exist in the form of annual audits, and in policies and procedures of the computer operations unit.

The trend from centralized processing to distributed processing and end-user computing has compounded the difficulty of achieving an acceptable level of control. Each CBIS subsystem imposes a unique control requirement.

Business Risks

Business operation requires that management take certain risks.[1] A **business risk** is the possibility of loss or injury that threatens the ability of the firm to operate as desired. Examples of business risks include a lack of customer acceptance of a new product, competition, business interruption, downturns in the economy, mistakes, and inefficient use of resources.

Management has three responsibilities in terms of risks:

1. To identify the risks

2. To implement controls that eliminate or minimize the impact of the risks

3. To follow up to ensure that the controls are correctly addressing the risks

The success with which management performs these tasks determines, to a large degree, the firm's ability to meet its objectives.

Business Controls

The systems that management implements to mitigate risks incorporate one or more controls. A **business control** is a mechanism that regulates or guides a business system, which can be the entire firm or a subsystem within the firm. The system can also be physical or conceptual.

We know from systems theory that a closed-loop system incorporates a feedback loop and a control mechanism that enable the system to control its own operation. Such a system is illustrated in Figure 12.1. The feedback loop consists initially of data, which is transformed into information. The information, in turn, is transformed into decisions. The control mechanism consists of three elements—the performance standards that the system is expected to attain, management, and the information processor. The information processor transforms the data into information, and management transforms the information into decisions.

Therefore, when business controls are applied in the broadest sense, they consist of a combination of the firm's standards of performance, management, and the information processor, working together to keep the firm on course.

[1]Much of the material in this chapter relating to the subject of risks and controls comes from The Institute of Internal Auditors Research Foundation's 1991 publication *Systems Auditability and Control Report*. For more information concerning the report, contact The Institute of Internal Auditors Research Foundation at 249 Maitland Avenue, Altamonte Springs, Florida 32701–4201.

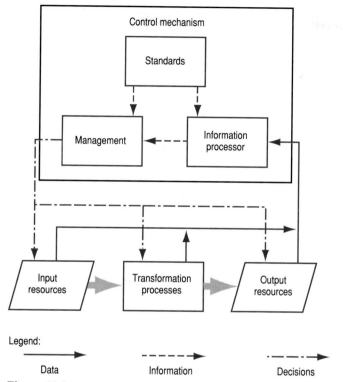

Figure 12.1
A system with a feedback loop and control mechanism.

The Firm's System of Internal Control

Each firm has a **system of internal control,** consisting of control procedures existing within a control environment. The system functions as a filtering device, eliminating events that threaten the firm's operations, and letting all events that support the firm's objectives flow through.

The Control Environment

The **control environment** is the setting established by management, within which systems controls are applied. It consists of such ingredients as the firm's organizational structure, the competence and integrity of the firm's human resources, the levels of authority and responsibility assigned to both managers and nonmanagers, and the availability of all resources necessary for applying and enforcing the controls.

Control Procedures

Control procedures are the routines that enable users to control the processes of their computer-based systems. There are two types of control procedures—general and application.

- **General Control Procedures** The controls that exist in the environment of the system are called **general controls.** They include the structure and policies of the information services organization, controls exercised during systems development, database and data communications controls applicable to all systems, and controls built into computer operations. General controls affect all applications.

- **Application Control Procedures** Controls built into individual applications, such as payroll and inventory systems, are called **application controls,** and are intended to ensure that the systems process the data as intended.

As systems become more complex and integrated, the dividing line between general and application controls becomes blurred. However, the tendency is toward greater reliance on general controls, a trend known as the **migration of controls.**

Computer-Related Risks

Each of a firm's computer-based systems faces three main types of risks:

- **Unauthorized Tampering** Persons who have no reason to do so are permitted access to the firm's information resources—using them, disclosing them, stealing them, modifying them, and inflicting destruction that denies service to rightful users. Direct contact with the hardware is not a prerequisite; access can be gained electonically from remote locations.

- **Loss of system integrity** The conceptual system does not accurately reflect the physical system.

- **Disruption** The computer-based system ceases to function.

Management relies on a combination of controls to eliminate these risks or minimize their effects.

Objectives of Computer-Related Controls

Controls of computer-based systems have three main objectives: (1) information integrity, (2) security, and (3) compliance.

Information Integrity

When a firm's data accurately represents its physical systems, this is called **information integrity.** Information integrity is achieved by controlling both data and processing. **Data controls** ensure that the data flows properly through the system, from its point of origin to its destination. **Processing controls** ensure that the data is processed as desired.

Security

Firms seek to keep information resources secure from all types of damage and misuse. Security threats include natural disasters as well as intentional and unintentional acts by persons both inside and outside the firm. Systems security is discussed in Chapter 13.

Compliance

Constraints are placed on an information system both by management and by elements in the firm's environment. The system must comply with these constraints. Management constraints consist of policies, procedures, and performance criteria. We have seen how the systems analyst and manager define a new system's performance criteria early in the life cycle. Environmental constraints consist of government laws and regulations, as well as standards from professional accounting and auditing organizations.

The Relationship Between Computer-Related Risks and Controls

When a computer risk is identified, the next step is to identify one or more controls that will either prevent the risk from occurring or minimize the damage should the risk occur. A particular risk might require a single control, or it might require more. By the same token, a particular control might be aimed at a single risk or more than one.

The Risk Matrix

A good way to systemically address risks is to organize them in a matrix. In a **risk matrix,** the components of a system are listed along the left side, and the various types of potential risks are listed across the top, as shown in Figure 12.2.[2] In this example, the types of risks concern threats to the information integrity of an order entry system—guarding against incomplete data, inaccurate data, and unauthorized transactions. Potential risks are entered into the appropriate cells.

---

[2]The idea for the risk and control matrices came from Dr. Jerry FitzGerald, a management consultant. For more information, see Jerry FitzGerald, *Business Data Communications,* 3d ed. (New York: John Wiley & Sons, 1990): 488–496.

System element	Risks		
	Incomplete data	Inaccurate data	Unathorized transactions
1. Log in sales orders	Missing data • Customer number • Customer order number • Customer order date	Wrong data • Customer number • Customer order number • Customer order date	No customer
2. Edit sales order data	Missing data • Item number • Quantity	Wrong data • Item number • Quantity	
3. Conduct credit check			Bad credit rating or credit limit exceeded

Figure 12.2
A risk matrix.

System element	Controls		
	Incomplete data	Inaccurate data	Unauthorized transactions
1. Log in sales orders.	Telephone customer for missing customer order number and order date.	Sight verify log entries.	Check customer number against master list.
2. Edit sales order data.	Use item description to look up missing item number in master list. Telephone customer for quantity.	Check item number against master list. Conduct reasonableness check on quantity.	
3. Conduct credit check.			Obtain current credit rating from credit bureau. Compare updated accounts receivable amount with credit limit.

Figure 12.3
A control matrix.

The Control Matrix

Once the risk matrix has been prepared, the task is to identify one or more controls for each risk. These controls can also be displayed in a matrix. In a **control matrix,** the cells contain controls that can be applied to the risks in corresponding cells of a risk matrix. Figure 12.3 identifies potential controls for the order entry system. These controls are then incorporated into the system design or into the system operation.

Areas of Computer Controls

Computer controls are applied in three main areas, all dealing with processes: systems development, systems design, and systems operation. These are illustrated in Figure 12.4.

Systems development controls are measures taken to ensure that computer-based systems are developed to solve identified problems, ensure that the use of such systems is justified and see to it that development proceeds according to schedule. Development controls are concerned with the *process* of systems development.

Systems design controls are routines and procedures incorporated into computer-based systems to ensure that the data is processed correctly. Systems design controls are a *product* of systems development.

Systems operation controls are policies and procedures governing the use of information resources intended to mitigate the risks these resources face daily.

Each of these control areas is described in the next three sections.

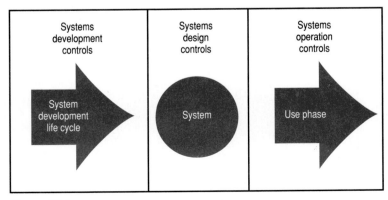

Figure 12.4
The three areas of computer controls.

Systems Development Controls

Management controls the development of computer-based information systems by implementing measures in a hierarchical manner. The hierarchy, pictured in Figure 12.5, includes information resources management, the strategic plan for information resources, the MIS steering committee, project management, and project controls.

- **Information Resources Management** At the top of the hierarchy is a commitment by the firm to a policy of information resources management. When a firm embraces the IRM concept, it recognizes information as a strategic resource and establishes information services as one of the major functional areas. The manager of information services, or CIO, and other executives on an

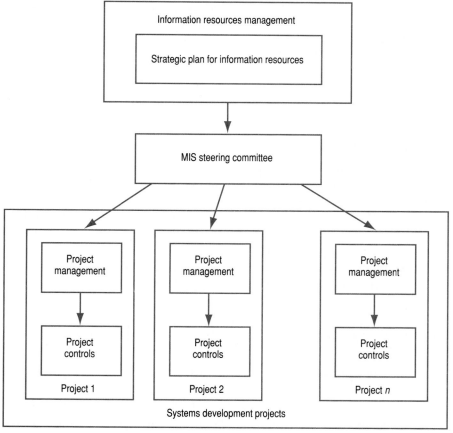

Figure 12.5
The hierarchy of systems development controls.

executive committee develop strategic plans for the firm and for the functional areas, including information services.

- **The Strategic Plan for Information Resources** This long-range plan identifies the information resources needed to develop new CBIS subsystems and maintain existing subsystems. As such, the plan provides the blueprint for all future systems development and maintenance.

- **The MIS Steering Committee** A group of top-level managers who represent a cross section of the firm's operations carries out the strategic plan for information resources. The group meets periodically to approve new systems development projects and to monitor the progress of ongoing ones.

- **Project Management** Each development project is managed by a project leader who makes periodic reports to information services management and the MIS steering committee.

- **Project Controls** Control points are built into each development project to ensure that the work is performed expeditiously and that the resultant system meets the defined needs.

The three upper-level controls (IRM, SPIR, and MIS steering committee) are put into place before any development effort begins. The two lower-level controls (project management and project controls) exist for the duration of each development project.

Controls are Built into Each Development Project

As each system is developed in accordance with the strategic plan for information resources, project controls are applied at key points. There are 15 control points in all, as illustrated in Figure 12.6.

Planning Phase Controls   Three controls are incorporated into the planning phase to ensure that only feasible development projects are launched and that the development proceeds efficiently.

- **The Feasibility Study** The systems analyst evaluates a new system idea in terms of various dimensions of feasibility.

- **The System Study Proposal** The systems analyst prepares written documentation of both the advantages and disadvantages of implementing a new system. The MIS steering committee makes a go–no go decision on continuing with the project.

- **The Control Mechanism** The final step of the planning phase calls for the MIS steering committee to establish a control mechanism consisting of a project plan, scheduled meetings, periodic reports, and graphic tools.

Analysis Phase Controls   Four analysis phase controls ensure that the information specialists and the users clearly understand the new system's expected performance.

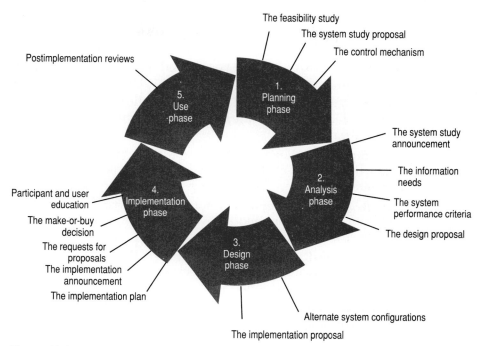

The feasibility study
The system study proposal
The control mechanism
The system study announcement
The information needs
The system performance criteria
The design proposal
Postimplementation reviews
Participant and user education
The make-or-buy decision
The requests for proposals
The implementation announcement
The implementation plan
The implementation proposal
Alternate system configurations

1. Planning phase
2. Analysis phase
3. Design phase
4. Implementation phase
5. Use phase

Figure 12.6
Project controls.

- **The System Study Announcement** Management announces the system study to the employees in order to obtain their cooperation.
- **The Information Needs** The users and the systems analyst jointly define the information output of the new system.
- **The System Performance Criteria** The users and analyst jointly specify the level of performance the new system must achieve.
- **The Design Proposal** The systems analyst provides the MIS steering committee with the basis for making a second go–no go decision concerning the project.

   Design Phase Controls   Two controls ensure that the firm does not embark on a process of assembling expensive information resources without a solid system design and the assurance that the design should be implemented.

- **Alternate System Configurations** The project team considers various ways to solve the problem as a means of ensuring that the best configuration is proposed.

- **The Implementation Proposal** The systems analyst provides a third opportunity for a go–no go decision.

  Implementation Phase Controls   Five controls are applied during implementation to ensure that the process goes smoothly, that resource acquisition decisions are sound, and that everyone is prepared when the system goes on the air.

- **The Implementation Plan** The implementation phase is kicked off by finalizing a detailed plan to ensure that all the information resources are made available at the right times.

- **The Implementation Announcement** Management makes a second announcement, this time to solicit the users' and participants' cooperation in the implementation effort.

- **The Requests for Proposals** Information services prepares RFPs for the acquisition of hardware and software to ensure that the best combination of resources is identified.

- **The Make-or-Buy Decision** Quantitative analyses are applied to the hardware and software acquisition decision to make certain that the payment method investment is sound.

- **Participant and User Education** All who will be involved with the new system receive education and training immediately prior to cutover.

  Use Phase Control   The final developmental control point exists in the use phase to ensure that the system functions as intended.

- **Postimplementation Reviews** Reviews and audits are conducted by internal auditors to validate system integrity, and by information specialists to ensure user satisfaction.

  In this way, projects are controlled by incorporating multiple controls at key points throughout the SLC.

Putting Systems Development Controls in Perspective

Management establishes an overall setting that consists of a commitment to use information as a strategic resource and a long-term plan that guides the assembly of the needed information resources. Within this setting, each development project receives management attention—from the MIS steering committee and the project leader. These managers rely on control points built into the life cycle at critical points to ensure that the development goes smoothly and that the desired end product is achieved.

Systems Design Controls

During the design phase, the systems analyst works with the users, other information specialists, and the internal auditor to develop the logical design of the new system. As this design unfolds, the designers are alert to the potential risks that the system will face and incorporate controls to mitigate those risks.

The Reality of Design Controls

As these specialists go about their task, they recognize that, regardless of the number of controls they incorporate, the system will never be completely protected from risks. With this in mind, the analyst and auditor seek to keep risks to a level that is acceptable to management.

Management bases its decision about how much to spend on controls on the expected amount of damage that could occur if the controls were not in place. The statistical concept of expected value captures this logic. The **expected value** of the potential loss is computed by multiplying the dollar amount of the loss by the probability that it will occur. For example, the expected value of a loss of $100,000 that has a 0.15 probability of occurring is $15,000. In this example, a control would not be implemented if its cost exceeded $15,000.

The Concept of Controls for Each System Element

So many potential control techniques can be incorporated into modern computer-based systems that system designers must systematically approach their task. Otherwise, the likelihood of control imbalance is high, with some areas being overcontrolled and other areas undercontrolled.

One technique that can contribute to a balanced application of controls is to view them in terms of the system elements that they affect. Figure 12.7 is a data flow diagram showing eight system elements found in all computer-based systems. The task of the designers is to incorporate as many controls into each element as is necessary to reduce risks to the desired level.[3]

System Elements

The system elements include:

- **Transaction Origination** Users originate transactions of various sorts. For example, customers order the firm's products, employees work a certain number of hours, and managers make decisions to be entered into a mathematical model. The transaction data is typically recorded on a written form called a **source document.** The main control objective for this element is to ensure that all transaction data is completely and accurately recorded. An example of a transaction origination control is the requirement that department supervisors sign employees' timecards before the cards are sent to information services.

- **System Input** Next, the transaction data is entered into the system. The main control objective is to ensure that the system input is accomplished in both a

---

[3]The idea of viewing controls in this manner was derived from *Systems Auditability and Control: Control Practices* (Altamonte Springs, FL: The Institute of Internal Auditors, 1977), 45–86.

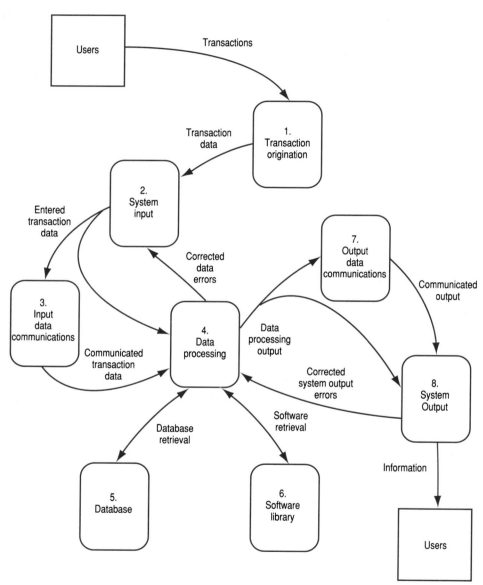

Figure 12.7
Controls are designed into system elements.

complete and an accurate manner. This is where data validation occurs; each entered data element is checked against predetermined characteristics, such as the type of data, field size, range of numeric values, and so on. Another example of a system input control is the use of prenumbered source documents to guard against loss or misuse. This second system element can be combined with the first when no source documents exist and the transaction data is entered directly into the system. An example of this form of system input is the taking of sales orders over the telephone by an order clerk sitting at a keyboard terminal.

- **Input Data Communications** The input data communications element exists when system input occurs some distance from the location of the processing. In this case, controls are implemented to ensure that data is completely and accurately transmitted from the sending nodes to the central computer, with no loss of security. An example of an input data communications control is the use of hardware and software that detects electronic transmission errors.

- **Data Processing** The transaction data enters the system either from the system input element or input data communications. The control objective is to ensure that the data is completely and accurately processed. An example of a data processing control is a program module that compares entered item numbers against a master list to ensure that the numbers are those of the firm's products.

- **The Database** Data processing often requires access to the database. Database controls have the objective of ensuring that the database is an accurate reflection of the physical resources and activities that it represents, and that its contents are made available to only authorized users. An example of a database control is the use of a user directory that identifies those persons authorized to retrieve data.

- **The Software Library** The **software library** is a collection of all the firm's computer programs. An example of a software library control is the policy of maintaining master copies of all approved programs in a vault and periodically comparing all operational programs to those masters. This policy is intended to detect instances when programs are illegally modified after they become operational.

- **Output Data Communications** The data processing output must be transmitted by means of output data communications when the output occurs in a location that is remote from the processing—for example, if a computer in Des Moines prints out a picking ticket in a warehouse in Cedar Rapids. The same controls can be used for output data communications that are used for input.

- **System Output** System output controls are intended to guarantee that the results of the processing are reported to only those who are authorized to receive them. An example of an output control is a routing sheet that accompanies a report mailed to a user. The user signs the sheet and returns it to information services as verification that the report was received.

Each of the eight system elements can be further subdivided into processes, and controls built into each process. The following sections describe each process and provide examples of the controls. The controls described here are those that would be found in a system such as payroll, where the using departments originate the transactions and transmit the data to information services by means of a data communications network. Information services processes the data and transmits the output to the using departments. Although a different set of controls would be devised for another type of system, such as a cash flow model used by a financial analyst, the same system elements would be present.

Transaction Origination Controls

Transaction origination consists of the four processes shown in Figure 12.8.

Originate Source Documents   For those systems triggered by source documents, controls can be established that relate to document design, document acquisition, and document storage. A system that computes overtime earnings is an example of a document storage control. The forms that the supervisors use to authorize the overtime are kept in a locked file cabinet until needed.

Authorize Processing   Transactions must be approved before they are processed. This approval is accomplished by setting limits on each manager's authority and requiring signatures on source documents. Some firms, for example, require that the purchasing director sign all purchase orders for large amounts.

Prepare Computer Input   Source documents are checked for accuracy and completeness. Documents that do not pass these checks are routed to error handling. For the documents that do pass the checks, one or more control files are established before the documents go to transaction entry. A **control file** is a collection of records created to establish a control over some process.

One type of control file is a **transaction log,** which captures certain key, identifying information about each transaction. For example, in an order entry system, each approved order is documented in an order log, which contains data such as customer number, customer order number, customer order date, and date received.

Another type of control file consists of a set of batch totals. A **batch total** is a total accumulated on a particular data field for a batch of source documents. Batch sizes of approximately 25 documents are common, and each batch may be represented by more than one batch total. Batch totals can be computed on amount fields, such as hours worked, unit price, and sales quantity. Some batch totals are accumulated on fields not normally totaled, such as employee number. These batch totals are called **hash totals;** they serve only a control function.

Handle Source Document Errors   Any errors detected during input preparation are corrected. The process of correcting errors in a computer-based system frequently entails the use of a suspense file. A **suspense file** is a temporary repository where records are retained until a certain action occurs, at which time the

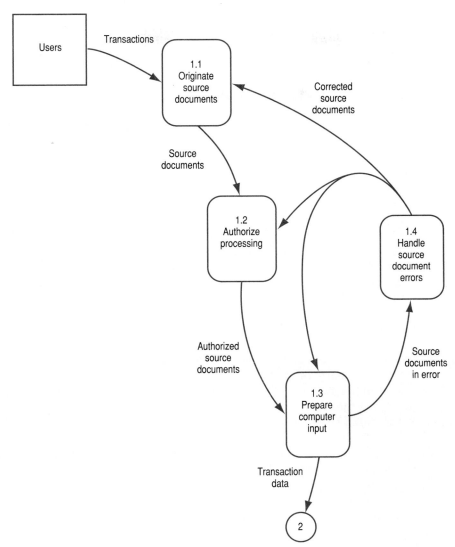

Figure 12.8
Transaction origination.

records are removed. In an order entry system, for example, sales orders are held in a Sales Order Suspense file until customers are contacted to clear up questions concerning item numbers, order quantities, and so on. As shown by the diverging data flows in the figure, the corrected source documents can reenter the system at any of three points, depending on the type of error.

The circle containing the 2 at the bottom of the figure indicates that the transaction data flows to system element 2.

System Input Controls

This is the point in the system where the input data is entered into the computer. Figure 12.9 shows the five processes involved.

Enter Data   Data entry operators or users use input devices to enter transaction data into the computer. Or, source data automation, in the form of OCR or MICR, can be used. A written procedure describing the processes the operator is to follow is an example of a data entry control.

Verify Data   When a high degree of accuracy is required of the input data, such as in financial applications, the data should be verified. This verification comes after data entry but before processing.

The simplest form of verification is for the data entry operator to visually verify data on the screen before it is entered. This is called **sight verification.** The main shortcoming of this approach is that the procedure is difficult to enforce. The operator might get in a hurry and not verify everything.

Another, more effective, approach is **key verification,** where a second person repeats the data entry, keystroke by keystroke, and the computer compares the two inputs. Any mismatch is an error in either entry or verification.

Any errors detected by the verification process are routed to error handling.

Handle System Input Errors   The system produces output—either a screen display or a hardcopy printout—that identifies the error records and error types.

For online systems, the data entry operator responds to the screen display by correcting the errors and reentering.

For batch systems, clerical personnel use the hardcopy to research the errors and determine the necessary corrective actions. The error documents are retained in a suspense file until they are returned to data entry, along with instructions for making the corrections. Data entry operators reenter the corrected data and the verification process is repeated.

Retain Source Documents   Source documents are not destroyed after data entry, but are held for a predefined time. Some documents are held for only a few days or weeks, and others for as long as seven years or more. The documents are filed in a **history file,** which is an accumulation of processed records held for future reference. The history provides an audit trail of the transactions. An **audit trail** is a record that enables each input transaction to be traced to its output, and each output to be traced to its input.

Balance Computer Input   Batch totals are accumulated for all transactions entered into the computer. These totals must balance with those established at transaction origination. When the two batch totals agree, it means that all the originated transactions for that batch have been entered into the computer. When the totals do not agree, the batch receives error handling, and corrected data is reentered. This process is repeated for a batch until its totals balance.

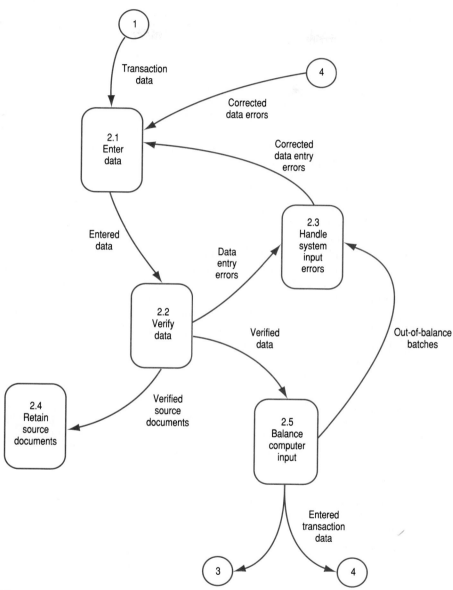

Figure 12.9
System input.

Input Data Communications Controls

All systems do not include data communications elements, but for those that do, the two main processes are message sending and message receiving, connected by a communication channel.[4] These processes are shown in Figure 12.10, along with a third process that maintains a log of all transmissions. These three processes are the same for both system input and output.

Send Messages   We use the term **message** to describe the data that comprises a single data transmission. The message can be a record of a transaction, several records, or a part of a record. Messages can be sent in several ways. The most common is to key the messages into the keyboard of a terminal. Higher speeds can be achieved by reading the messages from magnetic disk, magnetic tape, or the primary storage of a computer.

The most basic control over message sending is the physical security of the room containing the sending equipment, and the equipment itself. If possible, the room should be kept locked. It is also possible to lock the terminals or install them in lockable steel cabinets bolted to the floor.

The Communication Channel   The sender and receiver are linked by means of a communication channel. For long-distance transmission, this channel consists of circuits owned by a common carrier such as AT&T, GTE, or Sprint. The common carriers provide circuits in the form of telephone lines, coaxial cables, fiber-optic cables, and microwave equipment. For shorter-distance transmission, such as within a local area network, the circuits are owned by the firm and can consist of wires, coaxial cables, or fiber-optic cables.

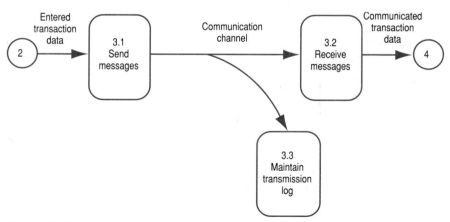

Figure 12.10
Data communications.

[4]For a more detailed discussion of risks and controls related to data communications, see *System Auditability and Control: Module 8 Telecommunications* (Altamonte Springs, FL: The Institute of Internal Auditors Research Foundation, 1991): 31–37, 53–56, 89–91, 94.

The most straightforward way to establish a control over the transmission is to either own the circuit or lease a private line from the common carrier. These circuits are more difficult to tap than a dial-up line shared with other users.

A second control can be established by converting the message to a coded form during the time it flows through the channel. The message is coded when it is sent and decoded when it is received. The term **encryption** describes the process of coding data so that it becomes meaningless to a viewer, such as a computer criminal, who somehow gains access to the channel.

Receive Messages   Controls over message receipt are meant to detect errors caused by the data transmission and to prevent unauthorized persons from gaining access to the computer system.

Error detection can be accomplished by means of extra data bits or characters added to the message. These bits, called **check bits,** and characters, called **check characters,** are computed by the sending equipment, based on the data that is sent. When the message is received, the receiving equipment computes a similar set of bits or characters and compares them with the ones in the message. When the bits or characters do not match, a transmission error is assumed. In that case, the receiving equipment corrects the error or causes the transmission to be repeated until a clear transmission is made.

Gaining unauthorized access to the computer system can be made more difficult by installing a **port protection device** as part of the receiving hardware. This device can verify the legitimacy of the transmission in several ways. One technique requires the sender to provide a password and then hang up the phone. The computer confirms that the password is acceptable by comparing it with a preauthorized list and then directs the port protection device to dial the user's number to establish the connection. This is an electronic version of the same technique that pizza restaurants often use when you order a pizza for delivery and they ask for your phone number. You hang up and the restaurant calls you back with the price. The technique ensures that the order originates from a customer and not a prankster. The port protection device protects the input ports of the computer from the same type of misuse.

Maintain Transmission Log   The physical security of the data communication network can be augmented by logical security. Examples of logical security measures are requirements that users provide passwords before being allowed to transmit data and that all transmissions be recorded in a log. A **transmission log** contains an entry for each message that is sent, identifying the sender, the particular terminal used, the date of the transmission, and the time of day. The transmission log provides an audit trail that is useful for identifying weak points in the network and for tracking computer criminals.

Data Processing Controls

As illustrated in Figure 12.11, only two control points exist within the data processing element: a data handling process and an error handling process.

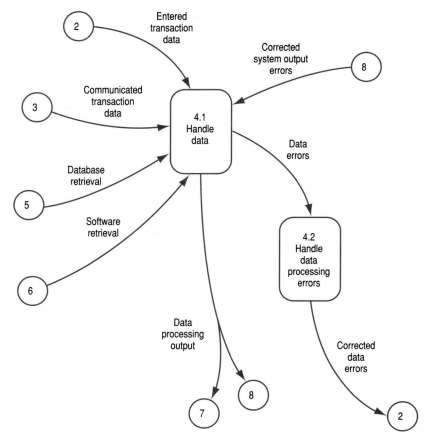

Figure 12.11
Data processing.

Handle Data   Data handling controls are used at the point in the system where the input data is transformed into the output information. Some data handling controls subject the data to various kinds of tests to detect errors that have escaped the previous system elements. For example, an error module can conduct a **data type check** to ensure that the data is the proper type—numeric or alphanumeric. A similar test is a **field size check** that limits fields to certain lengths. A third check, a **reasonableness check,** compares a data element with upper or lower limits, or perhaps a combination of the two, to ensure that the value is one that would normally be encountered. The reasonableness check is designed to prevent the processing of unintentional errors made by authorized users and intentional acts by computer criminals.

Handle Data Processing Errors   Any failures of the data to pass the various data processing checks are researched. An excellent example of this type of handling is the way that American Express establishes a reasonableness check on credit card charges for each card holder. When a transaction amount exceeds a predetermined limit, the card holder is called to verify that the transaction amount is correct.

If the error research calls for corrections, they are made, and the transaction data is returned to the transaction entry element.

Database Controls

We have seen how important data communications security is to maintaining the integrity of the system. If the data communications controls are somehow bypassed or made ineffective, the next major control point is the database. Because the database is the conceptual representation of the physical system of the firm, management takes every precaution to keep the data secure.[5]

The database is controlled in three basic ways. First, and most important, access is restricted to only authorized users. Second, a backup database is maintained, and third, statistics concerning database use are provided. Figure 12.12 shows these three processes.

Control Database Access   Database security has been likened to a series of barbed-wire fences, one inside another with the database in the center. To gain access to the database, a person must go through gates in each of the fences. These fences illustrate the hierarchical nature of database controls. This hierarchy includes, starting at the bottom, passwords, a user directory, a field directory, and encryption.

- **Passwords** Database management systems usually require the user to enter a special database password as the first step in gaining database access. This password is in addition to the one checked by the operating system when the user logs onto the system.

- **User Directory** A list of authorized database users is maintained in secondary storage and brought into primary storage when needed. A database user can be a human or a computer program.

- **Field Directory** Another list is maintained in secondary storage of the data fields in the database that the user is authorized to use and the type of processing permitted. Some users are allowed only to retrieve data elements. Others are allowed to change retrieved elements.

- **Encryption** When data is entered in the database, it is converted to a coded form, using an encryption key different from the one used in data communica-

[5]For more information on database controls, see *System Auditability and Control: Module 5 Managing Information and Developing Systems*, (Altamonte Springs, FL: The Institute of Internal Auditors Research Foundation, 1991): 40–48.

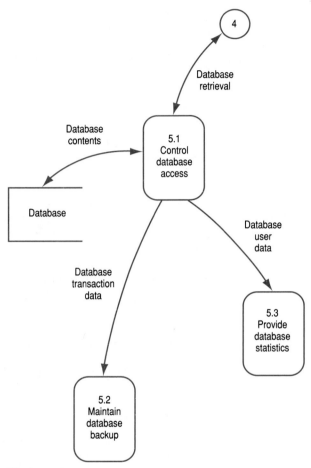

Figure 12.12
Database.

tions. Then, if an unauthorized user is somehow able to get through the three outer fences, the data is unusable. Authorized users have access to the encryption key that converts the encrypted data to its original form.

This hierarchy of controls is intended to make data available to only those persons who need it and to protect the data from damage or destruction by computer criminals.

Maintain Database Backup   A backup of the database should be maintained in the event that something happens to the original version. Microcomputer users provide this control by copying their files onto diskettes. For mainframe and mini-computer users, the database management software performs the task.

Provide Database Statistics   The larger-system DBMSs also provide statistics concerning database use. These statistics identify who is using the database, when it is being used, and which data elements are involved. This is useful information to the manager of database administration in monitoring database performance. The statistics often provide the audit trail that leads to the capture of a computer criminal.

Software Library Controls

A firm should be just as sensitive to the security of the programs that process its data as it is to the data itself. Someone intent on causing the firm harm or embezzling funds can accomplish this by modifying the software. One of the first tricks used by computer criminals was to modify banks' interest computation programs so that the fractions of cents left over from rounding operations were deposited in the criminals' accounts.

Figure 12.13 shows the two basic control points for the software library—control over access and control over changes.

Control Software Library Access   The same types of controls used for the database can also be used for the software library. Users can be required to enter passwords and pass user and field directory checks before gaining access to stored software. Finally, the software can be maintained in an encrypted form.

Control Software Library Changes   Management must ensure that unauthorized changes are not made to a program after it is put into use. A step toward

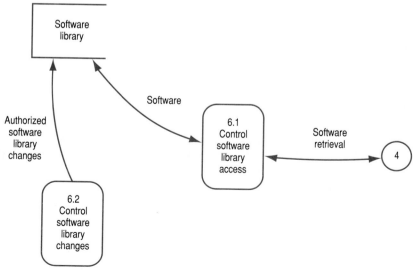

Figure 12.13
Software library.

achieving this objective is to create a **program change committee** that must approve all changes to operational programs. Another precaution is to periodically compare the code of operational programs with masters that were made when the systems first went on the air. The comparison will flag any changes that have been made.

Output Data Communications Controls

The same types of controls are used for communicating system output as are used for communicating system input.

System Output Controls

Controls are established on system output to ensure that all the input transactions are processed and that the output reaches the users. These controls are achieved by means of five processes, as shown in Figure 12.14.

Balance Computer Output   Batch totals for the computer output are compared to those accumulated during system input. Any out-of-balance conditions are handled as errors. This balancing operation is performed by information services to ensure that all the records they received from the users were processed by the computer.

Distribute Output   When the batch totals indicate that all the input records were processed, the output is distributed to the users. An example of a distribution control is the printing of the recipient's name and mail station number in the heading area of a report.

Balance System Output   The user generates a second set of batch totals to ensure that all records sent to information services were processed. These totals are balanced with those from transaction origination.

Handle Output Errors   Any errors detected by the two output batch balancing processes are corrected. The corrections reenter the system in the data processing element.

Retain System Output   Users retain copies of system output in a history file as an audit trail. After a specified length of time, the history file contents are destroyed. Shredding machines are the ideal way to dispose of hardcopy output that is no longer needed. The history of computer crime is full of cases where a printout taken from a wastebasket or a dumpster provided the means for breaching a firm's computer security.

Putting the Systems Design Controls into Context

All members of the project team are dedicated to developing a system that processes its data accurately and completely, provides its output to only authorized users, and keeps its resources secure from damage or destruction. Team members identify the risks for each system element, devise controls to guard against the

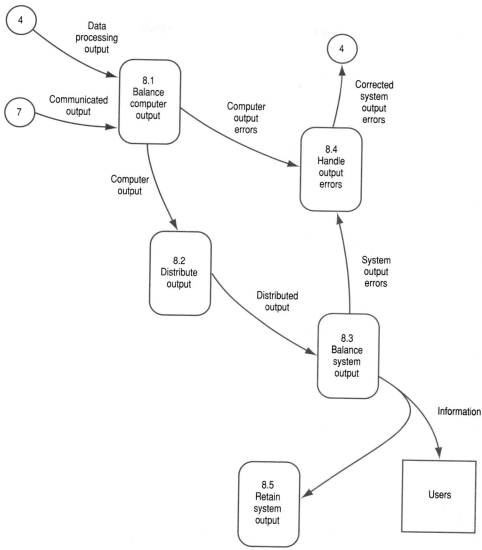

Figure 12.14
System output.

risks, and incorporate the controls into the system design. When the system becomes operational, these design controls combine with systems operations controls to achieve the level of information integrity, security, and compliance that
management desires.

Systems Operation Controls

Two main types of controls are applied to operational systems. One consists of the annual audits, and the other of computer operations policies and procedures.

Annual Audits

The firm's executives authorize annual audits of operational systems conducted by the internal auditors. At the same time, the CIO authorizes annual audits conducted by information specialists. As you will see in Chapter 15, the internal auditors validate system integrity, and the information specialists confirm user satisfaction.

These annual audits in the use phase, combined with the fifteen developmental control points, enable management to control all five phases of the SLC.

Computer Operations Policies and Procedures

Once the system passes its postimplementation reviews, the responsibility for the system is transferred from the project team to computer operations or the user. In the discussion that follows, the types of controls that can be installed by the manager of computer operations are described. Similar controls can be installed by the user but usually are less elaborate.

The operations manager achieves and maintains control of functioning systems in four different ways:

- Organizational structure

- Equipment maintenance

- Environmental control and physical security

- Contingency planning

The first two controls are discussed here. The third is discussed in Chapter 13. Contingency planning is a topic that is more appropriately included in a discussion of information systems management.

Organizational Structure   One of the fundamental tenets of control is the **segregation of duties,** which specifies that the overall workload is subdivided and allocated to more than one person. The rationale is that the greater number of people involved, the more difficult it is to violate the system, and the easier it is to detect a violation when one occurs. Violation requires collusion.

This segregation of duties is seen in banks, where one group of employees handles the opening of new accounts and another handles deposits and withdrawals. This arrangement makes it impossible for one person to open an account for the purpose of embezzling funds, transfer money into the new account from the accounts of depositors, withdraw the transferred funds, and skip town.

The organization of computer operations reflects this same segregation of duties, for the same general purpose. Segregation is intended to make it difficult for someone in operations to violate the system. Figure 12.15 shows a functional grouping of employees that forms the base for the organization of many operations units.

- **Scheduling** This group of operations personnel determines which jobs are run on the computer. These are batch jobs that are run according to a schedule, such as daily or weekly, or on an as-required basis. For a job to be run on the computer, it must be included in the daily run schedule.

- **Data Entry** These are the operations employees who enter data into the computer from source documents. This activity was centralized within computer operations during the early years of computing, but today it is dispersed throughout user areas in many firms. Computer operations has greater control over the centralized data entry unit than over those that are dispersed.

- **Input and Output Control** This group maintains control over source documents and data that enter information services from user areas and leave information services for user areas. The guiding principle of this group is "What flows in, must flow out." The basis for such control is a **batch log** that records the important data concerning each batch of transactions entering and leaving information services. The log identifies the batch, the date and time it was received, and the batch totals. Corresponding entries are made in the log when the system produces the batch output.

- **Media Library** Early computer systems made liberal use of magnetic tape. Over the years, firms accumulated thousands of these reels. Recently, the prac-

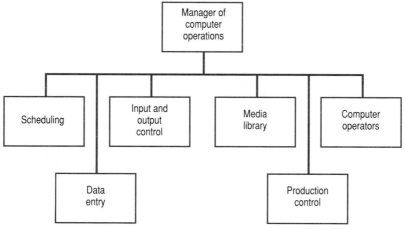

Figure 12.15
Organization of a computer operations unit.

Figure 12.16
A media library.

tice has been to use magnetic tape cartridges. Tape reels and cartridges are stored in a **media library** during the time that they are not mounted on online tape units. In addition to the tape, the media library also often is the storage area for disk packs—a stack of disks that can be mounted on a disk drive for use and then removed when the job is finished. Disk packs enjoyed their peak popularity during the 1970s. Figure 12.16 shows a typical media library.

The media library is operated in the same way as a book library. A librarian checks out reels and disk packs to operations personnel and checks them back in when the jobs are finished. The library procedures and records guard against unauthorized use of secondary storage media.

- **Production Control** This group is responsible for performing scheduled work. The manager of production control assigns computer operators the task of running particular jobs according to the schedule.

- **Computer Operators** These are the personnel who work with the mainframe or minicomputer, operating the console, loading and unloading tape units, changing paper in the printer, and so on.

By organizing the operations personnel in this manner, the manager of computer operations is able to maintain control over the daily activity in the firm's central computing facility.

Equipment Maintenance   In Chapter 14, you will learn about the various ways that a firm can pay for its hardware resources. When a firm rents or leases equipment, the maintenance is included in the rental or lease charge. However, when a firm purchases its own equipment, a separate maintenance contract must be negotiated.

During the early days of computing, all equipment maintenance was performed by employees of the hardware vendors. If you leased or purchased equipment from IBM, an IBM field engineer, or FE, took care of it for you. Over time, this changed. Large users, such as the federal government, formed their own corps of maintenance personnel. Also, other firms, sensing the potential profit in equipment maintenance, competed with hardware vendors for the maintenance business. Today's mainframe or minicomputer user, therefore, has a choice when it comes to maintenance. A firm can obtain it from a hardware supplier or a maintenance firm, or it can form its own staff.

The larger the computer installation, the more maintenance is required. In very large installations, the FEs have their own room next to the computer room, where they house their test equipment, maintenance manuals, spare parts, and so on. One or more full-time FEs can be assigned to maintain the firm's equipment.

The manager of computer operations sees to it that this maintenance is performed and that the system functions in an acceptable manner. The manager coordinates the maintenance schedule with the production schedule to ensure that the maintenance work does not interfere with production. Scheduled maintenance is often performed during odd hours—say 5:00 to 7:00 o'clock each weekday morning. Also, the manager receives reports that reflect the performance levels of the various pieces of equipment. These reports indicate when it might be appropriate to schedule additional maintenance or replace certain units.

The manager of computer operations also sees to it that the equipment performs satisfactorily. The maintenance contract, facilities, schedule, and reports enable the manager to carry out this responsibility.

Putting Operations Control in Perspective

As more and more information resources are distributed to user areas, the centralized computer operations unit loses control over those resources. From the standpoint of the firm, this is a major disadvantage of end-user computing. If control is to be maintained, the user areas must bear that responsibility. The users must apply the same types of control procedures found in the computer operations unit. Establishing responsibility for the control of distributed information resources is a key element in a firm's IRM policy.

The Unique Control Requirements of the CBIS Subsystems

The objective of the project team and computer operations is to protect each system element from its associated risks. However, certain elements are more critical than others, and certain applications expose the firm to greater risks than do others.

Table 12.1 identifies the critical control points for each of the five CBIS subsystems in terms of their system elements. Each cell of the table should receive the attention of the system designers and operations personnel, but the shaded cells represent the points where especially tight controls should be applied.

Data Processing System Controls

Each element of the data processing system is critical because data processing is the system that handles the firm's financial transactions. Anyone intent on embez-

Table 12.1   Key Control Points in the CBIS Subsystems

System Element	CBIS Subsystems				
	Data Processing System	Management Information System	Decision Support Systems	Office Automation Systems	Expert Systems
Transaction origination	▓				
System input	▓	▓	▓		▓
Data communications	▓	▓	▓	▓	▓
Data processing	▓				
Database	▓	▓	▓	▓	▓
Software library	▓	▓	▓		▓
System output	▓	▓	▓	▓	▓

zling funds by means of the computer will attack the data processing system. The data processing systems of financial institutions such as banks and credit card companies are especially high on the hit lists of computer criminals.

Management Information System Controls

The MIS makes problem-solving information available to managers throughout the firm. This information does not interest embezzlers, but it could have great value to anyone engaged in **industrial espionage,** the unethical gathering of data and information by a firm concerning its competitors.

To guard the MIS against industrial espionage, controls should make it difficult for an outsider to enter requests for information, transmit those requests to the central computer, gain access to the database, and receive the information. Many MIS designs, especially executive information systems, maintain critical information in the database in the form of preformatted reports. Access to such information could be especially helpful to competitors.

Decision Support Systems Controls

Although the DSS can produce the same types of information as the MIS, the DSS places more emphasis on mathematical modeling. Access to the modeling software could prove valuable to competitors because the software can represent logic that gives the firm a competitive advantage. For example, the logic incorporated in a DSS used by a construction firm to determine bid amounts would be extremely helpful to competitors.

Office Automation Controls

The critical element in office automation systems is the database where the electronic and voice mailboxes of the firm's executives and managers are stored. Access to these mailboxes, and the ability to retrieve their contents, could prove especially helpful to industrial spies.

Expert Systems Controls

Expert systems have control needs similar to those of DSSs. As with DSSs, the software used by expert systems is a competitive asset, for it represents the knowledge and experience of some of the firm's key specialists. Expert systems that are used in problem solving are targets for industrial spies. For example, a model used in setting prices for the firm's products would be helpful to competitors. In addition, expert systems used in data processing are targets for embezzlers. An expert system used to approve customers' credit purchases, for example, could be modified to commit fraud.

To date, few cases of firms attempting to gain strategic information, or intelligence, from their competitors' computer systems have been reported. However, this is not to say that such activity does not go on or that it is not a potential threat.

Using CASE for System Control

One of the problems encountered in traditional analysis with nonautomated tools is lack of consistency among diagrams. Almost all CASE tools offer support for consistency checking. Consistency checking ensures that the diagrams and data repository entries are complete, do not violate any methodological rules, and are consistent within and among diagrams.

Most CASE tools allow consistency checking for single diagrams or for the entire project. When each diagram is completed, it should be checked for consistency. The same holds true for the project; consistency checking should be conducted on the whole project as the first phase of testing. After the inconsistencies have been addressed, then testing with live and test data can be conducted.

Consistency checking can detect, but is not limited to, the following types of errors:

1. Unnamed or unnumbered modules and other constructs
2. Dangling modules, which have no inputs, outputs, or both
3. Incorrectly placed control structures such as looping structures
4. Lack of balance, such as a data flow that appears on several DFD levels, with different names and perhaps showing the data flowing in different directions
5. Lack of appropriate keys on data structures
6. Lack of data normalization in database designs
7. Lack of data repository entry for a structure that appears on a DFD
8. Lack of a DFD for an entry that appears in the data repository

All these controls are examples of design controls. When the project team takes advantage of the ability of a CASE tool to achieve consistency in system design, the chances of the resultant system's meeting the user's needs are increased.

Summary

Management identifies risks the firm faces, establishes controls to mitigate the risks, and then follows up to ensure that the controls work. When viewed in a systems context, a control consists of a feedback loop and a control mechanism. The feedback loop begins as data, is transformed first into information, and then into decisions. The control mechanism includes standards, management, and the information processor.

A firm's system of internal control consists of control procedures that exist within a control environment. The procedures can be general in nature or specific to particular applications. The trend is toward general controls.

Computer risks include unauthorized tampering, loss of system integrity, and disruption. Computer-related controls have the objectives of information integrity, security, and compliance. Information integrity is achieved through a combination of both data and processing controls.

Systems designers can array the various risks for a new system in a risk matrix. Then a control matrix can be prepared that identifies controls for each risk.

Controls are applied in three main areas—systems development, systems design, and systems operation. The controls in all three areas are intended to keep dynamic processes on course.

Systems development controls are arranged in a hierarchy. At the top is the firm's IRM policy and strategic plan for information resources. This plan is enforced by the MIS steering committee, which works through the project managers. During each SLC, development control is achieved by means of 15 project control points, beginning with the feasibility study and ending with postimplementation reviews.

So many different kinds of systems exist, and so many possible controls, that it is helpful for the systems designers to take a systems view. All systems have some combination of transaction origination, system input, input data communications, data processing, database, software library, output data communications, and system output. Within each of these system elements, controls can be incorporated into the various processes.

The manager of computer operations controls operational systems by means of an organizational structure that features segregation of duties and a thorough program of equipment maintenance. Managers of user areas have the control responsibility for the information resources available to them.

Of all the CBIS subsystems, the data processing system offers the greatest control challenge because it allocates the firm's financial resources. Each system element must be regarded as a potential entry point for computer criminals.

The other CBIS subsystems are targets for industrial spies who seek a competitive advantage. Preformatted reports in the EIS database, the software of DSSs and expert systems, and the mailboxes of office automation systems are the elements of the information-oriented CBIS subsystems that must receive special control attention.

The controls described in this chapter are combined with the security precautions discussed in the next chapter to produce systems designs that protect the firm's valuable information resources.

Key Terms

System of internal control

Control environment

Control procedures

General control

Application control

Migration of controls

Software library

Control file

Transaction log

Batch total

Hash total

Suspense file

Sight verification

Key verification

History file

Audit trail

Encryption

Transmission log

User directory

Field directory

Batch log

Media library

Industrial espionage

Key Concepts

The relationship between risks and controls, and how this relationship can be documented with risk and control matrices

How all computer controls are aimed at three dynamic processes—development, design, and operation

How control of the firm's information resources is a top-down process that begins with IRM and SPIR

Project control points in the SDLC

How the potential cost of a risk can be estimated by computing its expected value

Systems design controls as applied to system elements

Segregation of duties

Questions

1. What are management's three responsibilities in terms of risks?
2. What are the three elements in a firm's control mechanism? What function does each perform?
3. What are the main ingredients of a system of internal control?

4. What is meant by the term *migration of controls?*

5. What are the three main risks faced by a computer-based system?

6. What two types of controls are used to achieve information integrity?

7. What do the rows of the risk matrix represent? What about the columns? Do the rows and columns of the control matrix have the same meaning?

8. If a cell in a risk matrix contains a risk, what should the corresponding cell in the control matrix contain?

9. What are the three main areas where computer controls can be applied? Do these areas deal with dynamic or static processes?

10. Where does the control of systems development begin, in terms of time and place in the organizational hierarchy?

11. Why don't the systems analyst and internal auditor seek to provide 100 percent assurance against systems risks?

12. Which of the eight system elements is not directly involved with the computer?

13. Which system elements would be involved with an application in which a manager uses an electronic spreadsheet on a standalone micro? The manager enters data directly into the computer without recording them on a source document.

14. Give two examples of control files.

15. Would you find a hash total somewhere other than in a restaurant system? Explain.

16. What is the difference between a suspense file and a history file?

17. Should a user verify data entered into an end-user application? If so, how could it be done?

18. What else does an audit trail enable you to do in addition to tracing an output transaction back to its input?

19. Which manager in information services would be the most interested in the transmission log?

20. What are three types of checks that can be conducted on input data of a computer program?

21. What are the four "fences" that are built around the database? List them in order, beginning with the outer one.

22. Which person in information services would be especially interested in receiving a report of database statistics?

23. Name two ways management can ensure that improper changes are not made to operational programs.

24. When information services runs a job for a user, whose responsibility is it to balance the output?

25. What four ways can the manager of computer operations maintain control?

26. What is meant by *segregation of duties?*

27. Name three electronic recording media you would expect to find in a media library.

28. Which CBIS subsystem has the most critical control elements? Why?

29. Which system elements are critical for all CBIS subsystems? Explain why.

Topics for Discussion

1. What can a firm's management do to establish the proper control environment?

2. What do you think is the rationale for the migration of controls?

3. This chapter describes the role of the MIS steering committee in development controls. Does the committee play a role in design controls? If so, what is it? What about operation controls?

4. Are postimplementation reviews considered development controls or operation controls?

5. Suppose that a firm realizes that its output control of printing the recipient's name on the report is not adequate. What else could be done?

6. Could an error suspense file have any value in addition to providing a repository for error records? If so, explain why.

Problems

1. Redraw Figure 12.7 as a system flowchart. Assume that the computer configuration includes online keyboard input, magnetic disk storage, and hardcopy output. Use arrows to represent data communication.

2. Redraw Figure 12.7 as an action diagram. Use the selection construct to illustrate whether data communication is involved.

Case Problem
Cowpoke Creations (A)

You are the new manager of systems analysis for Cowpoke Creations, a large manufacturer of Western clothing located in Calgary, Alberta. You previously worked as a senior systems analyst for a Big Six accounting firm, where you specialized in auditing client's computer-based systems. Cowpoke was one of your clients, and Rae Summerfield, the CIO, liked your work so well that she offered you the management job when it opened up.

Cowpoke is a good example of information resources management. The information services organization is one of the major functional areas of the company, and Rae is a member of the executive committee. Rae works with the other executives in developing the strategic plan for Cowpoke as well as the plans for each of the functional areas. Rae is also chair of the MIS steering committee.

In your first meeting with Rae, she explains that the systems analysis section is in good shape, the previous manager having done a good job of building a staff. As a way for you to get your feet on the ground, Rae wants you to handle a complaint by the sales manager, Harold Hall, that the marketing information system is not functioning as intended. Because the system was implemented within marketing, Rae doesn't know the details. She wants you to call on Harold, find out what is going on, and report back to her.

You: Since I'm new to the organization, I would appreciate it, Mr. Hall, if you could give me a little background on your marketing information system.

Hall: Certainly. It all started about two years ago. I had a meeting with my regional managers, and we decided that we didn't have enough information for setting the annual sales quotas for the sales representatives. We needed historical data about what each rep had sold in past years, plus their projections for the coming year. Only some of the historial data, and none of the projection data, was in the computer. I checked with Rae and she told me that we would have to wait about 15 months before IS could get around to it; she always has a backlog. I kicked it around with my managers and we decided just to go on our own. End-user computing—I think that's what you call it.

You: Right. Do you have your own staff of information specialists?

Hall: We do. We have about a half-dozen systems analysts and an equal number of programmers. We don't have any operators because we always tie into the IS mainframe from terminals. And we rely on IS for database and data communications expertise.

You: Is this the only example of a functional area in Cowpoke implementing its own computer system independent of IS?

Hall: It sure is, but it might not be the last. Shortly after we got started, human resources began talking about doing the same. If we're successful, it could trigger a lot more activity of the same sort.

You: You say "*if*" you're successful. Is there any doubt?

Hall: There most certainly is. That was the reason for my call. I think we've failed miserably and I don't know whether anything can be salvaged.

You: Tell me what the problem is.

HALL: To begin with, none of us—the regional managers or myself—knows how to produce the information that we need. Our analysts did a poor job not only of designing the system but of teaching us how to use it. It is very complicated—user *un*friendly I would say. None of us are computer experts and really don't have time to invest in a big training program. So, every time we need to run the model, we have to ask our computer specialists to do it for us. It is very inconvenient, and not the way things were intended.

You: Did you receive any formal training?

HALL: None to speak of. Our lead analyst gave us a demo, but there are no written instructions.

You: You mention a model. I assume that the system is a mathematical model that computes the quotas. Is that right?

HALL: That's right, and the output is displayed on the screen in the form of a report. That's another thing. We can't get a hardcopy output. And, also, it would be nice if we could produce some graphs. But nobody asked us.

You: What do you mean: "Nobody asked us?"

HALL: Exactly what I said. Our specialists designed the system without checking with us. They used the old reports that we had been preparing manually and simply put them on the computer. Before we knew it, the system had been implemented. It would be too expensive to modify the system to give us what we need. At least, that's what our specialists say.

You: That's amazing. You didn't have any opportunity to approve the design as it unfolded or to exercise any kind of control over the development? For example, did the specialists sit down with you and ask you about your information needs? Or did they ask you how well the system must perform to satisfy your needs?

HALL: None of those things. We had a brief meeting where we mainly talked about the weather, and then they went their own merry way. The next thing we knew, the system had been designed and implemented on the mainframe.

You: Was there any kind of plan, like a schedule of the activities to take place, who would do each one, the start and completion dates, and so on?

HALL: Not to my knowledge. The specialists might have had one, but they never showed it to us.

You: Well, how would you summarize the situation right now?

HALL: In one word, grim. No, seriously, I would say that we have a system that we can't use because it is too complex, and even if we *could* use it, the output would not be in the format that we want.

You: Are you mainly unhappy about the format? Does the information itself appear to be accurate? Have you encountered any errors?

Hall: So far, everything looks accurate. We've spent a lot of time going over the reports, trying to understand them, and I'm convinced that the right numbers are there. It does appear that the specialists gave us a good database, and that was one of the main things we were after.

You: Well, that's good news. Maybe all is not lost. Exactly what do you want from me, and from IS?

Hall: First, I'd like to know where we went wrong so that if we ever do this again we won't make the same mistake. Also, I'd like your recommendation concerning what we should do. Should we try to live with the system as it is now, should we ask our specialists to reengineer it, or should we call on IS to bail us out?

You: Well, let me tell Rae about our conversation, and I'm sure you will hear from her shortly.

Assignments

1. Write a memo to Mr. Hall, for Rae's signature. Identify steps in the system life cycle where adequate controls were not applied. Just list the steps; there is no need to explain them. Use the description of the SLC in Chapter 8 as a guide. Keep in mind that in situations such as this, it is not always necessary to take all the steps. For example, because the intent was to use the IS mainframe, there was no need for a make-or-buy decision.

2. Write a second memo to Mr. Hall, also for Rae's signature, identifying which of his three alternatives you recommend marketing take. Assume that the IS backlog has been reduced to approximately three months. After stating your recommendation, list the advantages and the disadvantages of following such a program.

Selected Bibliography

Cloud, Avery C. "An EDP Control Audit With Teeth." *Journal of Systems Management* 41 (January 1990): 13ff.

Collier, Paul Arnold; Dixon, Robert; and Marston, Claire Lesley. "Computer Fraud: Research Findings from the UK." *Internal Auditor* 48 (August 1991): 49–52.

Crowell, David A. "Control of Microcomputer Software." *Internal Auditor* 48 (April 1991): 33–39.

Gallegos, Frederick. "Audit Contributions to Systems Development." *EDP Auditing* (Boston, MA: Auerbach Publishers, 1991): section 72-01-40, 1–14.

Gilhooley, Ian A. "Auditing Program Change Control Procedures." *EDP Auditing* (Boston, MA: Auerbach Publishers, 1991): section 75-04-20, 1–9.

Karabin, Stephen J. "Application Systems Control Standards." *EDP Auditing* (Boston, MA: Auerbach Publishers, 1991): section 74-04-40, 1–20.

Loch, Karen D.; Carr, Houston H.; and Warkentin, Merrill. "Why Won't Organizations Tell You About Computer Crime?" *Information Management Bulletin* 4 (February 1991): 5–6.

Mullen, Jack B. "Audit and Control of Program Changes: Library Structures and Access Scenarios." *EDP Auditing* (Boston, MA: Auerbach Publishers, 1991): section 74-03-20, 1–19.

Neumann, Peter G. "The Human Element: Inside Risks." *Communications of the ACM* 34 (November 1991): 150.

Perry, William E. "Ensuring the Integrity of the Database." *EDP Auditing* (Boston, MA: Auerbach Publishers, 1991): section 74-01-10, 1–15.

Wilkinson, Bryan. "Systems Development and Design Checklist." *EDP Auditing* (Boston, MA: Auerbach Publishers, 1991): section 74-04-10, 1–17.

CHAPTER 13

Systems Security

Learning Objectives

After studying this chapter, you should:

- Know the meaning of systems security and why the current level of interest in it is so high
- Be aware of the objectives of computer security and the threats that make those objectives difficult to achieve
- Know how a firm controls user access to its data and software resources
- Be aware of how a firm physically safeguards its facilities and hardware resources from both unauthorized acts and natural disasters
- Be familiar with the different types of malicious software threats and the security measures that can be taken to achieve prevention, detection, and recovery
- Know the guidelines systems designers can follow in building security into their systems

Introduction

Systems security consists of all activities that protect the firm's information resources from threats by unauthorized parties. It is necessary to identify system vulnerabilities and then incorporate countermeasures or safeguards that minimize potential damage. Firms are currently exhibiting an increasing interest in security because of the connectivity that characterizes modern systems. Security has three main objectives—confidentiality, availability, and integrity—but must also contend with threats of unauthorized use, disclosure and theft, destruction and denial of service, and modification.

The most effective means of protecting a firm's data and software resources from unauthorized acts is access control, achieved by means of user identification, authentication, and authorization. Users identify themselves by providing information, such as a password, or authenticate identity by using a device such as a smart card. Authorization is established by means of directories.

A firm's computing facilities and hardware are protected by physical safeguards implemented in a top-down manner. First the site is planned, next the

599

building is constructed, then physical access is controlled, and finally the proper room environment is provided.

Recently, systems security has focused mainly on malicious software threats, such as viruses, worms, and trojan horses. Firms enact safeguards intended to prevent this software from entering the system, detect entry when it occurs, and regain normal operation.

Although many people play important roles in systems security, systems analysts and users can apply several design techniques that will help ensure the integrity of the system.

This chapter concludes our description of systems analysis and design from an organizational perspective. The material you have learned will help you in your career as an information specialist or manager.

The Role of CASE in Systems Security

When a CASE tool is operating in a local area network (LAN) environment, controls and security take on very important roles. It is important to ensure that access to modules is controlled to prevent collisions and inconsistencies. Generally, only one user can access a diagram or data repository entry at a time. In a LAN environment, access to the LAN itself can be limited by the LAN management system. In addition, most CASE tools have access privilege levels that can be set by supervisors who have control over such privileges. In both of these situations, access is controlled by passwords and node identification. Node identification involves recognizing that a particular workstation has network privileges.

Access privileges can also be set for the target information system. Lists of full-privilege users who can read and write to databases and users with read-only privileges can be established to control access to the target system.

This type of access control is important during both the four developmental phases and the use phase. During development, many hours of labor can be lost if an unauthorized user gains access and either intentionally or inadvertently destroys or alters work that has been done. During the use phase, when CASE is being used to maintain or redesign the system, the need for the same type of security holds.

What Is Systems Security?

Systems security refers to the protection of a firm's information resources from threats by unauthorized parties. Systems security is not limited to hardware, but also includes physical facilities that house the hardware, software, data, users, information specialists, and administrators who manage these information resources. Security can be compromised by bad systems design, imperfect implementation, weak administration of procedures, or accidents.[1]

A firm implements an effective computer security program by first identifying the vulnerabilities of its information resources and then taking required countermeasures and safeguards. A **vulnerability** is some aspect of a system that leaves it open to attack. A **threat** is a hostile party with the potential to exploit that vulnerability and cause damage. A **countermeasure** or **safeguard** is an added step or improved design that eliminates the vulnerability and renders the threat impotent.[2]

Why the Current Interest in Systems Security?

Interest in systems security is higher now than ever before, because:[3]

- Firms are heavily dependent on information systems for their critical operations

- Most systems today feature online access from users located throughout the firm

- Applications that feature electronic data interchange (EDI) allow other organizations to have access to the firm's valuable information resources

In general, the increased importance of computer security is necessitated by the expanding connectivity of the firm's computing systems. **Connectivity** refers to the linkage of computing devices, allowing them to share data and information without human intervention. As shown in Figure 13.1, the computers located in a firm's central computing facility are only one piece of the security puzzle. Also included are computing facilities located in the firm's functional areas, remote workstations in practically every department, and both networks and personal computers in the firm's environment.

This interest in security is expected to increase. During the 1990s, certain technological trends will impose additional demands on systems control and security. Table 13.1 lists these trends, along with the risks that they impose.

Data communications networks will become more widespread, the trend to distributed databases will continue, firms will rely on prewritten software to a greater extent, expert systems will contain the knowledge and experience of the firm's most valuable employees, and CASE tools will provide a computer-based repository for system documentation. Anyone intent on harming a firm through its computer system can transform any of these technologies into vulnerabilities.

[1]*Computers at Risk.* (Washington, DC: National Academy Press, 1991): 7.
[2]*Computer at Risk,* p. 13.
[3]*Systems Auditability and Control: Module 1 Executive Summary* (Altamonte Springs, FL: The Institute of Internal Auditors Research Foundation, 1991): 2.

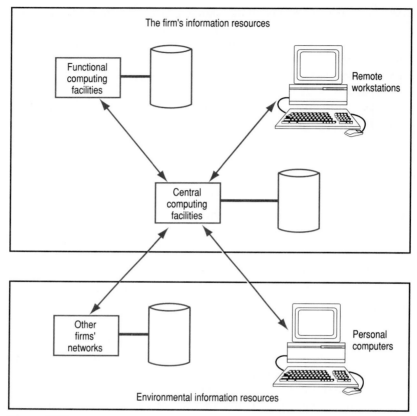

Figure 13.1
The expanded connectivity of computing devices increases the need for systems security.

Security Objectives

A firm installs security countermeasures or safeguards to achieve three main objectives: confidentiality, availability, and integrity. Figure 13.2 shows how security threats affect a firm's ability to realize these objectives.

Confidentiality   The firm seeks to protect its data and information from unauthorized persons, especially confidential material relating to the personal privacy of its employees.[4] The Privacy Act of 1974 outlines the major principles that the firm must follow:

---

[4]*Computers at Risk,* 66–67.

Table 13.1   Trends in Technology Will Stimulate Increased Interest in Systems Security

Technology Trend	Risks
Widespread data communications networks	Spread of viruses Unauthorized access Lack of control of end-user computing Environmental exposures due to electronic data interchange
Distributed databases	Lack of security Bad data due to lack of input controls
Prewritten software	Spread of viruses Inadequate controls
Expert systems	Over-reliance on the system Decisions made based on inaccurate, inconsistent, or outdated information
CASE tools	Uncoordinated tools used to develop applications, which may result in faulty and costly designs Opportunity for computer criminals to gain access to system documentation

Source: Systems Auditability and Control: Module 1 Executive Summary (Altamonte Springs, FL: The Institute of Internal Auditors Research Foundation, 1991): 11–12.

1. There must be no personal data recordkeeping system whose very existence is secret.
2. There must be a way for individuals to find out what information about them is in a record and how it is used.
3. There must be a way for individuals to prevent information obtained about them for one purpose from being used or made available for other purposes without their consent.
4. There must be a way for individuals to correct or amend a record of identifiable information about them.
5. Any organization creating, maintaining, using, or disseminating records of identifiable personal data must assure that data are used as intended and must take precautions to prevent misuse of the data.

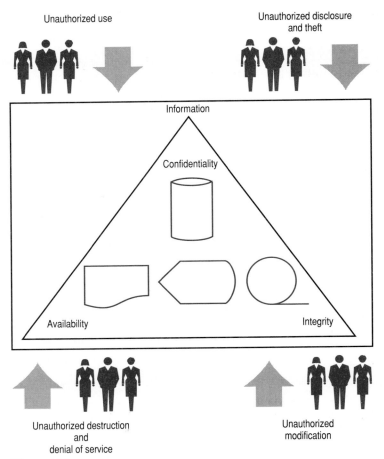

Figure 13.2
Unauthorized parties threaten systems security.

All firms are responsible for maintaining the confidentiality of personal data and information in their databases. Human resource information systems (HRISs) must safeguard information describing the firm's employees. Data processing subsystems, such as accounts receivable, purchasing, and accounts payable, must safeguard elements in the firm's environment.

Availability   The firm seeks to make its data and information available to those authorized to use it. This responsibility is given to the data processing system in terms of the firm's environment and to the other CBIS subsystems for the firm's employees.

Integrity   The firm seeks to maintain a conceptual information system that accurately reflects the physical system. This security objective is met by implementing controls in systems development, design, and use, as described in Chapter 12.

Historically, firms have spent the most money on maintaining confidentiality and the least on achieving integrity. However, availability is receiving increased management attention.[5]

Security Threats

As the firm goes about achieving its security objectives, four main threats stand in the way. The threats all take the form of unauthorized acts.

Unauthorized Use   Persons who are not ordinarily entitled to use the firm's resources are able to do so. For example, a computer criminal breaks into a firm's computer network, gains access to the telephone network, and makes unauthorized long-distance calls.

Unauthorized Disclosure and Theft   The contents of the firm's database and software library are made available to persons not entitled to have access. Industrial spies can gain valuable competitive information, and computer criminals can embezzle the firm's funds.

Unauthorized Destruction and Denial of Service   Persons can damage or destroy the hardware or software, causing a shutdown in the firm's data processing. This can be accomplished without setting foot inside the computer room. Computer criminals can log onto the firm's computer network from remote terminals and damage monitors, cause disks to crash, jam printers, and disable keyboards.

Unauthorized Modification   Changes can be made to the firm's data, information, and software, causing users to make the wrong decisions when using the system's output.

When one thinks of systems security, it is common to think of only attacks by computer criminals intent on stealing a firm's funds. As you can see, that threat is only one part of the security problem.

Logical and Physical Safeguards

Systems security can be achieved by means of both logical and physical safeguards. **Logical safeguards** are those built into the conceptual system to limit access to the database and software library. **Physical safeguards** are those incorporating physical security measures that limit access to the computing facilities and hardware.

[5]Dan White, Ernst & Young, in an address at the ACM Computer Security Seminar, Phoenix, Arizona, October 8, 1991.

The Role of Logical Safeguards in Systems Security

The key to protecting data and software by means of logical safeguards is access control. The reasoning is simple: If unauthorized persons are denied access, then the resources can be kept secure.

Access control is achieved in three ways—by user identification, by user authentication, and by user authorization. The position of these functions in the firm's overall security plan is illustrated in Figure 13.3.

- **User Identification** A person can be identified as one who is authorized to use the firm's information resources by providing information they *know*, such as a password. Identification can also include the user's *location*, such as a telephone number or network entry point.

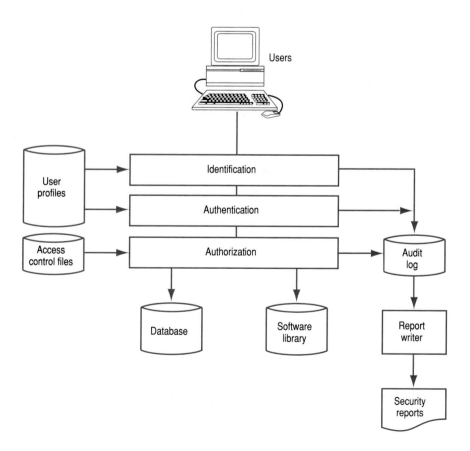

Figure 13.3
Access control functions.

- **User Authentication** Once a person has been identified, their authority to gain access to and use the information resources can be verified by a device they *have*, such as a smart card, token, or identification chip. User authentication can also be accomplished by a signature, or a voice or speech pattern.

- **User Authorization** With the identification and authentication checks passed, a person can then be authorized certain levels or degrees of use. This control is typically achieved by means of access control files stored in the computer's secondary storage. The access control files take the form of user and field directories. A **user directory** lists those persons authorized to have access to the system, and a **field directory** authorizes users to have access to and perhaps modify certain fields, or data elements, in the database.

These three access control functions are intended to limit system access to authorized users and to limit system use to only those processes to which the authorized users are entitled.

Passwords

Passwords have been widely adopted by both using firms and suppliers of prewritten software as the primary means of user identification. A **password** is a small number of characters, usually eight or fewer, typically selected by the user, which identifies that person as one who can have access to some portion of the firm's computing facilities.

Although passwords are extremely popular as a security measure, they do not provide foolproof protection from a resourceful computer criminal. Studies have shown, for example, that when choosing passwords, people have a tendency to select:[6]

- English words
- English words spelled backwards
- First names, last names, street names, city names
- The above with the first letters in uppercase
- Car license plate numbers
- Room numbers, social security numbers, or telephone numbers

When the computer criminal knows particular users, the chances are good that some of the users' passwords can be guessed.

Protecting Passwords

Because user passwords are, in effect, the keys to the computer system, they must be protected.[7] The firm can incorporate the following guidelines into its

[6]Ian H. Witten, "Computer (In)security: Infiltrating Open Systems." *Computers Under Attack,* edited by Peter J. Denning (New York: ACM Press, 1990): 110.
[7]Peter J. Denning, "The Internet Worm." *Computers Under Attack,* edited by Peter J. Denning (New York: ACM Press, 1990): 197–198.

overall security policy to achieve the greatest amount of security effectiveness from passwords.

- Assign a password to every account.
- Educate users concerning the proper selection of passwords. Users should be encouraged to design passwords so that they are difficult to determine. For example, passwords can contain punctuation marks. This education can be accomplished, in part, by including a section on computer security in user manuals.
- Check each new password to ensure that it cannot be easily guessed. Software routines are available to accomplish this.
- Keep passwords stored in an encrypted form, and the file containing the passwords inaccessible.
- Use an encryption algorithm that takes time to run—one second or more. This makes it expensive for criminals to engage in electronic password guessing.
- Augment passwords with authentication and authorization screens.
- Pay particular attention to networks that include mutually trusting computers. No computer outside the firm's network of processors should have unrestricted access, and none of the firm's computers should be able to give access to an outside computer.

Passwords have been used for approximately 30 years as a means of controlling computer access. By enacting policies and procedures governing password use, a firm can provide a sound basis for access control.

The Role of Physical Safeguards in Systems Security

Physical safeguards are appropriate for protecting computing facilities and hardware not only from unauthorized acts but also from natural disasters. Physical safeguards are achieved by proper site planning, building construction, physical access controls, and environmental controls.[8]

Site Planning

When the firm has the choice, it should locate its computing center in a remote area, away from large cities and airports. The site should make it inconvenient for a computer criminal to come into direct contact with the information resources. Also, the site should minimize the likelihood for natural disasters, such as earthquakes, floods, and hurricanes. However, the site should be one that is easily serviced by utility companies, particularly communications common carriers.

[8]This section is based on *System Auditability and Control: Module 4 Managing Computer Resources* (Altamonte Springs, FL: The Institute of Internal Auditors Research Foundation, 1991): 35–41.

Building Construction

The building that houses the hardware should have few windows and doors, with no access through ceiling crawl spaces, air vents, trash chutes, or mail slots. The outside grounds should be free of trees, shrubs, and other obstructions that could provide cover for intruders. The computer room should be located near the center of the building, and the walls should be heavy masonry rather than sheetrock so as to make penetration difficult.

Physical Access Controls

Access to the computer room should be controlled by both security guards and door locks. Security guards should utilize closed-circuit television to monitor activity outside the building, inside the building around the computer room, and in the computer room. Door locks can require the person wishing to gain access to insert a magnetically encoded ID card or enter a lock combination into a keypad. More sophisticated **biometric systems** identify persons by means of unique physical attributes, such as fingerprints, palm prints, voice patterns, vascular patterns on the backs of eyeballs, or even lip prints.

Environmental Controls

All rooms housing hardware and data should maintain the temperature and humidity within specified ranges. Different rooms can have different air pressures so as to control the flow of contaminants away from vital units, such as the CPU. An **uninterruptable power supply,** or **UPS,** should provide power long enough for the operations staff to execute an orderly shutdown in the event of an outage. In addition, an **emergency power supply,** or **EPS,** can provide electricity for longer periods by means of auxiliary generators. Policies should prevent smoking, eating, or drinking in the computer room. Combustible materials should be stored away from the hardware, and furnishings should be fire resistant. Fire detectors should be located not only on the ceilings and walls, but also under the raised floor.

Physical security is achieved by following an orderly, top-down strategy. First, a decision is made concerning the geographic area of the facility. Then, attention is paid to the building itself. Finally, details such as equipment and fixtures within the computer room are worked out.

Malicious Software Threats

Computer criminals often breach the security of a computer system by means of malicious software. **Malicious software** consists of complete programs or segments of code that perform functions not intended by the owners of the hardware. Most everyone has heard of computer viruses. In addition to this well-known type of malicious software, there are others, such as trapdoors, logic bombs, trojan horses, worms, and bacteria or rabbits.[9]

Trapdoors

A **trapdoor,** also called a **back door,** is code incorporated into an application program, usually by the programmer, to grant special privileges, such as bypassing user authentication or setup steps. The programmer inserts the code to facilitate debugging, and the code is invoked by some unique combination of input or a certain user password or identification code. The trapdoor becomes a security threat when it is left in the software after cutover and someone other than the initiator exploits it for criminal acts.

Logic Bombs

A **logic bomb** is code inserted into an application program, usually by a disgruntled programmer who seeks to harm her or his employer. The logic bomb looks for some particular signal and then executes processes that are not normally a part of the program. The signal might be a particular date or some combination of transaction data. When the bomb is triggered, it can modify or destroy data or software, halt the system, or cause other types of damage.

Trojan Horses

According to legend, a large, hollow, wooden horse was left outside the gates of Troy. After it was brought inside, Greek soldiers emerged from its belly during the night and opened the gates to their army, which destroyed the city. Today, anything that destroys an organization from within by means of misplaced trust is called a **trojan horse.**[10]

The key feature of a software trojan horse is that it is embedded in an application program and causes damage without giving any indication that the damage is taking place. The user sees the normal system interface and assumes that everything is going as planned; however, the code is accomplishing unauthorized processes, such as deleting or altering files. Table 13.2 lists a variety of trojan horses, along with brief descriptions.

All trojan horses are not harmful. One, called the Cookie Monster, displayed the message "I want a cookie" on the screen while someone was running a program. If the user did nothing, another message was displayed: "I'm hungry. I really want a cookie." This continued until the user typed the word *Cookie.*

An example of a trojan horse that caused damage is one that was left in a firm's software library by a terminated employee. The routine was implemented in a

---

[9]Most of the descriptions of these topics are based on Eugene H. Spafford, Kathleen A. Heaphy, and David J. Ferbrache, *Computer Viruses: Dealing with Electronic Vandalism and Programmed Threats* (Arlington, VA: ADAPSO, 1989).
[10]Witten, p. 117.

Table 13.2   Examples of Trojan Horses

DISKSCAN.EXE

This is a *PC Magazine* program to scan a (hard) disk for bad sectors; a joker edited the program to write bad sectors. Look for this under other names, such as SCANBAD.EXE and BADDISK.EXE.

EGABTR

The description says something like: "Improve your EGA display." But, when run, it deletes everything in sight and prints: "Arf! Arf! Got you!"

ELEVATOR.ARC

This poorly written program suggests in the documentation that you run it on a floppy. If you do not run it on a floppy, it chastises you for not reading the documentation. Regardless of which disk you run it on, ELEVATOR will erase your files and may format disks as well.

RCKVIDEO

This is another program that does what it's supposed to do, then wipes out hard disks. The program shows a simple animation of a rock star, then erases every file it can access. After about a minute of this, it will create three ASCII files that say: "You are stupid to download a video about rock stars," or something to that effect.

SEX-SNOW.ARC

This program deletes all the files in your directory and creates a gloating message, using those filenames.

SIDEWAYS.COM

A perfectly legitimate version of SIDEWAYS.EXE is circulating as well as the trojan horse version, SIDEWAYS.COM. Both advertise that they can print sideways, but the trojan horse version will trash a (hard) disk's boot sector instead. The trojan's ".COM" file is about three KB, whereas the legitimate ".EXE" is about 30 KB.

Source: Computers Under Attack, edited by Peter J. Denning (New York: ACM Press, 1990): 376–380.

graphics package and printed a small teddy bear somewhere on the paper, outside the area occupied by the graph. The trojan horse was so well hidden that it was found only when the terminated employee's replacement telephoned the employee and begged her to remove it.

Worms

A **worm** is a program that is transmitted from one computer to another on a network, replicating itself and overburdening the computers, denying service to authorized users. The idea of a software worm came from John Brunner's *The Shockwave Rider*.[11]

A worm does not contain code that alters data or software; however, it can transmit other code that does. The first worm was created in the early 1980s by John Shoch and Jon Hupp of Xerox PARC (Palo Alto Research Center) as a means of performing useful tasks using idle workstations on a network. Shoch and Hupp recalled reading Brunner's description and named their programs worms.

The most famous worm was the work of Cornell graduate student Robert Morris, who, on November 2, 1988, released a worm into the ARPANET network. ARPANET was the first major computer network, which was disbanded in 1989 but established standards that have been incorporated into many other worldwide networks.

In a period of eight hours, the Morris worm spread to between 2,500 and 3,000 VAX and Sun computers attached to the worldwide Research Internet. The worm, which has become famous as the Internet worm, caused many computers to be shut down by replicating itself and clogging the systems. No data files were destroyed, and most systems were restored to their original states within two or three days. Morris was suspended from the university, brought to trial, and convicted of violation of the Federal Computer Privacy Act of 1986.[12]

Bacteria or Rabbits

Bacteria or **rabbits** are similar to worms in that they clog the single system in which they are housed by replicating themselves. They are one of the oldest software means of taking down a system, having first been used on mainframe systems. Bacteria or rabbits generate multiple copies exponentially, quickly taking up all the available processor capacity, primary storage, or disk space.

Viruses

Finally we have the computer virus. A **virus** is a piece of code attached to a program, called a host, so that the virus code is executed whenever the host runs. The virus can be attached to any type of program—an operating system, an application program, a shell command file, a macro in a spreadsheet or word processing program, and so on. When the program is run, the virus code is executed and plays tricks or causes damage. In addition, the code can reproduce itself in another program, causing the virus to spread.

If the viral code copies itself only once a day, more than 50 copies will exist after a week because the copies can reproduce themselves. And, after a month,

---

[11]John Brunner, *The Shockwave Rider* (New York: Harper & Row, 1975): 257–260.
[12]Denning, p. 191.

about a billion copies will exist. If the virus reproduces once a minute, it takes only about half an hour to make a billion copies.[13]

Thus far, the primary targets of viruses have been personal computers, and the most popular means of dissemination has been diskettes. When a virus-infected diskette is loaded into a PC, the virus attacks that computer. Then, any diskettes produced by the computer can also contain the virus and be transported to other PCs. Viruses may also be disseminated via networks.

The first recorded instance of a computer virus occurred in the mid-1970s, when a program attacked the idle workstations in a Silicon Valley research center's computer network. The program displayed random patterns on the screens, disabled the keyboards, and clogged the network. It was years later, in 1984, that Fred Cohen of the University of Cincinnati first applied the term *virus* to software.

The first virus outside a computer laboratory setting was detected in January 1986.[14] The virus was called the Brain virus because it wrote the word *Brain* on the disk label of all the files it infected. In addition, it wrote two names and an address in Lehore, Pakistan. For this reason, the virus is also called the Pakistani virus.

In December 1987, a virus attacked the IBM internal communications network by drawing Christmas trees on the screens. It replicated itself many times and clogged the network, but it was detected and quickly removed before any damage was done. The significance of the IBM Christmas tree virus is the fact that it is believed to have been attached to a data file, whereas viruses are usually attached to programs.

In August 1989, fewer than 100 viruses existed.[15] By 1991, the number had reached over 1,000.[16] Each is identified by one or more unique names. Table 13.3 lists several IBM PC viruses that existed in 1989 and includes a brief description of each.

Virus Symptoms   You should suspect that your computer has been infected by a virus when:[17]

- The size of programs or data files suddenly increases or the amount of available disk space suddenly decreases.

- The operating system behaves oddly. Output is lost or garbled, or disk accesses fail. Also, system performance may slow down considerably as the virus traps service interrupts.

[13]*Computers at Risk*, p. 267.
[14]There is some disagreement concerning this date, which is reported by Spafford, Heaphy, and Ferbrache in *Computer Viruses*, p. 16. Harold Joseph Highland, in Denning, p. 293, identifies the date as October 22, 1987, and the place as the Academic Computer Center at the University of Delaware in Newark.
[15]Eugene H. Spafford, Kathleen A. Heaphy, and David J. Ferbrache, "A Computer Virus Primer," in *Computers Under Attack*, edited by Peter J. Denning (New York: ACM Press, 1990): 340–342.
[16]Ken Cutler, American Express, in an address at the ACM Computer Security Seminar, Phoenix, Arizona, October 8, 1991.
[17]"A Computer Virus Primer," pp. 329–330.

Table 13.3   Some IBM PC Viruses

Name	Identifier	Type
405	405 virus	Overwriting
Brain	Brain virus	Boot sector
Clone	Brain variant	
dBASE	dBASE virus	Memory resident
FuManchu	FuManchu virus	Memory resident
Israeli	Israeli virus	Memory resident
Italian	Italian virus	Boot sector
Pakistani	Brain alias	
Ping Pong	Italian alias	
Shoe	Brain variant	
South African	South African virus	Transient

Source: *Computers Under Attack*, edited by Peter J. Denning (New York: ACM Press, 1990): 340–341.

- You notice an unusually large number of disk accesses (as the virus attaches itself to new hosts). Sophisticated viruses do not show this symptom because they piggyback normal disk accesses.
- The application program behaves oddly. In some cases, the program behaves differently from machine to machine or from disk to disk.

Recovery from a virus is best accomplished by using **anti-viral software,** which deactivates a virus in a program and restores the program to its original condition. Table 13.4 identifies some examples of anti-viral software.

Security Measures Against Malicious Software

Firms are not at the mercy of computer criminals who use malicious software to breach systems security. They can take particular measures to prevent and detect malicious software, and to correct damage.[18]

Prevention   The best security measure is to prevent malicious software from entering the system. Access control aimed specifically at malicious software can be accomplished by taking the following measures:

- Educate employees concerning the potential damage that can be inflicted by malicious software, and establish procedures that can be followed to minimize the likelihood of such damage.

[18]*Computer Viruses*, pp. 31–42.

Table 13.4   Anti-Viral Software

VIRUSCAN

VIRUSCAN is designed to scan hard disks and floppies for infected programs. It can detect most IBM PC viruses.

FluShot+

FluShot+ installs itself as an interrupt monitor at boot time. It watches disk activity on IBM PCs and warns the user of suspicious activity. It monitors for some known virus signature activities and has options for encryption and checksum files.

Virus Rx

Virus Rx is a free program distributed by Apple Computer and is available from many merchants and bulletin boards. It detects INIT 29, nVIR, SCORES, and variants of nVIR on Apple Macintosh systems. Virus Rx has detected the presence of new viruses with its built-in self-checking feature. If it detects a change in itself while it is running, it alerts the user and changes itself into a document with a name asking the user to throw it away. This prevents Virus Rx from becoming a carrier if something infects it while running.

Source: Eugene H. Spafford, Kathleen A. Heaphy, and David J. Ferbrache, *Computer Viruses* (Arlington, VA: ADAPSO, 1989): 86.

- Establish policies that govern the use of (1) foreign software obtained from environmental sources such as electronic bulletin boards, and (2) anti-viral software.

- Establish procedures that prescribe the action to take in (1) reporting a malicious software attack, (2) obtaining help in disinfecting the software, (3) recovering from an attack by using backup files and memory dumps, and (4) assessing the damage that has been inflicted.

- Reduce the use of shared software to a minimum. Discourage users from using game software on the firm's computers, and obtain important software only from reliable sources.

- Establish a **quarantine station,** a system that is separated both physically and electronically from other systems and is used to check incoming software before it is put into use.

- Design networks that include **diskless nodes,** systems that do not have disk drives. In such networks, all software is housed in the file server, and new software is introduced only in a controlled environment, such as a quarantine station.

- Safeguard diskettes. Never boot a system using an original diskette; always use a copy. Keep original diskettes write protected and properly labeled. Keep all diskettes in a secure place, and do not share them with other users.

Detection   Additional safeguards can be enacted to detect malicious software that somehow enters the system:

- Be alert to malicious software symptoms. Because some software is triggered by a particular date or time of day, reserve a stand-alone system with its clock set a week ahead of time to detect the software before it has a chance to cause any real damage.
- Use checksums to detect added malicious code. **Checksums** are special totals computed from program code data. When a computer criminal adds malicious code, a different checksum is produced, waving a red flag that a security violation has occurred.
- Use anti-viral software designed to detect particular viruses. The software analyzes each program that is loaded from the software library.

Recovery   If malicious software gets through the prevention and detection barriers and is executed, recovery can be accomplished when you:

- Use disinfection utilities such as anti-viral software.
- Use backup files.

The policies and procedures that prescribe the firm's security against malicious software must be enacted at the executive level. This is because such software threatens systems throughout the firm, not only those of information services. The policies should be established by the executive committee and implemented and enforced by the MIS steering committee.

Achieving Security by Means of Systems Design

Exactly what is the relationship between security and systems analysis and design? As systems developers accomplish their tasks during the SDLC, they can build security into their systems by pursuing the following strategies:[19]

1. Strive for simplicity and smallness
2. Reduce exposure to failure of security
3. Restrict general access to software development tools and products

[19]This list is based largely on *Computers at Risk*, p. 35.

4. Develop generally available components with documented program-level interfaces

5. Provide excess memory and computing capacity

6. Include security criteria in selecting hardware configurations

7. Design software to limit the need for secrecy

8. Aim for building secure software by extending existing secure software

9. Use higher-level languages

10. Involve the internal auditor in all phases of systems development

11. Schedule more time and resources for quality assurance

Systems security must be planned well in advance—in a general sense for the entire firm, and in a particular sense for each system. The designers build security into the systems as they are developed.

Just as the American manufacturer says, "The quality goes in before the name goes on," systems designers should say, "The security goes in before the system goes on the air."

Summary

Systems security seeks to make computer-based systems less vulnerable to both internal and external threats by enacting countermeasures and safeguards. Interest in systems security is on the rise because the systems play such important roles in firms' operations, typically feature online input, and often connect multiple computers both inside and outside the firm.

Systems security has three main objectives—confidentiality, availability, and integrity. Confidentiality is closely related to personal privacy, and availability involves responsibilities to elements in the firm's environment as well as those inside the firm. Integrity is achieved through controls.

The main threats to systems security all deal with acts by unauthorized parties—use, disclosure and theft, destruction and denial of service, and modification.

Security of a computer-based system is achieved through a combination of logical and physical safeguards. Logical safeguards are based on the reasoning that someone cannot cause damage if they cannot gain access. Access control is aimed primarily at protecting the database and software library and consists of three screening functions—user identification, authentication, and authorization. Identification is proven by something that users know, or their locations; authentication is proven by something that users have or are; and authorization is based on entries in user and field directories. Passwords are widely used for identification, but the patterns that users follow in selecting them can provide clues to computer criminals. Firms can, however, follow several practices to keep their passwords secure.

Firms enact physical safeguards by first selecting a site that best insulates the computing facility from both computer criminals and natural disasters. Then the building is constructed, with particular attention paid to securing the computer room. Physical access to the room is controlled by guards and locks, and an appropriate room environment of temperature, humidity, and power is provided.

Much of the publicity concerning security violations has centered on malicious software. A firm can guard against all of the malicious software threats by implementing security measures designed to achieve prevention, detection, and recovery.

Systems developers can build certain features into the designs of their systems that contribute to security.

Key Terms

Systems security

Vulnerability

Threat

Countermeasure, safeguard

Connectivity

Confidentiality

Availability

Integrity

User directory

Field directory

Password

Biometric system

Uninterruptable power supply (UPS)

Emergency power supply (EPS)

Malicious software

Trapdoor, back door

Logic bomb

Trojan horse

Worm

Bacteria, rabbit

Virus

Anti-viral software

Quarantine station

Diskless node

Checksum

Key Concepts

The direct relationship between systems connectivity and interest in systems security

Security threats as unauthorized acts

How systems security is achieved by a combination of logical and physical safeguards

The three phases of access control—identification, authentication, and authorization

How damage from malicious software can be minimized by organizational policies and procedures

Questions

1. What types of resources does systems security protect?
2. List the three reasons for the current interest in systems security.
3. Which of the CBIS subsystems have special responsibilities in terms of achieving confidentiality?
4. Name four ways unauthorized persons can cause damage to a firm's information resources.
5. What is the difference between logical safeguards and physical safeguards?
6. What is the difference between user identification, authentication, and authorization?
7. Why is it easier for a computer criminal to gain access to a system when he or she knows one or more of the users?
8. What environmental conditions are necessary in a computer room that are not required for ordinary office space?
9. Programmers can create both trapdoors and logic bombs. How do they differ?
10. What is the key feature of a trojan horse?
11. How are a worm and a bacteria similar? How are they different?
12. How are computer viruses transmitted? What type of hardware do they most often affect?
13. How does a quarantine station work?

14. How could diskless nodes prevent malicious software?
15. Explain a checksum.

Topics for Discussion

1. Of the three objectives of systems security—confidentiality, availability, and integrity—which one would be of most interest to the manager of database administration? What about the data communications manager? The internal auditor? The manager of the human resources unit? The manager of the accounting department? The president?
2. What effect has end-user computing had on systems security?
3. Is all malicious software harmful?
4. Why locate a computer center away from an airport?
5. Why would a simple system be easier to secure than a complex one?
6. How would limiting access to CASE tools contribute to systems security?

Problems

1. The chapter lists 11 means by which systems developers can incorporate security into their designs. Classify each in terms of the phases of the SDLC. This can best be accomplished by listing the design strategies down the left margin and providing columns at the right for the four SDLC phases. Place a checkmark in the column or columns where each strategy would be applied. Which phase accounts for the largest number of strategies?
2. With the coordination of your instructor, visit a local firm and interview someone who is responsible for computer security. Compare what the firm is doing with what the chapter describes. Write a short memo to the person who is interviewed, outlining areas for possible improvement. Make an appointment with the person to discuss the memo.
3. Draw a floor plan of a computer room, and identify features that contribute to systems security. (The floor plan in Figure HB8.1 gives an idea of a general layout.)
4. Go to the library and find an article that deals with one of the types of malicious software. Write a short paper that summarizes the article. Your instructor will tell you how long the paper should be and its proper format.

(Use the Harcourt Brace scenario prior to Chapter 12 in solving this case problem.)

Pretend that you are Sandi Salinas or Russ March and that Mike has asked you for your appraisal of the Harcourt Brace security system. He wants a memo, identifying the various information resources and ranking them in order based on the degree of security measures that have been put in place. Recall that the resources include people (users and information specialists), hardware, software, data, information, and facilities. For any resources that you feel have received inadequate attention, make recommendations for beefing up their security.

Case Problem
Cowpoke Creations (B)

Recall from the Cowpoke Creations (A) case that you are the new manager of systems analysis. You did such a good job of handling the marketing information system assignment that Rae Summerfield, the CIO, wants you to take a look at the human resources division. They have indicated an intention to implement a human resources information system. Those efforts have been underway for about eight months, largely independent of any influence outside of HR. Rae is concerned that the HR people might be making some key decisions without the benefit of professional help, and she wants your insight. Rae tells you that HR is not represented on the executive committee or the MIS steering committee and that the HRIS is not a part of the strategic plan for information resources. The HRIS has been such a small-scale operation up to now that nobody has paid much attention to it, but it looks like the time has come to find out what is going on.

Rae asks you to meet with Bill Desmond, the HRIS director, and then report to her with your findings. You make an appointment with Mr. Desmond, spend a couple of hours talking with him in his office, and then accompany him on a tour through his area. When you think you've absorbed as much information as you can, you bid him farewell and tell him that you will stay in touch. Then you make an appointment to see Rae.

Rae: Well, what did you find out?

You: It's a much bigger operation than I anticipated. They have a local area network consisting of a network server, 12 micros, and a laser printer. Four of the micros are in managers' offices, and the rest are in an open work area. The HRIS database and software library are maintained on a hard disk in the network server, and the users download what they need.

RAE: What kind of security did you find?

You: There's good news and bad news. The network server is kept in a room that is locked at all times. Two operators work in the room, and they won't let anyone else in without Mr. Desmond's permission. That's the good news. The bad news is that the managers don't lock their offices at night, although there are locks on the doors. And all the micros have locks, but the users don't use the keys. I saw quite a few sets of keys lying about on the tables next to the keyboards. I also saw quite a few diskettes scattered around, especially out in the open work area.

RAE: Who has access to the micros? Can anyone operate them?

You: Well, the ones in the managers' offices are just for their own use, but the ones in the open area are available to any of the 25 or so clerical people who care to use them. I saw quite a few different people operating the micros just while I was in the area. I think the users all have unlimited access to the hardware, software, and data.

RAE: Is any of the data sensitive?

You: It most certainly is. It includes all the personnel data for every employee in the company. Plus, there are a lot of preformatted personnel reports stored in the database for quick retrieval—the reports that go to the board of directors and executive committee.

RAE: Do the users use passwords?

You: No. Mr. Desmond told me that they disabled the password feature on the DBMS to speed up the retrieval.

RAE: What did you see in the way of documentation?

You: I saw a lot of printed reports lying around. All in all, it was pretty messy. The wastebaskets were jammed full of old printouts. I think they must have just cleaned out their files.

RAE: Did Mr. Desmond say anything about their plans?

You: He certainly did. They eventually want to replace their LAN with a minicomputer that is networked to our mainframe. This will enable them to maintain the database on the mainframe and make it available to anyone in the company who needs it. Here is a copy of their long-range plan. [You hand Rae the planning report that Mr. Desmond gave you.]

RAE: [Glancing over the report] This doesn't look good. From what you say, they have an unsecured system. Before we let them tie into the mainframe, they are going to have to build in adequate controls and security measures. We had better let Mr. Desmond know what they must do if they expect to expand their network outside of HR. Why don't you draft a memo for my signature and we can discuss it?

Assignments

1. Write a memo to Mr. Desmond, for Rae's signature. Address the issue of *systems controls*. Base your suggestions on the material in Chapter 12. Suggest steps that HR can take to control each element of the HRIS. Separate the suggestions according to the eight systems elements. Assume that the application is essentially one of maintaining a database and that Cowboy Creations wants the data to be accurate, current, and secure from access by unauthorized persons.

2. Write a second memo to Mr. Desmond—also for Rae's signature. Address the issue of *systems security*. Base this memo on material in this chapter. Suggest some steps that HR can take in safeguarding the firm's information resources.

Selected Bibliography

Andrews, William C. "Contingency Planning for Physical Disasters." *Journal of Systems Management* 41 (July 1990): 28–32.

Baskerville, Richard. "The Developmental Duality of Information Systems Security." *Journal of Management Systems* 4 (Number 1, 1992): 1–12.

Berman, Alan. "Ensuring Dial-Up Security." *EDP Auditing* (Boston, MA: Auerbach Publishers, 1991): section 75-01-60.1, 1–9.

Burns, David C., and Sorkin, Horton Lee. "EDI Security and Controls." *Bank Management* 67 (February 1991): 27ff.

Coopee, Todd. "Backup Drill." *Corporate Computing* 1 (September 1992): 93ff.

Dear, Lorne A. "A Checklist for Auditing Local Area Network Security Controls." *EDP Auditing* (Boston, MA: Auerbach Publishers, 1991): section 75-01-45, 1–21.

FitzGerald, Jerry. "Detecting and Preventing Computer Viruses in the Microcomputer Environment." *EDP Auditing* (Boston, MA: Auerbach Publishers, 1991): section 75-01-50, 1–10.

"Four Steps for Building a Better Backup Plan." *Corporate Computing* 1 (September 1992): 117ff.

Haight, Nicholas, and Byers, C. Randall. "Disaster Recovery Planning: Don't Wait Until It's Too Late." *Journal of Systems Management* 42 (April 1991): 13–16.

Heath, Constance C. "How Healthy Is Your Power Protection?" *Modern Office Technology* 36 (November 1991): 35ff.

Janulaitis, M. Victor. "Creating a Disaster Recovery Plan." *Infosystems* 32 (February 1985): 42–43.

Joseph, Gilbert W. "Computer Viruses: How EDP Auditors Can Minimize the Risks." *EDP Auditing* (Boston, MA: Auerbach Publishers, 1991): section 75-01-55, 1–7.

Kerr, Susan. "Using AI to Improve Security." *Datamation* 36 (February 1, 1990): 57ff.

Logan, Andrew J. "Contingency Planning." *EDP Auditing* (Boston, MA: Auerbach Publishers, 1991): section 75-02-20, 1–18.

McAfee, John. "The Virus Cure." *Datamation* 35 (February 15, 1989): 29–40.

Moad, Jeff. "Disaster-Proof Your Data." *Datamation* 36 (November 1, 1990): 87ff.

Murray, W. H. "Security Considerations for Personal Computers." *IBM Systems Journal* 23 (Number 3, 1984): 297–304.

Romei, Lura K. "Power Protection: What Kind and Why?" *Modern Office Technology* 35 (May 1990): 66ff.

Schlack, Mark. "How To Keep Viruses Off Your LAN." *Datamation* 37 (October 15, 1991): 87ff.

Seeley, Donn. "Password Cracking: A Game of Wits." *Communications of the ACM* 32 (June 1989): 700–703.

Skupsky, Donald S. "Establishing an Effective Records Retention Policy." *Modern Office Technology* 36 (November 1991): 58ff.

Straub, Detmar W., Jr. "Organizational Structuring of the Computer Security Function." *Computers & Security* 7 (Number 2, 1988): 185–195.

Summers, R. C. "An Overview of Computer Security." *IBM Systems Journal* 23 (Number 4, 1984): 309–325.

Watts, Robert T., and Richards, Thomas C. "Microcomputer Software Backup: Legal and Ethical Considerations for the Individual and Small Business User." *Security Audit & Control Review* 9 (Spring 1991): 9–12.

Systems Implementation, Audit, and Maintenance

At this point in the system development life cycle, we have completed the systems analysis and design. We have considered alternate designs and selected the one that best meets the needs of both the organization and the users. In executing the design, we have also built in appropriate controls and security features.

We are now ready to execute the final phase of system development—implementation. Then, after cutover, we will audit the system to ensure that it continues to meet user needs. We will also maintain the system to extend its usability for as long as is practical.

Part 6 consists of two chapters. Chapter 14 deals with the implementation phase and Chapter 15 with use, audit, and maintenance. In addition, three technical modules relate primarily to the implementation phase. Technical Module H describes structured English that information specialists or users can use to document detailed processing logic. Technical Module I describes a similar tool called action diagrams. Technical Module J provides an example of a user manual that users can use to learn about the system prior to cutover and can refer to as needed during the use phase.

This part of the text marks the end of our journey. We have laid a foundation of systems theory, explained the main systems methodologies, and described in detail the phases of the system life cycle and the most popular tools of systems work. Knowledge gained from this journey puts you in an excellent position to pursue systems analysis and design, either as an information specialist or as a user.

Ongoing Case

Harcourt Brace & Company

IX—Implementing the Employee Scheduling System

(Use this scenario in solving the Harcourt Brace problem at the end of Chapter 14.)

With the completion of the system design, Ira Lerner and the rest of the order processing and billing group turned their attention to implementation. The system that existed on paper in the form of the project dictionary would have to be converted into a system consisting of hardware, software, procedures, data, and, most important, people.

Plan the Implementation

Ira called a group meeting for the purpose of developing the implementation schedule. There was no urgency in cutting over to the new system because the changes were a part of COPS2 and would not take effect until after successful implementation of COPS1 in November. However, it appeared that the order processing and billing group would have their COPS1 work completed by July 15 and could begin COPS2 projects. Ira was anxious to continue work on the employee scheduling system that Sandi and Russ had analyzed and the team had designed. The system documentation was thorough, and Ira wanted to provide Sandi and Russ with experience in completing an entire system development project. For these reasons, Ira made the employee scheduling system the top priority COPS2 project—a decision supported by Bill Presby, Mike Byrnes, and the COPS review committee.

IRA LERNER: We'll go about planning this project the same way we always do, only now we are a little more in control of our own destiny. We don't have any deadline dates imposed on us, so we're free to determine when the various activities will take place. Rather than fit the work to the schedule, we can produce a schedule that reflects the work to be done. Russ and Sandi, when we plan an implementation, we always anticipate nine main steps.

[Ira writes the steps on the chalkboard. The steps appear in Table HB9.1.]

Table HB9.1 **Implementation Steps**

1. Plan the implementation
2. Announce the implementation
3. Organize the project group
4. Select the hardware
5. Prepare the software
6. Prepare the database
7. Educate the users
8. Prepare the facilities
9. Cutover to the new system

Some of these will take more time to accomplish than others. What I'd like to do today is set some rough start and completion dates for each of the steps. We will do step 1, the planning, today. Then, we need to announce the project to the Bellmawr people so they can anticipate the change and be ready for cutover. The Bellmawr people who will be affected the most are the warehouse area supervisors, but the announcement should go through Dave Mattson since he is the Bellmawr manager.

RUSS MARCH: Ira, just how do you plan to do the announcing?

IRA LERNER: The best way is for someone to go to Bellmawr and review the final system design with Dave and the four area supervisors. You and Sandi are the logical choices since it has been your project all along. Meet with the people, review the system, and describe the implementation schedule. The review should be very brief since they have participated in the project from the beginning and approve the approach.

Next, we need to organize the project group. I've asked Mike Byrnes if our order processing and billing group can remain intact and do the work, and he's assured me that it can. But, we still need to decide who will do what work—who will code the programs, who will conduct the education, and who will coordinate the preparation of the physical facilities. We also need to agree on a reporting mechanism.

ANNE HOGAN: Russ, some of the assignments are very straightforward. Since Frank is our database guru, he will handle all of those issues. Since Phyllis is our network guru, she will do the same for data communications. Since Pam is the internal auditor, she will deal with the issues relating to system integrity.

IRA LERNER: And, since Suzanne volunteered to do the programming, I'm going to see that she gets her wish.

The selection of the hardware will pose no big problems. All we're talking about is a PC that will serve as the Bellmawr conference room terminal, along with the associated data communications gear. The PC will not only display the reports, but also produce hardcopy output that the supervisors can take with them as they return to their units. Since Harcourt Brace has established a formal procedure for obtaining PCs, a request will have to be submitted to the PC committee. Because the system is to provide management information, there is no doubt in my mind that the request will sail right through. Approval should be automatic.

ANNE HOGAN: The software involves little decision making as well. In cases such as this, where it is obvious that we need to do our own coding, we don't even bother to evaluate the prewritten software market. We have our own programmers and always code the application programs ourselves. So, we won't have to labor over a make-or-buy decision.

FRANK GRECO: The database will not be a big problem, either. We already have the data we need. It's only a matter of using it. So, the job of this guru will be pretty easy.

[Everybody appreciates Frank's break in the action, and then Ira calls for order.]

IRA LERNER: Education shouldn't be difficult. We'll have to focus on the area supervisors and make sure they know how to interpret the reports. Russ and Sandi, that can be your responsibility. You know what that means—a few more days in Bellmawr.

PHYLLIS STEINHAM: The physical facilities end of it won't be a big deal, either. We already have the data communications circuits to Bellmawr. It will only be necessary to tap into them and add another terminal in the conference room. I can probably do whatever I have to do over the telephone.

LLOYD HAMM: That just leaves cutover. What do you think, Ira, should we switch over immediately, or should we phase it in or run parallel?

IRA LERNER: Well, we really aren't replacing an existing system. This will be a completely new way to schedule the employees, so we won't have to worry about the new system matching the old one. That rules out a parallel cutover to compare the two systems' outputs. The new system is so simple that I see no reason to phase it in piece by piece. Does everybody agree that we can go with an immediate cutover? [Everybody indicates that the immediate cutover is the way to go.]

SANDI SALINAS: That covers all the points. What do we do now? Are we finished planning?

ANNE HOGAN: Oh, no. We have to assign responsibilities and set some dates. Ira, can we go ahead and do that now?

IRA LERNER: I don't see any reason why not. We seem to have a good feel for what has to be done. Does anybody have a problem with the assignments that I mentioned earlier—Sandi and Russ handling the systems work and Suzanne doing the programming? [Nobody indicates any problems.]
    All right, then. Let's take it from there.

[Ira goes to the chalkboard and adds the names next to the steps. He asks for help in estimating the start and completion dates for each step, assuming that work can begin August 15. The discussion produces the schedule illustrated in Table HB9.2.]

Table HB9.2   **Implementation Schedule**

Step	Responsibility	Dates Start	Complete
Plan the implementation	Team	6/21	6/21
Announce the implementation	Russ, Sandi	6/28	6/28
Organize the project group	Ira	6/21	6/21
Select the hardware	—	—	—
Prepare the software	Suzanne	8/15	10/15
Prepare the database	Frank	8/15	10/15
Educate the users	Russ, Sandi	12/2	12/6
Prepare the facilities	Ann	11/25	11/29
Cut over to the new system	Team	12/16	12/16

Announce the Implementation

The following week, Russ and Sandi flew to Bellmawr to advise management of the implementation plan. They first met with Dave Mattson and reviewed the portion of the new system's data flow diagrams that dealt with the employee scheduling. They explained the program that would use the standards to compute the area workloads. The program had been named the warehouse daily scheduling program. Then, they reviewed the report format that would be prepared for each of the unit supervisors. They had prepared hardcopy forms that could be left with Dave and the supervisors. The two junior analysts did not attempt to explain the system in detail; they realized that the educational portion of the implementation would accomplish that. Finally, they described the implementation plan, focusing on the projected dates leading to cutover on December 16. Dave could see no problem with the schedule.

After reviewing the system with Dave, Russ and Sandi met with the area supervisors as a group and repeated the description of the system and the schedule. The supervisors were also told that Sandi and Russ would return in December to conduct educational sessions for the supervisors and anyone else the supervisors would like to attend. With the announcement made, Sandi and Russ chalked up a few more frequent flyer miles on the way back to Orlando.

Organize the Project Group

Ira asked Sandi and Russ to prepare a report each week summarizing the progress. It would be nothing elaborate—probably a one-page memo. The report would inform Ira of any deviations from the schedule. In addition, Ira asked Russ and Sandi to drop in at least once a week for a chat. This was Ira's way of ensuring that the new analysts did not get in over their heads and that they experienced a successful development project.

Select the Hardware

Sandi and Russ worked with Phyllis Steinham in preparing the request for a PC. The PC committee approved it, as Ira had predicted. The PC was scheduled for delivery at Bellmawr in time for onsite education in early December.

Prepare the Software

Suzanne McNulla used Russ and Sandi's documentation as the basis for coding the warehouse daily scheduling program and the program used to display the supervisors' reports. The two programs that were Suzanne's responsibility are the rectangles in the Figure HB9.1 system flowchart.

Sandi and Russ had done such a good job on their documentation that the coding was very straightforward. Only occasionally did Suzanne have to check with the junior analysts to clear up a point.

Both programs were tested using data from the COPS database and a networked PC in the MIS department that was identical to the one to be installed in Bellmawr. Suzanne stepped through the processes with Sandi and Russ and then received Ira's OK that the software was ready to be implemented.

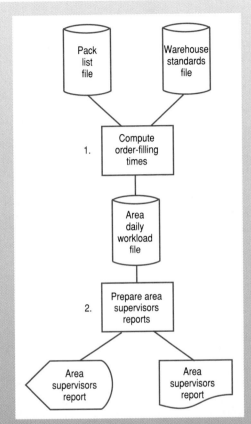

Figure HB 9.1
The system flowchart of the employee
scheduling system.

Prepare the Database

Frank Greco kept in contact with Suzanne as the programming was being done. Although Frank did not anticipate any problems in using the standards and pack list files, he wanted to make certain that everything went smoothly. It did.

Prepare the Facilities

Phyllis Steinham was right. She was able to direct the installation of the PC in the Bellmawr conference room by telephone. Some of the Bellmawr staff were experienced in data communications, and they were able to get the work done.

Educate the Users

Sandi and Russ returned to Bellmawr as planned and conducted a day-long session with the area supervisors. The morning focused on the reports and how to interpret them. The afternoon was a simulation of how the system would work. The supervisors retrieved their reports and conducted a mock discussion of the reassignments using the simulated data. The supervisors took the printouts back to their units and were able to clearly explain them to their order fillers. There were no problems. The Orlando analysts were confident that the system could be implemented without any difficulty.

Cutover

When December 16 rolled around, the system was ready. COPS had been successfully implemented the previous month. Everything had gone so smoothly with the scheduling system that Russ and Sandi decided not to travel to Bellmawr. If the system bombed, it would not be the end of the world. The supervisors could simply go back to their old approach until the bugs were worked out. But the system did not bomb. In fact, everything went exactly as planned.

Sandi and Russ felt good because they had played an important part in a system development project. They had been with Harcourt Brace only six months, yet they felt confident in their ability to do systems work. They had demonstrated a proficiency in

the use of the documentation tools, including CASE, and believed that they had satisfied the requirements for promotion. Ira thought so too, and told them that he would recommend them both for promotion to systems analyst after the system had been operational for 90 days.

Personal Profile—Suzanne McNulla, Programmer

Suzanne McNulla (shown on the previous page), a native of Alexandria, Virginia, earned a Bachelor of Science degree in management information systems from Old Dominion University. Her hobbies are modeling, boating, skiing, swimming, bowling, and traveling. She attributes her career successes to her dedication to striving for the best results and her positive attitude toward people. Suzanne believes the key to Harcourt Brace's continued success is an innovative product mix combined with excellent customer service.

CHAPTER 14

Systems Implementation

Learning Objectives

After studying this chapter, you should:

- Know how to develop a detailed implementation plan
- Be familiar with one way to motivate employees to cooperate with the implementation efforts
- Know how to keep up to date on advances in hardware technology
- Understand why it is important to include detailed design specifications in RFPs
- Know some criteria to use as the basis for selecting hardware and software suppliers
- Know how to go about verifying hardware and software suppliers' claims
- Know the different ways to pay for hardware resources
- Understand more about the make-or-buy decision as it relates to software
- Know the main avenues to custom software
- Be aware of the services offered by outsourcers
- Understand the process of preparing the database and know the conditions that influence the degree of difficulty
- Be better informed about the features that can be built into a modern computing facility to keep it functioning and secure
- Recognize the role of the systems analyst in the education of users and participants
- Know the main approaches to cutover and when each should be used
- Be aware of special demands that are placed on systems implementation by each of the CBIS subsystems

Introduction

The implementation phase of the SLC begins with detailed planning. Now that only one phase remains before cutover, the plan can be very specific—identifying exactly what will be done, who will do it, and when.

With the plan in place, management kicks off the activity by making an announcement to the employees. The purpose of the announcement is to let the employees know of the decision to implement the new system and ask for their cooperation in putting it into use.

The logical design that was prepared during the design phase provides the basis for selecting the hardware supplier(s). This selection involves evaluating supplier proposals and considering multiple dimensions of performance.

The basic software decision is whether the firm should prepare its own software or obtain prewritten software from suppliers—the make-or-buy decision. When the decision is to make, the programmer can use a programming language or a CASE tool. The user can use a more user-friendly language or application-development software. When the decision is to buy, the same general procedure used in obtaining hardware can be followed.

The preparation of the database is a two-step process. The first step is specification of the data in the form of the data dictionary, and the second is data entry.

The hardware, software, and data resources must be located in physical facilities. When the system is large, the facilities provide special capabilities for security, environmental controls, and emergency power. The manager of computer operations usually has the responsibility for putting the physical facilities in place.

The final task before cutover is education. Both users and participants must be checked out on their roles in the new system. The systems analyst can have the primary responsibility for the education, conducting classes or training sessions. Other sources, both inside and outside the firm, can also contribute.

When preparation is complete, the firm can cut over from the old system to the new one. Four cutover approaches are possible, each best suited for a particular application.

The composition of the resources expended in the implementation phase depends to a great extent on the system. Each CBIS subsystem exerts its own unique influence.

The Implementation Task

At the conclusion of the design phase, the new system is documented in the form of the logical design. Project team members who have the major implementation responsibility base their work on the logical design, contained in the project dictionary. Figure 14.1 shows the contents of the project dictionary as the implementation phase begins.

During the implementation phase, the systems analyst and the user relinquish their center-stage position to the programmer, who is the "star," and to the database and data communications specialists and the operations personnel, who play important supporting roles. However, the analyst and user continue to participate, helping the other team members as required and filling in any gaps in the documentation.

1. Problem definition*
2. System objectives and constraints*
3. Documentation of the current system
    3.1 Process documentation (a combination of the following)
        3.1.1 System flowchart
        3.1.2 Program flowcharts
        3.1.3 Data flow diagrams
        3.1.4 Hierarchy diagrams
        3.1.5 Warnier-Orr diagrams
        3.1.6 Structured English
        3.1.7 Other process documentation tools
    3.2 Data documentation
        3.2.1 Entity-relationship diagram
        3.2.2 Data dictionary entries
4. Performance criteria*
5. Documentation of the new system
    5.1 Alternate feasible system configurations
    5.2 Recommended system configuration
    5.3 Process documentation (same types of documentation as in section 3.1, with the exception of program flowcharts and the possible exception of structured English)
    5.4 Data documentation (same types of documentation as in section 3.2)
6. General implementation plan**
7. Appendix
    7.1 Transcripts of interview tapes
    7.2 Working papers
    7.3 Joint application design (JAD) session transcripts

* From the design proposal
** From the implementation proposal

Figure 14.1
The contents of the project dictionary at the beginning of the implementation phase.

The Implementation Phase of the System Life Cycle

Figure 14.2 shows the eight steps required to implement a new system. You can see that the MIS steering committee plays a more active role than in the previous phases. The committee is directly involved in each step. A good plan must be in place in order for the committee and other management to provide the necessary monitoring.

Step 1—Plan the Implementation

Much planning has already been accomplished during the SLC. The final step of the planning phase was the establishment of a control mechanism. When that mechanism was put in place, not much was known about the development effort, and the plan was very general. As the project unfolded, more was learned, and the plan became more specific. Now, with only a single developmental phase remaining, the implementation plan can be very detailed. Such detail is necessary because so many people will be involved.

The Implementation Plan

The project team prepares the implementation plan and presents it to the MIS steering committee for approval. The plan answers three important questions:

1. *What* activities will be required?
2. *Who* will perform the activities?
3. *When* will the activities be performed?

The questions are answered in this order. It is necessary to know the activities before the personnel can be assigned, because the assignments are based on the knowledge and skills required. For example, if the system will involve the inhouse development of a mathematical model, a programmer with modeling experience is assigned. Also, the time that the work will require depends on who will do the work since people work at different speeds. Figure 14.3 depicts a portion of an implementation plan.

Step 2—Announce the Implementation

Management triggers the implementation activity by again engaging in formal communication with the employees. The same process used to announce the system study is followed to accomplish two main objectives: (1) to inform the employees of the decision to implement the new system, and (2) to appeal to the employees for their cooperation.

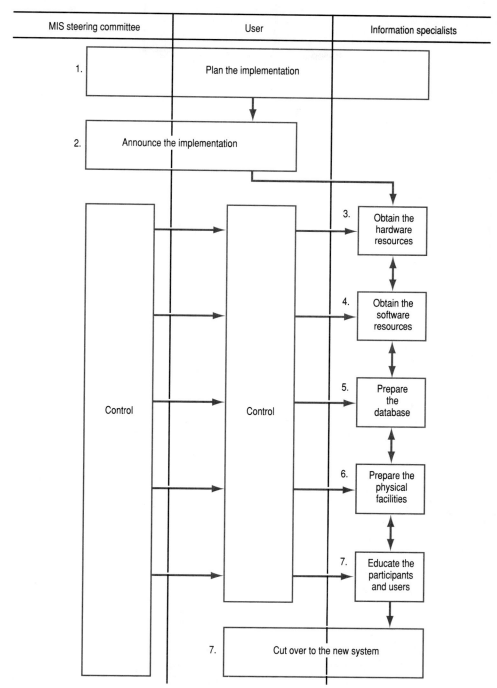

Figure 14.2
The implementation phase.

Functional system: Subsystem:	Marketing Product deletion model		
		Estimated	
Task	Responsibility	Start date	Stop date
Review system documentation	Estelle Williams Ronnie Wilkins	9/02	9/08
Prepare structured English	Estelle Williams	9/09	9/12
Code program	Estelle Williams	9/13	9/29
Prepare test data	Ronnie Wilkins	9/09	9/29
Conduct module tests	Estelle Williams Marcia White	9/30	10/05
Conduct integration tests	Estelle Williams Marcia White	10/06	10/16
Conduct system tests	Estelle Williams Henry Abernathy	10/17	10/24
Obtain user approval	Estelle Williams Henry Abernathy	10/25	10/30
Obtain MIS steering committee approval	Henry Abernathy Alberto Suarez MIS committee	11/01	11/01
Cut over to system	Operations staff	12/12	12/19

Figure 14.3
Example of a portion of an implementation plan.

In making the announcement, a good strategy for management is to link the employees' needs with the system objectives, as illustrated in Figure 14.4. Management first recognizes employee needs and then provides motivation for the employees to work toward meeting those needs. When employees work to satisfy their needs, the system meets its objectives.

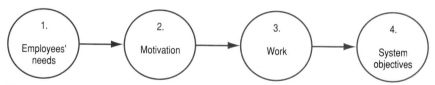

Figure 14.4
Satisfaction of employees' needs leads to achievement of system objectives.

The key to basing implementation on human factors considerations is correlating employee needs and system objectives. As an example of how this works, imagine that a new order entry system is planned. The supervisor of the order department knows that her workers have a need for a sense of accomplishment and that they take special pride in meeting challenges. With that understanding, management can emphasize to employees how the new system will relieve them of many of their clerical duties and give them responsibility for making more decisions. The computer will elevate their work above a clerical level and give them an opportunity to apply their experience and knowledge in a more meaningful, productive way.

The order department employees also benefit when the system enables the firm to be more successful. When the firm can do a better job of processing orders, it can compete better in its marketplace and increase its profits, and this allows the firm to provide a better work environment for the employees.

As users of the new system's output or participants in the system's operation, employees are important, and their cooperation is a must for a successful implementation.

Step 3—Obtain the Hardware Resources

The design phase produced a system configuration that the project team and the MIS steering committee believed to be best suited to meeting the system objectives. A configuration for a mainframe system is shown in Figure 14.5. This particular system will use online keydriven devices and OCR for input, DASDs and magnetic tape units for secondary storage, and a page printer for output. The task of the project team is to determine the best source, or sources, for these hardware units.

In the early years of computing, the practice was to acquire the entire system from a single hardware vendor, such as NCR or IBM. This logic prevailed until the market became flooded with peripheral devices that performed as well as, or better than, the units manufactured by the CPU supplier. A **peripheral device** is one that supports the CPU by performing input, secondary storage, or output functions.

Although the multi-supplier approach benefits the firm by ensuring that the best possible combination of units is assembled, it imposes a big responsibility on

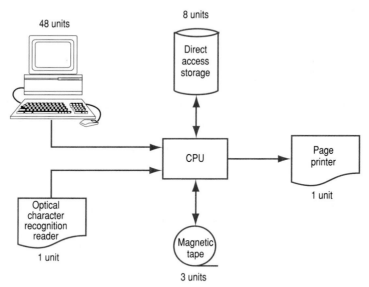

Figure 14.5
A system configuration.

systems developers: They must remain current on the products of a large number of hardware vendors.

Sources of Hardware Information

Three sources are especially effective in providing information about available hardware: the hardware vendors, computer magazines, and reference books.

Hardware Vendors   All hardware vendors, large and small, promote their products by means of advertising and personal selling.[1] You have seen computer ads on TV and in magazines. Other promotions, such as direct mailings and sales calls, are aimed at persons who are in a position to make or influence hardware decisions. If you become an information specialist, you will receive promotional material in the mail, informing you of new products and their capabilities. If you work for a large, mainframe user, you will also be called on in person by hardware suppliers' sales representatives. The reps will make certain that you remain current on their products by providing you with manuals, perhaps conducting onsite seminars, and offering courses on topics of special interest. When it is important

[1]In this text the term *supplier* is used for the organization that provides the firm with resources. However, in the computer field the term *vendor* is commonly used to describe suppliers of hardware and software.

for you to attend the courses, your employer will generally cover registration, travel, and other expenses. One of the advantages of working for a large computer user is the opportunity to remain on the cutting edge of new technology through vendor contacts.

Computer Magazines   The popular computer magazines that you find on newsstands and in libraries contain articles describing the latest hardware and software. Most magazines are focused on the microcomputer market. Examples are *Macworld, PC World, PC Computing,* and *PC Magazine.* Other magazines contain articles that review computer products of all sizes. Examples are *BYTE, Computerworld,* and *InfoWorld.* By reading these magazines, you can take a giant step toward staying current in computer technology.

Reference Books   Your college library most likely has one or more sets of reference books devoted to descriptions of computing hardware. Examples are *Data Pro 70: The EDP Buyer's Bible,* from McGraw-Hill, and *The Faulkner Report,* from Faulkner Technical Reports, Inc. Sets such as these provide descriptions of the units accompanied by performance statistics and prices. The books are kept current with revisions. Figure 14.6 is a table from *The Faulkner Report* comparing three models of Cray supercomputers.

One feature of the computer field that makes it so interesting is the constant change. It seems that every day something new is announced. This imposes a big responsibility on users, and an even bigger one on information specialists. Regardless of your career path, you must make an effort to stay current on computer technology. That will be one of the keys to your success. One of the advantages of a college education is the preparation it provides for conducting library research. By applying that skill after graduation you will be able to keep up with the ever-expanding computer field.

The Selection Process

The process of selecting hardware vendors was described in Chapter 8. Recall that the project team first prepares a request for proposal, or RFP, and that the vendors respond with both written and oral proposals. The RFP is illustrated in Figure 8.9, and a vendor proposal in Figure 8.10.

The contents of the RFP are the secret to successfully selecting a vendor. The RFP includes such details about the system design as (1) performance criteria, (2) equipment configuration, (3) transaction volumes, and (4) file sizes. In addition, the RFP specifies the installation schedule that the vendor must meet.

It is necessary to provide such details if each vendor is to propose a configuration that satisfies the logical design. The firm wants to be in a position to make an *apples-to-apples* comparison of the vendors' proposals, and that is possible only by specifying the design in detail. Otherwise, each vendor will recommend those units that they perceive to give them a competitive advantage, and the firm will be forced to make an *apples-to-oranges* comparison.

Model **Introduced**	**CRAY Y-MP2E** May 1990	**CRAY Y-MP4** March 1989	**CRAY Y-MP8** March 1989
Central Processor			
No. of CPUs	1 or 2	1–4	4–8
Technology	25K gate array ECL	25K gate array ECL	25K gate array ECL
Word Length (bits)	64	64	64
Cycle Time (nsec)	6	6	6
Peak Performance (FLOPS)	333M–666M	333M–1.3G	1.3G–2.7G
Memory			
Technology	64K-bit MOS	64K-bit MOS	64K-bit MOS
Min/Max Capacity (words)	16M/32M	16M/64M	32M/256M
No. of Memory Banks	64 or 128	128–512	512
Cycle Time (nsec)	15	15	15
Memory Protection	SECDED	SECDED	SECDED
Mass Storage			
Solid-State Disk Capacities (words)	32M or 128M	32M, 128M, 256M, 512M	32M, 128M, 256M, 512M
No. of Disk Storage Units	2–24	2–24	4–48
Capacity per Disk Unit (bytes)	1.2G, 5.2G	1.2G, 5.2G	1.2G, 5.2G

Model Introduced Central Processor	CRAY Y-MP2E May 1990	CRAY Y-MP4 March 1989	CRAY Y-MP8 March 1989
I/O and Data Communications			
No. of I/O Subsystems	1	1	1–2
No. of I/O Processors	1–4	1–4	4–8
No. of 6M-Byte/Sec Channels	2	4	8
No. of 100M-Byte/Sec Channels	2	4	8
No. of 1000M-Byte/Sec Channels	1	2	2
No. of Network Gateways	1–7	1–7	2–14
Networks	IBM SNA, Digital DECnet, Control Data CDCnet, Ethernet	IBM SNA, Digital DECnet, Control Data CDCnet, Ethernet	IBM SNA, Digital DECnet, Control Data CDCnet, Ethernet
Software			
Operating System	UNICOS, COS	UNICOS, COS	UNICOS, COS
Languages	FORTRAN 77, C, Pascal, CAL	FORTRAN 77, C, Pascal, CAL	FORTRAN 77, C, Pascal, CAL

Source: Faulkner Technical Reports, Inc. Reprinted with permission.

Figure 14.6
A table from a hardware reference set.

Evaluation Criteria

The selection of a hardware vendor is a semistructured problem. Some of the elements, such as the performance ratings of the equipment units, are highly structured and easily measured. Other elements, however, must be evaluated subjectively. Consider the following evaluation criteria:

- **Equipment Performance** How will the hardware perform in terms of its speeds, error rates, mean times to failure, and so on?

- **Equipment Maintenance** What service will the vendor provide to keep the hardware in running order?

- **Education** What does the vendor offer in terms of educational programs for users and participants? Many vendors have education centers where they offer special courses to their customers.

- **Industry Knowledge** How knowledgeable is the vendor in meeting the needs of the firm's particular industry? Most industries have unique data processing requirements, and a vendor who is aware of these needs can provide valuable consultation about how to best apply the hardware.

The project team can consider these evaluation criteria in a structured way by weighting the criteria and rating each on a quantitative scale. Chapter 11 described such an approach in evaluating the various system configurations, illustrated in Figure 11.8. The same approach can be used to evaluate the hardware vendors.

Verification of Vendor Claims

Always verify the hardware vendors' claims. The simplest way to do this is to read the hardware reviews in computer magazines. Two potentially more productive approaches are to seek out other users of the same equipment and to require the vendors to solve benchmark problems.

User Contacts   Persons connected with the project can contact a sampling of the vendors' customers in order to gauge vendor performance level. These contacts can be made by members of information services, the MIS steering committee, or the firm's purchasing department. Personal contacts are best, but the telephone can be used when the two parties already know each other—a common situation in many close-knit industries.

Benchmark Problems   When the firm's hardware order is extremely important to the vendors—for example, a multi-system order to be placed by a national chain—the firm can impose a benchmark problem. A **benchmark problem** is a process that each vendor must perform using its own hardware, and provides a major basis for evaluating that hardware. For example, all vendors could be required to use their equipment to update a master file with transaction data. The vendors would use data provided by the firm and the processing would be timed.

Figure 14.7
Education can be an important consideration in selecting hardware vendors.

In some cases, the firm supplies the software, and in others, the vendor is expected to do the programming.

Different Ways to Pay

The low prices of microcomputers have made outright purchase the most popular form of payment for hardware resources. Every day hundreds of individuals pay for their micros by writing checks or using credit cards. Business organizations purchase low-priced units the same way. For more expensive units, however, firms can consider three additional payment alternatives—rental, lease, and lease-purchase. Each payment method offers advantages and disadvantages.

Rental Plans   Some hardware vendors offer rental plans that enable their customers to make monthly payments for hardware and to cancel the agreement with short notice—generally 30 to 90 days. This is the most flexible payment method because it provides the firm with the ability to easily and quickly return unwanted equipment.

By renting, the firm does not have to worry about paying insurance premiums or property taxes on the equipment because these expenses are included in the rental amount. The firm is also entitled to equipment maintenance, provided by the vendor at no extra cost.

Large computer systems require both scheduled and unscheduled maintenance. **Scheduled maintenance** is work performed to prevent future equipment failure, much like changing the oil in your car every 4,000 miles. The term **preventive maintenance (PM)** is also used. **Unscheduled maintenance,** on the other hand, is work performed to repair a failure, like fixing a flat tire on your car. Computer maintenance must be performed by trained technicians, called **field engineers,** or **FEs.** Hardware vendors provide FEs to their rental customers.

The rental approach obviously offers significant advantages. However, it is the most expensive of the payment methods in terms of monthly cost. Another disadvantage is the fact that the rental payment does not enable you to use the equipment on overtime without incurring extra charges. The rental plan typically provides for 176 hours (8 hours per day for 22 days per month) or 200 hours of **prime shift usage** per month. Any additional usage, called **extra shift usage,** calls for an extra payment.

Lease Plans   Lease plans are similar to rental plans but bind the firm to a longer period of time, usually a minimum of one year. Leasing therefore lacks the flexibility of renting, but the monthly payments are lower. The same con-

Figure 14.8
Field engineers perform both scheduled and unscheduled maintenance.

tractual stipulations about maintenance, insurance, taxes, and extra shift usage apply.

Lease-Purchase Plans   Some leases include a clause providing the firm with an option to purchase the leased equipment. Some portion of the lease payments that have already been made—usually in the 50 to 70 percent range—can be applied to the purchase.

Purchase Plans   When a firm uses hardware for a long period—three to five years or more—outright purchase is more economical. The title passes from the hardware vendor to the firm, which assumes responsibility for maintenance, insurance, and taxes. However, because the equipment belongs to the firm, there is no extra shift charge.

Some larger firms have their own FEs, and other firms acquire FE services from the hardware vendor or from maintenance firms by means of a separate maintenance agreement.

A big deterrent to purchase is the large investment the firm must make at the outset. However, making that investment often entitles the purchaser to certain income tax advantages, depending on the laws in effect that year. Another deterrent to purchasing is the risk of investing in technology that will quickly become obsolete.

Selecting the hardware payment option is an important decision for a firm. Because the expenditure is significant, financial management often gets involved and uses such techniques as payback analysis and net present value calculations to determine which option is best. These techniques are described in Technical Module E.

The Significance of the Hardware Selection

When the evaluation of vendor proposals is completed and the best one identified, an order is placed to deliver the equipment according to a certain schedule. This hardware selection influences *all* the remaining implementation work, which specifies particular brands and models.

Putting Hardware Selection in Perspective

The two main decisions relating to hardware selection are the choice of the vendors and the payment method. The vendor decision is essentially a semistructured one and can be influenced by both politics and emotions. Some firms are dedicated to particular vendors, having built a strong relationship over the years. These firms will not consider other vendors, even though others may offer superior hardware or lower prices. The decision concerning the payment method is more structured and less susceptible to noneconomic influences.

The amount of attention given these decisions varies with their importance to both the firm and the vendors, and with the amount of money involved. Such decisions are typically *not* made by the project team. Rather, they are made by the

MIS steering committee or perhaps the executive committee. The CIO plays the key role.

Step 4—Obtain the Software Resources

The logical design of the new system processes exists in the form of DFDs, structured English, action diagrams, system flowcharts, and the like. This logical design must be transformed into a physical design consisting of executable computer code.

The Make-or-Buy Decision

Firms faced with the decision of how to best acquire the software needed for a new system have three choices:

1. **Custom Software** The firm's information services staff and users can create software especially tailored to the needs of the firm and the users.

2. **Software Vendors** The firm can purchase prewritten software from software vendors such as Lotus Development Corporation or Borland International.

3. **Outsourcers** The firm can contract with special suppliers, called outsourcers, who will perform some or all of the development of custom software for the firm.

Firm-produced Custom Software

The firm's custom software can be prepared by its own programmers or users.

The Programming Process   The majority of business data is being processed today by programs written by programmers in second-generation languages such as assembler, or third-generation languages such as FORTRAN, COBOL, PL/I, Pascal, BASIC, and C. In using these languages, the programmer employs the design documentation as a basis for the coding, as illustrated in step 1 of the system flowchart in Figure 14.9. The source program is entered into the computer in step 2, and the computer translates the source program into an object program in step 3. Testing is performed in steps 4 and 5, and debugging in step 6. Bugs are corrected by revising the code in step 1. When the program has been tested and debugged, it is put into operational use.

Fourth-Generation Languages   Second- and third-generation languages emerged during the 1950s and 1960s. During the 1970s a new breed of software came on the market that was intended to simplify the programming process. The key feature of the new software was user friendliness, making it possible for persons other than a programmer to generate their own code. The term **natural language** was used because the user phrased commands in the form of sentences. The term **nonprocedural language** was also used, because the user did not have to

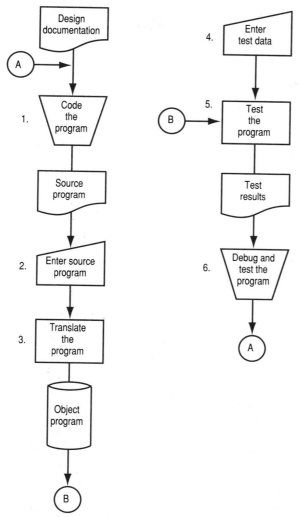

Figure 14.9
The traditional programming process.

spell out the desired processes step by step. Eventually, the term **fourth-genera-tion language,** or **4GL,** stuck.

Today, a large number of 4GLs are on the market, including such well-known products as Natural, Mark IX, FOCUS, IDEAL, and MANTIS. These packages can be used by either programmers or users.

Event-driven Languages   A more recent innovation in program translators is the **event-driven language,** which does not follow the typical top-down format of structured programming but is organized in terms of events. Perhaps a user employs a mouse to click on a button on the input screen to request a summary output. Figure 14.10 is an example of Visual BASIC code that produces the output in Figure 14.11.

An event-driven language is ideal for developing interactive programs featuring computer–user dialog. However, the syntax is such that the language appeals more to information specialists than to users.

CASE Tools   Many CASE tools include **code generators** that produce computer source code directly from the design documentation. Table 14.1 shows the source code languages generated by some popular CASE tools. The source code is then compiled, using the appropriate compiler to produce executable object code.

Automatic code generation is a major reason for the popularity of CASE. Much time-consuming and costly manual coding and debugging are eliminated. Also, by generating the code from the documentation, the information specialists

```
Sub Form_Load ()
   SummaryOutput.Left = (Screen.Width - SummaryOutput.Width) / 2
   SummaryOutput.Top = (Screen.Height - SummaryOutput.Height) / 2

   'Fill Output boxes with values from memory
   Summary(1).Caption = Format$(TotalInterestEarned, "#,###,##0")
   Summary(2).Caption = Format$(TotalInterestPaid, "#,###,##0")
   Summary(3).Caption = Format$(NetInterestEarned, "#,###,##0")
   If Val(Summary(3).Caption) > 0 Then
      Summary(3).BackColor = &HFF00&
   Else
      If Val(Summary(3).Caption) < 0 Then
         Summary(3).BackColor = &HFF&
      Else
         Summary(3).BackColor = &HFFFF&
      End If
   End If

   For A = 0 To 5
      If Index = 0 Or Index = 1 Or Index = 3 Then
         Text1(A).Caption = Format$(General(A), "#,###,##0")
      Else
         Text1(A).Caption = Format$(General(A), "##0")
      End If
    Next A
   For A = 0 To 2
      RecSched(A).Caption = Format$(ReceivePercent(A), "##0")
      PaySched(A).Caption = Format$(PayPercent(A), "##0")
      Next A
   For A = 2 To 13
      PayMonth(A).Caption = Format$(MonthlyPayment(A), "#,###,##0")
      SalesMon(A).Caption = Format$(SalesUnits(A), "#,###,##0")
      Next A
End Sub
```

Figure 14.10
An example of code written in an event-driven language to produce a summary report.

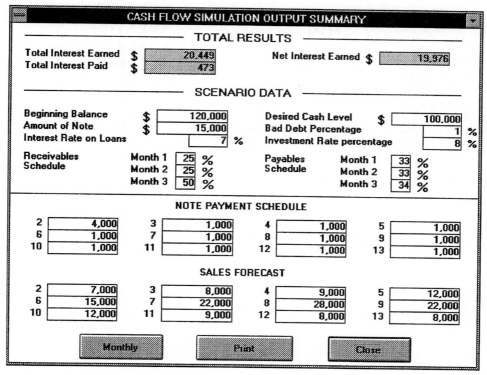

Figure 14.11
The summary report produced by the event-driven language code.

are assured that the program does what it was meant to do. Program maintenance is also simplified. Rather than maintain the code, the task is to maintain the documentation. The code generation capabilities of CASE tools represent the hope of the computer industry because of their ability to handle the large volume of application-development projects.

Application-Development Software   Although 4GLs have enabled users to engage in end-user computing, another class of software has had an even greater impact. The term **application-development software** is used to describe such systems as electronic spreadsheets, file management systems, word processors, desktop publishing systems, and graphics packages. These packages are commonly referred to as **productivity tools.** Most end-user computing today involves the use of application-development software.

Putting the Development of Custom Software in Perspective   Application-development software and, to a lesser extent, 4GLs have relieved information services of a great deal of software development pressure. Information specialists are freed up to develop the larger, more complex systems using the full-blown

Table 14.1   CASE Tool Code Generators Provide for a Variety of Source Program Languages

Tool Name	Supplier	Program Languages			
		Ada	C	COBOL	Other
EPOS	SPS Software Products & Services	●	●		FORTRAN, Pascal
IEF	Texas Instruments		●	●	
Microstep	Syscorp International		●		
Metavision	Applied Axiomatics		●	●	
MRC Productivity Series	Michaels, Ross, and Cole				RPG/400
Powerhouse	Cognos, Inc.				Powerhouse, 4GL, SQL
Powertools	Iconix Software Engineering	●	●	●	FORTRAN, Pascal, C++
Predict CASE	Software AG of North America				Natural
Teamwork	Cadre Technologies	●	●		
Telon	Pansophic Systems			●	PL/I

languages and CASE. Although CASE is potentially valuable for end-user computing, current tools appear too formidable to the users who are not interested in investing large amounts of time learning CASE and going through the steps of documenting their designs so as to generate the code. As of now, CASE is a tool of the information specialist.

Prewritten Software from Vendors

It may be obvious from the very beginning of the SLC that prewritten software will be required. In that case, the steps of the design phase that produce the new system documentation are bypassed. The assumption is that the software vendor will furnish the documentation with the software. However, a strong argument can be made for performing the rest of the SLC steps as described. The problem must be defined, and analysis and design work must be accomplished to specify the nature of the system to be used in solving the problem.

Obtaining Prewritten Software   Much prewritten software is obtained informally. Individuals and small firms often make purchases at computer stores, relying on salespersons for consultation.

However, when firms purchase expensive software or inexpensive software in large volumes, the tendency is to follow the same formal procedure described for the selection of hardware vendors. This involves preparing RFPs and evaluating proposals.

Software Evaluation Criteria   When evaluating the prewritten software, firms pay attention to factors such as performance capabilities, financial stability, documentation, and technical support.

- **Performance Capabilities** A firm can verify the software vendors' claims much as it did the hardware vendors'. Magazines that contain hardware articles also contain critical reviews of prewritten software systems. Also, reference books such as *The Software Encyclopedia,* from R. R. Bowker, and *Software Reviews On File,* from Facts On File, Inc., are devoted exclusively to software. In addition to the printed material, vendors are usually happy to demonstrate their packages, but the demos do not always provide a true picture of what the user can expect. A better approach is to have users in other firms demonstrate the packages and describe their experiences. For the most thorough

Figure 14.12
Computer stores can provide valuable service in selecting prewritten software.

validation, a firm should obtain a copy of the software from the vendor and use it for a trial period.

- **The Vendor's Financial Stability** The vendor's financial stability is important because the vendor will be expected to maintain the software over time—fixing bugs and bringing out improved versions. Too often, a firm will purchase software from a vendor who then goes out of business, leaving the firm "holding the bag."

- **Documentation** Firms also consider the documentation provided with the package. When the documentation is good, all is not lost even if something happens to the vendor. Someone, perhaps a software consultant, can use the documentation as a basis for maintaining the system.

- **Technical Support** It is important that the vendor provide technical support when it is needed. Software vendors usually have a hotline that users can turn to when they have questions or need help.

The performance capabilities provide a certain amount of structure to the task of evaluating the software, but the final decision is a semistructured one that requires considerable judgment.

The Significance of the Prewritten Software Decision   For a small firm with no information services staff, deciding on software is more important than deciding on hardware. Such firms generally *first* identify the best software and *then* acquire hardware that will run the software.

For a large firm with a staff of information specialists, the software decision can be just as important when the system has strategic value and the cost is high. One large, established firm recently went bankrupt due, in part, to an untimely decision to invest in an expensive software package.

Custom Software from Outsourcers

Until recently a firm's make-or-buy decision amounted to a choice between using company-developed software and software from vendors. Today, a third alternative exists, a special class of vendors called outsourcers. An **outsourcer** is a firm that assumes responsibility for some or all of a firm's computer-related activities.

Perhaps the most famous outsourcer is Electronic Data Systems, founded by Ross Perot. EDS entered into one of the largest outsourcing contracts in history when it agreed to a $2.1 billion deal with Continental Airlines.[2] The outsourcing arrangement that shocked the computer world, however, came in 1989 when Kodak decided to contract with Andersen Consulting to provide systems development, with IBM to operate its computer installations, and with Digital Equipment Corporation to manage its data communications networks. The Kodak action drove home the point that outsourcing is an option for even large, sophisticated computer-using firms.

[2]Clinton Wilder, "Giant Firms Join Outsourcing Parade," *ComputerWorld* 25 (September 30, 1991): 91.

Outsourcer Services   Outsourcers can offer any of the following services:[3]

- Data entry and simple processing
- Contract programming
- Facilities management
- Systems integration
- Support operations for maintenance, service, or disaster recovery

Facilities management (FM) means that the outsourcer operates the firm's computer center. **Systems integration (SI)** means that the outsourcer performs all the activities of the SDLC. When the system becomes operational, the SI outsourcer either turns it over to the user or operates it in accordance with an FM agreement.

Advantages of Outsourcing   Outsourcing has become a popular approach to systems development and use because it enables the firm to:

- Control costs. The firm's management believes that the outsourcer can do the contracted work for less or can keep the costs from getting out of hand. Management likes the ability to accurately predict future computing costs by entering into the outsourcing contract.

- Focus on new systems development. By ridding itself of resource-draining tasks, such as systems maintenance, the firm can concentrate its efforts on the development of new systems aimed at achieving greater competitive advantage.

- Gain access to leading-edge knowhow. The outsourcers may be able to offer needed knowledge and skills that do not exist within the firm. A good example of such knowledge is enterprise planning and strategic planning for information resources. Andersen Consulting served as an outsourcer for Harcourt Brace & Company in planning the COPS life cycle that is the subject of the ongoing case.

Outsourcing will no doubt continue to flourish, but there is a certain resistance to it. A 1990 survey of 21 large U.S. firms revealed that three-fourths were opposed to outsourcing.[4] One reason is the fear of losing control over the firm's information resources. Some firms have pursued the outsourcing strategy and later decided to resume doing their own work, a strategy called **insourcing.**

Software Testing

Chapter 8 explained that the structured approach applies not only to programming but also to the entire process of systems development. Figure 8.15

---

[3]Uday Apte and MaryAnne Winniford, "Global Outsourcing of Information Systems Functions: Opportunities and Challenges." *Managing Information Technology in a Global Society,* edited by Mehdi Khosrowpour (Harrisburg, PA: Idea Group Publishing, 1991): 58–59.
[4]Sid L. Huff, "Outsourcing of Information Services," *Business Quarterly* 55 (Spring 1991): 64.

illustrates how analysis, design, and implementation can all be done in a top-down manner.

During the implementation phase the program modules can be coded and tested in a top-down fashion, beginning with the upper-level modules and ending with those on the bottom.

Program testing is normally viewed as a step to be performed only on custom-generated code. However, it is a task that could be performed on software obtained from software vendors and oursourcers as well.

Step 5—Prepare the Database

Database preparation is a two-step process consisting of data description followed by data entry.

Data Description

The steps of describing the data are illustrated in Figure 14.13. The data dictionary specifies the format of the data to be entered in the database. Perhaps paper forms

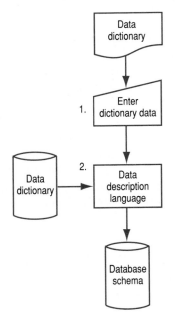

Figure 14.13
The data dictionary provides the basis for the data-base schema.

such as the data element dictionary entry—described in Chapter 10 and illustrated in Figure 10.15—are used. Or, the description can exist in secondary storage as a result of using a database management system (DBMS), a special data dictionary system, or a CASE tool. These software systems use a **data description language (DDL)** to describe the *format* of the data that will comprise the database. The description produced by the DDL is called the schema. The **schema** is a specification of each data element in terms of its name, data type (numeric or alphanumeric), size, and the number of decimal positions.

Data Entry

With the schema specified, the data can be entered into the database. There are three routes to the database, as illustrated in Figure 14.14. Existing data may need to be reformatted or cleaned up. Or, new data may have to be gathered.

- **Reformat Existing Data** The task of entering data into the database is easiest when the firm is converting from one computer database to another. Computer programs are written that transform the data in the existing format to the format required by the new system.

- **Clean Up Existing Data** A more difficult task is when the firm is converting to a computer system from a manual one. The data in the manual system most likely contains errors and inconsistencies that must be cleaned up. Much effort can be required to put manual files in a form acceptable to the DBMS.

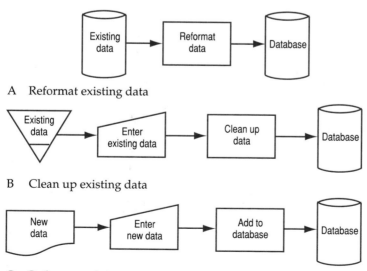

A   Reformat existing data

B   Clean up existing data

C   Gather new data

Figure 14.14
Three routes to the database.

- **Gather New Data** A potentially difficult situation is when data required by the database does not currently exist in the firm. However, it is not always necessary that the firm do the data gathering. More than 4,000 commercial databases are on the market.[5] Some are available for sale, and some, such as DIALOG and BRS (Bibliographic Retrieval Service), can be accessed online by paying a subscription fee.

Other Factors Influencing Database Preparation

In addition to the influence of the format and the availability of the data, the firm's type and size can also affect the database preparation task.

- **The Type of Firm.** Certain industries—insurance and banking, for example—require large databases of historical data. When firms in these industries convert to new computer systems, the databases must reflect past activities. For example, the record for a hospitalization policy issued in 1958 should reflect all activity since that date—changes in coverages, addition of dependents, claims history, and so on. Similarly, the record for a bank customer who is paying off a five-year automobile loan taken out three years ago should reflect all payments that have been made. Firms such as insurance companies and banks cannot simply create their databases as transactions occur; they must capture historical data, which can exist in large volumes and in various formats.

- **The Size of Firm** All else being equal, preparing a database for a large firm is more difficult than for a small one, strictly because of data volume. A small firm might be able to schedule its database preparation to begin one month prior to the planned cutover; but an industry giant might have to provide a lead time of six months, a year, or even longer.

Responsibility for the Database

The manager of database administration, assisted by a staff of DBAs, is responsible for making the database available. DBAs are members of the project teams during the SDLC and monitor system performance during the use phase.

Step 6—Prepare the Physical Facilities

The larger the computer system, the bigger the physical installation task.

Installing Large-scale Systems

Minicomputers and mainframes require special rooms, with some or all of the following features:

[5]John McMullen, "New Allies: IS and Service Vendors," *Datamation* 36 (March 1, 1990): 46.

- **Security** The firm protects its centrally located information resources from unauthorized use, damage, or destruction by keeping the computer room locked. Access is gained by opening the door with a key, inserting a magnetically encoded identification card into a reader, entering numbers into a combination lock, or satisfying more exotic screens, such as those employing fingerprints, palm prints, or even lip prints.

- **Raised Floors** Computer units are connected by cables the size of garden hoses and larger. Rather than lay the cables on top of the floor, where they would pose a hazard and be susceptible to damage, the cables are located under the floor. When the computer room is built, the floor is raised a distance of 2 feet or so above the regular floor. In addition to providing space for the cables, the raised floor also provides a plenum for directing air flow through the various units.

- **Temperature and Humidity Controls** Computer units operate best within certain temperature and humidity ranges. Computer rooms often have their own air conditioning and heating units to ensure that these environmental conditions are met.

- **Pollutant Controls** Air conditioning and heating units also remove pollutants such as dust and paper fibers from the air. When the computer facility consists of several rooms, each with its own type of equipment, separate air pressures can exist in each room to control the flow of pollutants. For example, the room housing the CPU can have a high pressure, and the room housing the printers can have a low pressure. When doors are opened between the rooms, any pollutants in the printer room will not flow into the CPU room.

- **Fire Controls** Smoke and fire detectors are located both on the ceiling of the computer room and under the raised floor. Fire suppression systems using chemicals or gas rather than water are installed. The use of gas requires that alarms sound to warn operators of the need to evacuate.

- **Uninterrupted Power** Power distribution units separate the computers from the source of municipal power in order to prevent them from being damaged by sudden fluctuations in voltage, such as when lightning strikes a transformer. In addition, many firms install auxiliary power generators to produce their own electricity when the municipal power goes out.

Installation Requirements of Networked Systems

When systems are networked together, installation consists of providing the necessary data communications circuits.

Networks that span relatively short distances, such as those in a department, building, or several adjacent buildings, are called **local area networks,,** or **LANs.** LAN circuitry, provided by the firm, can consist of twisted pairs, coaxial cables, or fiber-optical cables. **Twisted pairs,** consisting of four wires, are the type of circuitry used in telephones. This is the least expensive circuitry, but does not offer the high-quality transmission of the other choices. **Coaxial cables** are the type of circuitry used by cable TV services and can transmit the widest variety of signals, including television signals used in video conferencing. **Fiber-optical cables** offer the greatest security against unauthorized tampering because special equipment is required to tap into the circuit.

Networks that span large distances—across a state or nation or around the world—are called **wide area networks,** or **WANs.** WAN circuitry choices include fiber-optical cable and microwave transmission. WAN circuitry is provided by a common carrier, such as AT & T or Sprint.

Putting Construction of Physical Facilities in Perspective

The construction of physical facilities can be very expensive, and the decision of how much to spend is made by the executive committee and the MIS steering committee, with the CIO providing expertise.

Once approval is given to construct the physical facilities, the responsibility to accomplish the task ordinarily falls to the manager of computer operations, who works with the representatives of the firms supplying the hardware, security systems, and environmental controls, as well as the general construction contractor. Because of lengthy lead times, physical installation must be planned far in advance of the time when the computer is to be installed.

Step 7—Educate the Participants and Users

Shortly before cutover, participants and users must be educated and trained so that they are familiar with the new system. Participants must learn how to perform such tasks as data entry. Users must learn how to operate the computer and interpret output. A **user manual**—a written document created by project team members that explains the systems operation in terms the user can understand—can help educate users about the system. An example of such a manual can be found in Technical Module J.

When small numbers of participants and users are involved, they can be trained informally, in one-on-one sessions. Larger projects often require that the material be presented in classroom settings. The systems analyst is a logical choice for the teacher because she or he knows both the new system and many of the people who will be involved. However, in some firms, the teaching responsibility goes to a special educational group within the human resources unit. Representatives of hardware and software firms can also help. Many other organizations also offer various forms of computer training and education. For example, the Association for Computing Machinery (ACM) frequently holds workshops and seminars on such topics as CASE, artificial intelligence, and computer graphics.

Even if the systems analyst does not personally provide the education, he or she should identify the educational and training needs. It is best to schedule the sessions immediately prior to cutover so that the material will be fresh on the participants' and users' minds.

Step 8—Cut Over to the New System

With the physical facilities completed, the hardware and software in place, the data in the database, and the users and participants educated, it is time to put the new system into operation. **Cutover** is the act of replacing the existing system with the new one; the four approaches to cutover—pilot, immediate, phased, and parallel—are illustrated in Figure 14.15.

Pilot System A **pilot system** is one implemented in only one part of the firm's operations as a way to measure its impact. Once the pilot performs satisfactorily, the system is implemented throughout the firm using an immediate, phased, or parallel approach. An example of a pilot cutover is when a food wholesaler uses a mathematical model to determine the optimum layout of a particular distribution center. If all goes well, the firm uses the layout for the other distribution centers. If the pilot does not perform as intended, it might be necessary to rethink the system design.

Immediate Cutover An **immediate cutover** is when the entire existing system is discontinued and the entire new system is immediately put into effect. This total replacement occurs all at once, on a given day. Immediate cutover is feasible only for small firms or small systems because of the impact that cutover has on an organization. It is very disruptive to stop doing things the way they have been done, perhaps for years, and begin doing them a new way.

Phased Cutover When a firm carries out a **phased cutover,** the new system is put into use one part at a time. Each system part constitutes a phase. The parts can be:

- **Subsystems of the system.** Each subsystem is cut over separately. For example, when a manufacturer cuts over to a new accounting system, order entry is cut over first, inventory is cut over next, billing next, and so on.

- **Organizational units.** The system is put into use in only one part of the firm at a time. The parts can be organizational levels, functional areas, geographic sites, and so on. A good example is when the Air Force cuts over to a new aircraft maintenance system. The system is implemented airbase by airbase.

The phased approach is popular for large and complex systems that must be eased into use.

Parallel Cutover The most cautious approach is to follow a **parallel cutover** where the existing system continues to be used until the new one is fully checked out. Both systems run at the same time—in parallel. The main

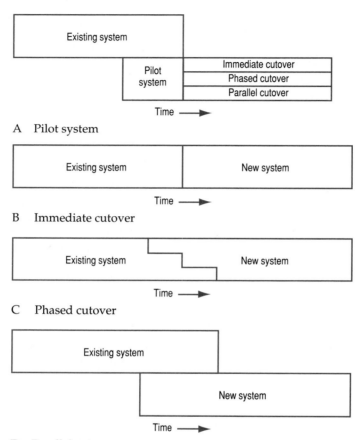

A   Pilot system

B   Immediate cutover

C   Phased cutover

D   Parallel cutover

Figure 14.15
The four approaches to cutover.

disadvantage is the cost. The firm must pay for two systems as long as the parallel operation continues. Also, it may be difficult to assemble two complete systems at the same time when they both require some of the same resources.

When the cutover is completed, the development portion of the SLC ends, and the use phase begins. The development has required much work over a long period—months, or perhaps a year or more. The firm hopes the use phase continues for an even longer period so that the developmental costs can be recovered and the benefits of the new system realized.

Using CASE for Systems Implementation

Many of the advantages of using CASE tools rather than relying on traditional techniques for systems development come during the implementation stage. Because CASE requires and encourages heavy user involvement in the development process, systems implementation is not as difficult due to user familiarity with the system.

Of all the implementation phase activities, CASE is especially helpful in coding, documenting, testing, and educating.

- **Coding**   When the CASE tool has a code generation capability, this step requires hardly any time.
- **Documenting**   The automatic documentation capability of CASE tools ensures that the documentation is done, is done correctly, and stays current with the code.
- **Testing**   Use of CASE reduces *unit testing,* the testing of separate programs or program modules, to an almost negligible level. However, *system testing*, the testing of complete systems, and integration with other systems is still needed.
- **Educating**   Heavy user involvement throughout the development effort reduces the amount of education necessary prior to cutover. In addition, using CASE tools to generate code ensures consistency in the user interface, which also reduces not only the time required for education but also its necessity.

These advantages can clearly be seen during implementation, but they also reappear during the use phase when maintenance and business process redesign become necessary.

Unique Implementation Requirements of the CBIS Subsystems

All CBIS subsystems do not pose the same implementation challenge. Some subsystems are easier to put into effect than others. Table 14.2 illustrates the relative complexity of implementing the five CBIS subsystems.

1. **Plan the Implementation** Data processing systems affect practically all the firm's clerical and administrative employees, and often production and sales personnel as well. The systems also have a direct effect on all the elements in the firm's environment except competition. For these reasons, the planning for data processing systems presents a major challenge. MIS, DSS, and expert system projects also require careful planning when those systems are especially complex or important. For example, when American Express implemented its expert system for validating customers' credit ratings, it planned the implementation with great care. Because of their simplicity, OA systems offer the least planning challenge.

2. **Announce the Implementation** The data processing system has the greatest potential impact on rank-and-file employees, making its announcement

Table 14.2   The Degree of Implementation Challenge Offered by the CBIS Subsystems

	Data Processing	MIS	DSS	Office Automation	Expert System
1. Plan the implementation	Major	Intermediate	Intermediate	Minor	Intermediate
2. Announce the implementation	Major	Intermediate	Minor	Intermediate	Minor
3. Obtain the hardware	Intermediate	Minor	Minor	Minor	Minor
4. Obtain the software	Intermediate	Major	Major	Minor	Major
5. Prepare the database	Major	Major	Intermediate	None	Intermediate
6. Prepare the facilities	Major	Major	Minor	Intermediate	Minor
7. Educate users and participants	Major	Major	Intermediate	Intermediate	Intermediate
8. Cut over to the new system	Major	Major	Minor	Intermediate	Minor

very important. MIS and OA projects also require special communications when their application is organization-wide. DSS and expert system projects usually affect small groups of problem solvers and can be more informally announced.

3. **Obtain the Hardware Resources** All projects do not require new or additional hardware, but when they do, the task can be the most difficult for data processing systems. This is because the computer–human interface can exert such a strong effect on the ultimate success of the system. You can imagine how much thought American Airlines put into selecting the keyboard terminals to be used around the world in its Sabre reservation system.

4. **Obtain the Software Resources** MIS, DSS, and expert system software is typically the most difficult to obtain because the programs solve problems

that are unique to the firm or to individuals in the firm. Because accounting and communications problems can usually be solved with off-the-shelf solutions, data processing and office automation software is the easiest to obtain. However, the wide variety of prewritten data processing software can complicate that selection task.

5. **Prepare the Database** By far the most difficult database to prepare is one required by a data processing system. Accounting data comprises a large portion of any firm's database. MIS projects also require tremendous effort when data must be gathered. DSS and expert system efforts can also impose database challenges, depending on their design, but OA does not.

6. **Prepare the Physical Facilities** When new or remodeled physical facilities are required, it is usually because of the need to revamp the data processing system or MIS. DSS and expert system projects typically involve only the installation of PCs or keyboard terminals. OA projects can require special preparation when audio and video conferencing facilities must be built.

7. **Educate the Participants and Users** Education and training are key to the success of all systems. However, this step demands the most attention for data processing and MIS projects because of the large number of people to be educated.

8. **Cut Over to the New System** Because of their organizational impact, data processing and MIS projects require the greatest amount of care in cutover. OA projects that affect many persons, such as word processing and electronic mail, can also pose cutover problems.

In most cases, data processing and management information sytems are the most difficult to implement because almost everyone in the organization is affected by the new processes. Office automation systems are usually the least difficult because of their turn-key nature.

Putting the Implementation Phase in Perspective

The programmer, DBA, data communications specialist, and operations personnel carry the implementation burden but they can do their work only when the analyst and the user have done theirs. Although the development of a computer-based system is a team effort, the systems analyst and the user set the course that leads to either success or failure.

Summary

The objective of the implementation phase is to convert the logical design into a physical design. The majority of this work is performed by information specialists other than the systems analyst. The MIS steering committee monitors each step of the process, using the detailed implementation plan as the basis. The plan specifies each step to be performed, who will do it, and when it will be done.

Because cooperation of both users and participants is necessary for a smooth implementation, management takes special care in announcing this final developmental phase. Employees' needs are linked to the systems objectives by means of individually tailored motivators, stimulating the employees' cooperation.

Selecting hardware is more difficult today than in the early years of computing because of the increased number of vendors. Information specialists keep up to date on hardware by taking advantage of educational and promotional material provided by the vendors, and by reading computer magazines and reference books.

Firms solicit proposals by providing vendors with RFPs, which provide details concerning the new system design so that all vendors propose hardware to do the same task. When evaluating the vendors' proposals, information specialists consider equipment performance and maintenance, education, and industry knowledge. Vendor claims are verified by checking published material, contacting other users, and, for especially large orders, requiring vendors to solve benchmark problems.

Selecting vendors is the first of two important hardware-related decisions. The second is determining which option to use to pay for the equipment—rental, lease, lease-purchase, or purchase. Financial management becomes involved when the stakes are high.

Software acquisition can take any of three paths. The firm can assign information specialists and users the task of preparing custom software, it can purchase prewritten software from vendors, or it can contract with outsourcers to prepare custom software for the firm.

When custom programming is to be produced inhouse, programmers can use second-, third-, or fourth-generation languages or event-driven languages. They can also use CASE tools that contain code generators. When users do the work, they use 4GLs or application-development software.

Selection of prewritten software can follow a formal process that involves RFPs and proposals. When selecting a software vendor, attention is given to performance capabilities, financial stability, documentation, and technical support. For small firms completely dedicated to prewritten packages, the software decision is more important than the hardware decision.

Contract programming is only one of several outsourcer services, which also include performing the entire SDLC (systems integration), managing the computing center after cutover (facilities management), and other operational tasks, such as data entry. Management is attracted to the outsourcer option because it enables them to control costs, focus on new systems development, and gain access to leading-edge knowhow.

Regardless of which software acquisition path is traveled, the end product should be tested. A top-down testing agenda is especially well suited to the inhouse development of custom code.

The preparation of the database begins with the data dictionary, which provides input to a data description language. The output is the database schema. Once the schema has been defined, the data can be entered. The entry process can vary in difficulty, depending on the extent to which the data must be gathered, cleaned up, and reformatted. The characteristics of the firm can also have an influence. Large firms in industries where historical data must be maintained offer the greatest database challenges.

The operations manager plays the main role in preparing the physical facilities, working with vendors and contractors. The facilities feature special accommodations, such as extra security; raised floors; separate temperature, humidity, pollutant and fire controls; and uninterrupted power. Networked systems require installation of the needed communications circuitry. When a LAN is involved, the circuit choices are twisted pairs, coaxial cables, or fiber-optical cables. A WAN usually involves fiber-optics or some type of microwave transmission.

The final step before cutover is the education and training of users and participants. The systems analyst can have this responsibility, or it can be assumed by others, such as instructors from the human resources unit or outside training organizations.

Cutover can be accomplished four basic ways. A pilot system can precede an immediate, phased, or parallel cutover. Immediate cutover can be considered for small firms and small systems. Phased cutover places less strain on the firm, whereas parallel cutover offers the greatest security against the devastating effects of new system failure.

All systems do not impose the same degree of implementation difficulty. As a rule, the data processing system and the MIS are the most difficult. Next come DSSs and expert systems. Office automation systems should be the easiest to implement.

With the system development completed, the users can begin to enjoy the benefits of the new system.

Key Terms

Benchmark problem

Scheduled maintenance, preventive maintenance (PM)

Unscheduled maintenance

Field engineer (FE)

Prime shift usage

Extra shift usage

Natural language, nonprocedural language, fourth-generation language (4GL)

Event-driven language

Code generator

Application-development software, productivity tool

Outsourcer

Facilities management (FM)

Systems integration (SI)

Insourcing

Data description language (DDL)

Schema

Local-area network (LAN)

Wide-area network (WAN)

User manual

Cutover

Key Concepts

The way that the spotlight shifts from the user and the systems analyst to the programmer and operations personnel during the implementation phase

The three dimensions of the detailed implementation plan—what, who, and when

The benefit of linking system objectives to employees' needs

The way that evaluation criteria seek to gauge multiple dimensions of hardware and software vendors' performance

The inverse relationship between the cost of a hardware payment plan and the flexibility that it offers

How the number of people affected by a computer-based system is related to the implementation difficulty

Questions

1. What documentation provides the basis for the work of the implementation phase? Where is this documentation contained?

2. What is the key to implementation control by the MIS steering committee?

3. What three questions, in order, are answered by the detailed implementation plan?

4. What are the two objectives of the employee implementation announcement?

5. Explain how the size of a firm can influence the difficulty of keeping up with advances in technology.

6. Since the hardware vendors are expert on their own products, why not let them propose the configuration they feel is best suited to a firm's needs?

7. List the four hardware vendor evaluation criteria. For each one that can be measured quantitatively, give an example.

8. Why would a firm go to the trouble of setting up a benchmark problem? Why would a vendor participate in it?

9. Which hardware payment plan should a firm choose when it is uncertain whether the system will succeed?

10. Which payment plans include equipment maintenance?

11. Name three sources of custom software. Hint: One is *not* software vendors.

12. How can a firm be certain that a software vendor's claims are true?

13. Why should a firm be concerned with the financial stability of a software vendor?

14. What outsourcer services, if any, are available during SDLC?

15. Why would a CIO be attracted to outsourcing?

16. What would attract the MIS steering committee to outsourcing?

17. What are the two main steps of database preparation?

18. What does the schema reveal about a data element?

19. Is it easier to prepare a database from computer files, such as magnetic tape and disks, or from manual files, such as paper documents? Explain why.

20. Name two characteristics of a firm that influence the difficulty of database preparation.

21. Name two purposes for a raised floor.

22. Which LAN circuitry is least expensive? Which offers the greatest security against tampering by a computer criminal?

23. Why is it advisable to wait until shortly before implementation to educate employees about the new system?

24. Which one of the four cutover strategies would normally be executed before the others?

25. Name two ways to divide the cutover into phases.

26. What are the disadvantages of parallel cutover?

27. Which of the CBIS subsystems is the most sensitive in terms of announcing its implementation to the employees? Explain why.

28. Which CBIS subsystem is the most demanding in terms of preparing the database? Explain why.

29. Under what circumstances would office automation pose an implementation problem?

Topics for Discussion

1. Assume that it will be your responsibility to announce the implementation of a sales reporting system to the firm's sales representatives. The system will distill data from the customer call reports submitted by the reps and produce preformatted reports that the reps can retrieve using their laptop computers. What will you emphasize in your announcement so that the reps will be motivated to cooperate with the implementation?

2. Explain how the selection of a hardware vendor affects each of the succeeding implementation steps—steps 4 through 8.

3. What features of application-development software account for its popularity with users?

4. The chapter does not mention evaluating financial stability when selecting a hardware vendor. Should this be considered?

5. Which feature, or features, of a mainframe computer room should also be considered in installing microcomputers?

6. A small firm has decided to purchase a microcomputer and a particular accounting package. Review the steps of the planning, analysis, design, and implementation phases, and identify those that could be skipped.

Problems

1. Go to your library and look over the selection of computer magazines. Pick one with an article on some piece of hardware that interests you and write a short summary.

2. Repeat problem 1, but pick a software article.

3. While in the library, find a hardware reference book. Select a review of a piece of peripheral equipment and write a one-page paper evaluating two vendors' products. Identify the vendor that you would recommend and give your reasons.

4. As directed by your instructor, visit a local firm and interview the person in charge of its computer unit. Ask the person to describe the unique characteristics of the industry that influence the way that data is processed. Summarize your findings in a short paper.

Consulting with

Harcourt Brace

(Use the Harcourt Brace scenario prior to this chapter in solving this problem.)

Pretend that you are Ira Lerner and want to use graphics to keep track of the implementation. Prepare the following documentation:

1. Draw a Gantt chart of the implementation schedule illustrated in Table HB9.2. If you have project management software, use it.

2. Read about network diagrams in Technical Module D and draw a CPM or PERT diagram of the Table HB9.2 schedule. Use your own judgment to determine the relationships among the activities. Use project management software if it is available.

3. Prepare a one-page memo that can be sent to Dave Mattson, announcing the implementation. The memo should briefly explain the system and include the implementation schedule illustrated in Table HB9.2. Dave can use the memo to announce the implementation to his area managers.

Case Problem
Splashdown (C)

In only a short time on the job you have made a name for yourself as a good forms designer. Your boss, Mildred Wiggins, who is the manager of systems analysis, first asked you to design a form for use in determining whether to prototype. That was in the case at the end of Chapter 8. Then she asked you for a form to be used in determining project feasibility. That was in Chapter 9. Now she wants a repeat performance in terms of the make-or-buy decision. The scene is Mildred's office.

You: So, I guess the big factor is the backlog of jobs waiting to go on the computer. If we have a lot waiting, we go with prewritten software. Right?

MILDRED: That's part of it, but we have to consider other factors as well. I'd like to hear your ideas.

You: Most of the prewritten software has to do with accounting applications. So the prewritten route is a more realistic option for data processing systems than for, say, MIS or DSS. Isn't that right?

MILDRED: That's correct. If an application is data oriented, we should give a lot of attention to prewritten software. On the other hand, if it's information oriented, such as MIS, DSS, or expert system, we had better get ready for a custom effort. However, there are exceptions to both of these rules. If any system has strategic value, then we might want to do it inhouse. And, if a lot of the type of prewritten software we need is available, we should consider buying it. But, and this is an important point, we should consider it only when the software vendors are stable companies with good track records for quality products, documentation, and support.

You: That's a good idea. Another thing we should include is the expertise that we have in MIS. If we have the necessary developmental skills, that's a strong reason to go custom. If not, we should go prewritten. Take, for example, someone who needs a math model. We have a lot of people with modeling experience and would probably want to do it ourselves. Right?

Mildred: You're right on target. Any other good ideas?

You: Well, I'm afraid I've about run out of gas. Let's see (thinking), the time frame might be a factor. If the user needs something right away, we might not have time to develop it inhouse.

Mildred: That's a good point. Any more gas in your tank?

You: No, only fumes. Have we covered everything?

Mildred: The only thing I can think of is the way the data is processed. If an application requires standard processing, there is a good chance that a prewritten package exists. Let's include that and come up with a prototype form. I'd like the form to use quantitative weights and rates for each of the criteria. We can vary both based on the situation. A high score would favor prewritten software. Work up a form and then test it out with a couple of our users. You won't have any trouble finding guinea pigs because there are so many potential users in the backlog. Try Harold Wu in purchasing and Laura Strong in finance.

(You rough out a form and head for Harold's office.)

Harold Wu: I'm glad to finally see someone from MIS. It's been six months since I asked for help. My needs are really quite simple. We've got this fantastic mainframe database of vendor information but I have a need for more detail. I buy only adhesives and I personally know hundreds of sales reps who work for the various vendors. I want to implement my own database on my micro and be able to quickly retrieve information about a particular sales rep. I might be talking to one on the phone, or one might walk in the door, and I will want to know what kind of service they gave on my last order. Did they go out of their way to meet my needs? Did they handle any credits quickly and cheerfully? Things like that. Am I going too fast? What can you do for me?

You: No, you're not going too fast. It seems to me that all you want to do is retrieve information from the database. No processing. Is that right?

Harold Wu: That's right.

You: Does your application have strategic value?

HAROLD WU: Well it does to me, but I don't think the company will go out of business if we don't do it. It should help me do my job better; I can have better relationships with the vendors.

YOU: We have a lot of database experts in MIS, but never develop the DBMS software ourselves. A lot of good software firms have a lot of products in that area. Tell me, when do you need this?

HAROLD WU: As soon as I can get it.

YOU: I see no real problem. I don't know why someone didn't get with you sooner. Listen, Harold, let me talk to my manager and I'll let you know what you can expect.

(Your next stop is finance.)

YOU: Tell me about your computer needs, Laura.

LAURA STRONG: Well, we need an expert system to help our financial managers determine whether to obtain long-term capital by borrowing money from the bank or issuing stock. It's a complex decision that depends on many factors.

YOU: Do you know of any packaged software that will do this? I guess if you did, you would be using it. Right?

LAURA STRONG: Right.

YOU: Is there anything special about the calculations? If every company were doing it, would they do it about the same way?

LAURA STRONG: I'm sure they would, but I don't think anybody has figured out how to do it yet.

YOU: Well, when you get into artifical intelligence, there are really no prewritten packages as such—just what we call expert system shells. They are prewritten systems but you still have to provide the knowledge base and it sounds like that would be a big job. MIS doesn't have anyone who is skilled in AI but we would like to learn. What's your deadline?

LAURA STRONG: The sooner the better. We've been waiting a long time. If we drag our feet much longer, one of our competitors might get the jump on us.

YOU: Well, we certainly don't want that to happen. Laura, let me bounce this off of my boss and I'll get right back with you.

Assignments

1. Design the form that Mildred has requested. Use your word processor or similar software. The form should consist of a series of scales such as the following:

The backlog of jobs awaiting computer processing is:

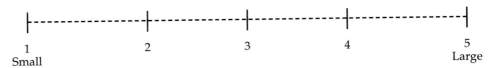

1   2   3   4   5
Small   Large

Each scale will have a weight. The weights should add to 1.00. For each scale, multiply the weight times the points to obtain the weighted points. Add the weighted points for an overall score. Hint: There should be at least eight scales. Make the form as user friendly and self-explanatory as possible. Print a blank copy.

2. Make two additional copies of your form and complete them using the findings of your data gathering.

3. Prepare a memo to Mildred, advising her of your forms design efforts and your recommendations for each of the two jobs. Do you recommend going the prewritten route, and, if so, how strong is your recommendation? Attach the blank form, followed by the two completed forms.

Selected Bibliography

Ameen, David A. "Evaluating Alternative Computer Acquisition Strategies." *Journal of Systems Management* 41 (September 1990): 15–20.

Apte, Uday, and Winniford, MaryAnne. "Global Outsourcing of Information Systems Functions: Opportunities and Challenges." *Managing Information Technology in a Global Society*, edited by Mehdi Khosrowpour. (Harrisburg, PA: Idea Group Publishing, 1991): 58–67.

Bostrom, Robert P.; Olfman, Lorne; and Sein, Maung K. "The Importance of Learning Style in End-User Training." *MIS Quarterly* 14 (March 1990): 101–119.

Drummond, Marshall E., and Reitsch, Arthur R. "Selection Criteria for Fourth Generation Languages." *Journal of Systems Management* 41 (September 1990): 24–27.

Gelperin, David, and Hetzel, Bill. "The Growth of Software Testing." *Communications of the ACM* 31 (June 1988): 687–695.

Gershkoff, Ira. "The Make or Buy Game." *Datamation* 36 (February 15, 1990): 73–77.

Huff, Sid L. "Outsourcing of Information Services." *Business Quarterly* 55 (Spring 1991): 62–65.

Johnson, James R. "Hallmark's Formula for Quality." *Datamation* 36 (February 15, 1990): 119.

Lederer, Albert L., and Prasad, Jayesh. "Nine Management Guidelines for Better Cost Estimating." *Communications of the ACM* 35 (February 1992): 51–59.

Li, Eldon Y. "Software Testing In a System Development Process: A Life Cycle Perspective." *Journal of Systems Management* 41 (August 1990): 23–31.

Loh, Lawrence, and Venkatraman, N. "Determinants of Information Technology Outsourcing: A Cross-Sectional Analysis." *Journal of Management Information Systems* 9 (Summer 1992): 7–24.

McFadden, Fred, and Discenza, Richard. "Confronting the Software Crisis." *Business Horizons* 30 (November–December 1987): 68–73.

McMullin, John. "New Allies: IS and Service Suppliers." *Datamation* 36 (March 1, 1990): 42ff.

McPartlin, John P. "Where's the Human Factor." *InformationWeek* (April 6, 1992): 26ff.

Nelson, R. Ryan, and Cheney, Paul H. "Training Today's User." *Datamation* 33 (May 15, 1987): 121–122.

O'Heney, Sheila. "Outsourcing Solutions to the DP Puzzle." *Bankers Monthly* 108 (July 1991): 26–28.

Ovedovitz, Albert. "Choosing Your Friends: Software Selection for Forecasting." *The Journal of Business Forecasting* 9 (Fall 1990): 15–17.

Schleich, John F.; Corney, William J.; and Boe, Warren J. "Pitfalls in Microcomputer System Implementation in Small Businesses." *Journal of Systems Management* 41 (June 1990): 7–10.

Stein, Elizabeth A. "How to Get What You Want." *Corporate Computing* 1 (September 1992): 146.

"Step-by-Step: 13 Ways to Pick the Winner." *Corporate Computing* 1 (September 1992): 148–149.

Taylor, Thayer C. "Choosing the Right Software Vendor: A Case of Natural Selection." *Sales and Marketing Management* 143 (October 1991): 46–50.

Violino, Bob. "Kodak's Next Step." *InformationWeek* (February 3, 1992): 10–11.

Weston, Rusty, and Cornwell, Dwight. "Micro Mini Mainframe." *Corporate Computing* 1 (September 1992): 62ff.

Structured English

Information specialists who develop business systems have generally recognized that good documentation is important because systems often have a life span of many years. Documentation is helpful not only for developing new systems, but also for maintaining and reengineering existing systems. For many years information specialists relied on system and program flowcharts as their primary documentation tools.

The Birth of Pseudocode

In nonbusiness applications, users have long done much of their own computing, but their programs often do not have the life expectancy of business systems. A mathematician, for example, might create a program to perform a series of calculations and then never use the program again. In such cases as this the need for thorough documentation is not as great as it is for a business organization. Nonbusiness users thus never felt a strong allegiance to flowcharting and began to look elsewhere for a more suitable tool. What they found was something called pseudocode. **Pseudocode,** a narrative documentation that looks like computer code but is not, is a shorthand way to jot down the main steps that must be performed. During the past 10 to 20 years pseudocode has been a popular documentation tool for those outside the business area.

The Birth of Structured English

Today, many businesses do not see the need to devote much time to documentation because their programs are often used once and then discarded. In firms where this is the case, the information services organization needs to provide users with a documentation tool suitable to their needs. Pseudocode seemed to be the right idea, but there were no established guidelines. One user's pseudocode might not even closely resemble that of another user. Because such individual differences in documentation were seen as a hindrance to use of the computer as an organizational resource, standardization was developed that could provide a basis for control.

Information services established pseudocode guidelines, and the result was called structured English. **Structured English** is a shorthand way to document processes using a narrative that conforms to structural and syntactical guidelines established by the firm.

A Structured English Example

Figure H.1 is an example of a program documented using structured English.

The first line identifies the program name. Next, comes the *driver module*, bounded by the words START and STOP. These words mark the logical beginning and end of the program.

In this example, the driver module contains three Perform statements, each referring to a subsidiary module. The subsidiary modules are listed below, with their names aligned on the left margin.

Notice that certain words are printed in uppercase, and there is quite a bit of indenting.

Guidelines for Using Structured English

Unlike flowcharting and, to a lesser extent, data flow diagramming, no generally accepted conventions for using structured English exist. Rather, each firm establishes appropriate guidelines that suit their particular needs. However, the following guidelines will give you some idea of a set that might be adopted.

- Conform to a structured format, consisting of a driver module and subsidiary modules arranged in a hierarchy.

- Bound the driver module with the words START and STOP.

- Label the first line of each subsidiary module with its name.

- Use statements consisting of a verb and object, such as *Read sales order record* and *Print detail line.*

- Assemble the statements in the same order in which their processes are to be performed.

- Use only structured programming constructs. Figure H.2 illustrates these constructs, using program flowchart symbols. The sequence construct groups statements that are executed one after the other. The selection construct represents IF/THEN logic, and the repetition construct controls looping.

- Indent the contents of the selection and repetition constructs. In Figure H.2, the selection constructs begin with an IF statement, and end with an END IF statement. The word THEN is on a separate line, as is the word ELSE, when it is present. The repetition construct is bounded by DO and END DO statements.

```
Screen Sales Orders
START
Perform Enter Sales Order Data
Perform Edit Sales Order Data
Perform Compute Order Amount
STOP
Enter Sales Order Data
      INPUT SALES.ORDER Data
Edit Sales Order Data
      Edit CUSTOMER.NUMBER by ensuring that it is a positive
            numeric field
      IF Edit Is Failed
            THEN WRITE REJECTED.SALES.ORDER Record
      END IF
      DO for Each Item
            EDIT ITEM.NUMBER to ensure that it matches a number
                  in the MASTER.ITEM.NUMBER.LIST
            IF Edit Is Failed
                  THEN WRITE REJECTED.SALES.ORDER Record
            END IF
            EDIT UNIT.PRICE to ensure that it matches the price
                  in the MASTER.PRICE.LIST
            IF Edit Is Failed
                  THEN WRITE REJECTED.SALES.ORDER Record
            END IF
            EDIT ITEM.QUANTITY to ensure that it is a positive
                  numeric field
            IF Edit Is Failed
                  THEN WRITE REJECTED.SALES.ORDER Record
            END IF
      END DO
Compute Order Amount
      IF no edit errors
            THEN DO for Each Item
                        COMPUTE ITEM.AMOUNT = ITEM.QUANTITY*
                              UNIT.PRICE
                        COMPUTE ORDER.AMOUNT = ORDER.AMOUNT sum
                  END DO
                  WRITE ACCEPTED.SALES.ORDER Record
      END IF
```

Figure H.1

The systems analyst uses narrative tools such as structured English to document systems design.

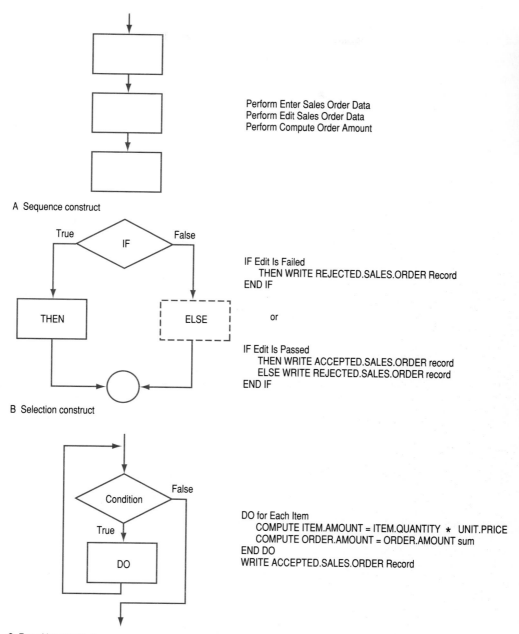

Perform Enter Sales Order Data
Perform Edit Sales Order Data
Perform Compute Order Amount

A Sequence construct

IF Edit Is Failed
    THEN WRITE REJECTED.SALES.ORDER Record
END IF

              or

IF Edit Is Passed
    THEN WRITE ACCEPTED.SALES.ORDER record
    ELSE WRITE REJECTED.SALES.ORDER record
END IF

B Selection construct

DO for Each Item
    COMPUTE ITEM.AMOUNT = ITEM.QUANTITY $\star$ UNIT.PRICE
    COMPUTE ORDER.AMOUNT = ORDER.AMOUNT sum
END DO
WRITE ACCEPTED.SALES.ORDER Record

C Repetition construct

Figure H.2
The three constructs of structured programming are incorporated into structured English.

- Use uppercase for all words typically found in a programming language—words such as READ, COMPUTE, PRINT, IF, THEN, ELSE, and DO.
- Use only data names described in the data dictionary, and print the names in uppercase. Examples are PAYROLL.FILE, CUSTOMER-RECORD, and ITEM_NUMBER.
- Keep in mind that you are *not* creating computer code and do not try to include every little detail. For example, in a computer program you might have to initialize primary storage locations before proceeding with processing, but in structured English you would not bother with such detail. The idea is to communicate the logic of a routine and not worry about how that logic will be implemented.

This list of guidelines may suggest the idea that the term *structured* comes not from the hierarchical, modular design but from the high level of discipline that can be imposed. This is a worthwhile observation. Structured English is just like pseudocode, except for the rules.

The Process Dictionary Entry

Although structured English can be recorded on any type of media, a special form has been designed. It is called the **process dictionary entry** and is illustrated in Figure H.3.

The *System Name* identifies the overall system, recognizing that structured English can be used at several points within a system to document the low-level detail. Several process dictionary entries may be required for a single system. The *Process* is the process within the DFD that documents the system. This identification, which includes the process number, links the process dictionary entry to a particular DFD process. The *Description* is a brief explanation of what the structured English accomplishes. The *Incoming Data Flows* identify the data flows that enter the DFD process, and the *Outgoing Data Flows* identify those that leave. The *Structured English* is the area where the narrative description is entered. Lengthy narratives are continued on the back of the form. The example in Figure H.3 includes most, but not all, of a complete structured English narrative.

Putting Structured English in Perspective

Structured English can be used to document a system in a summary as well as in detail, but it is particularly suited for providing the detail documentation that is not practical with DFDs. Therefore, structured English and DFDs work together to provide all the required process documentation.

As a firm goes about establishing its structured English guidelines, it should give thought to the needs of both users and information specialists. The guidelines should not expect too much from the users. For example, many of the guidelines

PROCESS DICTIONARY ENTRY

Use: To describe each use of structured English in a system.

SYSTEM NAME: *Order Entry*

PROCESS: *1.1 Verify item number*

DESCRIPTION: *Each item number on a sales order is compared to the Master Item Number list to ensure that the number is valid.*

INCOMING DATA FLOWS: *Sales orders*

Item numbers

OUTGOING DATA FLOWS: *Verified item numbers*

STRUCTURED ENGLISH:
Verify Item Number
   DO for Each Sales Order
     DO for Each Item
       Verify SALES.ORDER.ITEM.NUMBER by ensuring that
         it is in the MASTER.ITEM.NUMBER.LIST
     IF Verification Is Failed
       THEN Flag Record
     END IF
   END DO

(Continued on Back)

Figure H.3
An example of a process dictionary entry.

presented above assume that the user has some familiarity with structured programming, which is asking too much in most cases.

    Structured English is probably more appropriate for programmers than for systems analysts. Systems analysts might not be expected to document systems in such detail. Each firm must decide where to draw the line between the analyst's and programmer's responsibilities.

Problems

1. Use structured English to document the following processes, which the departmental supervisor performs when auditing time sheets.

   A. The supervisor examines the sheet and signs it if everything is in order.

   B. If any figures are unacceptable, the supervisor returns the sheet to the secretary.

   C. The acceptable sheets from all departments are sent in the company mail to the accounting department.

2. Use structured English to document the process that a bank's computer performs when editing a New Account application.

   A. The computer edits the new account data and prepares an error listing of all applications that contain errors.

   B. The error listing is sent to the new accounts department.

   C. New account data that does not contain errors is written onto a magnetic tape file named New Account file.

3. Use structured English to document the following system, which processes receipts from suppliers.

   A. Warehouse personnel remove packing lists from cartons.

   B. A data entry operator keys the packing list data into the computer and then files the packing lists in an offline Packing List History file.

   C. The computer program obtains the corresponding records from the Outstanding Purchase Order file, on magnetic disk, and uses that data to print a received purchases report, which is sent to the purchasing department. The same program writes a Supplier Payables file on magnetic disk.

   D. The Supplier Payables file is input to another program, which obtains supplier data from the disk-based Supplier Master file and prepares checks that are mailed to the suppliers. The program also creates a Supplier Check History file on magnetic tape.

Action Diagrams

A class of documentation tools known as **iteration tools** support decomposition by using a top-down approach to subdividing entities into smaller and smaller modules of detail. The lowest level contains the functional, primitive-level modules, whereas the upper levels contain control modules. A generic form of a decomposition tool is pictured in Figure I.1.

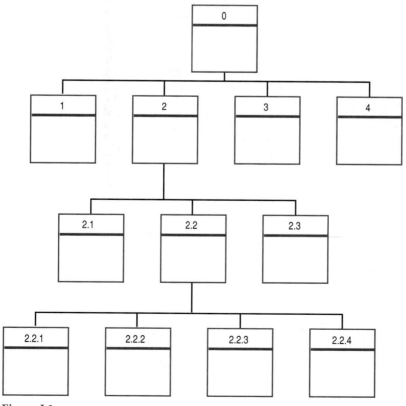

Figure I.1
A generic decomposition chart.

Some examples of decomposition tools are entity-relationship diagrams, Warnier-Orr diagrams, and action diagrams. We describe entity-relationship diagrams in Technical Module A. An illustration of a Warnier-Orr diagram appears in Figure 10.12, accompanied by a brief description. Action diagrams are the subject of this technical module.

An **action diagram** is a documentation tool that consists of a list of processes to be performed, with the pattern of the processing indicated by brackets and some accompanying titles. Figure I.2 is an action diagram that validates a sales order received from a customer.

The name *action diagram* is really a misnomer. Action diagrams are not very diagrammatic. Brackets are the primary symbols, used to depict levels and control structures. No boxes, lines, arrows, circles, or other such artwork is used.

Action diagrams were originally developed by James Martin to be used in conjunction with his information engineering (IE) methodology.[1] The main purpose of an action diagram is to clearly depict program logic in such a way that the diagrams can be translated by the computer into structured code. Action diagrams can also be database oriented and represent simple data-access tasks such as the creating, reading, updating, deleting, sorting, selecting, searching, and joining of data elements and tables.

When programs are to be coded by programmers, systems analysts can use action diagrams to convey program logic to the programmers. Additionally, Martin suggests that the diagrams are simple enough to be used by users engaged in end-user computing.

Not all CASE tools support action diagrams, but all I-CASE tools do. The CASE product of Texas Instruments, known as Information Engineering Facility (IEF), provides for the creation of action diagrams as the final step before code generation. The resultant code is structured and is easily maintained by modifying the CASE tool repository rather than manipulating the code itself. Adherence to a common tool, such as action diagrams, for specification of program logic provides a common language for all analysts, programmers, and users in a firm. This common understanding greatly enhances communication among professionals and the likelihood that a target system will function correctly and be easy to maintain.

The Use of Brackets

The brackets of action diagrams enclose or delineate related sets of statements or instructions. The brackets can be used to delineate all levels of actions, from high-level functions such as menu choice all the way down to the primitive level logic such as an IF/THEN/ELSE/ structure.

All brackets, regardless of level, follow the same rules:

[1]James Martin and Carma McClure, *Action Diagrams* (Englewood, NJ: Prentice-Hall), 1985.

TM

I

```
        VALIDATE ORDER
        CHECK GENERAL FORMAT
              ┌─IF ERROR
              └─WRITE ERROR
              ┌─WHEN NEW CUSTOMER
              │  ENTER NAME AND ADDRESS
              │  ENTER ZIP CODE
              │  ENTER TERMS
              │  ENTER PAYMENT
              │        ┌─IF ERRORS
              │        │  SET INVALID INDICATOR
              │        │  ELSE
              │        └─SET VALID INDICATOR
              │  WHEN OLD CUSTOMER
              │  CHECK FOR VALID TERMS
              │  CHECK FOR PAYMENT
              │        ┌─IF ERRORS
              │        │  SET INVALID INDICATOR
              │        │  ELSE
              │        └─SET VALID INDICATOR
              │  WHEN CANCEL
              └─SET CANCEL FLAG
        IF INVALID INDICATOR IS SET
        WRITE ERROR MESSAGE
```

Figure I.2
Action diagram for validation of ORDER.

1. Brackets are entered from the top.
2. Once entered, sequential execution proceeds from top to bottom.
3. Brackets are exited from the bottom.

Titles may be used at the right of the top of the bracket. When used, the titles serve the same function as comment lines in program code. In the example in Figure I.2, the outer bracket is titled *Validate Order*, which is the name of the entire routine. The other brackets do not contain titles.

TM
I

Support for the Three Basic Structured Programming Constructs

Action diagrams support the three basic constructs of structured programming. These constructs are illustrated in Technical Module H, in Figure H.2. The constructs include sequence, repetition, and selection.

Sequence Construct

In the **sequence construct,** the sequential progression is from one module to another or one statement to another, with no branches or GOTOs.

One or more sequential actions may be included within a bracket. The actions are listed, in verb-object form, one after another and are executed in the order in which they are listed. Figure I.3 illustrates the generic use of action diagrams to support sequential processing.

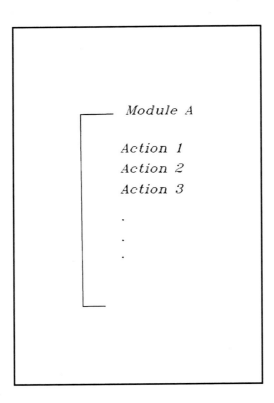

Figure I.3
The sequence construct in action diagrams.

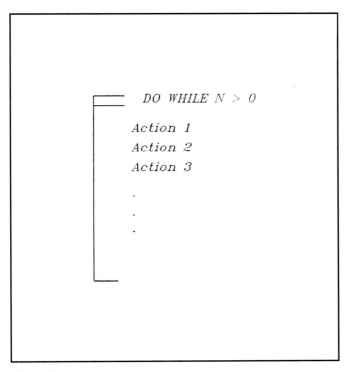

Figure I.4
The repetition construct in action diagrams.

Repetition Construct

In the **repetition construct,** also called the **iteration construct,** processes are repeated a certain number of times. Structured iteration without GOTOs means use of such logic as DO WHILE, REPEAT UNTIL, and PERFORM UNTIL.

Action diagrams use a double bar at the top of the bracket to indicate iterative processing or looping. The condition that controls the iteration is written out at the right of the double bar. In REPEAT UNTIL iterative control structures, the *UNTIL* condition is written at the bottom of the bracket to the right. Figure I.4 illustrates a DO WHILE and DO UNTIL repetition structure.

Selection Construct

In the **selection construct,** also called **branching,** program logic follows one of two or more paths, depending on a condition that may or may not exist. This is the IF statement logic of a programming language.

There are two acceptable selection control structures. The first is the IF/THEN/ELSE structure, and the second is the CASE structure. In action diagrams, the IF/THEN/ELSE structure has a bracket that goes from the IF statement to the last action and a **partition** (a short line going off to the right and connecting to the bracket), which delineates the ELSE portion of the statement. Figure I.5 illustrates a generic IF/THEN/ELSE control structure in action diagram form.

In the CASE type of selection, there is an overall bracket, and at each value that the CASE conditional value takes on, there is a partition indicating mutually exclusive actions. Only one of the CASE partitions is executed. An example of a CASE construct is menu selection. If the menu contains five choices, a partition occurs when the conditional variable takes on the values 1 through 5. Only one choice can be executed at a time. Figure I.6 illustrates a generic use of the CASE construct in action diagram form.

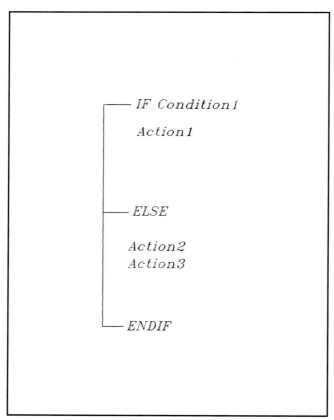

Figure I.5
The selection construct in action diagrams.

TM

I

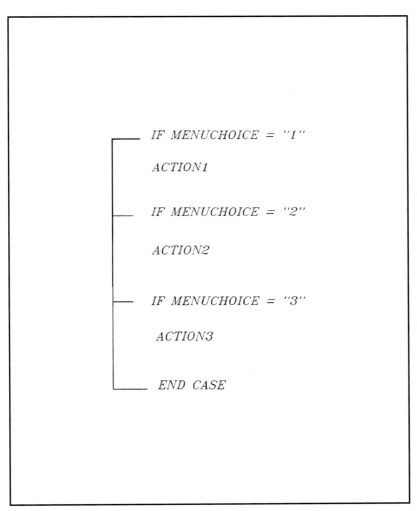

Figure I.6
The CASE selection construct in action diagrams.

You can see from the examples how the brackets add to the readability of the diagrams, making them easier to understand than purely narrative documentation, such as pseudocode and structured English. The brackets make the diagram less intimidating and more appealing to users than other types of documentation.

TM

I

Action Diagram Capabilities

Action diagrams facilitate several capabilities that are characteristic of structured systems. These capabilities relate to tree structures, nesting, program stubs, module inputs and outputs, and database actions.

Tree Structures

If you turn an action diagram so that the length of the bracket is at the top, you will see that the action diagram can be read like a tree chart. See Figure I.7, which illustrates an action diagram translated into tree form.

Nesting

Brackets can also show nesting, which is related to hierarchical ordering. When programs contain several levels of nesting, understanding this complexity becomes difficult. The brackets in action diagrams allow the nesting levels to be clearly identified. Every entry point and exit point of each construct is easy to see. Figure I.8 illustrates program logic with nesting five levels deep.

Program Stubs

In top-down design and coding, it is sometimes necessary to indicate **program stubs,** which are planned modules that as yet have not been designed or coded. Action diagrams can indicate the existence of a program stub by using a bracket drawn with dashed lines and an asterisk before the bracket title, as shown in Figure I.9.

Module Inputs and Outputs

At the primitive level, action diagrams can be used to illustrate the inputs and outputs of program modules. To do this, the brackets are used to form a rectangle. The title of the module is inserted in the top bar of the rectangle. Inputs are noted at the upper right corner of the rectangle, and outputs are noted at the lower right corner. The rectangles can be nested in the same manner as the brackets; however, nesting of rectangles can become cumbersome at more than three levels. Figure I.10 illustrates a nested primitive-level action diagram with input and output.

Database Actions

When action diagrams are used to illustrate database actions, the data name is enclosed in a rectangle, as shown in Figure I.11. The rectangle signals the reader that the boxed information is a record or data structure name. If the action is a data

TM

I

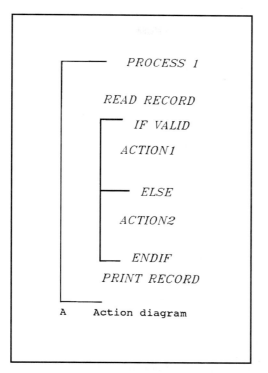

A   Action diagram

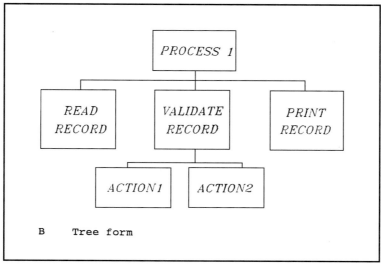

B   Tree form

Figure I.7
An action diagram translated into tree form.

TM
I

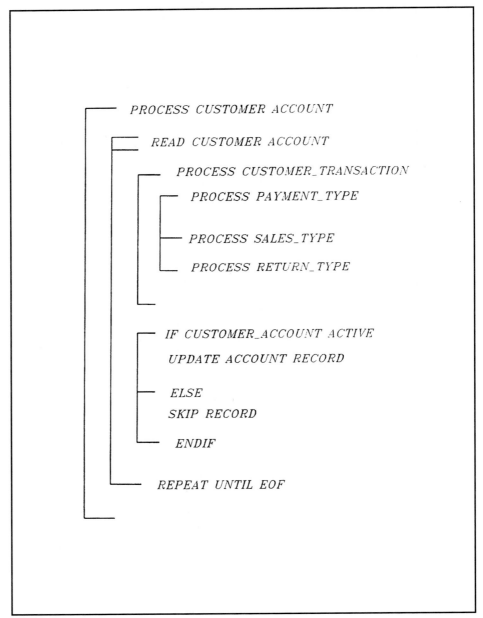

Figure I.8
An action diagram with nesting.

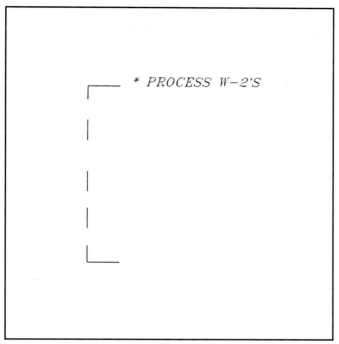

Figure I.9
A program stub in action diagram form.

access action such as SORT, JOIN, or PROJECT, the compound record or data structure name is boxed in by a double rectangle. In action diagram SELECT statements, the double rectangle indicating the compound record or data structure can be followed by a conditional statement that indicates the bases on which the SELECT is performed.

A Sample Action Diagram

In this technical module, all the figures are drawn in a generic form. Figure I.12A lists BASIC code from a fairly simple program. Figure I.12B shows the code in action diagram form. If this diagram were given to a programmer, he or she could easily translate it into the language to be used in the target system.

TM

I

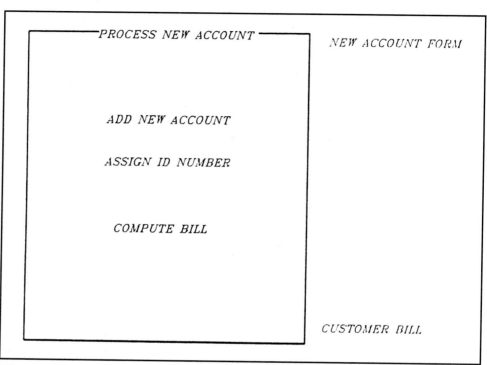

Figure I.10
An action diagram module with input and output.

Putting Action Diagrams in Perspective

Action diagrams, like Warnier-Orr diagrams, can document systems at all levels. Therefore, you can completely model a process with a series of action diagrams. Unlike Warnier-Orr, however, action diagrams are receiving some degree of support from CASE tools, especially I-CASE. This support should ensure their continued use.

Systems analysts should be proficient in action diagramming as a way to document program logic for either programmers who will code the logic in program languages, or CASE tools that will generate the code directly from the documentation. Used in this way, action diagrams have a place in systems projects when information specialists develop systems for users.

Perhaps the real strength of the action diagram, however, is its appeal to users engaged in end-user computing. The reason for the appeal is simplicity. It would be difficult to conceive of a way to document structured processes in a less intimidating way. Any documentation tool that has appeal to end users is a pearl of great price.

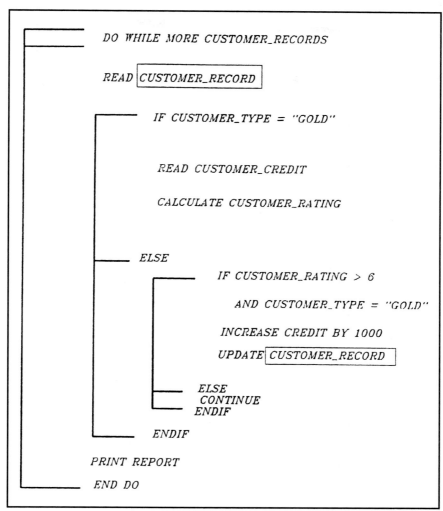

TM
I

Figure I.11
Database actions in action diagram form.

Problems

1. Draw generic action diagrams to illustrate the following:

   A. Nesting

   B. Module inputs and outputs

   C. Iteration

D.   IF/THEN/ELSE selection

E.   CASE selection

F.   Program stub

2.   Draw an action diagram using the following code:
DO WHILE NOT EOF
READ TRANSACTION
CHECK GENERAL FORMAT
IF ERROR WRITE ERROR MESSAGE ELSE CONTINUE
IF CUSTOMER_TYPE = "NEW"
CREATE CUSTOMER_NAME
CREATE CUSTOMER_ADDRESS
SET CUSTOMER_BALANCE = 0
ELSE IF CUSTOMER_TYPE = RENEWAL
CHECK VALID TERM
ENDIF
ENDIF
IF ERRORS SET INVALID INDICATOR ELSE SET VALID INDICATOR END
ENDIF
WRITE REPORT
END DO

```
4010 WHILE RES$ <> "Q"              'KEEP GOING UNTIL Q ENTERED
4020    LET NUM2 = 1                'START DRILL AT 1
4030    WHILE NUM2 <= 12 AND RES$ <> "Q"
4040       GOSUB 5000               'SET UP PROBLEM
4050       GOSUB 6000               'INPUT THE ANSWER
4060       GOSUB 6500               'VALIDATE ANSWER
4070       IF ANS = NUM1 + NUM 2
               THEN GOSUB 7000
               ELSE GOSUB 7500
4080       GOSUB 8000               'PRINT BOTTOM MESSAGES
4090       LET RES$ = INPUT$(1)     'GET ONE CHAR FROM KEYBOARD
4100       LOCATE 12,35
4110       PRINT SPACE$(9)          'CLEAR OUT MESSAGE
4120       GOSUB 8500               'CLEAR OUT BOTTOM MESSAGES
4130    WEND
4140    LOCATE 9,35
4150    PRINT SPACE$(15)            'CLEAR OUT PROBLEM
4160    IF RES$ <> "Q"
           THEN GOSUB 2000          'INPUT A NEW NUMBER FOR DRILL
4170 WEND
```

A BASIC programming code

Figure I.12

An action diagram of BASIC programming code.

Continued

TM

I

```
            ┌──────┐   DO WHILE RES$ <> "Q"
            │      │   LET NUM2 = 1
            │      │  ┌─────── DO WHILE NUM2 ,+ 12 AND RES$ <> "Q"
            │      │  │        GOSUB 5000
            │      │  │        GOSUB 6000
            │      │  │        GOSUB 6500
            │      │  │    ┌────── IF ANS = NUM1 + NUM2
            │      │  │    │       THEN GOSUB 7000
            │      │  │    │────── ELSE
            │      │  │    │       GOSUB 8000
            │      │  │    └────── ENDIF
            │      │  │        LET RES$ = INPUT$(1)
            │      │  │        LOCATE 12,35
            │      │  │        PRINT SPACE$(9)
            │      │  │        GOSUB 8500
            │      │  └─────── END DO
            │      │          LOCATE 9,35
            │      │          PRINT SPACE$(15)
            │      │          IF RES$ <> "Q"
            │      │  ┌─────── THEN GOSUB 2000
            │      │  │        ELSE
            │      │  │        CONTINUE
            │      │  └─────── ENDIF
            └──────┴───── END DO
```

B Action diagram

Figure I.12   Continued

3. Draw an action diagram for the following database actions:
   SELECT SUPPLIER FROM QUOTATION WHERE RATING > 3
   SORT SUPPLIER BY LAST_NAME
   PROCESS NEW EMPLOYEE
   CREATE EMPLOYEE RECORD
   CREATE PAYROLL RECORD
   CREATE PERSONNEL RECORD
   IF EMPLOYEE_TYPE = "SALARY" ENTER SALARY_AMOUNT ELSE
   EMPLOYEE_TYPE = "HOURLY" ENDIF

4. Draw an action diagram for the following menu. The selection condition is the numeric value of the menu choice.

```
                    PAYROLL SYSTEM
                      MAIN MENU
                       MM/DD/YY
              1.   FILE MAINTENANCE
              2.   PAYROLL CALCULATION
              3.   REPORT GENERATION
              4.   EXIT

      ENTER THE NUMBER OF THE ACTION YOU
      WISH TO TAKE  [        ]
```

5. Draw a high-level action diagram of the following procedure:

   When a customer returns an item to MAXIMART, he or she must go to the customer service department in the rear of the store. The item is returned to a customer service representative along with the sales receipt. If the customer does not have the receipt, the item cannot be returned. The service representative then follows one of these procedures:

   For Cash Sales   The service representative writes up a cash voucher for the returned item, which is signed by the customer. The cash voucher is attached to the original sales receipt and placed in the cash register drawer. The amount of the credit is rung up on the cash register, and the cash is removed from the drawer and given to the customer. The returned item is placed in a bin.

TM

I

For Credit Sales   The service representative writes up a credit voucher for the returned item, which is signed by the customer. One copy of the credit voucher is given to the customer, one copy is attached to the sales receipt and placed in a bin with other such documents, and one copy is retained by the service department. The copy with the sales receipt will be used later as a source document for data entry. The returned item is placed in a bin.

For Exchanged Goods   The customer may exchange the undesired item for an item with the same stock number and price. If this is the case, no paperwork is required. The customer retains the original sales receipt.

In all cases, if the item is damaged, it is placed in a bin with other items that will be sent back to the manufacturers. If the item is not damaged, it is placed in a bin with other items to be retagged and returned to the sales floor.

The credit vouchers are sent to data entry at the end of each day. The data entry personnel credit the appropriate customer accounts using keyboard terminals. These entries are combined to make up a daily credit voucher transaction file. The customer accounts are also part of the customer master file, which is updated after all the transactions have been entered. Also, the appropriate daily sales logs must be updated.

Finally, a summary report of all returned items is generated. The data for this report is pulled from both the credit voucher transaction file and the daily cash report, which shows all types of sales, credits, and allowances from the cash register activity for the day. A four-digit code, which signifies type of transaction, separates cash sales, credit sales, cash refunds, and credit refunds.

6. Draw a hierarchy diagram similar to Figure I.1 from the action diagram that you drew in Problem 5.

Pricing Model Version 2.0
User's Manual

Re-engineered by

TEXAS REENGINEERING CONSULTANTS

Chris Byrne
Cynthia Caldarola
Steven Gray
Mir Asad Ali Khan
Gautam H. Mudunuri
Jeff Rush

TM

J

Table of Contents

TM

J

Installation

This section describes the content of the *TRC Pricing Model*. In addition, it describes the computer requirements you must have to use the *Pricing Model*. Please read these instructions carefully before beginning the installation process.

Checking Your Package

The *Pricing Model version 2.0 User Manual* should contain the following:

1. Brief introduction of the program and its purpose
2. Installation instructions
3. Information on the user interface, including menus
4. Report layouts
5. Tutorial
6. Glossary of terminology
7. Error message catalog

The *TRC Pricing Model* should include a 3 1/2-inch high density disk and a 5 1/4-inch double-sided, double density disk.

System Requirements

To run the *TRC Pricing Model* on your IBM PC or compatible (80386 or higher), you should have:

1. 2 megabytes (MB) of random access memory (RAM)
2. A hard disk and either a 3 1/2-inch or 5 1/2-inch floppy disk drive
3. 1 megabyte (MB) of available hard disk storage
4. Disk Operating System (DOS) version 5.0 or higher
5. *Microsoft Windows 3.0* or higher
6. Mouse (optional)

TM

J

Installing the Model

Follow these steps to install the system:

1. Make sure you have the 3 1/2-inch or 5 1/4-inch disk containing the *TRC Pricing Model*.

2. Insert the installation disk in the appropriate drive (A: or B:)

3. Change to this drive by typing
    a:
    if the disk was put into drive A or
    b:
    if the disk was put into drive B.

4. Execute the installation program from the disk by typing
    install c:\windows
    The parameter passed to the install program is the drive and directory where *Microsoft Windows* is installed on your system.

5. Next, set up the *TRC Pricing Model* in *Microsoft Windows*. Start by opening the **Applications** program group in the Program Manager under *Windows*. Choose the **File-New...** menu option from the Program Manager menu.

6. Select the **New Program Item** option and click on the **OK** button.

7. Enter details of the program item as follows:
    under *Description,* enter **Pricing Model** and
    under *command line,* enter **c:\windows\price.exe**.

    Note: The drive and path of price.exe depends on the *Windows* installation on your machine.

8. Set the *working directory* to a directory where you plan on keeping all of your *Pricing Model* data files. After this information has been entered, click on the **OK** button.

9. An icon for the *Pricing Model* program will be created under the *Applications* program group. The program can be run by double clicking on the Pricing icon with the mouse.

ii

CHAPTER 1
What is the Pricing Model ?

What does the model do ?

The model simulates a market where a firm is constrained in its decisions by what its competition is doing and the economic and seasonal fluctuation. The model considers projected changes in competitors' prices and marketing budgets from the previous quarter to the next. The model also takes into consideration the economic environment and seasonal sales fluctuations.

By considering the prior quarter's environment, plus that of the next quarter, the model is able to project changes in the demand for a firm's product.

Who can use this model ?

This model can be used by managers in the marketing, finance, or manufacturing areas.

A marketing manager would use the model to set the price and the marketing budget.

A finance manager would use the model to make the price decision in those firms where that decision is not assigned to marketing.

A manufacturing manager would use the model to determine the level of plant investment and research and development.

How does the model work ?

This model is used dynamically to find the net profit after tax for a firm. It is deterministic and non-optimizing. You use the model in a "what-if" manner to determine the best solution. The best solution is the one that maximizes the after-tax profit.

The model starts by accepting your input describing the previous quarter's situation and the projected situation for the upcoming quarter. Then, it accepts your decisions and projects your revenue and expenses for the upcoming quarter. The four key decision variables are: price, plant investment, marketing, and research and development. When making the projection, the model considers how the economy will change and what changes your competition is likely to make in pricing and marketing its products.

In order to find the best solution, you must continue changing the decision variables and running the simulation until you obtain the highest profit. This model is designed so that you, the user, optimize the profit.

TM

J

CHAPTER 2
How to Use the System

FUNCTION KEYS

These function keys are used on the data input and report screens in the system. Also, wherever you see a key word with a letter underlined, you can press ALT and the letter to execute that function.

<TAB>	Next field
Shift-<TAB>	Previous field
Ctrl-F1	Help for current field
F2	Load input data
Shift-F2	Save input data
F3	Load results
Shift-F3	Save results
F4	Simulate the next quarter
F5	View simulation reports
Ctrl-X	Exit the system

MENU OPTIONS

FILE MENU OPTIONS

This screen contains a pull-down menu for the file options, providing the user with the capability to manipulate files, generate simulation data, and print reports.

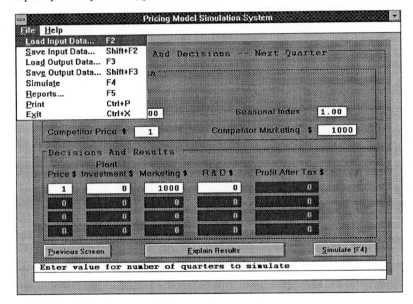

Figure 2.1

To obtain the File Menu, click on the name "File" with the mouse, or press **Alt+F**. This menu contains eight options:

Load Input Data (F2)
This option allows the user to load an input data file from a disk drive or hard disk. The user can select this option by clicking on the option in the menu bar with the mouse, pressing **F2,** or pressing **Alt+L**.

2-1

TM

J

Save Input Data (Shift+F2)
This option allows the user to save an input data file to a disk drive or hard disk.
The user can select this option by clicking on the option in the menu bar with the mouse,
pressing **Shift+F2,** or pressing **Alt+S**.

Load Output Data (F3)
This option allows the user to load an output data file from a disk drive or hard disk.
The user can select this option by clicking on the option in the menu bar with the mouse,
pressing **F3,** or pressing **Alt+D**.

Save Output Data (Shift+F3)
This option allows the user to save an output data file to a disk drive or hard disk.
The user can select this option by clicking on the option in the menu bar with the mouse,
pressing **Shift+F3,** or pressing **Alt+E**.

Simulate (F4)
This option allows the user to initiate a simulation by clicking on the option in the menu
bar with the mouse or pressing **F4**.

Reports (F5)
This option takes the user to the main report menu *PRICING MODEL SIMULATION
FOR FOUR QUARTERS*, where the user can look at all of the reports generated from the
simulation.

Print (Ctrl+P)
This option prints the current screen. The default printer is LPT1. These options can be
changed inside *Microsoft Windows*. The user can select this option by clicking on the
option in the menu bar with the mouse, pressing **Ctrl+P**, or pressing **Alt+P**.

Exit (Ctrl+X)
This option exits the user from the entire system. A message will be displayed on the
screen to warn the user before actually exiting. The user can select this option by clicking
on the option in the menu bar with the mouse, pressing **Ctrl+X**, or pressing **Alt+X.**

TM

J

LOAD INPUT DATA

This screen allows the user to load an input data file. All of the input files use the
extension ".prd."

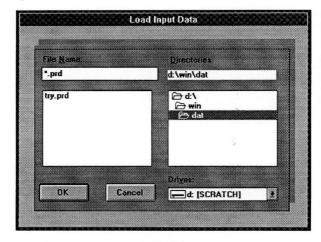

Figure 2.2

This screen contains three fields:

File Name
In the File Name box is a list of the files found in the current directory. The user may type
in a file name with the ".prd" extension or click on the file name with the mouse.

Directories
The directory box shows a list of the directories under the current directory in a "tree
structure." The user may either type in the directory name, specifying the drive name first,
or click on the directory with the mouse.

Drives
The drive can be changed by clicking on the "down arrow" on the right side of the drive
box. A list of drives will be displayed. When the user reaches the bottom of the list, the
arrow will change to an "up arrow."

2-3

TM

J

This screen also contains two buttons:

<u>OK</u>
By selecting the OK button, the file specified on the screen will be loaded into the program and the user will be returned to either the *INTERNAL FIRM AND ENVIRONMENTAL DATA -- LAST QUARTER* screen or the *ENVIRONMENTAL DATA AND DECISIONS -- NEXT QUARTER* screen, depending on where the user selected this option.

<u>Cancel</u>
By selecting the Cancel button, the user is returned to the original screen without any action being taken on the file.

SAVE RESULTS

This screen allows the user to save the output data in a file with the ".pro" extension.

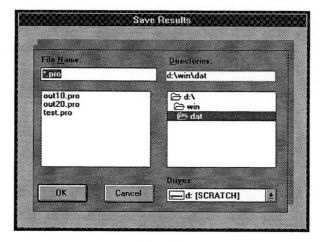

Figure 2.3

This screen contains three fields:

File Name
In the File Name box is a list of the files found in the current directory. The user may type in a file name with the ".pro" extension or click on the file name with the mouse.

Directories
The directory box shows a list of the directories under the current directory in a "tree structure." The user may either type in the directory name, specifying the drive name first, or click on the directory with the mouse.

Drives
The drive can be changed by clicking on the "down arrow" on the right side of the drive box. A list of drives will be displayed. When the user reaches the bottom of the list, the arrow will change to an "up arrow."

This screen also contains two buttons:

OK
By selecting the OK button, the output data will be saved in the file specified on the screen and the user will be returned to either the *CURRENT PERIOD DECISIONS* screen, the *OPERATING STATEMENT* screen, or the *INCOME STATEMENT* screen, depending on where the user selected this option.

Cancel
By selecting the Cancel button, the user is returned to the original screen without any data being written to the file.

TM

J

HELP MENU OPTIONS

This screen contains a pull-down menu with the help options, providing the user with the capability to view a screen with a list of all of the function keys that can be used in the system and information on how to get help on a particular field.

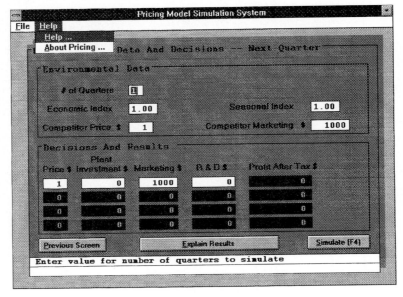

Figure 2.4

To obtain the Help Menu, click on the name "Help" with the mouse or press **Alt-H**. This menu contains two options:

Help
This option displays a screen with a list of all of the function keys and their actions. These function keys can be used anywhere in the system. On this screen, there is also information on how to find help on any input field. The user can select this option by clicking on the option in the menu bar with the mouse or pressing **Alt+H**.

TM

J

About Pricing

This option displays a screen with information about *Texas Re-engineering Consultants*. The user can select this option by clicking on the option in the menu bar with the mouse or pressing **Alt+A**.

HELP SCREEN

This screen displays a list of the function keys for use on any of the screens in the system. For information on a particular input field, the user must click on the word with the mouse or refer to the *Glossary* in this manual.

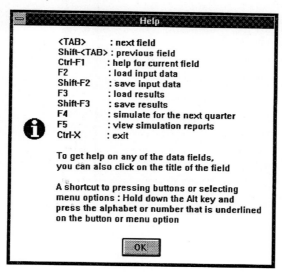

Figure 2.5

OK

By clicking on the OK button or pressing the <ENTER> key, the user will be returned to the original screen from which this option was chosen.

DATA ENTRY FORMS

INTERNAL FIRM AND ENVIRONMENTAL DATA --LAST QUARTER

This screen allows the user to enter the decision variables for the last quarter of the user's firm and the user's competitor.

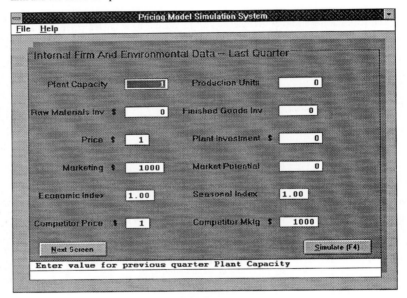

Figure 2.6

This screen contains twelve fields and two buttons which are:

Plant Capacity
This is the production capacity of the plant, in units. The plant can produce this capacity working only a single shift (8 hours per day; 5 days per week). Enter a positive value ranging from 1 to 9,999,999. Do not enter the commas.

Production Units
This is the production volume, in units. Enter a positive value ranging from 1 to 9,999,999. Do not enter the commas.

2-9

Raw Materials Inventory

This is the dollar value of raw materials inventory. You can estimate the number of units that can be produced by dividing this value by the unit materials cost of $6.25. Enter a positive value ranging from $0,000,000 to $9,999,999. Do not enter leading zeros or the dollar sign or commas.

Finished Goods Inventory

These are the unsold units of finished goods from the previous quarter. Enter a positive value ranging from 0 to 999,999. Do not enter the comma.

Price

This is the price that you charged during the previous quarter, or the price that you wish to charge during the upcoming quarter. Enter any positive value ranging from $1 to $999. Do not enter the dollar sign or the comma.

Plant Investment

This is the amount of money that you invested in improving your plant last quarter, or the amount that you wish to invest this quarter. Money invested in the plant does not take effect during the period of the investment, but in the following period. Plant investment tends to counteract the effect of depreciation, which reduces production capacity by a factor of 2.5% (.025) per quarter. Production capacity for the next quarter is increased one unit for each $70 invested in plant this quarter. Enter any positive value from $0 to $9,999,999. Do not enter the dollar sign or commas.

Marketing

This is the amount of money invested in all types of marketing activity (personal selling, advertising, sales promotion, etc.) last quarter, or the amount that you want to invest during the upcoming quarter. The higher the investment in marketing, the higher the market potential. Enter a positive value ranging from $1,000 to $999,999. Do not enter the dollar sign or the comma.

Market Potential

This is the **demand** for your product. It is influenced by the economic and seasonal indexes, as well as the price that you and your competitors charge and the marketing investment by your firm and your competitors. This value is computed by the model, and can range from 0 to 9,999,999. Do not include the comma.

Economic Index

This is an index that represents the health of the economy. The base for the index is a value of 1.00. Values higher or lower represent the percentage increases or decreases. For example, an index of 1.10 means that the economy is 10% healthier than normal. A healthy economy (index values over 1.00) has the effect of increasing the market potential

TM

J

for the product. A weak economy (index values less than 1.00) has the effect of decreasing the market potential. Enter any value from 0.00 to 9.98. You do not have to enter leading or trailing zeros. The value must be positive.

Seasonal Index
The seasonal index represents the fluctuation in sales of the product throughout the year. The average quarter's sales is represented by the value 1.00. A strong sales quarter (index value more than 1.00) has the effect of increasing the market potential for the product. A weak sales quarter (index value less than 1.00) has the effect of decreasing the market potential. Enter any value from 0.00 to 9.98. You do not have to enter leading or trailing zeros. The value must be positive.

Competitor Price
This is the average price charged by all the firm's competitors. Enter any positive value ranging from $1 to $999. Do not enter the dollar sign.

Competitor Marketing
This is the average amount of money invested in all types of marketing activity (personal selling, advertising, sales promotion, etc.) by all of the firm's competitors. Enter a positive value ranging from $1,000 to $999,999. Do not enter the dollar sign or the comma.

At the bottom of the screen are two buttons:

Next Screen
By clicking on this button or pressing **Alt+N**, the user may proceed to the *ENVIRONMENTAL DATA AND DECISIONS-- NEXT QUARTER* screen.

Simulate (F4)
By clicking on this button, pressing **F4**, or pressing **Alt+S**, the user may initiate the simulation, once the appropriate data has been entered into the fields.

Messages are displayed on the bottom of the screen to remind the user what to do in the fields.

TM

J

ENVIRONMENTAL DATA AND DECISIONS - NEXT QUARTER

This screen allows the user to input data for the next quarter. The user is able to simulate the data on this screen and receive an interpretation of the data. Each time a quarter is simulated, the row of four decisions on the lower half of the screen is moved down one row. The four quarters' activity is displayed in a reverse sequence (4-3-2-1) reading from top to bottom.

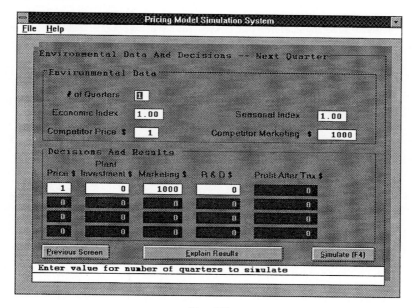

Figure 2.7

This screen contains ten fields and three buttons which are:

Number of Quarters to Simulate (1 or 4)
This is the amount of time for which you want to run the simulation. The only acceptable values are 1 quarter or 4 quarters. Enter either a <1> or a <4>.

Economic Index
This is an index that represents the health of the economy. The base for the index is a value of 1.00. Values higher or lower represent the percentage increases or decreases. For example, an index of 1.10 means that the economy is 10% healthier than normal. A

2-12

healthy economy (index values over 1.00) has the effect of increasing the market potential for the product. A weak economy (index values less than 1.00) has the effect of decreasing the market potential. Enter any value from 0.00 to 9.98. You do not have to enter leading or trailing zeros. The value must be positive.

Seasonal Index
The seasonal index represents the fluctuation in sales of the product throughout the year. The average quarter's sales is represented by the value 1.00. A strong sales quarter (index value more than 1.00) has the effect of increasing the market potential for the product. A weak sales quarter (index value less than 1.00) has the effect of decreasing the market potential. Enter any value from 0.00 to 9.98. You do not have to enter leading or trailing zeros. The value must be positive.

Competitor Price
This is the average price charged by all the firm's competitors. Enter any positive value ranging from $1 to $999. Do not enter the dollar sign.

Competitor Marketing
This is the average amount of money invested in all types of marketing activity (personal selling, advertising, sales promotion, etc.) by all of the firm's competitors. Enter a positive value ranging from $1,000 to $999,999. Do not enter the dollar sign or the comma.

Price
This is the price that you charged during the previous quarter, or the price that you wish to charge during the upcoming quarter. Enter any positive value ranging from $1 to $999. Do not enter the dollar sign or the comma.

Plant Investment
This is the amount of money that you invested in improving your plant last quarter, or the amount that you wish to invest this quarter. Money invested in the plant does not take effect during the period of the investment, but in the following period. Plant investment tends to counteract the effect of depreciation, which reduces production capacity by a factor of 2.5% (.025) per quarter. Production capacity for the next quarter is increased one unit for each $70 invested in plant this quarter. Enter any positive value from $0 to $9,999,999. Do not enter the dollar sign or commas.

Marketing
This is the amount of money invested in all types of marketing activity (personal selling, advertising, sales promotion, etc.) last quarter, or the amount that you want to invest during the upcoming quarter. The higher the investment in marketing, the higher the market potential. Enter a positive value ranging from $1,000 to $999,999. Do not enter the dollar sign or the comma.

TM

J

Research and Development

Research and development is an investment in the future. It serves to increase the market potential for coming quarters by an unknown amount. You can assume that money invested in this quarter will have only a minimal effect next quarter, somewhat more effect two quarters from now, and the maximum effect three quarters from now. R&D has no effect on the current quarter's market potential. Enter any positive value ranging from $0 to $9,999,999. Do not enter the dollar sign or the commas.

Profit After Tax

This is the final result after one quarter of the simulation has been run. This result cannot be changed. After it is displayed on the screen during a four quarter simulation, you can change the next quarter data and decisions.

At the bottom of the screen are three buttons:

Previous Screen

By clicking on this button the user is able to move to the *INTERNAL FIRM AND ENVIRONMENTAL DATA -- LAST QUARTER* input screen.

Explain Results

By clicking on this button or pressing **Alt+E**, the user is able to move to the *EXPLANATION OF RESULTS* screen, which tells the user which factor had the greatest effect on the simulation for each quarter.

Simulate (F4)

By clicking on this button, pressing **F4**, or pressing **Alt+S**, the user may initiate the simulation once the appropriate data has been entered into the fields.

Messages are displayed on the bottom of the screen to remind the user what to do in the fields.

EXPLANATION OF RESULTS

This screen displays an explanation of the results obtained from the simulation. There are three factors: market potential, plant capacity, and raw materials inventory. The limiting factor is displayed in a table for each quarter of the year.

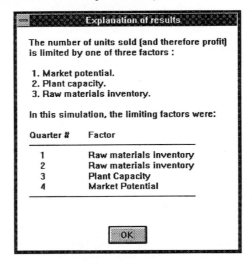

Figure 2.8

OK
By clicking on the OK button or pressing the <ENTER> key, the user will be returned to the *ENVIRONMENTAL DATA AND DECISIONS - NEXT QUARTER* screen.

TM

J

REPORTS

PRICING MODEL SIMULATION FOR FOUR QUARTERS

This is the main report menu that allows the user to view the *CURRENT PERIOD DECISIONS* report, the *OPERATING STATEMENT* report, or the *INCOME STATEMENT* report for all four quarters of the simulation.

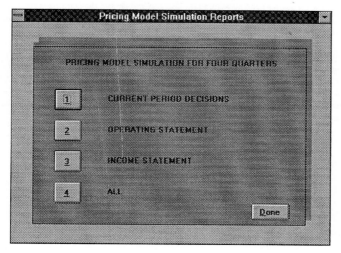

Figure 2.9

This screen contains four options and one button which are:

Current Period Decisions
By selecting this option, the user is able to view a screen with the four decision variables for each of the quarters. The user may select this option by clicking on the "1" button with the mouse or by pressing **Alt+1**.

Operating Statement
By selecting this option, the user is able to view a screen with the market potential, the sales volume, the production units, the finished goods inventory, and the plant capacity units for each of the quarters. The user may select this option by clicking on the "2" button with the mouse or by pressing **Alt+2**.

Income Statement
By selecting this option, the user is able to view a screen with an income statement, listing the expenses of all of the elements used in the simulation and the profit after tax. The user may select this option by clicking on the "3" button with the mouse or by pressing **Alt+3**.

All
By selecting this option, the user is able to view all of the screens. The first screen that appears is the *CURRENT PERIOD DECISIONS* screen and the last one is the *INCOME STATEMENT* screen. The user may select this option by clicking on the "4" button with the mouse or by pressing **Alt+4**.

Done
By selecting this button, the user is returned to either the *INTERNAL FIRM AND ENVIRONMENTAL DATA -- LAST QUARTER* screen or the *ENVIRONMENTAL DATA AND DECISIONS -- NEXT QUARTER* screen, depending on the screen that was used to execute this option. The user may select this button by clicking on the mouse or by pressing **Alt+D**.

TM

J

CURRENT PERIOD DECISIONS REPORT

This report lists the value of each of the decision variables in dollars for all four quarters in the fiscal year. This report also allows the user to view a histogram for each decision variable over a one year period.

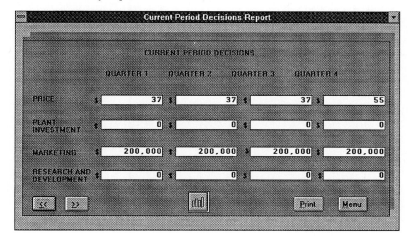

Figure 2.10

This report contains five buttons located on the bottom of the screen which are:

By clicking on this button or pressing **Alt+<**, the user may return to the previous screen, which is the main reports menu *PRICING MODEL SIMULATION FOR FOUR QUARTERS.*

By clicking on this button or pressing **Alt+>**, the user may go to the next screen, which is the *OPERATING STATEMENT REPORT.*

By clicking on this button, the user may view the histogram for each of the decision variables in this screen.

2-18

TM

J

Print
By clicking on this button or pressing **Alt+P,** the user may send a copy of the Current
Period Decisions Report to the printer. The default printer is LPT1. The user may change
this inside *Microsoft Windows.*

Menu
By clicking on this button or pressing **Alt+M,** the user is taken back to the main reports
menu *PRICING MODEL SIMULATION FOR FOUR QUARTERS.*

TM

J

OPERATING STATEMENT REPORT

This report lists the value of each of the elements: market potential, sales volume, production units, finished goods inventory, and plant capacity in units for all four quarters in the fiscal year.

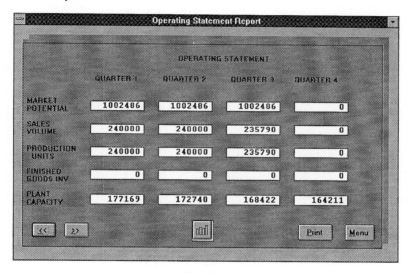

Figure 2.11

This report contains five buttons located on the bottom of the screen. The buttons are used in the same manner as those on the *CURRENT PERIOD DECISIONS REPORT* screen.

INCOME STATEMENT REPORT

This report is an accumulation of total expenses and receipts for all of the quarters in the fiscal year. At the bottom of the report is the net profit after tax.

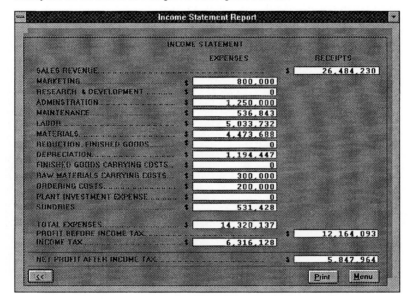

INCOME STATEMENT

	EXPENSES	RECEIPTS
SALES REVENUE		$ 26,484,230
MARKETING	$ 800,000	
RESEARCH & DEVELOPMENT	$ 0	
ADMINSTRATION	$ 1,250,000	
MAINTENANCE	$ 536,843	
LABOR	$ 5,033,732	
MATERIALS	$ 4,473,688	
REDUCTION, FINISHED GOODS	$ 0	
DEPRECIATION	$ 1,194,447	
FINISHED GOODS CARRYING COSTS	$	
RAW MATERIALS CARRYING COSTS	$ 300,000	
ORDERING COSTS	$ 200,000	
PLANT INVESTMENT EXPENSE	$ 0	
SUNDRIES	$ 531,428	
TOTAL EXPENSES	$ 14,320,137	
PROFIT BEFORE INCOME TAX		$ 12,164,093
INCOME TAX	$ 6,316,128	
NET PROFIT AFTER INCOME TAX		$ 5,847,964

Figure 2.12

This report contains three buttons at the bottom of the screen.

By clicking on this button or pressing **Alt+<**, the user may return to the previous screen, which is the *OPERATING STATEMENT REPORT*.

Print
By clicking on this button or pressing **Alt+P**, the user may print a copy of the Income Statement Report.

Menu
By clicking on this button or pressing **Alt+M**, the user is taken back to the main reports menu *PRICING MODEL SIMULATION FOR FOUR QUARTERS*.

TM

J

CHAPTER 3
How to Interpret the Reports

CURRENT PERIOD DECISIONS REPORT

This report lists the value of each of the decision variables in dollars for all four quarters in the fiscal year.

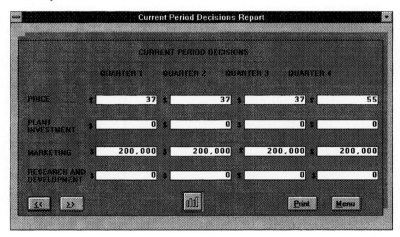

Figure 3.1

Below is a description of the meaning of each of the decision variables:

Price
This variable is the amount of money you charged for your product. It takes effect for the current quarter.

Plant Investment
This variable is the amount of money spent to increase the production capacity of your plant. The capacity decreases at a rate of 2.5% per quarter due to depreciation. Plant investment is used to: (1) counteract the depreciation and/or
(2) increase the future capacity.
You increase your plant capacity by one production unit for every $70 invested. Money spent on plant investment has no effect on the current quarter's capacity, but it takes effect the next quarter.

3-0

Marketing

This variable is the money spent for personal selling, advertising, and sales promotion. It takes effect for the current quarter. There is no carryover from one quarter to the next.

Research and Development (R&D)

This variable is a long-term investment that increases your market potential by an unknown amount during future quarters. The effect does not begin until the next quarter and increases each quarter after that one. The effect of each quarter's investment is cumulative.

For example, the market potential for Quarter Four is increased by the influence of the investment in Quarter One plus the investment in Quarter 2 plus the investment in Quarter 3. It is not necessary to invest in each quarter since the decision, whether to invest, is your decision.

TM

J

OPERATING STATEMENT REPORT

This report lists the value of each of the elements: market potential, sales volume, production units, finished goods inventory, and plant capacity in units for all four quarters in the fiscal year.

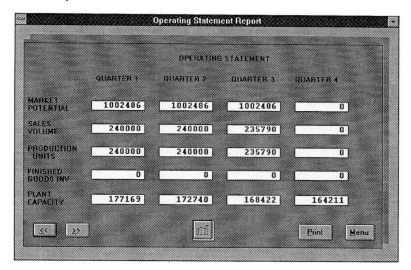

Figure 3.2

Below is a description of the meaning of each of the elements:

Market Potential
This is the number of units you can sell. It is a function of:
- the previous quarter's market potential
- the economic situation
- the competitive situation
- your pricing and marketing decisions.

The market potential for the upcoming quarter is increased (or decreased) from the previous quarter in the following manner:

Market potential is <u>increased</u>:
- 1.4% for each percentage increase in the economic index
- 1.0% for each percentage increase in the seasonal index
- 5.0% for each percentage <u>decrease</u> in your price
- 2.5% for each percentage increase in the competitors' price

TM

J

- 0.3% for each percentage increase in your marketing budget
- 0.2% for each percentage decrease in the competitors' budget

Sales Volume
This is a combination of the finished goods inventory carried over from last quarter plus this quarter's production.

Production Units
This is the difference between your <u>finished goods inventory</u> left over from the previous quarter and the <u>market potential</u>.

Finished Goods Inventory
This is the number of units available at the end of the current quarter.

Plant Capacity
This is the production capacity of the plant in units. The plant can produce this capacity working only a single shift (8 hours per day for 5 days a week).

INCOME STATEMENT REPORT

This report is an accumulation of total expenses and receipts for all of the quarters in the fiscal year. At the bottom of the report is the net profit after tax.

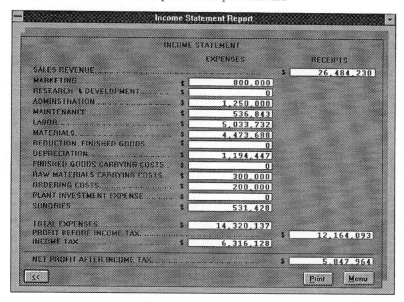

Figure 3.3

Below is a description of each of the expenses and receipts:

Sales Revenue
This is the sales volume times the price.

Marketing and Research Development (R&D)
These are your input variables, prior to the simulation.
Refer to the Operating Statement for a complete description.

Administration
This expense is $275,000 if production units do not exceed plant capacity.
Otherwise, it is $325,000.

TM

J

Maintenance
This is calculated at $0.75 per unit of production.

Labor
This is calculated as production units times $6.18 (labor cost per unit), when production units do <u>not</u> exceed plant capacity. When "overtime" production is necessary, labor is calculated at plant capacity times $6.18 plus the overtime percentage ($6.18 times 1.5).

Materials
This expense is $6.18 times the number of production units.

Reduction in Finished Goods Inventory
This is the cost of reducing the value of the finished goods inventory, based on a standard production cost of $12 per unit. The number of reduced units is calculated by subtracting the number of production units from the number of sales units.

Depreciation
This expense is calculated at 2.5% of the current plant capacity times $70. The $70 figure is the plant value per unit of production. Although investment in plant can offset the effect of depreciation on plant capacity, the investment does not eliminate or reduce the depreciation expense.

Finished Goods Carrying Costs
These are determined by multiplying $2 times the quantity of finished goods remaining at the end of the current quarter. The quantity remaining should be zero in most instances since the model schedules production to result in a zero ending balance. The ending balance is calculated by adding last quarter's finished goods to production units, and subtracting this sum from sales units.

Raw Materials Carrying Costs
These are 5% of the value of the raw materials from the previous quarter. When the model runs for four quarters, it is assumed that the initial value of raw materials (for Quarter 1) also is available for each of the remaining quarters.

Ordering Costs
These are always $50,000. This is the cost of the repetitive ordering strategy.

Plant Investment Expense
This is a penalty for investing large amounts in plant. Large amounts indicate poor long-range planning, resulting in expensive "crash programs." The expense is the plant investment squared, divided by 10,000,000.

Sundries
These are a constant of $10,000 plus $0.72 times current plant capacity.

TM

J

Profit Before Income Tax

This is the total expenses subtracted from sales revenue. A loss is indicated by a minus sign at the right of the *profit-before-tax* figure.

Income Tax

This is computed at a rate of 21% on the first $12,500 profit and 48% on the excess.

Net Profit After Income Tax

This appears on your input screen after a simulation is completed and at the bottom of the *Income Statement*.

CHAPTER 4
Tutorial

Step 1.   Start the *Pricing Model* program be double-clicking on the pricing icon or type **price.exe** inside of *Windows*.

Step 2.   The *INTERNAL FIRM AND ENVIRONMENTAL DATA -- LAST QUARTER* form will appear. Type the following data into each of the fields (you can <TAB> or use the mouse to move between fields):

- Plant Capacity - 800,000
- Production Units - 975,000
- Raw Materials Inventory - $1,450,000
- Finished Goods Inventory - 85,250
- Price - $39
- Plant Investment - $0
- Marketing - $450,000
- Market Potential - 975,000
- Economic Index - 1.15
- Seasonal Index - 0.95
- Competitor Price - $40
- Competitor Marketing - $425,000

Step 3.   Enter the following projections for the next four quarters. Click on the *Next Screen* Button or press **Alt-N.**

Type the following data into each of the fields (you can <TAB> or use the mouse to move between fields):

	Qtr 1	Qtr 2	Qtr 3	Qtr 4
Economic Index	1.35	1.37	1.41	1.45
Seasonal Index	1.05	1.25	1.35	0.95
Competitor Price	$34	$33	$32	$31
Competitor Marketing	--- 500,000 each quarter ---			

Step 4.   Now run the simulation by clicking on the *Simulate* button, pressing **F4**, or pressing **Alt-S**. The profit after tax will be blue for a profit and red for a loss. If you want an explanation of the results, click on the *Explain Results* button or press **Alt-E**. This will tell you what factor limited the number of units sold during a particular quarter.

Step 5.   View the reports generated by clicking on the *File* pull-down menu option, located at the top of the screen. Select *Reports* on this menu by clicking on the name, pressing **F5**, or pressing **Alt-R**. Next you will see the *PRICING*

TM

J

MODEL SIMULATION FOR FOUR QUARTERS screen. Select "1" by clicking on it or pressing **Alt-1**.

Step 6.   Now the *CURRENT PERIOD DECISIONS REPORT* will be displayed. You may print a copy of this report by clicking on the *Print* button or pressing **Alt-P**.

Step 7.   Next, click on the icon of the graph. This will cause a screen to pop up with a histogram for each of the four quarters. When you are finished looking at it, click on the *Exit* button or type **Alt-E**.

Step 8.   If you want to go to the next report, *OPERATING STATEMENT REPORT,* then click on the ">>" button or press **Alt->**. Otherwise, click on the "<<" button or press **Alt-<** or click on the *Menu* button or press **Alt-M** to return to the main reports menu. When you get to the Operating Statement, you may repeat steps 6 and 7. The Operating Statement will take you to the *INCOME STATEMENT* when you select ">>" and return you to the *CURRENT PERIOD DECISIONS REPORT* when you select "<<".

Step 9.   To exit the reports menu, click on the *Done* button or press **Alt-D**.

Step 10.   To exit the entire system, select the *File* pull-down menu by clicking on it or pressing **Alt-F**. Next select *Exit* by clicking on it or pressing **Ctrl-X** or **Alt-X**. A confirmation box will pop up to make sure you want to exit. Select yes by clicking on it or pressing **Y**.

4-1

APPENDIX A

GLOSSARY OF TERMS AND ABBREVIATIONS

Administration - $275,000 if production units do not exceed plant capacity, otherwise it is $325,000.

Competitor Marketing - the average amount of money invested in all types of marketing activity (personal selling, advertising, sales promotion) by all of the firm's competitors.
RANGE: 1,000 - 999,999

Competitor Price - the average price charged by all the firm's competitors.
RANGE: 1 - 999

Economic Index - an index that represents the health of the economy. The base for the index is a value of 1.00. Values higher or lower represent the percentage increases or decreases.
EXAMPLE : An economic index of 1.10 means that the economy is 10% healthier than normal. A healthy economy has the effect of increasing the market potential for the product.
RANGE: 0.00 - 9.98

Finished Goods Inventory - the number of units available at the end of the current quarter.
DEFAULT: 0
RANGE: 0 - 999,999

Income Statement - contains the following items: *sales revenue, marketing and research and development, administration, maintenance, labor, materials, reduction in finished goods inventory, depreciation, finished goods carrying costs, raw materials carrying costs, ordering costs, plant investment expense, sundries, total expenses, profit before income tax, income tax,* and *net profit after income tax.*

Labor - calculated as production units times $6.18 (labor cost per unit). This is when the production units do not exceed plant capacity. When "overtime" production is necessary, labor is calculated as plant capacity times $6.18 and is added to the difference between production units and plant capacity times $6.18 times 1.5.
EXAMPLE: Plant capacity   300,000
              Production units   400,000

Labor = 300,000 * $6.18 = $1,854,000
          100,000 * $6.18 * 1.5 = 927,000
Total labor expense = $2,781,000

Maintenance - calculated at $0.75 per unit of production.

Marketing - money spent for personal selling, advertising, and sales promotion. It takes effect the current quarter. There is no carryover from one quarter to the next.
DEFAULT: 1,000
RANGE: 1,000 - 999,999

Market Potential - the number of units you can sell. It is a function of:
    1) the previous quarter's market potential,
    2) the economic situation,
    3) the competitive situation,
    4) your pricing and marketing decisions.
Market potential increases from the previous quarter by:
    5.0% for each percentage decrease in your price.
    2.5% for each percentage increase in the competitor's price.
    1.4% for each percentage increase in the economic index.
    1.0% for each percentage increase in the seasonal index.
    0.3% for each percentage increase in your marketing budget.
    0.2% for each percentage decrease in the competitor's marketing budget.
Market Potential is decreased in a similar manner by the percentage changes mentioned above, except in the opposite direction.
DEFAULT: 0
RANGE: 0 - 9,999,999

Materials - expense is $6.18 times the number of production units.

APPENDIX B
ERROR MESSAGES

Error messages for the *TRC Pricing Model* are displayed promptly upon the detection of an error. There are two types of error messages documented: <u>Data Field Entry Messages</u> and <u>General Error Messages.</u>

<u>Data Field Entry Error Messages</u>

These messages are displayed in a line at the bottom of the main price form screens.

Too Many Characters.
This message is displayed when the user enters too many characters in an input field.

Decimal Point Not Allowed.
This message is displayed when the user enters a decimal point in an input field that should not have one.

Too Many Decimal Points.
This message is displayed when the user enters more than one decimal point in a field that allows a decimal point.

No Spaces In Between.
This message is displayed when a space is entered within a numeric data field.

Invalid Character.
This message is displayed when an invalid character (alphabetic) is entered in a numeric field.

Field must be entered. Press <ESC> to restore old value.
This message is displayed when the user blanks out a data field that *has to be entered,* and tries to navigate to another field or execute a menu option.

Value too small. Minimum value is : ......
This message is displayed when a value less than the minimum allowed is typed into a field and the user tries to exit the field or execute a menu option.

Value too big.   Maximum value is : .....
This message is displayed when a value greater than the maximum allowed is typed into a field and the user tries to exit the field or execute a menu option.

Can simulate for either 1 quarter or 4 quarters.
This message is displayed when a value other than 1 or 4 is entered into the *# of quarters* field. Simulation can be performed for either 1 quarter or a sequence of 4 quarters, <u>only</u>!

TM

J

General Error Messages

These messages are displayed in pop-up dialog boxes. After reading the error message, the dialog box can be closed by clicking on the **OK** button.

Nothing to report !
Run a simulation first or load previous results
This message is displayed when the user tries to use the *Reports* menu option and there is no output to report. A report can be viewed <u>only</u> after a simulation is performed for 1 or more quarters. Reports for old simulations can be viewed by loading the results from a file.

Help file price.hlp not found
On-line help will not be available
When the program loads, it tries to locate and read the help file (which contains help information for all the data fields). If the file is not found, the above message is displayed.

This quarter's marketing cannot be
more than 15% over last quarter's marketing
There is a restriction that the user cannot change the amount spent on marketing by more than 15%.

This quarter's marketing cannot be
more than 15% below last quarter's marketing
There is a restriction that the user cannot change the amount spent on marketing by less than 15%.

Simulation in progress.
Do you want to abort simulation
and continue with load input data
This message is displayed when the user tries to load an input data file while a simulation is running.

..... : file not found.
This message is displayed when an non-existent file name is entered in either the **Load Input Data...** or the **Load Output Data...** option.

Simulation in progress.
Do you want to abort simulation
and continue with load results
This message is displayed when the user tries to load an output data file while a simulation is running.

TM

J

No output to save
This message is displayed when the user tries the **Save Output Data...** option when there are no results to save. A simulation should be performed first or previous output results should be loaded before the **Save Output Data...** option is used.

..... exists !
Do you want to overwrite ?
This message is displayed when the user tries to overwrite an old input or output data file.

Nothing to explain !
Run a simulation first or load previous results
This is displayed when the user tries the **Explain Results** option, when there is no output data to be explained.

Cannot Explain Results !
You have loaded results from a file that was
  saved by an old version of this program.
The old program did not support this
  EXPLAIN RESULTS feature and therefore
  the file does not contain necessary
  information to explain results.
The old version of this program did not have the *Explain Results* feature. Therefore, when a *Pricing Model version 1.0* output file is loaded, the *Explain Results* feature is not available and this error is displayed.

X—AUDITING THE EMPLOYEE SCHEDULING SYSTEM

(Use this scenario in solving the Harcourt Brace problem at the end of Chapter 15.)

During the three months following implementation, Russ and Sandi kept in contact with Dave Mattson and the area supervisors at Bellmawr. The analysts called each week to make certain that things were going well. The Bellmawr managers were pleased with the system and quickly became dependent on it. One remarked to Russ on the phone, "I don't know how we ever got along without it." As far as the users were concerned, the system was a success, and the users' view was the one that mattered most to Russ and Sandi, and to the rest of the MIS staff.

But simply because users are satisfied with a system is no reason to assume that the system is functioning properly. A system must provide an accurate reflection of the physical system, and that reflection is of prime concern to accountants and auditors.

Harcourt Brace has a policy of conducting an impartial audit of a new system 90 days after implementation to prove the system's integrity. In the case of the new Bellmawr employee scheduling system, the audit was scheduled for March 16. For large systems, Harcourt Brace contracts with a consulting or auditing firm, such as Andersen Consulting, to conduct the audit. For smaller systems, such as the Bellmawr scheduling system, the audit is conducted by the firm's internal auditing staff.

Ann Johnston, an internal auditor, was assigned to audit the new employee scheduling system. She spent several days in the Orlando MIS department, talking with Russ and Sandi, and reviewing the documentation in the project dictionary. Then, on March 16, Ann traveled to Bellmawr to conduct a postimplementation evaluation. Although Pam Robinson served on the development team as an internal auditor, company policy is to use a different employee for the postimplementation audit.

Ann talked with the Bellmawr managers and reviewed copies of the hardcopy reports the supervisors had kept on file. She also examined productivity figures on selected order fillers to confirm that they had been efficiently scheduled. The evidence confirmed that the system was performing as designed. Ann returned to Orlando and filed a written report with her supervisor, who forwarded a copy to Mike Byrnes. Mike routed the report to Bill Presby and Ira Lerner.

A similar audit would be conducted annually for the remainder of the system's life.

Personal Profile—Ann Johnston, Internal Auditor

Ann Johnston grew up in Merritt Island, Florida, and graduated from the University of Florida with a Bachelor of Science degree in Computer Science. Her hobby is music—and anything to do with it. She relates her success to two lessons taught to her by her father: the importance of religion and how to work. Although her parents were always there to help, she learned at an early age to provide for herself, paying for her own car insurance and extras. Ann sees the keys to Harcourt Brace's success as providing a good product, people who believe in that product, and management innovation.

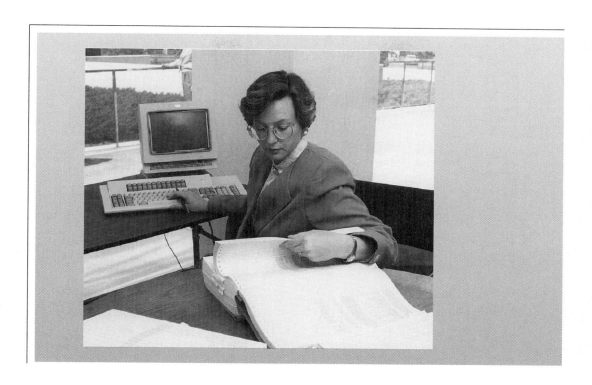

CHAPTER 15

Systems Use, Audit, and Maintenance

Learning Objectives

After studying this chapter, you should:

- Understand the purpose of the postimplementation review and annual system audits and know the roles played by the systems analyst and the internal auditor
- Know more about the firm's internal auditing unit and how internal auditors conduct their systems audits
- Appreciate the burden systems maintenance imposes on the firm's information resources and know how firms are coping with that burden
- Be familiar with the special maintenance needs of the CBIS subsystems

Introduction

The cutover to a new system marks the beginning of the use phase of the SLC. The users use the system and derive the benefits that provided the justification for the developmental effort.

Shortly after cutover, when the system has had an opportunity to prove itself, a postimplementation review is conducted to verify user satisfaction and ensure that system controls are intact. Separate reviews are conducted by the systems analyst and the internal auditor, a process repeated annually throughout the life of the system. When the system fails to pass a review or audit, efforts are made to correct the deficiencies.

The information system audit is only one of three types that the internal auditor conducts. The other two are financial and operational audits. Management monitors the activity of the internal auditors by establishing an audit committee and by having the manager of internal auditing report directly to a top-level executive.

The internal auditor is mainly concerned with incorporating controls into systems development, design, and use. In conducting audits of computer-based systems, the auditor uses special auditing software, including knowledge-based systems.

The longer a system is in use, the greater the need to maintain it by correcting errors, bringing it up to date, and incorporating improvements. Each CBIS subsystem imposes unique maintenance requirements, and firms are responding with tactics aimed at transforming maintenance into a strategic weapon.

The Use Phase of the System Life Cycle

Three main activities occur during the use phase of the SLC. Users use the system, information specialists and internal auditors audit it, and information specialists, working with users, maintain it. Figure 15.1 illustrates these activities, which are controlled by the MIS steering committee.

The Need for Management Follow-Up

As long as a computer-based system remains in operation, management must constantly evaluate it to ensure that it performs as intended. This follow-up responsibility is a vital part of the systems approach to problem solving described in Chapter 7. Recall that the systems approach consists of the 10 steps illustrated at the top of Figure 15.2. Step 10, *Follow up to ensure that the solution is effective,* is represented by formal audits that are conducted during the use phase of the system life cycle.

There are two types of audits—the postimplementation review and the annual audits.

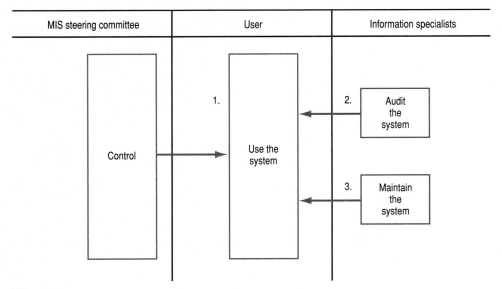

Figure 15.1
The use phase.

I. Preparation effort

    1. View the firm as a system

    2. Recognize the environmental system

    3. Identify the firm's subsystems

II. Definition effort

    4. Proceed from a system level to a subsytem level

    5. Analyze system parts in sequence

III. Solution effort

    6. Identify the alternate solutions

    7. Evaluate the alternate solutions

    8. Select the best solution

    9. Implement the selected solution

    10. Follow up to ensure that the solution is effective

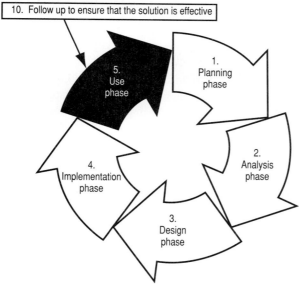

Figure 15.2
Follow-up activity is a vital part of the systems approach.

    The **postimplementation review** is a study of a newly implemented system to ensure that it meets both the needs of the user and the performance specifications established by the user and information services. The postimplementation review should be conducted as soon as the system bugs have been ironed out and the users and participants are familiar with system operations. This usually occurs about 90 days after cutover. For simple systems, the review can come earlier; for complex ones, later.

An **annual audit** is a study of an operational system to ensure that it continues to meet users' needs and that its controls continue to work effectively. An annual audit can trigger a new system life cycle when management concludes that the system cannot be kept alive through maintenance.

Both the postimplementation review and the annual audits are conducted by two types of specialists—the systems analyst and the internal auditor.

The Auditing Responsibility of the Systems Analyst

The work of the developmental project team is not completed until it is clear that the user is satisfied with the new system. This assurance can be achieved by conducting a postimplementation review. The project leader is responsible for conducting such a review but can assign the task to a team member. The systems analyst is a good choice because he or she is the most familiar with the system and its purpose. The analyst can carry out the review either informally or formally.

Informal Review

An **informal review** is one conducted without following an established procedure. Such a review is usually associated with a small system or a small firm. The analyst typically contacts the user or users in person. When more than a few users are involved or they are located some distance away, contact can be made by telephone. The analyst asks appropriate questions, such as "How do you feel about the system in general?" and "Is there anything you would like to change?" Any indication of user dissatisfaction causes the analyst to probe deeper to learn the details. The informal review is a good example of unstructured data gathering.

The analyst describes the findings of the informal review in a memo, which is reviewed with the project leader. When the project leader approves the memo, it is sent to the CIO, often over the project leader's signature. Copies of the memo also go to each member of the project team.

When the systems analyst encounters situations that call for improvement, the project leader directs the team to perform additional work. The team is not officially disbanded until the postimplementation review indicates that the user is satisfied with the system.

Formal Review

A **formal review** is conducted in accordance with a written policy established by the CIO or the MIS steering committee. In a formal review, the analyst uses a list of questions prepared by information services. The questions are intended to measure the level of user satisfaction in two main areas—system procedure and system output. Questions relating to the procedure essentially concern the system's user friendliness. Questions relating to output measure system capabilities according to the four dimensions of information—accuracy, timeliness, relevance, and completeness. Figure 15.3 is an example of such a question list in a written questionnaire format. The analyst can ask the questions in person, over the telephone, or by mail.

When all sampled users and participants have been questioned, the analyst analyzes the data and prepares a report organized along the following lines:

- Summary description of the system
- Summary of findings
- Recommendations
- Supporting data

The first two sections provide a management overview, and the recommendations section identifies and explains any follow-up action that might be required. The supporting data consists of tabulations of responses to the objective questions, plus any especially informative comments the users might have made.

The formal report follows the same distribution pattern as the informal one. The project leader approves it and sends it to the CIO and project team members. The CIO makes the report available to the MIS steering committee for those systems that have strategic value to the firm.

The Auditing Responsibility of the Internal Auditor

The other specialist who evaluates computer-based systems is the internal auditor.[1] We have included an internal auditor as a member of the project team throughout the developmental period. However, general practice is to use a different internal auditor for the postimplementation review and annual audits in order to ensure objectivity.

Types of Auditors

Firms usually use two types of auditors—internal and external. **Internal auditors** are employees of the firm and conduct in-house audits on an ongoing basis as directed by the firm's management. For example, management might be concerned about a particular area of the firm's operations and direct that an audit be conducted. **External auditors,** on the other hand, are employees of accounting firms such as Arthur Andersen and Price Waterhouse. External auditors conduct the annual audits required by law to confirm that the information in annual reports to stockholders accurately reflects the financial status of the firms.

[1]Much of the material relating to the internal auditor was provided by The Institute of Internal Auditors Research Foundation's 1991 publication *Systems Auditability and Control Report.* This report is based on a study of 260 firms throughout the world conducted in 1990 by Price Waterhouse and sponsored by IBM. For more information concerning the report, contact The Institute of Internal Auditors Research Foundation, 249 Maitland Avenue, Altamonte Springs, Florida 32701-4201.

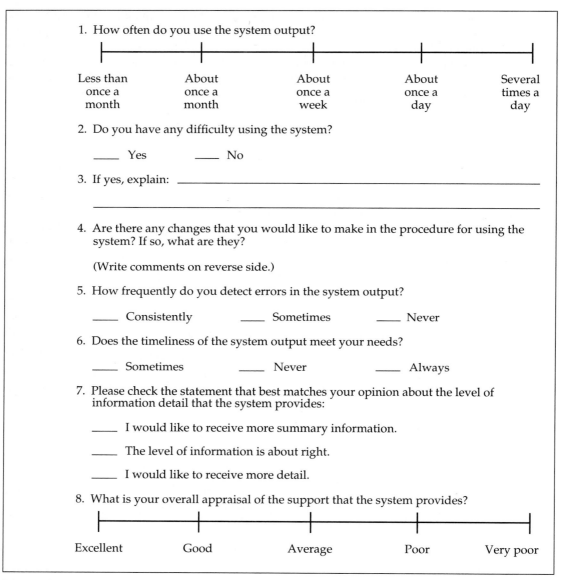

1. How often do you use the system output?

| Less than once a month | About once a month | About once a week | About once a day | Several times a day |

2. Do you have any difficulty using the system?

_____ Yes   _____ No

3. If yes, explain: _____

_____

4. Are there any changes that you would like to make in the procedure for using the system? If so, what are they?

(Write comments on reverse side.)

5. How frequently do you detect errors in the system output?

_____ Consistently   _____ Sometimes   _____ Never

6. Does the timeliness of the system output meet your needs?

_____ Sometimes   _____ Never   _____ Always

7. Please check the statement that best matches your opinion about the level of information detail that the system provides:

_____ I would like to receive more summary information.

_____ The level of information is about right.

_____ I would like to receive more detail.

8. What is your overall appraisal of the support that the system provides?

| Excellent | Good | Average | Poor | Very poor |

Figure 15.3
Some questions that the systems analyst can ask during the postimplementation review.

Not all firms have internal auditing units. Very small firms typically rely on external auditors for their in-house audits. As a firm grows in size, management generally establishes an auditing staff consisting of employees with the same backgrounds, capabilities, and professional certifications as the external auditors. The larger the firm, the more internal auditors are required. General Electric, for example, has an internal auditing staff of approximately 135. The staff at IBM is even larger—around 170.

The Position of Internal Auditing In the Organization

Auditors must have the complete endorsement of top management if they are to gain access to the various areas of a firm's operations and their systems and data. This endorsement is achieved by means of an audit committee and organizational positioning of the internal auditing unit.

The Internal Audit Committee   The **audit committee** consists of selected members of the firm's board of directors and is responsible for the integrity of the firm's conceptual systems. The audit committee serves the stockholders or owners by overseeing the activities of the firm's internal auditing unit.

The Internal Auditing Unit   The manager of the internal auditing unit typically reports either directly to the CEO or to a top-level executive who reports to the CEO. At General Electric, for example, the vice-president of the corporate audit staff reports to the senior vice-president of finance, who reports to the CEO. The internal auditing manager is responsible for keeping both top management and the audit committee informed about the status of the firm's conceptual systems and makes periodic reports, both oral and written.

Types of Internal Audits

The internal auditing unit is responsible for conducting three types of audits—financial, information systems, and operational. A **financial audit** is a study of a firm's accounting system to ensure that it includes accounting controls. An **information systems (IS) audit** reviews all the components of a firm's computer-based systems to ensure that they are being used as intended. IS audits consist of both the postimplementation review and annual audits. An **operational audit** is concerned with the efficiency and effectiveness of the firm's activities. In conducting an operational audit, the auditor might compare the firm's performance with industry standards.

Internal Auditors with Computer Expertise

The term **EDP auditor** is often used to describe the internal auditor who conducts IS audits. EDP, for **electronic data processing,** is the term used during the early years of the computer to distinguish that technology from others, such as punched-card machines. During this period, the internal auditors who understood computers became known as EDP auditors. However, since computing expertise has become so common among internal auditors, the term EDP auditor is

Figure 15.4
Auditors gather data throughout a firm's facilities.

less appropriate. We use the term *internal auditor*, a practice that is becoming more and more prevalent.

The Internal Auditing Process

In conducting the postimplementation review and annual audits, the internal auditor follows the same procedure as the systems analyst, obtaining feedback from the users. In addition, the auditor studies the performance of the system, usually in extreme detail, to determine the extent to which system controls are functioning as intended. When auditing an inventory system, for example, the auditor might ride a boom in the warehouse to count cartons stacked to the ceiling.

The auditor prepares a written report of the findings for the manager of the internal auditing unit, who, in turn, makes the report available to management. The CIO receives a copy, as do other executives, including, possibly, members of the board of directors. Failure to meet internal audit standards commits information resources to correcting the problems.

Computer-Based Auditing Tools

As internal auditors conduct an audit, they interview persons working in the area and analyze the system input, processing, and output. When auditing a computer-based system, the internal auditor uses software that falls into four main categories—retrieval and analysis programs, system software examination tools, integrated transaction testing techniques, and knowledge-based systems.[2]

- **Retrieval and Analysis Programs** This software enables the auditor to retrieve records from the database, perform certain arithmetic and logical processes, and print or display the results. Retrieval and analysis software is typically designed for use on a microcomputer. When auditing a mainframe or minicomputer system, the software downloads the data onto the micro, where the analysis is performed.

- **System Software Examination Tools** This commercially available software enables the auditor to read and analyze system software data, such as operating system parameters and control and security data.

- **Integrated Transaction Testing Techniques** This software is designed to audit application software by processing test data and reporting the results. The procedure the auditor follows is generally the same as that followed by the programmer when testing a program.

- **Knowledge-Based Systems** This software functions as an expert system, guiding the auditor in gathering the data and interpreting the results. To date such software has been relatively rare, but its use is increasing.[3]

The first audit software was designed to evaluate centralized, mainframe-based systems. The trend toward decentralized information resources in the form of end-user computing and microcomputers is imposing new demands on the auditing process and is creating a need for a new set of tools.

The Internal Auditor as a Team Member

The internal auditor is a member of the project team during the developmental period. During this time the auditor shares in the responsibility of identifying new system risks and designing controls that will address those risks. During the use phase, beginning with the postimplementation review and continuing with annual audits, the internal auditor has the responsibility of ensuring that the controls are working properly and that the integrity of the firm's data is maintained.

The internal auditor is an expert in identifying risks and designing controls. It is best to have that expertise on the project team as the system is being developed, rather than waiting until after implementation. If inherent weaknesses exist within a system design, the earlier they are discovered and corrected, the better. Figure 15.5 shows that the cost of correcting an error after design is 4,000 times greater than the cost of correcting the error during the early stages of design.

---

[2]*System Auditability and Control* (Altamonte Springs, FL: The Internal Auditors Research Foundation, 1991): 2–38.
[3]Carol E. Brown and Mary Ellen Phillips, "Expert Systems for Internal Auditing," *Internal Auditor* 48 (August 1991): 23–28.

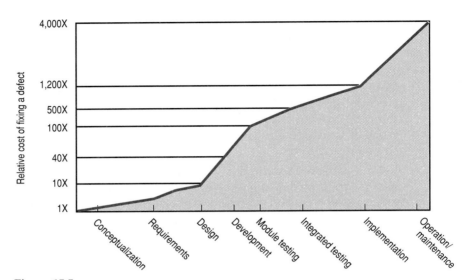

Figure 15.5
Errors and weaknesses in a system design cost less to correct early in the system life cycle.

Because of the contribution the internal auditor can make to system development, information specialists should not view the auditor as an outsider who imposes unrealistic constraints. By cooperating with the auditor and allowing that person to contribute her or his full range of knowledge and skills, information specialists increase the opportunity for developing a system that satisfies the user's needs.

Systems Maintenance

Business computing is over 35 years old. During this time, firms have implemented thousands of computer-based systems, some lasting only a few months before being replaced or scrapped, and others proving to be almost indestructible. Some firms are still using programs written in the 1950s.

Throughout the use phase of the SLC, work is performed to keep the systems alive. This work is called **systems maintenance.**

In Chapter 8 it was noted that at some point systems maintenance becomes impractical and the systems must be completely overhauled. We used the term **business process redesign (BPR)** for this overhaul activity, which is achieved by means of restructuring, reverse engineering, or reengineering. This chapter describes the systems maintenance that keeps systems functioning until BPR becomes necessary.

Types of Systems Maintenance

Three types of systems maintenance exist—corrective, adaptive, and perfective.[4]

Corrective Maintenance   The type of maintenance intended to correct errors is called **corrective maintenance.** The programmer tries to do a thorough job of debugging prior to cutover, but some errors invariably remain and do not surface until months or even years later. When that happens, corrective maintenance is performed.

Some think corrective maintenance represents the biggest part of a firm's total maintenance activity. However, such is not the case. It has been estimated that the average is in the 20 percent range.

Adaptive Maintenance   A firm's conceptual systems represent physical systems that change over time. Some of these changes are triggered by actions within the firm, such as the addition of new products and the opening of new sales offices. Other changes are triggered by actions within the firm's environment. For example, if new union contracts go into effect and local governments change the method of computing sales taxes, these changes in the physical systems must be reflected in the conceptual systems. These changes are called **adaptive maintenance.** Adaptive maintenance accounts for the smallest amount of the total maintenance activity—only about 10 percent.

Perfective Maintenance   After users use a computer-based system for a while, they see ways to improve it. For example, a credit manager may realize that it would be easier to follow up on unpaid accounts if the cutomers' telephone numbers were added to the past-due receivables report. Improvements such as these are called **perfective maintenance** and account for most maintenance activity—over 50 percent.

The large amount of perfective maintenance being done emphasizes an important but often overlooked point concerning maintenance. It is easy to regard maintenance as something that must be done to keep systems performing at the same level. However, most maintenance is initiated by users seeking improvements.

The Burden of Systems Maintenance

The volume of systems maintenance activity for computer-using firms has increased as new systems were added. This increase is illustrated in Figure 15.6. In large firms, with fully developed computer-based information systems, it is not uncommon for systems maintenance to account for 60 to 80 percent of the information resources. In this setting, it is almost impossible for a firm to expand the scope of its computer applications.

[4]This discussion of maintenance activity, and the percentages quoted, are taken from Efraim Turbin, "Putting Out Fires," *Decision Line* 22 (December–January 1991): 8–9.

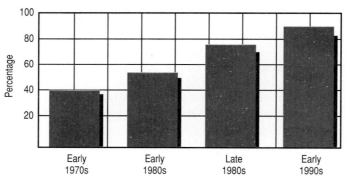

Figure 15.6
The portion of software resources dedicated to maintenance.

The Stigma of Systems Maintenance

Because of the burden that it imposes, many firms regard systems maintenance as a second-class activity. They require new hires to prove their worth as maintenance programmers before being considered for promotion to other positions, such as development programmer. Also, these firms assign their less efficient programmers to maintenance projects. The morale of the information specialists assigned to systems maintenance suffers in this environment.

Solutions to the Maintenance Problem

The best solution to the maintenance problem is a long-term one, aimed at making the maintenance activity less labor intensive. Other, more immediate results can be achieved by providing the maintenance team with development expertise and building a more positive attitude toward maintenance work.

Make Maintenance Work Less Labor Intensive   The best solution to the maintenance burden is the one that reduces the amount of time that the information specialists must devote. This reduction in maintenance labor can be achieved by taking advantage of existing software tools.

CASE tools can be used in the development of all new systems. Such tools allow the systems to be maintained through the documentation rather than the code. This strategy will ease the eventual maintenance burden for all new systems being added to the portfolio.

Business process redesign tools can put existing systems in a form that allows them to be more easily maintained. Restructuring can put the systems in a more structured form, and reverse engineering can provide needed documentation. Reengineering can upgrade the functionality of the systems.

The CASE and BPR tools let the computer, rather than the information specialists, do most of the maintenance work.

Provide Development Expertise   Some firms have created a position in information services called maintenance liaison. The **maintenance liaison** is a member of a developmental project team who joins a maintenance team once the system becomes operational. The maintenance liaison provides the maintenance team with an insight into the system that comes only from participating in its development. When the maintenance liaison joins the maintenance team, a member of that team fills the vacancy on the development team, providing for a rotation of duties between development and maintenance.

Build Positive Attitudes   Information specialists must cease to regard maintenance negatively. Sid L. Huff, an MIS professor at Western Business School, suggests a combination of name change, education, and reward structure.[5] Rather than distinguish between new systems development and systems maintenance, the terms *future systems* and *installed systems* should be used. The different terminology should accompany a recognition that the installed systems constitute a substantial part of the overall systems activity.

Information services staffs should be educated about the positive aspects of installed systems work in terms of career advancement. Work on installed systems demands just as much creativity and professionalism as that on future systems, and it offers the opportunity to apply leading-edge tools such as CASE and BPR software.

Education is most effective when accompanied by a change in reward systems that recognizes contributions to installed systems. Salaries, bonuses, and recognition for installed systems work should be equal to that of future systems work.

U.S. West, for example, which spends about 85 percent of its software dollar for maintenance, gives special awards and bonuses to maintenance staff members. Management also solicits the maintenance programmers' suggestions for improvements and their preferences for benefits, such as flexible hours and freedom to transfer among groups. Bank of Canada follows a similar proactive strategy by using maintenance bonuses for motivation, paying programmers up to $3,000 per year for meeting performance levels.[6]

By recognizing the importance of maintenance and giving it as much attention as systems development, firms such as these transform an activity that many regard as a liability into a strategic tool.

The Unique Maintenance Requirements of the CBIS Subsystems

We have seen how the different CBIS subsystems present unique opportunities and constraints during systems development. The same applies to maintenance. Each subsystem offers a unique challenge in the three types of maintenance.

[5]Sid L. Huff, "Information Systems Maintenance," *Business Quarterly* 55 (Autumn 1990): 30–32.
[6]These examples are from Jeff Moad, "Maintaining the Competitive Edge," *Datamation* 36 (February 15, 1990): 61ff.

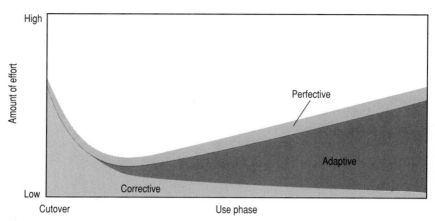

Figure 15.7
The data processing system requires a high level of adaptive maintenance.

Data Processing System Maintenance

The data processing system has the highest level of corrective maintenance after cutover because errors in accounting software often are not detected until live data is used. If a firm has created its own custom data processing software, maintaining the software is its responsibility. In the case of prewritten software, the supplier assumes the maintenance responsibility.

Figure 15.7 indicates that the biggest increase in maintenance as the use phase wears on is in the adaptive area.[7] The programs are updated to reflect changes imposed by both the firm's environment and its management. Once the data processing system becomes operational, less perfective maintenance is required for it than for any of the other subsystems. The problems the data processing system addresses are the most structured, and there is less room for variation in how the tasks are performed.

Management Information System Maintenance

The MIS offers a pattern of corrective maintenance similar to that of data processing, as shown in Figure 15.8. However, the corrective maintenance level is lower because much of the MIS output is in the form of reports prepared using DBMS query languages or 4GLs. Much of this software is maintained by software suppliers, and the custom software tends to be relatively simple in design. However, mathematical models do not fit this picture and can require formidable corrective maintenance.

[7]As with the other interpretations of unique CBIS subsystem characteristics that have been made in this text, this discussion of maintenance patterns is largely subjective, and little or no supporting data exists. Also, as previously emphasized, differences such as these are expected to apply in a general way, but features of specific systems will produce different effects.

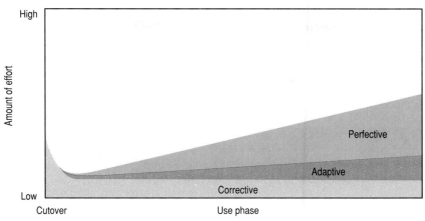

Figure 15.8
Most of the maintenance of the management information system is perfective in nature.

Because the MIS is aimed at meeting general information needs, the need for adaptive and perfective maintenance is relatively modest.

Decision Support Systems Maintenance

The maintenance pattern for DSSs is similar to that of the MIS. The main difference is in perfective maintenance, which, as Figure 15.9 shows, increases more because the DSSs are tailored to information needs of individuals and small groups.

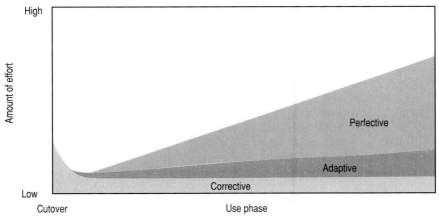

Figure 15.9
Decision support systems demand the most in terms of perfective maintenance.

Because of employee turnover, the composition of the user groups is constantly changing. New users frequently have different views on how the information should be produced and reported and require that the software be revised.

Expert Systems Maintenance

Expert systems require a high level of adaptive maintenance because they are designed to provide consulting advice conerning particular areas of business operations. As requirements in the areas change, the expert systems must be modified. Take, for example, expert systems that internal auditors use to audit computer-based systems. Changes in auditing laws and regulations require that new auditing techniques be employed. Those changes must be reflected in the expert systems. Much perfective maintenance must also be performed on expert systems as the experts learn more about the problem domains. Figure 15.10 shows that the overall maintenance level of expert systems can be the highest of all the CBIS subsystems.

Office Automation Systems Maintenance

The only CBIS subsystem that does not require maintenance is office automation. All OA software is prewritten, and the maintenance is performed by the suppliers.

Putting CBIS Subsystem Maintenance in Perspective

The differences in the maintenance requirements of the CBIS subsystems influence the manner in which systems modifications are made. Both corrective and adaptive maintenance may be accomplished with little or no direct contact with the users. Users notify the information specialists of errors and changes in the systems' environment, and the modifications are made without a great deal of user

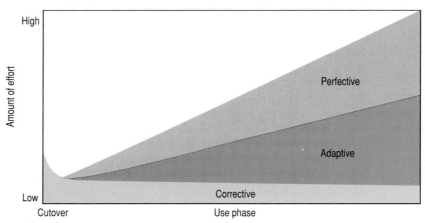

Figure 15.10
Expert systems require the highest levels of both adaptive and perfective maintenance.

involvement. Perfective maintenance is another matter. When users want to improve their systems, the maintenance team might have to spend as much time with the users learning about their changed needs as was spent initially when the system was developed.

Using CASE for Systems Documentation

One of the greatest benefits of using a CASE product is that, as the SDLC unfolds, the documentation is automatically generated. In addition, because maintenance is performed from the documentation rather than from the code itself, the documentation is regenerated during maintenance and will always match the operational system.

The following is a partial list of the types of automatic documentation that are generated through the use of CASE:

1. Data repository reports, which are complete or partial listings of all the data elements and structures contained in the system

2. Various diagrams, including but not limited to:
   a. Leveled sets of data flow diagrams, serving as both logical and physical models of the existing and proposed system
   b. Systems flowcharts
   c. Systems architecture
   d. Structure charts
   e. Entity-relationship diagrams
   f. Organization charts
   g. Hierarchy charts
   h. Dependency diagrams
   i. Action diagrams
   j. Menu structure diagrams

3. Various project management tools, including but not limited to:
   a. Network diagrams such as PERT charts and CPM diagrams
   b. Milestone schedules
   c. Resource histograms
   d. Work breakdown charts
   e. Gantt charts
   f. Budgets
   g. Tracking charts

4. Source and object code listings

5. Consistency-checking reports

6. Requirements specifications

7. Other documentation related to the CASE tool of choice

Much of the effort of systems work consists of documentation. CASE assumes much of this workload, increasing both the quality and quantity of the systems analyst's work.

A Final Word

Before you leave the subject of systems analysis and design and turn your attention to other areas, you should realize that we presented the topics in a prescriptive fashion. We described things as they *should be,* rather than as they always *are.* It is important that you understand this now because when you graduate and begin work, you may find that your firm does not measure up to all the yardsticks of this text. To be sure, a large number of firms are doing all the things that we have talked about—and more. But, many, many others still have a way to go.

Why have we taken this prescriptive approach? Because college education should prepare you for the future rather than the present. It is the responsibility of instructors, textbook authors, and all other developers of educational materials to give you the tools with which to be successful—one, two, five, ten years down the road. No doubt, the scene we have painted for systems work is the wave of the future, and the majority of firms are headed in this direction. With the knowledge and skills gained from this text and this course, you will not only ensure a successful career for yourself but will make a major contribution to the growth and success of your firm.

Although we have passed along a large number of *do's* and *don'ts* in the process of describing modern systems work, two points are extremely important.

- First, computer-based systems are developed to meet the needs of the firm, not individuals within the firm—no matter how lofty their titles. Because the systems are for the firm, it exerts a strong influence on their design, implementation, and use. This influence comes in a top-down fashion, beginning with activities of enterprise planning and strategic planning for information resources. No matter where you are located in the organization, all your systems work will be influenced by this setting.

- Second, practically everybody in the firm engages in systems work to some degree at one time or another. The systems analyst has this job full-time and can develop an expertise that can benefit users throughout the firm. Managers certainly do systems work as well, but they must balance this activity with their many other responsibilities. Nonmanagers also continually analyze their own systems and design and implement improvements, very often as a matter of daily routine. The bottom line is that everyone is a systems analyst.

If you become a professional systems analyst, you must view the trend to end-user computing not as a constraint but as an opportunity. To be sure, the *proportion* of systems developed by the systems analyst is shrinking. But there will never come a day when there is absolutely no need for systems analysts. Regardless of how capable the users become, they will always prefer that others—the information specialists—build many of their systems.

As you work as an information specialist in the user environment, you should develop expertise in those areas where you will be expected to carry the major share of the load. Become knowledgeable about organizational systems, such as data processing and MIS, and sophisticated systems, such as complex DSSs and expert systems. Also, cultivate the human relations skills that enable you to work with users, including the ability to communicate both orally and in writing.

The computer field may seem old to you. However, compared with most other fields of human endeavor, it is just starting. We have tried to distill the important material that has evolved in the relatively brief history of business computing and give you the understanding that you need in order to continue the evolution. We believe that we have done this and have given you something valuable that you can carry with you along the way.

Summary

Once a computer-based system is implemented, follow-up activity by management is necessary to ensure that the system is performing as planned. Management does not personally evaluate the system but delegates that responsibility to specialists.

There are two kinds of evaluations. The first, which immediately follows cut-over, is called the postimplementation review. All succeeding evaluations, which are generally performed once a year, are called annual audits. Specialists who conduct the postimplementation review and annual audits include the systems analyst and the internal auditor. The analyst seeks to confirm user satisfaction, and the auditor seeks to ensure system integrity.

The systems analyst can conduct the postimplementation review either informally or formally. In an informal review, the analyst determines the questions to ask, asks them in person or over the telephone, and writes up the findings in a memo. In a formal review, the analyst uses a predetermined list of questions, which can be presented in person, over the telephone, or through the mail, and documents the responses in a report. The findings of both types of reviews go to the CIO, and, in the case of systems with strategic value, to the MIS steering committee.

An internal auditor performs basically the same work as an external auditor but is an employee of the firm. Management supports internal auditing activities by establishing an audit committee on the board of directors and by positioning the internal auditing unit high in the organization. The manager of internal auditing reports to a top-ranking executive, such as the CEO, and makes written and oral reports to the audit committee.

Internal auditors, often called EDP auditors, conduct three types of audits. A financial audit is also conducted by an external auditor. An information systems audit consists of a postimplementation review or annual audit. An operational audit has the goal of making procedures more efficient and effective.

Internal auditors use several computer-based auditing tools. Retrieval and analysis programs allow the auditor to retrieve database contents, system software examination tools enable the audit of system software, integrated transaction testing techniques validate the integrity of programs with test data, and knowledge-based systems provide the auditor with expert advice.

During the SDLC the internal auditor identifies risks the new system might face and is instrumental in incorporating features into the system design that mitigate such risks. To achieve an unbiased evaluation of the new system when it becomes operational, a different internal auditor is used for the postimplementation review and annual audits.

Systems maintenance is conducted during the use phase to keep the system running. When it becomes obvious to top management that maintenance is no longer feasible, it authorizes business process redesign, following strategies of restructuring, reverse engineering, or reengineering.

Three types of systems maintenance exist. Corrective maintenance fixes errors, adaptive maintenance keeps systems current, and perfective maintenance incorporates improvements. This maintenance activity not only drains information resources but can also have a demoralizing effect on information specialists.

The best long-term solution to the maintenance challenge is to use CASE in all new system development work and to use BPR software to document and restructure existing systems. More immediate results can be achieved by making development expertise available to the maintenance team in the form of a maintenance liaison and by building positive attitudes through education and rewards.

Each CBIS subsystem offers a unique maintenance challenge. Data processing systems feature high levels of corrective and adaptive maintenance but modest perfective maintenance. Management information systems impose relatively balanced demands on all types of maintenance. Decision support systems place heavy demands on perfective maintenance, as do expert systems. Expert systems also require much adaptive maintenance. Office automation systems make few, if any, maintenance demands because that responsibility is assumed by software vendors.

Key Terms

Postimplementation review

Annual audit

Informal review

Formal review

Internal auditor

External auditor

Audit committee

Financial audit

Information systems audit

Operational audit

EDP auditor

Systems maintenance

Maintenance liaison

Key Concepts

How the postimplementation review and annual audits enable management to ensure that systems solutions are working

How systems maintenance is performed until it becomes ineffective, at which time a strategy of business process redesign is pursued

The three types of systems maintenance and the variation in their mix from one CBIS subsystem to another

Questions

1. Where do the postimplementation review and annual audits appear in the systems approach?

2. When is a postimplementation review scheduled? Why is this time chosen?

3. Which information specialist typically conducts the postimplementation review? Why is this person a good choice?

4. What is the difference between the systems analyst's informal and formal review in terms of how the data is gathered? What is the difference in terms of how the findings are reported?

5. What sort of supporting data would you find in a systems analyst's formal report?

6. What is the difference between an external and an internal auditor?

7. Where in the organizational structure would you find the internal auditing unit? Why is this location appropriate?

8. What is an audit committee? What function does it serve?

9. What three types of audits does an internal auditor perform?

10. Why is the term *EDP auditor* no longer in vogue?

11. How are the results of the internal auditor's audit reported to management?

12. Which type of auditing software requires the use of a test deck? Which type has been rare in the past but is increasing in use?

13. Why include the internal auditor as a member of the project team from the beginning of the SLC?

14. Why can't all the program bugs be identified and corrected prior to cutover?

15. Does systems maintenance seek to keep systems at their current level of performance or to improve them? Explain.

16. What software tools can make systems maintenance less labor intensive?

17. What is a *maintenance liaison?*

18. How can a more positive attitude toward systems maintenance be achieved among information specialists?

19. Which CBIS subsystem demands the highest level of corrective maintenance?

20. Why do DSSs require so much perfective maintenance?

21. Why do office automation systems require little or no maintenance?

22. Which type of maintenance demands the most analyst-user interaction?

Topics for Discussion

1. The board of directors is composed of both the firm's executives and prominent outside persons such as educators, bankers, and lawyers. In forming the audit committee, would it be a better choice to include only the outsiders, only the executives, or some combination?

2. What is the relationship, if any, between systems maintenance and BPR?

3. Comment on a CIO's statement that "We don't have a big maintenance problem because our users seldom ask for improvements to their systems."

4. Should a college restructure its MIS program to better prepare its graduates for systems maintenance? If so, how?

5. Can you see any possible problems caused by giving bonuses to systems maintenance personnel? If so, what are they and how could they be prevented?

Problems

1. Go to the library and look through magazines on internal auditing. Use the bibliography at the end of this chapter to identify magazine titles. Select an article that interests you and write a short paper summarizing your findings.

2. Visit a local computer-using firm and interview an internal auditor. Ask questions to determine the type of computer background the person has and the extent of his or her involvement with computer-based systems. Ask what the auditor's views are on the working relationships with information services and how they might be improved. Write a short paper that summarizes your findings.

Consulting with
Harcourt Brace

(Use the Harcourt Brace scenario prior to this chapter in solving this problem.)

Postimplementation Review

Pretend that you are Ann Johnston. Write a memo to your supervisor, Albert Guenther, advising him of the successful postimplementation review. Briefly describe the system, explain how the system was audited, and state that the system satisfied the audit.

Assume that Mike Byrnes has requested that Albert Guenther use the same internal auditor for both system development and the postimplementation audit. In the case of the Bellmawr system, Pam Robinson would have performed both jobs. Mike rationalizes that using the same person would save the time a new auditor must spend in learning the system. This procedural change would also reduce the time of the MIS staff in orienting the new auditor. Pretend you are Albert Guenther and write a memo to Mike, responding to his request.

Critique of the Harcourt Brace Computer Operation

Now it is time to be yourself. Write a memo to Mike Byrnes that provides him with your evaluation of the Harcourt Brace computer operation. Identify the things that you believe the firm is doing right and the things that need improving. Organize the memo as you see fit, but provide Mr. Byrnes with five suggestions for improvement. Your instructor will advise you whether there is a maximum length for the memo.

Case Problem
Anderson & Anderson of Arizona, Incorporated

Anderson & Anderson, a large Phoenix insurance agency, serves as selling agent for many property and casualty insurance companies. Bryan Anderson, the president, presides over an executive committee that includes all the vice-presidents. Kay Salmon, vice-president of management information systems, and Avery Barneby, vice-president of internal auditing, are members.

The company uses the computer in almost every facet of its operations. A mainframe located in the Phoenix headquarters is networked to microcomputers in sales offices throughout the state. Applications range from data processing to expert systems, putting Anderson & Anderson on the cutting edge of computing technology. The MIS division is regarded as one of the top computer staffs in the state.

MEMO

TO: Kay
COPY: Avery
FROM: Bryan
SUBJECT: Responsibility for System Audits
DATE: October 6

Yesterday Avery and I were talking about how we audit our computer systems. As you know, we require duplicate audits—one by MIS and one by internal auditing. In some instances, they go on concurrently. Isn't this overkill? Why don't we just do a single audit and give Bryan's group the responsibility? I would like your reaction.

Assignment

Pretend that you are Kay Salmon. Write a memo to Bryan, with a copy to Avery, evaluating both the single-audit and dual-audit approaches. Conclude the memo with your recommendations.

Selected Bibliography

Alschuler, David. "Beating Maintenance Madness." *System Builder* 4 (April–May 1991): 43–44.

Burch, John. "The Postimplementation Audit." *EDP Auditing* (Boston, MA: Auerbach Publishers, 1991): section 73-01-20, pp. 1–16.

Chandra, Satish, and Menezes, Dennis. "An Examination of the Software Development Backlog Problem." *Journal of Information Technology Management* 2 (Number 3, 1991): 1–7.

Cloud, Avery C. "An EDP Control Audit With Teeth." *Journal of Systems Management* 41 (January 1990): 13ff.

Courtemanche, Gil. "Outsourcing the Internal Audit Function." *Internal Auditor* 48 (August 1991): 34–39.

Crowell, David A. "Control of Microcomputer Software." *Internal Auditor* 48 (April 1991): 33–39.

Durant, Jerry E. "Applying Systematic Testing to Application Development Audits." *Internal Auditor* 48 (February 1991): 38–44.

Gibson, Virginia R., and Senn, James A. "System Structure and Software Maintenance Performance." *Communications of the ACM* 32 (March 1989): 347–358.

Graham, Lynford E.; Damens, Jeffrey; and Van Ness, George. "Developing Risk Advisor: An Expert System for Risk Identification." *Auditing: A Journal of Practice & Theory* 10 (Spring 1991): 69–96.

Klenk, Robert. "Identifying and Testing MIS Controls." *EDP Auditing* (Boston, MA: Auerbach Publishers, 1991): section 72-03-60, 1–11.

Kumar, Kuldeep. "Post Implementation Evaluation of Computer-based Information Systems: Current Practices." *Communications of the ACM* 33 (February 1990): 203–212.

Landry, Terry. "The EDP Audit Role On MIS Steering Committees." *EDP Auditing* (Boston, MA: Auerbach Publishers, 1991), section 72-01-45, 1–9.

Moad, Jeff. "Asking Users To Judge IS." *Datamation* 35 (November 1, 1989): 93ff.

Moeller, Robert R. "Auditing End-User Computing." *EDP Auditing* (Boston, MA: Auerbach Publishers, 1991): section 74-05-01, 1–14.

Vasarhelyi, Miklos A., and Halper, Fern B. "The Continuous Audit of Online Systems." *Auditing: A Journal of Practice & Theory* 10 (Spring 1991): 110–125.

Action diagram: a documentation tool that uses brackets and titles to show the hierarchical structure of processes to be performed.

Activity: an arrow on a network diagram.

Actual activity: work that is done and accomplishments that are made. Actual activity is compared with desired activity as a means of engaging in management by exception.

AD/Cycle: the IBM methodology that provides a framework for the integration of CASE tools from many software vendors.

Adaptive maintenance: see **Systems maintenance.**

Analysis: see **Definition effort.**

Analysis phase: the second phase of the system life cycle, consisting mainly of the system study. See **System study.**

Annual audit: the audit of a computer-based system conducted each year by the systems analyst or the internal auditor. The annual audit follows the postimplementation review. See **Postimplementation review.**

Anti-viral software: software that can restore a program to its original condition after it has been infected by a virus.

Anonymity: the ability of a group DSS session to keep each participant's identity secret as they enter their comments, observations, and suggestions.

Application: a job performed by a computer. Examples are inventory, payroll, and billing.

Application backlog: jobs awaiting implementation on the computer.

Application control: control applied to a particular application, as opposed to applications in general. See **Migration of controls.**

Application software: programs that process a firm's data. Examples are payroll, inventory, mathematical models, and statistical packages.

Application-development software: the name used in this text to describe the new breed of software such as electronic spreadsheets and database management systems. These are not prewritten packages in the true sense; the user must tailor them to fit particular applications.

Artificial intelligence (AI): the activity of providing machines such as computers with the ability to display behavior that would be regarded as intelligent if it were observed in humans. AI includes work in the areas of perception, learning, automatic programming, and expert systems.

Attribute: a distinguishing characteristic of a data entity. For example, an attribute of an employee is employee number. Particular values are assigned to each attribute for the purpose of identifying and explaining each occurrence.

Audit committee: a committee composed of members of the board of directors that oversees the firm's internal auditing operation.

Audit trail: a history of some aspect of a firm's activity. The history might be a detail report or a magnetic tape file containing a record of each transaction. The audit trail enables a transaction to be traced from its source to its conclusion, and vice versa.

Availability: one of the objectives of systems security; to make information resources available to users.

Back door: see **Trojan horse.**

Backup plan: the part of a firm's contingency plan that provides backup computing facilities.

Bacteria: a type of malicious software that primarily attacks mainframe computers by replicating itself so many times that it clogs the system. The term **rabbit** is also used.

Balanced DFD: a hierarchical arrangement of data flow diagrams that reflects a top-down continuity in terms of data names and structures.

Batch log: a record maintained by computer operations of all users' batch jobs flowing into and out of the department.

Batch processing: one of the two basic ways to process data characterized by grouping transactions so that all are handled at one time. It is the most efficient way to use the computer, but its main disadvantage is that files are not kept current as transactions occur. Contrast with **Online processing.**

Batch total: a total that is accumulated from a numeric field in a batch of documents. Unlike a hash total, the batch total can have some informational value. See **Hash total.**

Benchmark problem: one or more of the firm's programs that hardware vendors are expected to run on their equipment in order to provide a common basis for evaluation.

Biometric system: a physical security system, such as a door lock that can be opened only by successfully passing a test unique to a particular individual— for example, a voice pattern or fingerprint.

Bottom-up approach: a way of accomplishing a particular task by beginning with the details and gradually synthesizing them into a whole. This was the original approach to systems analysis and design, but has given way to the top-down approach. See **Top-down approach.**

Brainstorming: the process a group follows in discussing a topic to increase the level of understanding. Members of the group are encouraged to present their ideas, which are then discussed. The approach is especially useful when the group has had little opportunity to become familiar with a problem area.

Break-even analysis: a method of economically justifying a new system by comparing its costs with those of the existing system.

Break-even point: the point at which the costs of the new system equal those of the existing system.

Business process redesign (BPR): the activity of changing a basic process of the firm for the purpose of making the process more effective or efficient.

Checksum: a total computed from program code, which is used to detect instances where the code has been changed without authorization.

Chief information officer (CIO): the person in charge of the firm's information resources, who participates equally with other executives in mapping out corporate strategy. The CIO can have the title of vice-president of information services.

Choice activity: the term used by Herbert Simon for the selection of the best alternative.

CIO concept: the idea that the person in charge of information services is an executive who participates with other executives in solving top-level problems affecting all the firm's operations.

Circuit: the transmission facility that provides one or more channels in a data communications network. Commonly used circuits are telephone lines, coaxial cables, fiber-optical cables, and microwave signals. Also called a **line.**

Client/server computing: the currently popular design of data communications networks that allows users to do their own computing on their microcomputers or terminals, and information services to process its jobs on the central computer.

Closed system: a system that does not interface with its environment.

Closed-loop system: a system with a feedback loop.

Code generator: a lower-CASE tool that produces program statements in a language such as COBOL from specifications supplied by the programmer.

Combination key: multiple data elements used to identify a data entity.

Command level user: an end user who can communicate with the computer by means of software command language.

Communication chain: the oral and written communications linking the user, information specialists, and computer.

Competitive advantage: an edge the firm enjoys over its competitors in meeting the needs of its customers.

Competitive intelligence (CI): information a firm gathers on its competitors.

Computer criminal: someone who seeks to cause some type of harm to the firm by engaging in unauthorized computer operations.

Computer literacy: knowledge of the computer—generally how it works, its terminology, its capabilities and limitations, and so on.

Computer-aided software engineering (CASE): the various software-based documentation tools, code generators, and prototyping tools that facilitate systems development and maintenance.

Computer-based information system (CBIS): the term used in this text to describe all the computer applications in a firm—data processing, MIS, DSS, OA, and expert systems.

Conceptual resource: a resource that represents a physical resource. Examples are data and information.

Conceptual system: a system that represents a physical system. The representation is accomplished by the storage of data reflecting conditions (such as level of inventory) and activities (such as work flow).

Confidentiality: one of the objectives of systems security; to keep certain data confidential.

Connectivity: this term has two meanings in the computer field. First, it refers to the ability of a system to exchange data and programs with other systems using a data communications network. Second, it refers to the possible relationships that can exist between entities in an entity-relationship diagram. See **Entity-relationship diagram.**

Constraint: any limitation on the firm's ability to operate in the desired manner.

Consumer: the name that marketers use for a customer. See **Customer.**

Context diagram: the highest-level data flow diagram. It presents the system in context with its environmental interfaces.

Context-sensitive help: a help message that explains a particular part of a program, usually a point on the screen where the cursor is located when help is requested. Contrast with **Functional help.**

Contingency planning: the activity that outlines what the firm will do in the event its information resources are disabled in some way. Originally called **disaster planning.**

Control environment: the setting a firm strives to establish for the purpose of achieving and maintaining control over its operations.

Control file: a file established for the purpose of achieving control over the system. The file can contain error records, batch totals, or other similar data.

Control matrix: a grid that identifies the controls for each of the risks identified in a risk matrix. See **Risk matrix.**

Control mechanism: most often used to describe the portion of a closed-loop system that adjusts the system so that the objectives can be achieved. Also used to describe the means by which management controls the system life cycle.

Control procedures: software routines incorporated into systems for the purpose of controlling them. Two types exist: gene:al and application. See **General control** and **Application control.**

Conversion: see **Cutover.**

Corrective maintenance: see **Systems maintenance.**

Cost avoidance: a strategy for economic justification of a computer system whereby existing expenses are not reduced, but future expenses are not increased.

Cost reduction: the traditional strategy for economic justification of a computer system whereby existing expenses are reduced.

Cost-benefit analysis (CBA): the study of a computer system that seeks to provide an economic justification by comparing the expected costs and benefits. For the system to be justified, the benefits must exceed the costs.

Countermeasure: an action taken for the purpose of thwarting a breach in systems security.

Critical path: the sequence of activities from the beginning to the end of a network diagram that represents the longest lapsed time. This path contains no slack and determines the time necessary to complete the work represented by the diagram. See **Slack.**

Critical path method (CPM): a type of network diagram that includes a single time estimate for each activity.

Current state: the condition of an entity at the present time. The entity, such as the firm or one of its operations, is monitored by comparing the current state with the desired state. See **Desired state.**

Customer: an individual or organization that purchases the firm's products or services.

Cutover: the process of halting the use of the existing system and beginning the use of the new system. Also called **conversion.**

Data: facts and figures that are relatively meaningless to the user. Data—the raw material of information—is transformed into information by an information processor.

Data analysis: see **Normalization.**

Data communications: the transmission of data from one geographic location to another. Although the terms **teleprocessing** and **telecommunications** are often used interchangeably with data communications, the former terms encompass a wider variety of media, such as voice communications.

Data communications network: the interconnection of computing equipment using data communications circuits, which allows data to be transmitted from one location to another.

Data description language (DDL): the syntax used to specify a database schema or subschema.

Data dictionary: a description of all data elements used by all the firm's computer programs. The description includes such specifications as the data element name, the type of data (numeric, alphabetic, alphanumeric), the number of positions, and how the ele-

ment is used. Some data dictionaries are maintained in computer storage.

Data dictionary system (DDS): the software system that converts the data dictionary into the schema.

Data element: the smallest unit of data in a record. Examples are name, age, and sex. Also called a **data item,** and a **field.**

Data element dictionary entry: the format for describing all important details concerning a data element.

Data flow: an arrow in a data flow diagram. Can be viewed as *data on the move.*

Data flow diagram (DFD): a top-down, structured analysis and design tool that consists primarily of symbols representing processes and arrows representing flows of data between the processes.

Data flow dictionary entry: the format for describing all important details concerning a data flow on a data flow diagram.

Data processing (DP): operations on data that transform it into a more usable form, such as sorted data, summarized data, or stored data. The term is also used to describe accounting applications as opposed to those of a decision-support nature. Also called **transaction processing.**

Data processing system: the group of procedures concerned primarily with processing a firm's accounting data. The term **accounting system** is also used.

Data store: the symbol in a data flow diagram that represents a repository of data, such as a file or database. Can be viewed as *data at rest.*

Data store dictionary entry: the format for describing all important details concerning a data store in a data flow diagram.

Data structure dictionary entry: the format for describing all important details concerning a data structure.

Data validation: the activity of seeking confirmation of gathered data, such as by comparing it with data from other sources.

Database: in the broadest sense, all the data existing within an organization. In a narrower sense, a database consists of only the data stored in the computer's storage in such a manner that retrieval is facilitated. The narrower view is most common.

Database administrator (DBA): a person in a firm who has responsibility for computer-based data.

Database concept: a relatively new way of thinking about an organization's data resource that regards multiple files as comprising a single reservoir. Files continue to exist separately in the computer's storage, but can be integrated logically using a variety of techniques. The logical integration makes it possible to quickly and easily extract the contents from several files for processing.

Decision: the selection of a course of action.

Decision making: the process of making a decision. A problem solver makes multiple decisions in the process of solving a single problem.

Decision support system (DSS): a concept originating in the early 1970s that focuses on the decisions necessary to solve single problems, usually of a semistructured nature. A DSS is usually designed to meet the needs of a single decision maker, or a small group of decision makers. Contrast with **Management information system.**

Definition effort: the portion of the systems approach to problem solving that consists of a definition of the problem—where it is located and what is causing it. Also called **analysis,** the **analysis phase,** and **diagnosis.**

Delphi method: an approach to gathering nonquantitative data that involves successive responses from a panel. For each round, the panel leader provides feedback from the previous round that serves to bring divergent views together. This recursive process continues until the responses reflect a single position.

Descriptor: an attribute of a data entity that provides information about that entity but does not identify it. Contrast with **Identifier.**

Design: see **Solution effort.**

Design activity: the term used by Herbert Simon for the identification and evaluation of alternatives.

Design phase: the third phase of the system life cycle, in which the logical design is developed. See **Logical design.**

Design proposal: the document prepared by the systems analyst that provides justification for proceeding to the design phase.

Desired activity: work that is to be done and accomplishments that are to be made. Desired activity is compared with actual activity as a means of engaging in management by exception.

Desired state: the state or condition of the entity when it is meeting its objectives. A problem exists when the desired state is not the same as the current state. See **Current state.**

Detail report: a report containing detail, as opposed to summary, information.

Diagnosis: see **Definition effort.**

Direct access storage: one of the two basic types of storage, which allows individual records to be accessed as opposed to sequentially searching the entire file. The other type of storage is **sequential storage.**

Direct access storage device (DASD): a secondary storage unit that has the capability of sending the access mechanism directly to a certain location where data can be written or read.

Direct file organization: the means of storing data in a direct access storage device so that records can be accessed directly by using the record keys as the basis for the storage addresses.

Disaster planning: see **Contingency planning.**

Diskless node: a networked microcomputer that does not have disk drives. This design technique contributes to security, making it difficult for persons to enter unauthorized data into the system.

Distributed processing: a data communications network consisting of multiple computers. Also called **distributed data processing (DDP).**

Documentation tool: a means of describing an existing or new system with graphics, words, or a combination of the two. The documentation can be prepared manually or with the use of a computer, and the medium can be paper or magnetic storage.

Drill down: a term usually associated with executive information systems, whereby the user is able to retrieve successively more detailed displays.

DP programmer: an end user who can communicate with the computer by means of programming language.

DSS: see **Decision support system.**

Dummy activity: an activity on a network diagram that does not represent work to be done but, rather, is used to provide activities with unique node numbers.

Economic feasibility: the characteristic of a system that produces financial benefits in excess of its costs.

Economic justification: support for expenditures, expressed in monetary terms.

EDP auditor: an internal auditor who is computer literate.

Electronic data interchange (EDI): the flow of data from one firm to another by means of some type of data communications network.

Electronic data processing (EDP): the term used initially to mean computer processing. Sometimes used as a synonym for **data processing.**

Emergency power supply (EPS): a source of electrical power for a computer that is used when the regular power is lost. The EPS, which ordinarily consists of one or more generators, is used until regular power is restored.

Encryption: the coding of data stored in a computer or transmitted over a data communications channel for the purpose of making the data meaningless to an unauthorized viewer.

End-user computing (EUC): development and use of a computer application with some degree of independence from assistance provided by information specialists.

End-user computing support personnel: an end user who is assigned to information services but who is dedicated to assisting users in developing their applications.

End-user programmer: an end user who is assigned to information services but performs work for users on a contract basis.

Enterprise data model: A description of the data that will be required by the firm in the future in order for the firm to meet its strategic objectives. Also called the **corporate data model.**

Enterprise data planning: a component of enterprise planning that determines the data needs of the firm. See **Enterprise planning.**

Enterprise model: an output of enterprise planning that provides a description of the basic processes that a firm is to perform. See **Enterprise planning.**

Enterprise modeling: see **Enterprise planning.**

Enterprise planning: the activity that is conducted by the firm's executives for the purpose of defining the firm's strategic objectives and appraising the ability of the firm to meet them. Outputs include the enterprise model and the enterprise data model. See **Enterprise model, Enterprise data model.**

Entity: a term used in relation to both mathematical modeling and the database. When used in modeling, the entity is the condition or process that is represented. When applied to the database, an entity is a person, organization, place, object, or event that is included in an entity-relationship diagram. See **Entity-relationship diagram.**

Entity-relationship diagram (ERD): a documentation tool that describes the firm's data in a general way. The diagram shows the entities represented with data, and their relationships. See **Entity.**

Entropy: the degradation of a system as it runs out of resources.

Environment: everything outside of a firm, or outside of a system.

Environmental constraint: an influence on a firm's operations caused by an element in the environment.

Environmental data: data that describes activity or conditions in the firm's environment. In order for the data to be meaningful, it must be transformed into information. See **Environmental information.**

Environmental information: information that describes activity or conditions in the firm's environment.

Ergonomics: the study of the effect of the computer and its associated equipment and furnishings on the user's physical condition.

Error suspense file: a file that contains error records. The records are held in the file until they are corrected or removed.

Ethical feasibility: the ability of a system to operate within ethical boundaries.

Evaluation criteria: the factors used in measuring each alternate solution to a problem. The criteria are intended to identify the solution that best enables the system to meet its objectives.

Event-driven language: a program language that is structured based on events that occur, such as the pressing of a button displayed on the screen.

Exception principle: see **Management by exception.**

Exception report: an application of the principle of management by exception to a report. The report calls the problem solver's attention to only the exceptional situations—variations above or below the acceptable range.

Executive: a manager on the top level of the organizational hierarchy.

Executive committee: the formal group of executives who regularly make the key decisions in a firm.

Expert system: a computer program that can function as a consultant to a problem solver by not only suggesting a solution but also explaining the line of reasoning that leads to the solution. Such a program is an example of artificial intelligence.

External auditor: a person from an outside organization, such as an accounting firm, who evaluates a firm's conceptual systems.

External total: a control total that is created from data before it is entered into the computer. External totals are compared with internal totals to make certain that all data enters the computer. See **Internal total.**

Extra shift usage: the charge levied by the supplier for use of leased or rented equipment in addition to the contracted amount. The extra shift usually occurs after usage exceeds eight hours a day, or 176 or 200 hours a month. Contrast with **Prime shift usage.**

Facilities management (FM): a service provided by an outsourcer, which consists of operating the firm's computing facilities.

Feasibility: the inherent ability of a possible solution to be implemented and to solve the problem. There are different kinds of feasibility—technical, economic, noneconomic, legal and ethical, operational, and schedule.

Feasibility study: the activity of determining the feasibility of a system or a project. See **Feasibility.**

Feasible solution: see **Feasibility.**

Feedback: information the system uses to regulate itself. Feedback can also refer to information "fed back" to the firm from other elements in the distribution channel, such as wholesalers, retailers, and customers. Contrast with **Feedforward information.**

Feedback loop: the portion of a system that enables the system to regulate itself. Feedback is obtained from the system and transmitted to the control mechanism. The control mechanism makes adjustments to the system as required. A system with such a capability is called a closed-loop system. See **Closed-loop system, Open-loop system.**

Feedforward information: information provided to elements in a firm's channel of distribution. Examples include information the firm provides to wholesalers, retailers, and customers.

Field: the portion of a record reserved for a single data element. See **Data element.**

Field directory: a list used by the DBMS to determine the data elements a user can retrieve, and the operations permitted on those elements.

Field engineer (FE): the person who keeps the computing equipment in working order.

Figure n diagram: the term used in this text to describe a data flow diagram that is on any level below the Figure 0 diagram in the top-down hierarchy. See **Figure 0 diagram.**

Figure 0 diagram: the data flow diagram just below the context diagram in the top-down hierarchy. See **Context diagram.**

File: a group of records relating to a particular subject.

File maintenance: the process of keeping a file up-to-date by adding, deleting, and modifying records.

Financial audit: an audit conducted by an external or internal auditor for the purpose of verifying the accuracy of the firm's accounting figures.

Financial community: those elements in a firm's environment that influence the flow of money to and from the firm. Examples of such elements are banks, insurance companies, and governmental agencies such as the federal reserve system.

Float: the time that elapses between purchase by a customer until receipt of payment.

Flowchart: a schematic diagram of a process, using standardized symbols. When the diagram represents an entire system, it is called a **system flowchart.** When it represents only a single program within the system, it is a **program flowchart.**

Form-filling technique: one of the three ways to enter data and instructions into a computer from an online keydriven device. The screen is designed to resemble a printed form so that you can move the cursor from field to field as the data elements are entered. Contrast with **Menu-display technique** and **Prompting.**

Formal review: a review of an operational system conducted in accordance with written policy.

Formal system: a system described by a procedure, or used according to a schedule. Examples are programs that print periodic reports, and scheduled meetings.

Forward engineering: the component of reverse engineering that consists of the developmental phases of the system life cycle taken in the normal sequence. See **Reverse engineering.**

Fourth-generation language (4GL): see **Natural language.**

Functional area: see **Functional organization structure.**

Functional help: a help message that explains how the user can perform certain functions, such as printing output or reading from a diskette. Contrast with **Context-sensitive help.**

Functional organization structure: segregation of a firm's resources based on the major functions performed. The main functional areas are finance, human resources, information services, manufacturing, and marketing. Functional organization structure is reflected in information services when all systems analysts are assigned to one unit, all programmers to another, and so on.

Functional support personnel: an end user who is assigned to a user area but possesses the same level of knowledge and skills as an information specialist in information services.

Functionality: what a system does—the operations it performs.

Gane-Sarson methodology: a technique used in drawing data flow diagrams, where processes are represented with upright rectangles.

Gantt chart: a horizontal bar chart that uses the bars to illustrate how activities span time.

General control: a control that applies to all systems. Contrast with **Application control.**

Go/no-go decision point: a step in the system life cycle when a decision is made whether to continue with the project or to terminate it.

Graphic report: a report that conveys information using primarily pictorial techniques, such as charts, graphs, and diagrams. Contrast with **Tabular report.**

Group decision support system (GDSS): a DSS used by several people who jointly make a decision.

Hardcopy: a paper document.

Hash total: a total accumulated for data that is meaningless except for use as a control. The data, such as birth dates or item numbers, is not normally totaled.

Help message: a screen display intended to assist the user in overcoming a particular difficulty in using the computer.

History file: a data file that describes some aspect of the firm's past activity. The file may serve as an audit trail. See **Audit trail.**

Human factors considerations: attention to the human element in systems designs. The concern is for the individual to successfully work within the system. Also called **behavioral considerations.**

Identifier: an attribute of a data entity that identifies that entity.

Immediate cutover: the replacement of the old system with the new system without any overlap.

Impact printer: a printer that functions by causing a print element with raised characters to strike an ink ribbon, which, in turn, strikes the paper.

Implementation: the activity involved with the transformation of the logical design into the physical design. See **Logical design, Physical design.**

Implementation phase: the fourth phase of the system life cycle where implementation occurs. See **Implementation.**

Implementation proposal: the document prepared by the systems analyst to provide the basis for proceeding with the implementation phase.

In-depth personal interview: a type of face-to-face interview used to gather detailed information from a small number of people.

Indexed sequential file organization: the means of storing data in a direct access storage device so that records can be accessed by means of an index.

Industrial engineer (IE): the person who studies a physical system for the purpose of making it more efficient. The IE's work also involves establishing conceptual systems to control the physical system.

Industrial espionage: unethical techniques a firm employs to gather information on its competitors.

Informal review: a review of an operational system that intentionally does not follow written policy.

Informal system: a system that is not described by a procedure or used according to a schedule. Examples are programs that produce one-time reports, and unscheduled meetings.

Information: processed data that is meaningful to the user.

Information center: an area in a firm reserved for hardware, software, and support personnel, which are made available to the firm's employees who want to engage in end-user computing.

Information economics: the approach to justification of computing expenditures that recognizes all possible values and all possible risks.

Information engineering (IE): James Martin's top-down approach to information systems development, which begins with strategic planning, includes business area analysis, and considers both data and activities.

Information literacy: an understanding of how to use information in problem solving.

Information overload: the situation when a problem solver is presented with more information than is needed.

Information processor: the unit that transforms data into information. It can be a human, a computer, or some other device.

Information resources: all the resources used to transform the firm's data into information—including hardware, software, data, information specialists, users, participants, and facilities.

Information resources management (IRM): a firm's formal program for utilizing its information resources so that they provide maximum user support.

Information services: the term used in this text to describe the organizational unit established to provide computer support. The term is often shortened to IS, which can also mean *information systems.*

Information specialist: any person whose primary occupation concerns providing computer-based systems. Examples are systems analysts, programmers, operators, network managers, and database administrators.

Information systems audit: a postimplementation review or an annual audit of a computer-based system to determine whether it is performing as intended. The audit is performed by an internal auditor or systems analyst.

Input bottleneck: the situation that exists when a system cannot handle its volume of input data.

Insourcing: the act of recapturing work previously contracted to an outsourcer.

Integrated application generator: a prototyping tool that can provide all of the desired elements in a new system.

Integrated CASE (I-CASE) tool: a CASE tool that provides support throughout the top-down development of a system, ranging from strategic planning through implementation and maintenance.

Integrated services digital network (ISDN): a data communications circuit that has the ability to transmit voice, data, text, and image signals at high speeds.

Integrity: the characteristic of a system that enables it to accomplish what is intended.

Intelligence: a term that has two meanings in the computer field. First, intelligence can be information that describes activities or conditions within a firm's environment. Second, intelligence can be information that enables problem-solvers to look into the future.

Intelligence activity: the term used by Herbert Simon for the analysis phase.

Internal auditor: an employee whose main responsibility is to ensure the integrity of the firm's conceptual systems.

Internal constraint: an influence on a firm's operations, which originates within the firm.

Internal data: data that describes activities or conditions within the firm. Internal data must be transformed into internal information in order to be meaningful.

Internal information: information that describes activities or conditions within the firm.

Internal total: a total accumulated from data in the computer's storage, which is compared with an external total to ensure that all data enters the system. See **External total.**

Interorganizational system (IOS): a cooperative effort between a firm and its suppliers and members of its distribution channel to work together as one coordinated unit.

Joint application design (JAD): an approach to the design of a new system that calls for participants to have a meeting in order to work out details. The meeting, called a **JAD session,** can span several days.

Key verification: a control over a system's input requiring a second keying operation after the data has been entered. The data from the key verify step is compared to that from the data entry step to ensure that it matches.

Knowledge acquisition: the identification of the thought processes applied by an expert in solving a problem. Once acquired, the thought processes are incorporated in the knowledge base of an expert system.

Knowledge engineer: the name used to describe a person who is capable of working with an expert in developing an expert system.

Knowledge-based system: see **Expert system.**

Lease plan: a means of acquiring hardware resources by paying a monthly fee for a set period of time.

Lease-purchase plan: a means of acquiring hardware resources whereby some portion of the monthly lease payments can be applied to eventual purchase.

Legacy system: a system that has been in use for a long time and represents a drain on information resources but cannot be eliminated because it performs an essential function.

Legal feasibility: the ability of a system to function within the law.

Leveled DFD: a data flow diagram that exists on several hierarchical levels.

License agreement: the contract between the supplier and the user of prewritten software that specifies any limitations in how the software can be used, such as the authorization to make backup copies.

Life cycle organization structure: a way to organize information services personnel based on the tasks they perform during the system life cycle, such as new system development and maintenance. Contrast with **Functional organization structure.**

Likert scale: a scale used in a questionnaire, usually consisting of a line with five points that represent conditions such as "Strongly agree," "Agree," and so on.

Line: see **Circuit.**

Line item: an entry in a listing. An example is a detail line on an invoice describing a product that the customer has purchased.

Local area network (LAN): a network of computers connected by circuitry owned by a firm. In most cases the network is restricted to a small area, such as a building.

Logic bomb: a type of malicious software that is activated by a particular event such as a date.

Logical design: the documentation of a new or improved system, prepared by the systems analyst during the design phase. The documentation exists in the form of either paper-based or computer-based descriptions.

Lower-CASE tool: a CASE tool used in the lower levels of the hierarchy as a firm follows a top-down system development approach. A good example is a code generator. Contrast with **Upper-CASE tool, Middle-CASE tool.**

Maintenance liaison: a member of the developmental team who is given maintenance responsibilities after cutover. The purpose of the liaison is to provide continuity between development and maintenance.

Make-or-buy decision: the choice that a firm faces of developing its own software or acquiring prewritten software from a vendor.

Malicious software: software that has an inherent ability to harm the firm in some manner. Examples are viruses, worms, and trojan horses.

Management by exception: a technique whereby a manager is concerned only with activities falling outside an area of acceptable performance.

Management control system: the name that Robert Anthony gave to systems used by middle-level managers.

Management function: a basic task that all managers perform, such as plan, organize, staff, direct, and control. Henri Fayol is credited with originating the concept.

Management information system (MIS): a system that provides information for decision making. The term was originally used to distinguish such a computer application from the traditional accounting jobs. The text uses the term to describe information systems that are intended to meet the general information needs of all managers in the firm, or of all managers in a specific organizational area.

Manager: a person who directs the activity of others.

Managerial role: an interpersonal, informational, or decisional activity performed by a manager, as defined by Henry Mintzberg.

Market analysis: a study of user perceptions about the tasks they perform, intended to ensure user support should the computer be applied to those tasks.

Master file: a file containing data of a fairly permanent nature. Master files typically are maintained for a firm's customers, personnel, inventory, and so on. The files form the conceptual resource.

Mathematical model: any formula or set of formulas that represents an entity.

Media library: the place in the computer operations area where magnetic media, such as disks and tapes, are stored.

Menu: a list of choices displayed on a screen.

Menu-display technique: one of the basic ways to enter data and instructions into a computer from an online keydriven device. The computer displays a menu, and the user selects the choice that instructs the computer what to do next. For other ways, see **Form-filling technique** and **Prompting.**

Methodology: a recommended way of performing a task. A systems development methodology such as the system life cycle provides the setting within which the various documentation tools are used. See **Documentation tool.**

Middle-CASE tool: a CASE tool that can be used in the middle levels of the top-down development hierarchy. Tools that draw data flow diagrams and create data dictionary entries are examples.

Migration of controls: the trend away from controls that are unique to particular applications, and toward controls that apply to all applications. See **Application control, General control.**

MIS steering committee: the group in an organization responsible for establishing policy for information resources and overseeing the development of computer-based systems.

Model: a representation of some phenomenon. Various types of models exist—physical, graphic, narrative, and mathematical.

Narrative report: a report that provides information in written form—sentences and paragraphs.

Natural language: a new breed of software intended to facilitate end-user computing with user-friendly syntax. The name **fourth-generation language,** or **4GL,** is also used because of the advancements beyond third-generation

languages such as COBOL. The name **nonprocedural language** is also appropriate, because the processes do not have to be spelled out in a particular order, as is the case with a programming language.

Negative entropy: the ability of a system to keep from running down by storing resources for future use.

Net present value (NPV): a comparison of the costs and benefits of a new system whereby future amounts are discounted to produce current values.

Network: see **data communications network.**

Network diagram: a drawing that uses arrows to represent activities or jobs to be done. The arrows are connected to show the interrelationships of the activities.

Network manager: the person responsible for a data communications network.

Node: a circle on a network diagram used to connect two or more arrows, or activities.

Noneconomic feasibility: the characteristic of a system that produces benefits that cannot be measured in monetary terms.

Nonimpact printer: a printer that functions without impact of a print element on an ink ribbon. Examples are ink jet printers and laser printers.

Nonprocedural language: see **Natural language.**

Nonprogrammed decision: the term used by Herbert Simon to describe a decision made without the benefit of a precise understanding of the elements involved and their relationships. Contrast with **Programmed decision.**

Nonprogramming end user: an end user whose capability for communicating with the computer is restricted to menus.

Normalization: the process of converting the data elements in a file or database into a series of normal forms, in order to make the organization of the data elements more efficient.

Object-oriented system: a system with a structure based on the objects that the system represents.

Object-oriented system development: the system life cycle that is followed by an object-oriented system. See **Object-oriented system.**

Objective: what a system is intended to accomplish, usually stated in broad terms. More specific standards are used to guide the system toward its objectives.

Objective question: a question that requires a specific answer. True-false and multiple-choice questions are examples. Contrast with **Subjective question.**

Observation: the gathering of data and information by viewing activity as it occurs, or by viewing evidence that the activity has occurred.

Office automation (OA): all the electronic technologies used to facilitate the flow of communications within the firm, and between the firm and its environment. Examples are word processing, electronic mail, and teleconferencing.

Offline: the situation when something is not connected directly to the computer. For example, if a cash register terminal is used to prepare a disk for processing on a mainframe, the terminal is offline. Contrast with **Online.**

Online: the situation when something is connected directly to the computer. For example, if a terminal is used to enter data into a mainframe, the terminal is online to the mainframe. Contrast with **Offline.**

Online processing: one of two basic ways to process data requiring the computer configuration to include some means of entering transactions as they occur, plus direct access storage. The main advantage of online processing is that it enables the conceptual system to stay up-to-date with the physical system. Contrast with **Batch processing.**

Open system: a system that interfaces with its environment.

Open-ended question: see **Subjective question.**

Open-loop system: a system without a feedback loop.

Operational audit: an inspection of a conceptual system for the purpose of improving it. The audit is normally conducted by an internal auditor.

Operational control system: the name that Robert Anthony gave to systems used by lower-level managers.

Operational feasibility: the ability of a system to function as intended. The cooperation of users and participants is usually a key element.

Operator-directed dialog: the communication between the computer and the user that is controlled by the user. For example, the user enters commands that tell the computer what to do. Contrast with **Program-directed dialog.**

Outsourcer: an organization that performs some or all of a firm's computing operations, based on a long-term agreement.

Parallel communication: the situation in a group decision support setting where multiple participants can enter comments at the same time.

Parallel cutover: the dual operation of both the old and the new system until the performance of the new system is proven.

Participant: the term used in this text to describe a person who is not a user of a system's output, but makes the system work. Examples are clerical personnel and data entry operators.

Password: the unique combination of characters entered by the user of a system in order to gain access to all or part of the system—its programs and data.

Payback analysis: an approach to the economic justification of a new system that determines when a new system will pay for itself. See **Payback point.**

Payback point: the point when the cumulative benefits from using a new system equal the cumulative costs of developing and using it.

Perfective maintenance: see **Systems maintenance.**

Performance criteria: the standards a new system must meet in order to satisfy its users.

Periodic report: a report prepared on a certain schedule, such as monthly. Also called a **repetitive report** or **scheduled report.**

Personal interview: the gathering of data and information through face-to-face questioning.

Phased cutover: the gradual introduction of a new system, one subsystem at a time.

Physical design: the product of the implementation phase of the system life cycle, in the form of hardware, software, and data, which has the capability of processing the data in the desired way.

Physical resource: a resource that exists physically. Personnel, material, machines, and money are examples. Contrast with **Conceptual resource.**

Physical system: a system that exists physically. Examples are humans, computers, and firms. Contrast with **Conceptual system.**

Pilot cutover: an approach to putting a new system into use that calls for a trial system. If the trial system, or **pilot,** performs satisfactorily, the cutover to the entire system is accomplished.

Planning horizon: the future time period for which a manager has a planning responsibility.

Portfolio approach: a means of justifying computer use by considering all of the applications in a composite way.

Postimplementation review: a formal evaluation of a system conducted shortly after the system is implemented. Both the systems analyst and the internal auditor can conduct separate reviews.

Preparation effort: the portion of the systems approach to problem solving involving a systems view of the problem area.

Preventive maintenance (PM): see **Scheduled maintenance.**

Prime shift usage: the charge levied by the supplier for use of rented or leased equipment during the contracted period. The prime shift usually consists of eight hours of use a day, or 176 or 200 hours a month.

Problem: a condition or event that damages or threatens to damage the organization in some negative way, or improves or promises to improve the organization in some positive way.

Problem avoider: a person who dislikes problems and will not attempt to solve them even when they become evident.

Problem cause: the underlying reason why a problem exists. Often called a **root cause.** In order for the problem to be solved, the problem cause must be isolated from the symptoms.

Problem definition: see **Definition effort.**

Problem identification: the step in the problem-solving process in which the problem solver becomes aware of a problem or potential problem.

Problem seeker: a person who enjoys the challenge of solving problems and seeks them out.

Problem signal: see **Problem trigger.**

Problem solver: a person who will not make a special effort to uncover problems, but will not back away when they become evident.

Problem solving: all the activity leading to the solution of a problem.

Problem statement: a specific description of a problem, which includes its cause and effect.

Problem trigger: something that signals a problem or impending problem.

Problem understanding: the step of the problem-solving process whereby the problem solver separates the problem symptoms from the root cause and identifies the problem location.

Process dictionary entry: the format for describing a process, using structured English.

Program change committee: the group who must approve each change to an operational program.

Program evaluation and review technique (PERT): a type of network diagram that includes three time estimates for each activity—optimistic, pessimistic, and most likely.

Program-directed dialog: the communication between the computer and the user, which is controlled by the computer. For example, the computer displays a menu and the user makes a selection. Contrast with **Operator-directed dialog.**

Programmed decision: the term used by Herbert Simon to describe a decision that is made by following a prescribed routine. Contrast with **Nonprogrammed decision.**

Programmer-analyst: a person who performs both programming and systems analysis duties.

Project control mechanism: the means of controlling a project. Examples include scheduled meetings, graphics, and written reports.

Project dictionary: the repository for all documentation prepared during a system development project.

Project management: the act of managing a project such as the development of a computer-based system. The MIS steering committee performs this activity in a general fashion, whereas the project leader has responsibilities to his or her specific project team.

Project management system: the special group of software used for project planning and control.

Project schedule: the timetable followed in developing a system.

Project team: the group that has responsibility for developing a system.

Prompting: one of main ways to enter data and instructions into a computer from an online keydriven device. The computer displays a prompt, in the form of a question or command, and the user makes the desired response. For other ways, see **Form-filling technique** and **Menu-display technique.**

Proposal: a suggestion that a particular action be taken concerning system development. The systems analyst prepares system study, design, and implementation proposals. Suppliers of hardware and software make proposals that their products be selected. The proposal usually is in a written form but can be accompanied by an oral presentation.

Prototype: a system that is developed not with the intention of completely meeting a user's needs, but of providing the user with an idea of how the system ultimately will appear and be used. Over time, the prototype is modified until it either serves as the blueprint of the op-

erational system or becomes the operational system.

Prototyping: the act of using a series of prototypes as a means of defining users' needs.

Prototyping toolkit: a collection of software tools that can be used for prototyping.

Pseudocode: a relatively informal way of stating the detailed logic of a program, developed as an alternate to program flowcharts. A formalized outgrowth of pseudocode used in a business setting is called **structured English.**

Purchase plan: a means of acquiring hardware whereby the title passes to the purchaser.

Quarantine station: a stand-alone computer used to screen incoming software to ensure it is not malicious. See **Malicious software.**

Questionnaire: a form containing questions to be asked in a survey. The questionnaire can be printed on paper or displayed on a computer screen.

Rabbit: see **Bacteria.**

Rapid application development (RAD): James Martin's methodology for quickly developing computer-based systems. It is a top-down approach that uses modern tools, such as fourth-generation languages and integrated CASE tools, and SWAT teams. See **Fourth-generation language, Integrated CASE tool,** and **SWAT team.**

Realtime processing: the response by a conceptual system to signals from a physical system that is sufficiently fast to control the physical system as actions and transactions occur. The processing performed by an online credit-approval system in a department store is an example.

Record: a collection of data elements that relate to a certain subject. Multiple records comprise a file.

Record layout: the earliest form of data documentation, consisting of a form that shows the location and size of a record's fields.

Record search: the use by the systems analyst of historical records as a source of information during the analysis phase.

Reengineering: an approach to business process redesign consisting of a backward progression through the system life cycle, using restructuring or reverse engineering, followed by a forward progression. See **Business process redesign, Restructuring, Reverse engineering,** and **Forward engineering.**

Relationship: the association between two data entities in an entity-relationship diagram, represented with a diamond symbol.

Repetition construct: one of the three main structured design constructs, dealing with the repeated execution of processes.

Repetitive report: see **Periodic report.**

Repository: a place where something is stored. This term is most often used in conjunction with CASE to describe the magnetic storage of the system documentation. In the IBM AD/Cycle methodology, it includes such documentation in addition to data and software.

Request for proposal (RFP): the formal notification by a firm to prospective suppliers that a proposal is desired. See **Proposal.**

Requirements analysis: see **Performance criteria.**

Requirements specification: see **Performance criteria.**

Respondent: a person who supplies information and data in response to a survey.

Restructuring: an approach to business process redesign that calls for converting a nonstructured system into one that reflects a structured design. See **Business process redesign.**

Reverse engineering: an approach to business process redesign consisting of a backward progression through the phases of the system life cycle in order to producing needed documentation. See **Business process redesign.**

Review activity: the term used by Herbert Simon for follow-up after a decision is made.

Right to privacy: one of an individual's basic rights that has prompted government legislation aimed at preventing harm caused by inaccurate or irrelevant data in a computerized database, or inappropriate use of the data.

Risk matrix: a table that classifies risks in some way. It provides the basis for identifying the need for corresponding controls. See **Control matrix.**

Safeguard: see **Countermeasure.**

Schedule feasibility: the ability of a system to be implemented in the specified time period.

Scheduled maintenance: repair work carried out at specified intervals. Also called **Preventive maintenance.**

Scheduled report: see **Periodic report.**

Schema: a description of all data elements in the database.

Security: see **Systems security.**

Selection construct: one of the three main structured design constructs, dealing with the logical selection of alternatives.

Selective dissemination of information (SDI): providing information to only those persons who should receive it.

Semistructured problem: a problem that includes some identifiable variables whose composition and interrelationships are understood. This is the type of problem that the DSS is intended to address.

Sequence construct: one of the three main structured design constructs, dealing with the execution of a series of processes.

Sequential file organization: the storage of records in secondary storage, one after the other.

Sequential storage: a type of secondary storage in which records can only be accessed one after the other, not directly.

Sight verification: the visual inspection of data before it is entered into the system in order to ensure the accuracy of the data entry operation.

Simulation: the process of using a model to represent some phenomenon.

Slack: the amount of time for an activity not on the critical path of a network diagram, which does not affect the overall duration of the project.

Software engineering: the application of scientific principles to the development of computer programs.

Software library: the accumulation of computer programs in secondary storage.

Solution criterion: the level of performance that must be achieved in order to solve a particular problem.

Solution effort: the portion of the systems approach to problem solving that includes the identification of the best solution, its implementation, and follow up. Also called **design** and the **synthesis phase.**

Source data automation (SDA): the design of a source document so that its data can be entered into a computer without the need for manual keying. Magnetic ink character recognition and optical character recognition are examples.

Source document: the document that contains input data to a system.

Special report: a report prepared in response to a special request or event as opposed to one prepared on a regular schedule. Special reports are often prepared by querying the database.

Specification document: the output of a JAD session that defines the system design that has been agreed upon by the participants. See **Joint application design.**

Standard: a measure of acceptable performance.

Standards manual: a written collection of the standards to be followed in the development and use of a firm's information resources.

Steady state: the condition when a system is in equilibrium in relation to its environment.

Strategic objective: a long-term objective that a firm, or a subsidiary unit within the firm, is expected to achieve. Strategic objectives are established by a firm's executives.

Strategic business plan: the process a firm intends to follow in achieving its strategic objectives.

Strategic planning system: the name that Robert Anthony gave to systems used by top-level managers.

Strategic planning for information resources (SPIR): the development of long-range plans for the use of the firm's information resources.

Structured English: a relatively disciplined narrative description of a system or procedure that evolved from pseudocode. See **Pseudocode.** Structured English is excellent for supplementing data flow diagrams.

Structured interview: a method for gathering data and information directly from a respondent by asking only prearranged questions.

Structured systems analysis: a top-down study of the firm's operations, first focusing on systems and then on successively lower-level subsystems.

Structured systems design: the method followed in systems design that consists of initially specifying the system in general terms and gradually making the description more detailed.

Structured systems implementation: a top-down approach to implementing a system, where the major capabilities are first made operational, and then gradually refined by including more detail.

Structured problem: a problem consisting entirely of identifiable variables whose composition and relationships are understood. The problem of how much of an item to order (the economic order quantity) is an example.

Study project proposal: the document prepared by the systems analyst that proposes a system study be conducted.

Subjective question: a question that a respondent can answer in any way that he or she likes. Often called an **open-ended question.** Subjective questions are usually used when you do not have a thorough understanding of the subject area and are looking to the respondent to give you insight. Contrast with **Objective question.**

Subsystem: a system within a system.

Supersystem: the environment within which a system exists. Also called **Suprasystem.**

Suprasystem: see **Supersystem.**

Supplier: an organization that provides the firm with needed resources. Also called a **vendor.**

Survey: a study that involves the gathering of data by means of personal, telephone, or mail interviewing.

Surveyor: the person who conducts a survey.

Suspense file: a file containing records that are awaiting further processing. An example is a file of error records that are being corrected.

SWAT team: the name that James Martin gives to a project team that specializes in performing certain specialized tasks during system development, such as conducting feasibility studies, providing economic justification, and so on. The letters stand for *Skilled With Advanced Tools.*

Symptom: a result of a problem rather than its root cause. See **Problem cause.**

Synthesis: see **Solution effort.**

System: an integration of elements designed to accomplish some objective.

System development life cycle (SDLC): the phases of the system life cycle prior to cutover—planning, analysis, design, and implementation. See **System life cycle.**

System documentation: a description of a system using any combination of narrative and graphics. The documentation medium can be either paper or electronic storage.

System flowchart: see **Flowchart.**

System integrity: see **Integrity.**

System life cycle (SLC): the phases of developing and using a computer-based system.

System of internal control: all the control procedures a firm uses to ensure that activities are performed as they should. The procedures exist within a control environment. See **Control environment.**

System software: software required to use a computer. Examples are operating systems, translators, and utilities.

System study: the study of the existing system conducted by the systems analyst during the analysis phase of the system life cycle.

System study proposal: the document prepared by the systems analyst to provide justification for proceeding to the analysis phase of the system life cycle.

System trustworthiness: the ability of a system to inspire confidence in its users. Trustworthiness is achieved through a combination of safety, reliability, and security.

Systems analysis: the process of studying an existing system for the purpose of making improvements or replacing it with a new system.

Systems analyst: the person who analyzes and designs business systems. The systems can use either computer or non-computer technology.

Systems approach: a problem solving methodology consisting of understanding the problem before a solution is attempted, and evaluating multiple feasible solutions.

Systems design: the process of developing a description of how a new or improved system will function. The term logical design is used to describe the documentation prepared by the systems analyst in the design phase of the system life cycle that serves as a basis for the physical design in the implementation phase. See **Logical design, Physical design.**

Systems implementation: the process of converting the logical design of a new system into the physical design.

Systems integration (SI): a service provided by an outsourcer who performs all the required steps of developing a new system.

Systems maintenance: all the activity related to keeping software in a usable state. There are three types: **adaptive maintenance** reflects changes in the system's environment, **perfective maintenance** improves the system's output, and **corrective maintenance** corrects errors.

Systems orientation: see **Systems view.**

Systems planning: all work that comprises the planning phase of the system life cycle. The purpose of the planning is to determine whether to embark on a formal system study.

Systems security: the protection of a firm's information resources from harmful unauthorized acts.

Systems view: the ability to see phenomena as systems. Such a view enables someone to better understand the phenomena by recognizing the roles of the various elements in achieving the overall objective.

Tabular report: the traditional report format, consisting of data arrayed in rows and columns. Contrast with **Graphic report.**

Task analysis: a study of a task conducted by asking the user to describe her or his feelings as the task is performed.

Technical feasibility: the characteristic of a system that incorporates available or attainable technology in the form of hardware and software.

Terminator: the symbol in a data flow diagram that represents an environmental element or entity.

Threat: a person or organization that has the capability for breaching a firm's computer security.

Timesharing: the use of a single computer by multiple users.

Top-down approach: the currently popular way to study, design, and implement computer-based systems. This approach is characterized by an initial concern with the major system elements, which are then gradually subdivided into lower levels of subsystems.

Transaction file: a file containing descriptions of transactions, such as product sales. The transaction file is used to update a master file.

Transaction log: a file of data that describes the details of the firm's transactions. The transaction log provides an audit trail of the firm's activities. See **Audit trail.**

Transaction processing: see **Data processing.**

Transmission log: a file of data that describes each message transmitted over the firm's network. The log includes data such as the sender, the sender's location, the date, and the time, and serves as a control over unauthorized use of the network.

Trapdoor: a code incorporated into a program by the programmer as a means to facilitate easy debugging, which is exploited by a computer criminal. Also called a **back door.**

Trojan horse: the name given to any malicious software that causes damage without giving any indication of that damage to the user.

Turnaround document: a slip of paper or a card customers enclose with their payments so a firm can avoid the input bottleneck. This is an example of source data automation. See **Input bottleneck, Source data automation.**

Type I prototype: a prototype that eventually becomes the operational system.

Type II prototype: a prototype that serves as a blueprint for the operational system.

Uninterruptable power supply (UPS): a source of short-term power, such as batteries, that can be used to complete processing when regular power is lost. Unlike an emergency power supply, it is not intended for long-term use. See **Emergency power supply.**

Unscheduled maintenance: unanticipated repair work on a computer.

Unstructured interview: a face-to-face data gathering session where there is no predetermined list of questions, and the respondent can answer in any way that he or she likes.

Unstructured problem: a problem consisting of variables and their relationships that are not identified.

Upper-CASE tool: a CASE tool used in the upper levels of the hierarchy when a firm follows a top-down system development approach. A good example is a tool that facilitates strategic planning for information resources. Contrast with **Lower-CASE tool, Middle-CASE tool.**

Use phase: the fifth and final phase of the system life cycle, where the user uses the system outputs.

User: a person who uses the output of a computer-based system. The term **end user** is often used interchangeably, but is more appropriate in those situations where the person does some or all of the developmental work. See **End-user computing.**

User directory: a list of authorized users employed by the database management system as a means of maintaining database security.

User factor stage: a special step added at various points in the system development life cycle to ensure that the new system takes into account human factors considerations.

User friendly: the name given to hardware or software that is easy to learn and use.

User manual: a document that explains how to use a system.

Value chain: the term used by Michael Porter and Victor Millar to describe the benefits achieved when a firm links the physical and informational activities that are related to providing its products.

Value system: a linkage of a firm's value chain with those of its suppliers, distribution channel members, and customers. See **Value chain.**

Vendor: see **Supplier.**

Virus: a type of malicious software that is spread from one system to another by means of diskettes or network transmissions. The most popular targets are microcomputers. The software may or may not cause serious harm to the computer and its applications.

Vulnerability: a weak spot in a system that leaves it open to attack by a computer criminal.

Warnier-Orr diagram: a systems analysis and design tool that documents a system using a hierarchical arrangement of brackets. Each bracket includes notation that reflects a structured programming construct—sequential, selection, or repetition.

Wide-area network (WAN): a data communications network that spans a large geographic area. The circuits are provided by a common carrier such as AT&T or Sprint. Contrast with **Local area network.**

Worm: a type of malicious software that replicates itself geometrically to clog the computers on a network.

Yourdon-Constantine methodology: the technique used in drawing data flow diagrams where processes are represented with circles.

Index

Credits and Sources

Figure 1.1: Metropolitan Museum of Art, Ford Motor Company Collection.

Figure 1.2: © Gabe Palmer/The Stock Market.

Figure 1.8: © Peter Steiner/The Stock Market.

Figure 1.17: © Randy Duchaine/The Stock Market.

Figure 2.1: American Airlines photo.

Figure 2.13: © Steonen Derr/The Image Bank.

Figure 3.1: © Gabe Palmer/The Stock Market.

Figure 3.2: © John McDonough/Sports Illustrated.

Figure 3.3: © Tom Tracy/The Stock Market.

Figure 3.9: (a) © Comstock. (b) © Charles Gupton/The Stock Market. (c) © Comstock.

Figure 3.16: adapted from Per O. Flaatten, Donald J. McCubbrey, P. Declan O'Riordan, and Keith Burgess, *Foundations of Business Systems* (Chicago: The Dryden Press, 1989), pp. 125-126.

Figure 4.9: © Joel Gordon.

Figure 4.10: © Larry Hamill.

Figure 4.11: © Spencer Grant/PhotoBank.

Figure 4.12: Courtesy USAA.

Figure 4.13: Courtesy Dow Jones.

Figure 4.14: © Pete Saloutos/The Stock Market.

Figure 5.2: © PhotoBank.

Figure 5.7: © Spencer Grant/Stock Boston.

Figure 5.8: © Bob Daemmrich/Stock Boston.

Figure 6.12: Computer screen shot courtesy of Comshare, Inc., Ann Arbor, Michigan.

Figure 6.13: University of Georgia Management Department.

Figures 6.14 and 6.15: adapted from Raymond McLeod, Jr., *Management Information Systems*, 5th editon (New York: Macmillan, 1993). Reprinted with permission.

Figure 6.18: adapted from Raymond McLeod, Jr., *Management Information Systems*, 5th edition (New York: Macmillan, 1993). Reprinted with permission.

Figure 6.19: Courtesy AT&T Archives.

Figures 6.20 and 6.21: adapted from Raymond McLeod, Jr., *Management Information Systems*, 5th edition (New York: Macmillan, 1993). Reprinted with permission.

Figure 7.1: Courtesy Coca-Cola.

Figure 7.2: © Joel Gordon.

Figure 8.21: James Martin, *Rapid Application Development* (New York: Macmillan, 1991): p. 355.

Figure 8.22: James Martin, *Rapid Application Development* (New York: Macmillan, 1991): p. 127.

Figures 9.4 and 9.5: © Mary Ann Fittipaldi.

Figure 9.6: © Greg Nelson.

Figure 9.7: © Terry Wild Studio.

Figure 10.3: © Comstock.

Figure 10.4: Rollerblades Co.

Figure 10.20: Jay Prakash, "How Europe Is Using CASE," *Datamation* 36 (August 1, 1990):p. 80.

Figure 11.2: Per O. Flaatten, Donald J. McCubbrey, P. Declan O'Riordan, and Keith Burgess, *Foundations of Business Systems,* second edition (Fort Worth, TX: The Dryden Press, 1992): p. 217.

Figure 11.4: Courtesy CalComp.

Figure 11.6: Texas Instruments.

Figure 11.16: Sirkka Jarvenpaa and Gary W. Dickson, "Graphics and Managerial Decision Making: Research Based Guidelines," *Communications of the ACM* 31 (June 1988): p. 770.

Figures 11.17-11.20, Tables 11.1-11.2: Jean-Marc Nerson, "Applying Object-oriented Analysis and Design, " *Communications of the ACM* 35 (September 1992: pp. 64, 69, 70, 71.

Figure F.1: adapted from Raymond McLeod, Jr., *Decision Support Software for the IBM Personal Computer: dBASE III Plus, Lotus, WordPerfect* (Chicago: Science Research Associates, 1988): p. 176.

Figure F.2: adapted from *Macintosh User's Guide for Desktop Macintosh Computers* (1991), pp. 18-19.

Figure F.3: Courtesy of Comshare, Inc. University Support Program, Austin, TX.

Figures F.8-F.11: Aaron Marcus, "Designing Graphical User Interfaces: Part I," *UnixWorld 7* (August 1990): pp. 108, 110, 113, 114.

Figure F.12: Aaron Marcus, "Designing Graphical User Interfaces: Part II," UnixWorld 7 (September 1990): p. 121.

Figure F.14: Aaron Marcus, "Designing Graphical User Interfaces: Part II," UnixWorld 7 (September 1990): p. 122.

Figure F.15: Paul A. Davis, Computer Science, Texas A & M University.

Figure 12.16: © Joel Gordon.

Table 13.1: *Systems Auditability and Control: Module 1 Executive Summary* (Altamonte Springs, FL: The Institute of Internal Auditors Research Foundation, 1991): pp. 11-12.

Tables 13.2-13.3: Peter J. Denning, ed., *Computers Under Attack* (New York: ACM Press, 1990): pp. 376-380, 340-341.

Table 13.4: Eugene H. Spafford, Kathleen A. Heaphy, and David J. Ferbrache, *Computer Viruses* (Arlington, VA: ADAPSO, 1989): p. 86.

Figure 14.6: Faulkner Technical Reports, Inc.

Figure 14.7: © Masterson/PhotoBank.

Figure 14.8: Courtesy of IBM.

Figure 14.12: © Spencer Grant/Stock Boston.

Figure 15.4: Inland Steel Co. photo.

Figure 15.5: Frederick Gallegos, "Audit Contributions to Systems Development." In *EDP Auditing* (Boston: Auerbach Publishers, 1991), section 72-01-40, p. 9.

Figure 15.6: Jeff Moad, "Maintaining the Competitive Edge," *Datamation* 36 (February 15, 1990): p. 66.